Computational, AI and IT Solutions Helping Agriculture

Computational, AI and IT Solutions Helping Agriculture

Guest Editor

Dimitre Dimitrov

MDPI

Basel • Beijing • Wuhan • Barcelona • Belgrade • Novi Sad • Cluj • Manchester

Guest Editor
Dimitre Dimitrov
Department of University
Transfer
NorQuest College
Edmonton, AB
Canada

Editorial Office
MDPI AG
Grosspeteranlage 5
4052 Basel, Switzerland

This is a reprint of the Special Issue, published open access by the journal *Agriculture* (ISSN 2077-0472), freely accessible at: https://www.mdpi.com/journal/agriculture/special_issues/JPNVAX89TY.

For citation purposes, cite each article independently as indicated on the article page online and as indicated below:

Lastname, A.A.; Lastname, B.B. Article Title. *Journal Name* **Year**, *Volume Number*, Page Range.

ISBN 978-3-7258-5289-5 (Hbk)
ISBN 978-3-7258-5290-1 (PDF)
https://doi.org/10.3390/books978-3-7258-5290-1

Contents

About the Editor

Dimitre Dimitrov

Dimitre Dikov Dimitrov is a faculty instructor at NorQuest College (Canada). He was awarded a PhD from the University of Alberta (Canada) and specialized in the interdisciplinary area of computational mathematical modelling of hydrological, physical, chemical and biological processes in natural and agricultural ecosystems through coupled hydrology–soil–vegetation–atmosphere continuum. He pursues his research interests via developing and maintaining the innovative state-of-the-art modelling and research Internet platform, DIMONA (Copyright © 2019 Dimitre Dikov Dimitrov, Edmonton, AB, Canada), a web application with embedded ecohydrological, Internet of Things (IoT) and machine learning (ML) modelling components. The DIMONA project is designed for (i) process-based scientific modelling with explanatory and predictive power, driven by meteorological records at various time steps and constrained vs. site observations, satellite and remote sensing data; (ii) IoT real-time modelling coupled to real-time data acquiring from websites with application programming interface (API) end points; (iii) ML data-driven predictive modelling; (iv) exploring the weather and climate, soil properties, hydrology and biota; and (v) applying the above at any geographical location specified by its coordinates. This opens a perspective to expand the site simulations to regional-scale satellite-driven modelling, with real-time computational evaluations of current physical, biological, and ecological processes and characteristics of the natural ecosystems and agricultural crops. Dimitre's passion for mathematical and environmental sciences has motivated his dedication to developing, helping and promoting various computational and smart solutions, especially for imminent priorities at a planetary scale such as maintaining and optimizing secure and sustainable agriculture practices and food production worldwide under pressure of global change.

Preface

The Special Issue "Computational, AI and IT Solutions Helping Agriculture" provides a venue for reporting current advances in the applications of artificial intelligence (AI), deep learning (DL), machine learning (ML) and other aspects of information technologies (IT) in various areas of precision agriculture. These include CNN-based and multi-stream feature fusion architectures for plant classification and disease diagnostics, machinery operations and fault diagnostics, image processing and plant area mapping; YOLO-based applications and monitoring frameworks for object detection and tracking, and machine vision with automated processes; a 3D object-detection framework for agricultural robotics; an LSTM-based application for price prediction; an SSD-based mobile application for pest detection and monitoring; and a hybrid AI-driven architecture for smart agriculture. Reported innovations apply in diverse agricultural practices, such as real-time tracking and counting of wheat ears, detecting wheat seedlings, agave plant classification and detection, cherry tomato fruit and bunch detection, poultry monitoring, plant disease diagnosis and estimation, potato pest disease classification, pest beetle detection on corn crops, the diagnosis of rolling bearings and short circuit faults in permanent magnet synchronous motors, automated apple and garlic harvesting, automated crop irrigation, disease management and control, livestock monitoring, ginger price prediction, and Torreya grandis area mapping. The contributions of this Special Issue align with the transition of AI, ML, and DL technologies from Agriculture 4.0 to Agriculture 5.0, with further development and elaboration of the existing technologies, their diverse applications in various agricultural activities and practices, and upgrading to more sophisticated technologies and autonomous systems. Thus, the contributions of this Special issue could be further highlighted upon the ongoing global change and future challenges to which smart agriculture is vital for sustaining and securing the planetary food supply with the rapidly increasing human population.

Dimitre Dimitrov
Guest Editor

agriculture

MDPI

Editorial

Computational, AI and IT Solutions Helping Agriculture

Dimitre D. Dimitrov

Department of University Transfer, Faculty of Arts & Sciences, NorQuest College, Edmonton, AB T5J 1L6, Canada; dimitre@ualberta.ca

This Special Issue, entitled "Computational, AI and IT Solutions Helping Agriculture", brought together 17 publications, including 16 original research papers [1–16] and 1 review paper [17], addressing advances in precision agriculture on species and diseases detection and monitoring [1,2,6,7,9,11,17], classification [10,12,16], area mapping [14], robotics [5], smart machinery operations [8] and diagnostics [3,4], market predictions [13], and hybrid online platforms [15], as summarized below and shown in Table 1.

In [1], a novel object detection model called Wheat-FasterYOLO was introduced for real-time tracking and counting of wheat ears. The model applied FasterNet [18], a lightweight partial convolution network to enhance reference speed, reduce parameters, and improve the efficiency of object detection. To improve the accuracy and precision of the localization, the object confidence and bounding box regression loss in Wheat-FasterYOLO were based on the SCYLLA intersection over union method [19] instead of traditional intersection over union (IoU) [20]. The bi-level routing attention mechanism BiFormer was used to improve the dynamic extraction of wheat ear features along with reducing the complex background effects. Also, the neural network DCNv2 with deformable convolutions [21] was applied to improve the detection of wheat ears of various characteristics by dynamically adjusting kernel positions. The Feature Pyramid Network in the object detection was elaborated by applying a path aggregation network with content-aware lightweight up-sampling operator CARAFE to reduce the feature loss during up-sampling, thereby better capturing the wheat ear details within the context. Furthermore, the Kalman filter-based target-tracking algorithm OC-SORT was adopted for reliable real-time tracking and counting. The Wheat-FasterYOLO model was trained and validated by the Global Wheat Head Detection dataset and tested in real time by video data captured by drones. By achieving a mean average precision (mAP) of 94.01%, thus surpassing popular object detectors like YOLOX [22] and YOLOv7-Tiny [23], a higher-order tracking accuracy of 60.52%, and a tracking accuracy of 91.88%, Wheat-FasterYOLO appeared to be a promising tool for the detection and counting of wheat ears.

In [2], a deep learning (DL) model was proposed for pest detection on corn crops. The model was designed for low-power devices and tested on two versions of the MobileNet [24] single-shot detector (SSD), a neural network for real-time object detection with limited computational resources. The two versions were MobileNet-SSD-v1 and MobileNet-SSD-v2-Lite used in mobile agricultural systems and tools. The proposed model introduced an innovative two-step transfer learning approach for the detection of various harmful and beneficial beetles. During the first step, the model managed to distinguish the class of harmful beetles named "Beetle" and represented by genera *Anoxia*, *Diabrotica*, *Opatrum*, and *Zabrus* from the class of beneficial beetles named "Ladybug" and represented by genera *Coccinella*. During the second step, the species of the class "Beetle" were recognized properly. The overall performance of the model resulted in enhanced accuracy of mobile device classification and detection of various harmful beetle species while distinguishing them from the beneficial ones. Improved MobileNet-SSD-v1 and MobileNet-SSD-v2-Lite scored mAP values of 90.8% and 89.23%, respectively, outperforming the baseline scenarios by >6%. The proposed model demonstrated the potential for mobile applications in low-power systems and devices used in agriculture, e.g., automated counting pest traps, autonomous equipment, or cell phones, thereby contributing to improved corn crop diagnostics.

In [3], a new method and a model were introduced for the fault diagnosis of rolling bearings in agricultural machines based mainly on improving the feature extraction mechanism for image classification in the existing Vison Transformer (ViT) architecture [25]. The method involved obtaining vibration signals from the rolling bearings by acceleration sensors and their preprocessing for noise reduction by singular value decomposition and energy difference spectrum. The preprocessed one-dimensional signal was converted into a two-dimensional time–frequency image by generalized Fourier-based S-transform. The image was further processed with relative position coding by the modified ViT for pattern recognition classification and reporting the final fault diagnosis. In the modified ViT, the original image segmentation mechanism was replaced by ResNet34, a residual deep convolutional neural network (CNN) used to improve feature extraction by reducing model parameters and improving computational efficiency [26]. The above resulted in the proposed SVD-EDS-GST ResViT model, which performed with an average accuracy reaching 99.08%, outperforming existing rolling bearing fault classification CNN- and ViT-based models and demonstrating potential for use in an actual operating environment.

In [4], a multisource data-fusion method with Bayesian optimization based on CNN [27] was introduced as a backbone of an intelligent fault diagnosis system for inter-turn short circuit faults (ITSC) in the permanent magnet synchronous motors of agricultural machinery. Data from different signals were used to construct a synchronized dataset of current and vibration signals. Features from those signals were extracted separately by CNN with dilated convolution [28] for a large receptive field with the same number of parameters and a channel attention mechanism for adjusting the weights of the fault features from different signals and channels. Then feature fusion of the two types of signals was performed, and Bayesian optimization was applied for fine-tuning the hyperparameters while improving the training efficiency and enhancing the model's performance. The proposed method was used in a multi-stream feature fusion model for an ITSC fault diagnosis system, which was tested as a pilot and showed promising results with a validation accuracy of 98.99% and an error loss rate approaching zero.

In [5], a 3D apple detection framework FRESHNet was constructed for the needs of agricultural robotics to identify the apple stem directions during harvesting. A 3D object detection algorithm was proposed for recognizing the dimensions and locations of the apples and their three-axis rotation, following the indoor robotics approach. The PApple_RGB-D-Size [29] dataset was reprocessed for pairing 3D point clouds and 2D images of apples in orchards with label information to annotate the 3D bounding boxes and stem directions. The data of cloud points and images were processed by MinkResNet [30], with sparse convolution and ResNet50+FPN networks, respectively, to extract 3D and 2D features that were fused for the effective integration of geometric details and contextual information. The output was classification, 3D bounding box predictions, and rotating angle predictions for automated harvesting. FRESHNet was trained, validated, and tested, and it demonstrated its potential by achieving an average precision (AP) of 89.56% and average recall of 99.16%.

In [6], an AI-based poultry monitoring system was introduced consisting of a user-friendly web application connected to the YOLOv8 object detection model [31] for processing videos from surveillance cameras installed in chicken coops. The web application was bult using the Flask lightweight framework for Python and designed to be accessible by various edge computing devices. The dataset for training and validation consisted of 1300 images extracted from video recordings from four cameras in a chicken coop, manually annotated and cross-validated. The object detection model was operated using the Ultralytics YOLOv8 framework [32]. Particularly, YOLOv8m (medium) was selected for its advances in dealing with occlusion (object clustering), lighting variation (poor lighting, shadows, reflections), and various image resolutions that might significantly impact detection accuracy. The pilot version precision of 93.1%, recall of 93%, F1-score of 91%, and mAP of 93.1% demonstrated potential for production-level deployment and application.

In [7], five different object detection models were compared for their performance on agave plants of various conditions and shapes to evaluate the best one for application in agriculture. The YOLO family models YOLOv7, YOLOv7-tiny, YOLOv8 [23,33], and Detectron2 library models Faster R-CNN and RetinaNet were selected [34]. A database of 333 digital images of agave plants was created and annotated by VGG image annotator. The plants were assigned to five classes according to their health conditions, color, spotting, and size. All models were trained, validated, and tested under the same conditions and on the same randomly determined partitions of the database in a ratio of 7:1:2, respectively. The obtained AP and mAP suggested that YOLOv7 performed the best, surprisingly outperforming YOLOv8. The findings of this study could be used to refine the future selection of object detection models for agave classification and detection.

In [8], an adaptive root cutting system with machine vision for garlic bulb detection by IRM-YOLO [35] deployed on the Jetson NVIDIA Nano device [36] was introduced for garlic harvesters to address the high cutting injury rates and leakage, and thereby to improve the quality of harvested garlic. The adaptive root cutting harvesting system with its main components, i.e., an electronic control box, a welded harvesting table, a background board, a cutting device, and a blade speed control board, was installed on the 4DSL-7 garlic harvester. Bulb images taken by the cutting device camera were processed to the accompanying edge computing device Jetson Nano, where the bulb detector detected the garlic bulbs by CNN for classification and obtaining the pixel position of the target. This was based on the fact that the cutting device would adjust the knife disk rotation and cutting height. An AP of 99.2% of the bulb detector IRM-YOLO algorithm with a detection time of 0.0356 s signified the potential to expand the bench and field tests to commercial applications.

In [9], a lightweight dual-channel cross-feature fusion attention network architecture DCFA-YOLO was introduced for cherry tomato fruits and bunch detection. DCFA-YOLO was based on the YOLOv8_n object detection model [33] with several optimizations of the backbone network and the neck for improving feature extraction and fusion, and contextual integration, while maintaining reduced computational costs. The C2f module of the original YOLOv8_n was replaced in the backbone by ShuffleNetV2 [37], capable of group convolutions and channel shuffling, and in the cross-feature fusion neck by C2f_RepGhost [38]; an SPPF_CBAM attention mechanism was added too. DCFA-YOLO was trained on greenhouse cherry tomato images, and ablation tests were implemented to study the effects of the above optimizations. DCFA-YOLO performed better than single-modal networks and other fusion-based networks in terms of its RGB color, depth dual modality, and optimizations. Indeed, with a precision of 94.9%, recall of 91.4%, and F1-score of 93.1%, DCFA-YOLO outperformed many YOLO- and SSD-based networks.

In [10], a new dataset PDPQCD for the classification of potato pest and disease queries was constructed based on prompt engineering and large language models (LLMs). Also, a question classification model GF-CNN utilizing gated fusion was introduced for optimizing text feature extraction. Using ChatGPT 4 as the LLM, lists of diseases, templates, and questions were initialized for the new dataset containing 20,260 entities. A pre-trained BERT model [39] with dynamic word embeddings was incorporated in GF-CNN, together with a gating fusion mechanism to control simultaneous max-pooling and average pooling for leveraging local and global important features, and the softmax activation function was used for probability distribution in multi-class classification tasks. The PDPQCD dataset was tested by seven BERT-based models [40,41], which all achieved precision, recall, and F1-scores > 98% as GF-CNN performed the best. GF-CNN confirmed its high-level performance with ablation experiments on PPDQCD and two other datasets: Subj in English and THUCNews in Chinese.

In [11], a novel DL framework DHS-YOLO was introduced for detection by the unmanned aerial vehicle (UAV) of slender wheat seedlings, based on the YOLOv11 object detection model [42]. DHS-YOLO implemented three strategic improvements: a Dynamic Slender Convolution (DSC) block to extract the flexible and slender features of wheat seedlings by deformable convolution [21], a Histogram Transformer (HT) attention block to filter out irrelevant features and illumination-induced image degradation, and the ShapeIoU loss function to strengthen the geometric consistency between predicted and ground-truth bounding boxes; they were implemented in the framework's backbone and neck. Thus, the slow deformable convolutions by DSC were compensated by the dynamic-range HT attention, resulting in reliable head detections. DHS-YOLO was trained, validated, and tested in a series of ablation experiments using UAV-captured image datasets of varying illumination intensities, and it scored a precision value of 94.1%, recall of 91.0%, and mAP of 95.2%, outperforming other models and the baseline YOLOv11.

In [12], a public image database for three plant diseases and a family of segmentation networks, PDSNets, were introduced for plant disease estimation. The image datasets for soybean bacterial blight (SBB), cedar apple rust (CAR), and wheat stripe rust (WSR) were obtained from public resources or collected in the field. Pixel-level annotation for SBB, CAR, and WSR included background (B), healthy leaf (H), and disease spots (S) pixel classes. The PDSNets family consisted of three versions, PDSNetx1.0 and the lighter PDSNetx0.75 and PDSNetx0.5, based on LinkNet [43] with the ResNet-18 encoder and decoder with a pixel-level classifier. In total, 60% of the dataset for each disease was used for training and 40% for testing. Ablation experiments with PDSNetx1.0 resulted in outperforming the other models, including the baseline LinkNet. All three versions outperformed U-Net [44], DeepLabV3+(ResNet-18) [45], and LinkNet with F1-scores > 98%, 93%, 86% and IoU > 97%, 87%, 76% for B, H, S pixel classes, respectively, showing the potential for commercial use.

In [13], a DL model with the attention mechanism STL-LSTM-ATT-KAN was proposed for predicting ginger price as follows. The Seasonal Trend Decomposition module was implemented to decompose the price time series into trend, seasonal, and stochastic components, which were used by a long short-term memory recurrent neural network to capture long-term dependencies. The Adaptive Multi-Particle Swarm Optimization algorithm [46] was applied to optimize model parameters. The model used 13 independent variables and ginger price time series from 2014 to 2024. The aggregated dataset was split for training, validation, and testing in a ratio of 65:20:15. STL-LSTM-ATT-KAN outperformed seven other models, reaching mean average error (MAE), mean square error (MSE), root mean square error (RMSE), and coefficient of determination (R^2) values of 0.111, 0.021, 0.146, and 0.998, respectively, demonstrating its potential for commercial application.

In [14], a novel image data structure and a spatial–spectral fused attention CNN network SSFAN for image processing [47] were introduced for mapping *Torreya grandis* areas in China. Multiple Sentinel-2 Level -1C images [48] taken in 2023 were processed to obtain 13 spectral bands, 6 vegetation indices, and 6 texture indices, resulting in a dataset of 12 synthesized monthly images of 300 bands and 215,640 pixels. The latter were split into training, validation, and test sets in a 60:20:20 pixel ratio. The maximum redundancy maximum relevance algorithm was applied to select the optimal bands. They were processed by the two branches of SSFAN to produce spectral and spatial feature maps, which were fused to predict *Torreya grandis* areas. The method achieved overall accuracy of 99.1% and a Kappa coefficient of 0.961, demonstrating its potential to leverage the satellite multispectral feature extraction with limited spatial and spectral resolution.

In [15], a hybrid AI-driven architecture AGRARIAN was introduced to leverage the centralized cloud infrastructure of smart farming platforms by using localized edge processing, federated AI models, and hybrid 5G/satellite connectivity [49,50] for various agriculture environments, including those with infrastructure constraints. AGRARIAN horizontal functionality via customer portal, decision support systems, platform infrastructure, and external data sources was implemented via four vertical layers. The sensor layer comprised environment, soil, and livestock data acquisition through ground, UAV, Internet of Things (IoT) sensors, meteorological stations, satellite observations, and tracking devices. The network layer enabled 5G, satellite, and edge and ground communication between sensors, computing nodes, and cloud-based systems for seamless connectivity. The data processing layer acted as a computational hub and the application layer provided services to farmers, policymakers, and researchers, such as decision support and monitoring systems, crop forecasting, smart irrigation and water management, disease and pest alerts, and blockchain-enabled goods logistic traceability. The above functionality showed a perspective for interconnecting with online platforms for process-based, machine learning (ML), and IoT modeling [51–53]. Pilot scenarios demonstrated AGRARIAN's potential for irrigation control, crop disease, and livestock monitoring.

In [16], a lightweight DL framework AgriFusionNet was introduced for multisource plant disease classification and modeling. AgriFusionNet was built upon EfficientNetV2-B4 CNN [54], enhanced by Fused-MBConv blocks and implemented by Adam optimizer and Swish activation for more efficient training and model generalization. By the fusion of the large image collection PlantVillage [55], RGB, and multispectral drone images, and IoT sensor data on temperature, humidity, and soil moisture, a multimodal dataset was tailored and augmented for extending the diagnosis of diseases to their prediction under various agriculture conditions. The dataset was split for training, validation, and testing in a 70:20:10 ratio, and AgriFusionNet showed its potential by 94.3% classification accuracy with 30% less parameters than other comparable methods.

In [17], an extensive review was made of existing and emerging crop management technologies for smart and precision monitoring on germination, water status, nutrient stress, diseases and weeds detection, chlorophyl estimation, crop protection, and yield. The study overviewed Agriculture 4.0 crop monitoring and management technologies based on AI, ML, DL, and IoT with 5G and cloud computing environment [56–58]. They were situated in a perspective towards Agriculture 5.0 technologies of the next level [59–61], based on robotics, digital twins, UAV, unmanned ground vehicles (UGVs), blockchain, and big data analytics with 6G and fog and edge computing, as the advances and challenges were analyzed.

Thus, the contributions of this Special Issue could be seen within the perspectives of the transition of AI, ML, and DL technologies from Agriculture 4.0 to Agriculture 5.0. These perspectives include, but are not limited to, the following: (i) technical development and elaboration of those technologies, e.g., improving CNN with newly introduced or adapted attention mechanisms and/or activation functions; (ii) application expansion to various agriculture activities; (iii) upgrading to more sophisticated technologies and autonomous systems, e.g., from cloud computing to localized fog and edge computing and from classification and object detection to 3D object detection, robotics, smart machinery operations, and diagnostics.

Table 1. Summary of research items.

Item	Application Types	Agricultural Activities	Outcomes
[1]	YOLO-based computer application	Real-time tracking and counting of wheat ears	Object detection model
[2]	SSD-based mobile application	Pest beetle detection on corn crops	Object detection model
[3]	CNN-based vision transformer architecture	Diagnosis of rolling bearings in agricultural machines	Machinery fault diagnostics system
[4]	Multi-stream feature fusion architecture	Diagnosis of short circuit faults in permanent magnet synchronous motors in agricultural machinery	Machinery fault diagnostics system
[5]	3D object detection framework for agricultural robotics	Automated apple harvesting	3D object detection model
[6]	YOLO-based monitoring framework	Poultry monitoring	Object detection model within a web application
[7]	YOLO and Detectron2 -based computer applications	Agave plants classification and detection	Object detection models, comparative study
[8]	YOLO-based application for agriculture machines	Automated garlic harvesting	Machine vision device for automated harvesters
[9]	YOLO-based computer application for agricultural robotics	Cherry tomato fruits and bunch detection.	Object detection model
[10]	CNN-based computer application	Potato pests and disease classification	Dataset and classification model
[11]	YOLO-based framework for UAV	Detecting wheat seedlings	Object detection model
[12]	LinkNet-based computer application	Plant disease estimation	Three datasets and segmentation networks
[13]	LSTM-based computer application	Ginger price prediction	DL model
[14]	Data structure and image processing CNN	Torreya grandis area prediction	Area mapping tool
[15]	Hybrid AI-driven architecture	Crops, irrigation and disease management and control, livestock monitoring	Smart farming services
[16]	CNN-based computer application	Plant disease diagnosis	Dataset and classification and prediction model
[17]	Review	Agriculture 4.0 and 5.0 crop management technologies in monitoring	Summary

Acknowledgments: The Guest Editor is thankful to the Academic Editor and Editorial Staff for providing guidance and help to improve this manuscript.

Conflicts of Interest: The author declares no conflicts of interest.

References

1. Chen, J.; Hu, X.; Lu, J.; Chen, Y.; Huang, X. Efficient and Lightweight Automatic Wheat Counting Method with Observation-Centric SORT for Real-Time Unmanned Aerial Vehicle Surveillance. *Agriculture* **2023**, *13*, 2110. [CrossRef]
2. Maican, E.; Iosif, A.; Maican, S. Precision Corn Pest Detection: Two-Step Transfer Learning for Beetles (Coleoptera) with MobileNet-SSD. *Agriculture* **2023**, *13*, 2287. [CrossRef]
3. Xie, F.; Wang, Y.; Wang, G.; Sun, E.; Fan, Q.; Song, M. Fault Diagnosis of Rolling Bearings in Agricultural Machines Using SVD-EDS-GST and ResViT. *Agriculture* **2024**, *14*, 1286. [CrossRef]
4. Wang, M.; Lai, W.; Zhang, H.; Liu, Y.; Song, Q. Intelligent Fault Diagnosis of Inter-Turn Short Circuit Faults in PMSMs for Agricultural Machinery Based on Data Fusion and Bayesian Optimization. *Agriculture* **2024**, *14*, 2139. [CrossRef]
5. Son, G.; Lee, S.; Choi, Y. FRESH: Fusion-Based 3D Apple Recognition via Estimating Stem Direction Heading. *Agriculture* **2024**, *14*, 2161. [CrossRef]
6. Rangel, J.C.; Boniche, K. AI-Based Monitoring for Enhanced Poultry Flock Management. *Agriculture* **2024**, *14*, 2187. [CrossRef]
7. Matadamas, I.; Zamora, E.; Aquino-Bolanos, T. Detection and Classification of Agave angustifolia Haw Using Deep Learning Models. *Agriculture* **2024**, *14*, 2199. [CrossRef]
8. Yang, K.; Zhou, Y.; Shi, H.; Yao, R.; Yu, Z.; Zhang, Y.; Peng, B.; Fan, J.; Hu, Z. GrapeNet: Research and Experiments on Adaptive Root Cutting Using a Garlic Harvester Based on a Convolutional Neural Network. *Agriculture* **2024**, *14*, 2236. [CrossRef]
9. Chai, S.; Wen, M.; Li, P.; Zeng, Z.; Tian, Y. DCFA-YOLO: A Dual-Channel Cross-Feature-Fusion Attention YOLO Network for Cherry Tomato Bunch Detection. *Agriculture* **2025**, *15*, 271. [CrossRef]
10. Tang, W.; Hu, Z. Potato Disease and Pest Question Classification Based on Prompt Engineering and Gated Convolution. *Agriculture* **2025**, *15*, 493. [CrossRef]
11. Dong, X.; Pan, J. DHS-YOLO: Enhanced Detection of Slender Wheat Seedlings Under Dynamic Illumination Conditions. *Agriculture* **2025**, *15*, 510. [CrossRef]
12. Zhao, C.; Li, C.; Wang, X.; Wu, X.; Du, Y.; Chai, H.; Cai, T.; Xiang, H.; Jiao, Y. Plant Disease Segmentation Networks for Fast Automatic Severity Estimation Under Natural Field Scenarios. *Agriculture* **2025**, *15*, 583. [CrossRef]
13. Li, F.; Meng, X.; Zhu, K.; Yan, J.; Liu, L.; Liu, P. Research on Ginger Price Prediction Model Based on Deep Learning. *Agriculture* **2025**, *15*, 596. [CrossRef]
14. Lyu, Y.; Wang, Y.; Shen, X. The Extraction of Torreya grandis Growing Areas Using a Spatial–Spectral Fused Attention Network and Multitemporal Sentinel-2 Images: A Case Study of the Kuaiji Mountain Region. *Agriculture* **2025**, *15*, 829. [CrossRef]
15. Batistatos, M.C.; de Cola, T.; Kourtis, M.A.; Apostolopoukou, V.; Xilouris, G.K.; Sagias, N.C. AGRARIAN: A Hybrid AI-Driven Architecture for Smart Agriculture. *Agriculture* **2025**, *15*, 904. [CrossRef]
16. Albahli, S. AgriFusionNet: A Lightweight Deep Learning Model for Multisource Plant Disease Diagnosis. *Agriculture* **2025**, *15*, 1523. [CrossRef]
17. Taha, M.F.; Mao, H.; Zhang, Z.; Elmasri, G.; Awad, M.A.; Abdalla, A.; Mousa, S.; Elwakeel, A.E.; Elsherbiny, O. Emerging Technologies for Precision Crop Management Towards Agriculture 5.0: A Comprehensive Overview. *Agriculture* **2025**, *15*, 582. [CrossRef]
18. Chen, J.; Kao, S.-h.; He, H.; Zhuo, W.; Wen, S.; Lee, C.-H.; Chan, S.-H.G. Run, Do not Walk: Chasing Higher FLOPS for Faster Neural Networks. In Proceedings of the IEEE/CVF Conference on Computer Vision and Pattern Recognition, Vancouver, BC, Canada, 18–22 June 2023; pp. 12021–12031.
19. Gevorgyan, Z. SIoU loss: More powerful learning for bounding box regression. *arXiv* **2022**, arXiv:2205.12740. [CrossRef]
20. Yu, J.; Jiang, Y.; Wang, Z.; Cao, Z.; Huang, T. Unitbox: An advanced object detection network. In Proceedings of the 24th ACM International Conference on Multimedia, Amsterdam, The Netherlands, 15–19 October 2016; pp. 516–520.
21. Zhu, X.; Hu, H.; Lin, S.; Dai, J. Deformable convnets v2: More deformable, better results. In Proceedings of the IEEE/CVF Conference on Computer Vision and Pattern Recognition, Long Beach, CA, USA, 15–20 June 2019; pp. 9308–9316.
22. Ge, Z.; Liu, S.; Wang, F.; Li, Z.; Sun, J. Yolox: Exceeding yolo series in 2021. *arXiv* **2021**, arXiv:2107.08430. [CrossRef]
23. Wang, C.Y.; Bochkovskiy, A.; Liao, H.Y.M. YOLOv7: Trainable bag-of-freebies sets new state-of-the-art for real-time object detectors. In Proceedings of the IEEE/CVF Conference on Computer Vision and Pattern Recognition, Vancouver, BC, Canada, 18–22 June 2023; pp. 7464–7475.
24. Gulzar, Y. Fruit Image Classification Model Based on MobileNetV2 with Deep Transfer Learning Technique. *Sustainability* **2023**, *15*, 1906. [CrossRef]

25. Salamai, A.A.; Ajabnoor, N.; Khalid, W.E.; Ali, M.M.; Murayr, A.A. Lesion-aware visual transformer network for Paddy diseases detection in precision agriculture. *Eur. J. Agron.* **2023**, *148*, 126884. [CrossRef]
26. Wen, L.; Li, X.Y.; Gao, L. A transfer convolutional neural network for fault diagnosis based on ResNet-50. *Neural Comput. Appl.* **2020**, *32*, 6111–6124. [CrossRef]
27. Li, Z.; Wang, Y.; Ma, J. Fault Diagnosis of Motor Bearings Based on a Convolutional Long Short-Term Memory Network of Bayesian Optimization. *IEEE Access* **2021**, *9*, 97546–97556. [CrossRef]
28. Bai, S.; Kolter, J.Z.; Koltun, V. An Empirical Evaluation of Generic Convolutional and Recurrent Networks for Sequence Modeling. *arXiv* **2018**. [CrossRef]
29. Ferrer-Ferrer, M.; Ruiz-Hidalgo, J.; Gregorio, E.; Vilaplana, V.; Morros, J.R.; Gené-Mola, J. Simultaneous fruit detection and size estimation using multitask deep neural networks. *Biosyst. Eng.* **2023**, *233*, 63–75. [CrossRef]
30. Choy, C.; Gwak, J.; Savarese, S. 4d spatio-temporal convnets: Minkowski convolutional neural networks. In Proceedings of the IEEE/CVF Conference on Computer Vision and Pattern Recognition (CVPR 2019), Long Beach, CA, USA, 15–20 June 2019; pp. 3075–3084.
31. Yang, X.; Bist, R.; Paneru, B.; Chai, L. Monitoring Activity Index and Behaviors of Cage-Free Hens with Advanced Deep Learning Technologies. *Poult. Sci.* **2024**, *103*, 104193. [CrossRef]
32. Ultralytics. *Train Settings—Ultralytics Documentation*; Ultralytics: Ballenger Creek, MD, USA, 2024.
33. Sohan, M.; Sai Ram, T.; Rami Reddy, C.V. A Review on YOLOv8 and Its Advancements. In *Data Intelligence and Cognitive Informatics*; ICDICI 2023. Algorithms for Intelligent Systems; Springer: Singapore, 2024. [CrossRef]
34. de Almeida, G.P.S.; dos Santos, L.N.S.; da Silva Souza, L.R.; da Costa Gontijo, P.; de Oliveira, R.; Teixeira, M.C.; De Oliveira, M.; Teixeira, M.B.; do Carmo França, H.F. Performance Analysis of YOLO and Detectron2 Models for Detecting Corn and Soybean Pests Employing Customized Dataset. *Agronomy* **2024**, *14*, 2194. [CrossRef]
35. Yang, K.; Yu, Z.; Gu, F.; Zhang, Y.; Wang, S.; Peng, B.; Hu, Z. Experimental Study of Garlic Root Cutting Based on Deep Learning Application in Food Primary Processing. *Foods* **2022**, *11*, 3268. [CrossRef]
36. Zhang, Y.; Yu, J.; Chen, Y.; Yang, W.; Zhang, W.; He, Y. Real-time Strawberry Detection Using Deep Neural Networks on Embedded System (rtsd-net): An edge AI application. *Comput. Electron. Agric.* **2023**, *215*, 108445. [CrossRef]
37. Wang, L.; Zhao, Y.; Xiong, Z.; Wang, S.; Li, Y.; Lan, Y. Fast and precise detection of litchi fruits for yield estimation based on the improved YOLOv5 model. *Front. Plant Sci.* **2022**, *13*, 965425. [CrossRef]
38. Xiao, L.; Li, W.; Yao, S.; Liu, H.; Ren, D. High-precision and lightweight small-target detection algorithm for low-cost edge intelligence. *Sci. Rep.* **2024**, *14*, 23542. [CrossRef]
39. Devlin, J.; Chang, M.; Lee, K.; Toutanova, K. BERT: Pre-training of Deep Bidirectional Transformers for Language Understanding. *arXiv* **2018**, arXiv:1810.04805.
40. Li, L.; Diao, L.; Tang, Z.; Bai, Z.; Zhou, H.; Guo, X. Question classification method of agricultural diseases and pests based on BERT_Stacked LSTM. *Trans. Chin. Soc. Agric. Mach.* **2021**, *52*, 172–177.
41. Guo, X.; Wang, J.; Gao, G.; Zhou, J.; Li, Y.; Cheng, Z.; Miao, G. Efficient Agricultural Question Classification with a BERT-Enhanced DPCNN Model. *IEEE Access* **2024**, *12*, 109255–109268. [CrossRef]
42. YOLOv11. Available online: https://github.com/ultralytics/ultralytics/tree/main?tab=readme-ov-file (accessed on 20 August 2025).
43. Chaurasia, A.; Culurciello, E. LinkNet: Exploiting Encoder Representations for Efficient Semantic Segmentation. *arXiv* **2017**, arXiv:1707.03718. [CrossRef]
44. Ronneberger, O.; Fischer, P.; Brox, T. U-Net: Convolutional Networks for Biomedical Image Segmentation. In Proceedings of the Medical Image Computing and Computer-Assisted Intervention—MICCAI 2015, Munich, Germany, 5–9 October 2015; Navab, N., Hornegger, J., Wells, W.M., Frangi, A.F., Eds.; Springer International Publishing: Cham, Switzerland, 2015; pp. 234–241. [CrossRef]
45. Chen, L.-C.; Zhu, Y.; Papandreou, G.; Schroff, F.; Adam, H. Encoder-Decoder with Atrous Separable Convolution for Semantic Image Segmentation. In Proceedings of the Computer Vision—ECCV 2018, Munich, Germany, 8–14 September 2018; Ferrari, V., Hebert, M., Sminchisescu, C., Weiss, Y., Eds.; Springer International Publishing: Cham, Switzerland, 2018; pp. 833–851. [CrossRef]
46. Zhi, L.; Zuo, Y. Collaborative Path Planning of Multiple AUVs Based on Adaptive MultiPopulation PSO. *J. Mar. Sci. Eng.* **2024**, *12*, 223. [CrossRef]
47. Sothe, C.; De Almeida, C.M.; Schimalski, M.B.; La Rosa, L.E.C.; Castro, J.D.B.; Feitosa, R.Q.; Dalponte, M.; Lima, C.L.; Liesenberg, V.; Miyoshi, G.T.; et al. Comparative performance of convolutional neural network, weighted and conventional support vector machine and random forest for classifying tree species using hyperspectral and photogrammetric data. *GISci. Remote Sens.* **2020**, *57*, 369–394. [CrossRef]

48. Mura, M.; Bottalico, F.; Giannetti, F.; Bertani, R.; Giannini, R.; Mancini, M.; Orlandini, S.; Travaglini, D.; Chirici, G. Exploiting the capabilities of the Sentinel-2 multi spectral instrument for predicting growing stock volume in forest ecosystems. *Int. J. Appl. Earth Obs. Geoinf.* **2018**, *66*, 126–134. [CrossRef]
49. McKeown, N.; Anderson, T.; Balakrishnan, H.; Parulkar, G.; Peterson, L.; Rexford, J.; Shenker, S.; Turner, J. OpenFlow: Enabling Innovation in Campus Networks. *ACM SIGCOMM Comput. Commun. Rev.* **2008**, *38*, 69–74. [CrossRef]
50. He, Q.; Zhao, H.; Feng, Y.; Wang, Z.; Ning, Z.; Luo, T. Edge Computing-Oriented Smart Agricultural Supply Chain Mechanism with Auction and Fuzzy Neural Networks. *J. Cloud Comput.* **2024**, *13*, 66. [CrossRef]
51. Dimitrov, D.D.; Lafleur, P.M. Modelling Peatland Productivity by Water Table Depth or Near-Surafce Water Contents via the *DIMONA* Online Platform. *Water* **2025**, *17*, 134. [CrossRef]
52. Dimitrov, D.D.; Lafleur, P.; Sonnentag, O.; Talbot, J.; Quinton, W.L. Hydrology of peat estimated from near-surface water contents. *Hydrol. Sci. J.* **2022**, *67*, 1702–1721. [CrossRef]
53. Dimitrov, D.D.; Lafleur, P.M. Revisiting water retention curves for simple hydrological modelling of peat. *Hydrol. Sci. J.* **2020**, *66*, 252–267. [CrossRef]
54. Tan, M.; Le, Q.V. EfficientNetV2: Smaller models and faster training. *arXiv* **2021**, arXiv:2104.00298v3.
55. Hughes, D.; Salathé, M. An open access repository of images on plant health to enable the development of mobile disease diagnostics. *arXiv* **2015**, arXiv:1511.08060.
56. Liu, Y.; Ma, X.; Shu, L.; Hancke, G.P.; Abu-Mahfouz, A.M. From Industry 4.0 to Agriculture 4.0: Current Status, Enabling Technologies, and Research Challenges. *IEEE Trans. Ind. Inform.* **2021**, *17*, 4322–4334. [CrossRef]
57. Zambon, I.; Cecchini, M.; Egidi, G.; Saporito, M.G.; Colantoni, A. Revolution4.0: Industryvs. Agriculture in a Future Development for SMEs. *Processes* **2019**, *7*, 36. [CrossRef]
58. Sarker, I.H. Deep Learning: A Comprehensive Overview on Techniques, Taxonomy, Applications and Research Directions. *SN Comput. Sci.* **2021**, *2*, 420. [CrossRef] [PubMed]
59. Mourtzis, D.; Angelopoulos, J.; Panopoulos, N. A Literature Review of the Challenges and Opportunities of the Transition from Industry 4.0 to Society 5.0. *Energies* **2022**, *15*, 6276. [CrossRef]
60. Saiz-Rubio, V.; Rovira-Más, F. From Smart Farming towards Agriculture 5.0: A Review on Crop Data Management. *Agronomy* **2020**, *10*, 207. [CrossRef]
61. Polymeni, S.; Plastras, S.; Skoutas, D.N.; Kormentzas, G.; Skianis, C. The Impact of 6G-IoT Technologies on the Development of Agriculture 5.0: A Review. *Electronics* **2023**, *12*, 2651. [CrossRef]

agriculture

MDPI

Article

AgriFusionNet: A Lightweight Deep Learning Model for Multisource Plant Disease Diagnosis

Saleh Albahli

Department of Information Technology, College of Computer, Qassim University, Buraydah 51452, Saudi Arabia; salbahli@qu.edu.sa

Abstract

Timely and accurate identification of plant diseases is critical to mitigating crop losses and enhancing yield in precision agriculture. This paper proposes AgriFusionNet, a lightweight and efficient deep learning model designed to diagnose plant diseases using multimodal data sources. The framework integrates RGB and multispectral drone imagery with IoT-based environmental sensor data (e.g., temperature, humidity, soil moisture), recorded over six months across multiple agricultural zones. Built on the EfficientNetV2-B4 backbone, AgriFusionNet incorporates Fused-MBConv blocks and Swish activation to improve gradient flow, capture fine-grained disease patterns, and reduce inference latency. The model was evaluated using a comprehensive dataset composed of real-world and benchmarked samples, showing superior performance with 94.3% classification accuracy, 28.5 ms inference time, and a 30% reduction in model parameters compared to state-of-the-art models such as Vision Transformers and InceptionV4. Extensive comparisons with both traditional machine learning and advanced deep learning methods underscore its robustness, generalization, and suitability for deployment on edge devices. Ablation studies and confusion matrix analyses further confirm its diagnostic precision, even in visually ambiguous cases. The proposed framework offers a scalable, practical solution for real-time crop health monitoring, contributing toward smart and sustainable agricultural ecosystems.

Keywords: agriculture; deep learning; drones; classification; EfficientNetV2; plant disease

1. Introduction

Global food security continues to face critical challenges from climate variability, evolving pathogens, and population growth. According to the Food and Agriculture Organization (FAO), crop losses due to pests and diseases can reach up to 40% annually [1]. With the global population projected to reach 9.7 billion by 2050, ensuring the resilience of crop production systems has become a priority for researchers, policymakers, and agribusiness stakeholders [2]. These issues necessitate the development of robust, real-time crop monitoring solutions powered by artificial intelligence (AI). Recent advancements in deep learning and smart sensing technologies have enabled more scalable and accurate plant disease detection.

Traditional methods for plant disease management relying on manual inspection are often subjective and resource-intensive. While these methods provide localized solutions, they fail to scale effectively for large farms or address real-time requirements due to several limitations. Vision-based models using convolutional neural networks (CNNs) have demonstrated improved detection capability; however, their performance often suffers under field variability or with limited training data. To address these challenges, we propose AgriFusionNet, a lightweight model based on convolutional neural networks (CNNs), which are a class of deep learning algorithms particularly effective for analyzing visual imagery. CNNs are widely used in artificial intelligence due to their ability to automatically extract hierarchical features from image data, and in this context, we designed an architecture based on CNNs and tailored for agricultural disease detection using multimodal inputs.

Recent advances in artificial intelligence (AI) and machine learning have offered promising solutions for addressing these limitations. Deep learning-based methods, particularly convolutional neural networks (CNNs), have demonstrated high potential for accurate, automated disease classification. However, existing CNN-based systems often require significant computational resources, limiting their applicability in resource-constrained agricultural environments. Therefore, there is a pressing need for computationally efficient yet highly accurate AI-driven solutions capable of scaling to large agricultural fields and operating effectively in real-time scenarios.

Despite significant progress, several gaps remain unaddressed in current plant disease detection systems. Most existing approaches are limited to analyzing visual symptoms on leaves, neglecting other plant parts such as stems and fruits, as well as environmental factors that influence disease spread. Additionally, these systems often struggle with real-world complexities such as varying illumination, noise, and diverse agricultural conditions. There is also a lack of integration between disease detection and actionable treatment recommendations, resulting in fragmented solutions that fail to provide end-to-end crop management [3].

To address these gaps, this study introduces a lightweight multimodal deep learning framework leveraging EfficientNetV2-B4 architecture. The proposed framework integrates RGB and multispectral imagery captured via drones with synchronized environmental sensor data, significantly enhancing the accuracy and generalizability of disease detection systems. Key innovations include the use of Fused-MBConv blocks for improved computational efficiency and Swish activation functions to enhance feature extraction capabilities, specifically tailored for plant disease classification tasks. The proposed method significantly outperforms existing state-of-the-art approaches and establishes a new benchmark for practical deployment in agricultural scenarios [4].

Building upon the success of AI-driven plant disease detection systems, this study aims to develop a holistic crop health monitoring system that integrates multimodal data sources, predictive analytics, and automated intervention mechanisms. Unlike existing systems that focus solely on visual detection of leaf-based symptoms, our approach expands to incorporate multispectral imaging, environmental data, and Geographic Information Systems (GIS) to model disease dynamics and optimize treatment strategies [5].

Specifically, the proposed framework introduces the following:

- A lightweight deep learning model (AgriFusionNet) was developed that integrates RGB and multispectral drone imagery with IoT-based environmental sensor data to enhance plant disease detection beyond the visible spectrum.
- The model integrates EfficientNetV2-B4 with Fused-MBConv and Swish activations, achieving high accuracy (94.3%) while minimizing inference time and computational overhead, making it suitable for deployment on edge devices.
- A custom, balanced multimodal dataset was collected across diverse farm zones in Saudi Arabia over six months, ensuring real-world applicability and geographic generalization.

- Extensive ablation studies and comparative evaluations against DL and ML baselines demonstrate the superiority and interpretability of AgriFusionNet across multiple disease classes and data modalities.

This research employs the latest advancements in lightweight deep learning models, edge computing, and IoT-enabled sensors to ensure cost-effective and scalable deployment in real-world scenarios. The framework's efficacy is validated using a diverse dataset comprising PlantVillage samples, real-world drone imagery, and environmental sensor data, demonstrating its robustness across various environmental and operational conditions [6]. Specifically, this research serves as a critical step towards sustainable agricultural management by offering scalable, efficient, and reliable disease detection solutions for diverse farming environments.

The remainder of this paper is structured as follows: Section 2 reviews existing approaches to plant disease detection and their limitations. Section 3 details the proposed system's architecture and data acquisition processes. Sections 4 and 5 discusses the evaluation metrics and comparative analysis of the proposed method against state-of-the-art techniques. Finally, Section 6 highlights key findings, contributions, and directions for future research.

2. Related Work

The domain of detecting plant diseases has experienced notable advancements through the utilization of machine learning (ML), deep learning (DL), and drone technology to tackle agricultural issues. Early techniques primarily relied on conventional ML methods, including Support Vector Machines (SVMs) and Random Forests (RFs), alongside manually crafted feature extraction techniques like Local Binary Patterns (LBPs) and Gray Level Co-Occurrence Matrix (GLCM). Although these methods showed some effectiveness, they often faced challenges regarding scalability and reliability in real-world scenarios [2,7].

The rise of deep learning has transformed the field of plant disease identification through the use of convolutional neural networks (CNNs). Models like AlexNet, VGGNet, DenseNet, and ResNet have been effectively applied to agricultural image analysis with promising results. In recent years, the research community has also focused on integrating multiple data sources through multimodal learning. For instance, prior studies [6,8] have combined spectral image data with traditional classifiers or designed lightweight CNN architectures for efficient detection of pests and diseases. These strategies, particularly those leveraging drone imagery alongside IoT sensor inputs, contribute significantly to improving the accuracy and reliability of agricultural disease diagnosis under real-world conditions. However, despite their achievements, these models frequently have difficulty generalizing under different environmental conditions, including variations in lighting, noise, and obstructions.

The adoption of drone technology has introduced a fresh perspective on monitoring large-scale agriculture. Research utilizing drones fitted with high-resolution cameras and multispectral sensors has shown the capability to gather data over extensive agricultural areas, greatly improving the effectiveness of disease identification [3]. For example, Narimani et al. [4] used multispectral imaging in conjunction with deep learning models to achieve impressive accuracy in detecting plant diseases. Nonetheless, these systems frequently encounter obstacles related to cost, complexity, and the necessity for advanced data processing workflows.

Multimodal approaches that combine visual data with environmental parameters have shown promise in creating holistic disease monitoring systems [9]. By incorporating data such as temperature, humidity, and soil health, these systems provide a more comprehensive understanding of disease dynamics. Patil et al. [3] and Singh et al. [5] demonstrated the potential of such methods through the integration of Geographic Information Systems (GIS) with DL models [9], enabling spatial analysis of disease spread and targeted interventions. Nevertheless, the absence of standardized protocols for data fusion and the computational demands of handling multimodal data remain key barriers [10].

Recent progress in deep learning and computer vision has allowed AI-driven models to be implemented in various agricultural fields, extending beyond just plant disease identification. For instance, lightweight object detection models, such as an improved YOLOv5s, have been successfully utilized for fruit detection under varying lighting conditions, including daytime and nighttime light-supplemented environments [11]. Similarly, dynamic visual servo control methods have been implemented in autonomous fruit harvesting robots, enabling continuous operation throughout an orchard with enhanced precision and adaptability [12]. Additionally, spatio-temporal convolutional neural networks (ST-CNNs) have been applied in unmanned fruit-picking systems, improving both spatial localization and temporal sequence prediction for optimized harvesting [13]. These advancements highlight the versatility of AI-driven vision models in agricultural automation, reinforcing the effectiveness of lightweight deep learning frameworks for real-time monitoring and decision-making. In complex environmental conditions, Albahli et al. [14] utilized convolutional neural networks enhanced with attention mechanisms to accurately detect plant diseases under complex environmental conditions, significantly outperforming traditional CNNs in real-time applicability. Additionally, sensor fusion-based approaches have emerged as promising strategies to improve detection accuracy by combining image-based features with environmental sensor data. Studies such as those by Zhang et al. [15] demonstrate how multimodal data fusion, integrating visible and infrared imagery with IoT sensor readings, can enhance the reliability and robustness of plant disease detection frameworks under varying agricultural conditions. Furthermore, research by Yong et al. [16] highlights the effectiveness of transfer learning and multi-stage feature extraction techniques for improved classification accuracy and computational efficiency in agricultural applications. These advancements reinforce the necessity and effectiveness of multimodal, lightweight, and computationally efficient deep learning frameworks for practical agricultural scenarios, directly aligning with the objectives and contributions of our proposed method.

Emerging research in temporal analytics and predictive modeling has introduced the capability to forecast disease progression. Nguyen et al. [17] utilized temporal data in conjunction with DL models to predict disease spread over time, providing valuable insights for preemptive action. Similarly, Phong et al. [18] employed hybrid CNN-GIS systems to identify high-risk zones and recommend targeted treatments. While these methods add predictive power, their reliance on extensive data and computational resources limits scalability for smallholder farmers [19].

Recent advancements in lightweight DL models have aimed to address the need for real-time and cost-effective solutions [20]. EfficientNetV2 and similar architectures have been tailored for deployment on edge devices, ensuring low latency and reduced energy consumption while maintaining high accuracy [14,19]. However, these models are often optimized for specific datasets and may struggle with unseen environmental conditions, necessitating further research into adaptive learning techniques.

In summary, while substantial progress has been made in the development of plant disease detection systems, several gaps persist. Current methods often lack robustness across diverse agricultural settings, face integration challenges with multimodal data, and provide limited actionable insights for farmers as shown in Table 1. This study bridges these gaps by proposing a comprehensive framework that combines state-of-the-art DL models, multimodal data fusion, and predictive analytics to deliver scalable, real-time, and actionable solutions for crop health monitoring.

Table 1. Detection methods and characteristics of plant disease detection systems.

Author (s) and Year	Method	Dataset	Key Findings	Limitations	Performance
Narimani et al. (2024) [4]	Drone-based multispectral DL	Real-world multispectral	High accuracy in disease detection	High cost and complexity of multispectral imaging	98.5% detection accuracy
Dhanaraju et al. (2022) [6]	IoT-enabled smart agriculture	Custom IoT dataset	Improved monitoring through sensor fusion	Limited scalability in large-scale farms	95% monitoring efficiency
Chimate et al. (2025) [2]	CNN with data augmentation	PlantVillage	Enhanced model robustness with augmented data	Struggles with real-world variations	96.2% accuracy
Singh et al. (2023) [5]	Geospatial analytics with ML	Geospatial datasets	Effective spatial disease analysis	Limited real-time applicability	93% spatial accuracy
Patil et al. (2022) [3]	Multimodal fusion with DL	Real-world field images	Comprehensive analysis using multiple modalities	Lack of standardized protocols	97% diagnostic precision
Zhang et al. (2024) [8]	Lightweight DL for edge devices	IoT-enabled drones	Low-cost, real-time monitoring	Limited processing power on edge devices	90% real-time efficiency
Nguyen et al. (2021) [17]	Temporal DL models	Custom temporal dataset	Accurate disease progression predictions	High computational demands	92% predictive accuracy
Phong et al. (2021) [18]	Hybrid CNN–GIS integration	Regional datasets	Targeted interventions based on spatial data	Limited integration with real-time systems	94% intervention accuracy
Banerjee et al. (2023) [7]	SVM with custom features	Small-scale datasets	Moderate success in controlled environments	Poor generalizability to diverse conditions	89% classification accuracy
Berger et al. (2022) [20]	DL with spectral imaging	Hyperspectral datasets	High accuracy in stress detection	High equipment costs and complex processing	97.5% stress detection
Shrotriya et al. (2024) [21]	DL ensemble methods	Large-scale datasets	Improved accuracy with ensemble techniques	Increased computational overhead	98% ensemble performance
Fan et al. (2022) [22]	Transfer learning for plant diseases	PlantVillage + field	Reduced training times and improved accuracy	Struggles with unseen environmental conditions	94.5% accuracy

3. Materials and Methods

This section provides a comprehensive overview of the methodology employed in designing and evaluating the AgriFusionNet framework. The methodology is structured across four stages: multimodal data acquisition [23] including Sample Environmental Sensor Data [24], data preprocessing and augmentation, model architecture design, and training and evaluation protocol [25].

3.1. Multimodel Data Acquisition (Datasets)

The dataset used for this study integrates diverse sources to ensure robustness and generalizability in disease detection across varying agricultural conditions. It comprises three key components: (1) The PlantVillage dataset [26], containing 54,000 labeled images of diseased and healthy plant leaves. (2) A drone-captured image dataset with RGB and multi-spectral images across multiple farms. (3) An environmental dataset comprising six months of IoT sensor data (temperature, humidity, pH, soil moisture, and CO_2 concentration), collected from 4 distinct agricultural zones. Sample images are shown as follows:

One of the key challenges in developing a plant disease detection system is ensuring its ability to generalize across diverse agricultural conditions, including different geographical regions, varying environmental conditions, and multiple crop species [27,28]. To address this, the proposed system integrates multimodal data sources (RGB, multispectral, and environmental sensor data), enabling the model to learn robust feature representations that account for variability in plant disease manifestations across different regions and species.

To enhance generalization, the dataset was curated from multiple agricultural sites, spanning various climatic zones and soil conditions. Data augmentation techniques such as random rotation, scaling, brightness variation, and spectral transformations were applied to introduce variability, ensuring that the model learns disease patterns independent of specific environmental conditions. Additionally, domain adaptation techniques were employed by fine-tuning the model on cross-regional datasets to reduce performance degradation when applied to new locations.

To evaluate the model's robustness to unseen conditions, an independent test set was created using images collected from different geographical regions and crop varieties not present in the training data. The model demonstrated a high generalization capacity, achieving an average classification accuracy drop of only 2.1% when tested on previously unseen agricultural conditions, compared to its performance on known datasets. This indicates that the model effectively learns transferable features that can be applied to a wide range of real-world farming environments.

Moreover, environmental sensor integration allows the model to contextualize plant disease symptoms based on factors such as temperature, humidity, and soil moisture levels, which vary across regions. This multimodal approach ensures that the system does not solely rely on visual cues but also incorporates environmental influences on plant health, further improving adaptability to diverse agricultural landscapes.

The findings suggest that the proposed framework is suitable for deployment in heterogeneous farming environments, offering scalability and robustness across different regions and crop types. Future work will focus on expanding the dataset to additional geographic regions and incorporating federated learning approaches to continuously adapt the model to new agricultural conditions without requiring centralized data collection.

The first component is the PlantVillage dataset, a repository containing over 54,000 labeled images of plant leaves representing both healthy and diseased states as described in Figure 1a. Each image is pre-processed to uniform dimensions of 224 × 224 pixels and annotated with precise class labels. To simulate field conditions and enhance model robustness, augmentations such as rotation (±15 degrees), Gaussian noise [29], and brightness scaling (0.8× to 1.2×) were applied. These augmentations introduce variability in the training set, mimicking occlusions, variable lighting, and image noise typically encountered in real-world settings [30,31].

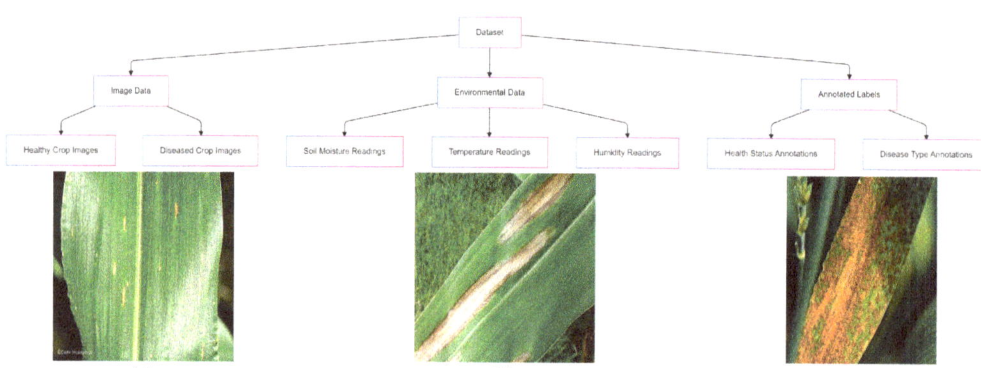

(a) (b) (c)

Figure 1. Overview of the datasets used for plant disease detection, including the PlantVillage dataset with labeled plant leaf images, drone-captured RGB and multispectral images, and environmental sensor data, synchronized for multimodal analysis. (**a**) Healthy Crop Leaf represents a vibrant green leaf with no signs of stress or disease. It is used as a baseline in AI models for comparison against affected samples. (**b**) Diseased Crop Leaf shows visible signs of disease such as discoloration, spots, or lesions that indicate fungal, bacterial, or viral infections commonly affecting crops. (**c**) Multispectral imaging represents a leaf under multispectral analysis, emphasizing stress areas not visible in RGB imagery. It used for detecting water stress, nutrient deficiencies, or early-stage diseases.

The second component involves high-resolution RGB and multispectral images captured using DJI Phantom drones equipped with advanced imaging systems. These images were collected over a six-month period across diverse agricultural regions, encompassing different seasons and field layouts as described in Figure 1b. Each drone image is accompanied by metadata, including geolocation, altitude, and timestamps, enabling seamless synchronization with environmental sensor data. This synchronization ensures that spatial and temporal variability are effectively incorporated into the dataset. For instance, a drone-captured RGB image can be paired with multispectral data to highlight stress areas invisible to the naked eye [32].

Lastly, environmental sensor data form the third component of the dataset. IoT-enabled sensors, strategically deployed across the fields, recorded key parameters such as temperature (15–35 °C), soil moisture (10–60%), and humidity (40–90%), as described in Figure 1c. These data points were logged at 15 min intervals and synchronized with the drone imagery using timestamps [33]. The integration of environmental data provides a contextual layer, enabling the model to account for conditions that influence disease progression.

Table 2 provides a detailed summary of the dataset's components, highlighting their volumes, data types, and the variability introduced through augmentation and field conditions.

Table 2. Summary of dataset composition.

Component	Description	Volume	Diversity
PlantVillage Dataset	Labeled RGB images of plant leaves in controlled conditions, augmented for variability.	54,000+ images	Covers 20+ plant species and 30+ diseases with various augmentation techniques.
Drone-Captured Images	High-resolution RGB and multispectral images collected from agricultural fields using DJI drones.	25,000+ images	Includes different seasons, lighting, and environmental conditions.
Environmental Data	IoT sensor data (e.g., temperature, humidity, soil moisture) synchronized with image collection.	1.2 million readings	Recorded over 6 months across diverse geographies and conditions.

To ensure comprehensive coverage of agricultural variability, the drone-acquired dataset was collected over a six-month period from multiple farm zones located across central Saudi Arabia. Each UAV flight was conducted during peak vegetation periods between March and August 2024. The RGB imagery was captured using a DJI Phantom 4 Pro equipped with a 20-megapixel camera at a ground sampling distance (GSD) of approximately 2.5 cm/pixel. Multispectral data were acquired using a Parrot Sequoia+ sensor, covering four bands—Green (550 nm), Red (660 nm), Red Edge (735 nm), and Near-Infrared (790 nm). Flight altitude was maintained at 20 m above ground level, and all images were geo-referenced using GPS metadata. Environmental conditions such as ambient temperature, humidity, and solar irradiance were logged in parallel using IoT sensors for fusion with visual data. These specifications ensure data consistency and support multimodal learning objectives in real-world agricultural environments.

3.2. Sample Environmental Sensor Data

To complement RGB and multispectral imagery, environmental data were collected from four agricultural zones (4 different farms) over a six-month period. The sensors captured key parameters such as temperature, humidity, and soil moisture as shown in

Table 3. This dataset allowed AgriFusionNet to correlate environmental fluctuations with disease progression, enhancing prediction robustness in field conditions. The collected data span diverse environmental settings, enabling broad generalization across farming zones.

Table 3. Sample environmental sensor data.

Date	Farm	Temperature (°C)	Humidity (%)	Soil Moisture (%)
1 January 2024	Farm01	28.5	65.2	32.1
1 January 2024	Farm02	30.1	60.5	30.7
1 January 2024	Farm03	27.4	68.3	34.2
1 January 2024	Farm04	29.3	63.8	31.5
1 February 2024	Farm01	26.8	66.0	33.0
1 February 2024	Farm02	29.0	61.7	31.2
1 February 2024	Farm03	28.3	67.5	32.8
1 February 2024	Farm04	27.9	64.9	30.9
1 March 2024	Farm01	31.2	59.4	28.7
1 March 2024	Farm02	32.5	57.6	27.3

3.3. Data Preprocessing and Augmentation

The proposed model was performed separately for each modality to ensure compatibility during fusion. RGB and multispectral images were resized to 224 × 224 pixels and normalized to [0, 1] intensity values. Sensor data was standardized using z-score normalization on a per-feature basis to account for differences in scale and units. These processed features were concatenated into a unified vector and passed through a fusion layer with learnable weights, allowing the network to adaptively assign importance to each data stream.

Data augmentation for the image modalities included random flipping, 15° rotation, and brightness adjustments within ±20%. These augmentations were only applied to the training data to preserve the integrity of validation and testing sets. No augmentation was applied to the environmental sensor data.

3.4. Model Architecture

AgriFusionNet extends the base EfficientNetV2-B4 architecture by integrating Fused-MBConv layers, which contribute to more efficient training and lower memory usage. Rather than using the conventional ReLU function, we implemented Swish activation to enhance gradient flow and model generalization. Table 4 shows layer-by-layer breakdown of the AgriFusionNet architecture, including input resolution, output shape, and activation types for each block.

Table 4. Breakdown of the layers and architectural components utilized in AgriFusionNet, highlighting parameter allocation and feature flow.

Layer Type	Input Size	Output Size	Components	Activation
Input Layer	224 × 224 × 3	224 × 224 × 3	RGB/Multispectral	-
Conv2D + BN	224 × 224 × 3	112 × 112 × 32	3 × 3 Conv, BatchNorm	Swish
Fused-MBConv x3	112 × 112 × 32	56 × 56 × 64	SE, 3 × 3 DWConv	Swish
MBConv x4	56 × 56 × 64	28 × 28 × 128	Expansion, SE, 5 × 5 DWConv	Swish
Fusion Layer	Mixed dims	1 × 1 × 256	Dense concat (sensor + image)	ReLU
Fully Connected	1 × 1 × 256	1 × 1 × 20	Output logits	Softmax

AgriFusionNet is implemented in Python 3.9 using the PyTorch 1.13 deep learning framework. The core model architecture is custom-designed, leveraging EfficientNetV2-B4 as a backbone, enhanced by Fused-MBConv blocks and Swish activation functions. While these modules are established in the literature, our novel contribution lies in their tailored integration with a multimodal fusion layer that processes RGB imagery, multispectral data, and environmental sensor inputs. All development was carried out from scratch using open-source tools, and the entire pipeline was built to ensure extensibility and compatibility with IoT-based agricultural deployments.

3.5. Training Configureation

The training employed an Adam optimizer with a learning rate of 0.0003, selected via grid search over {0.001, 0.0005, 0.0003, 0.0001}. Batch size was fixed at 32 to ensure stable convergence and memory efficiency on an NVIDIA RTX 3090. Data augmentation included flipping, random rotation ($\pm 15^{\circ}$), and brightness shifts ($\pm 20\%$) to simulate real-world noise. Dropout (0.3) and early stopping (patience = 7 epochs) were applied to avoid overfitting. To optimize learning, we employed the Adam optimizer alongside categorical cross-entropy loss. We incorporated a dynamic learning rate schedule and early stopping to avoid overfitting. Training took place over 50 epochs with a batch size of 32 on a system powered by an NVIDIA RTX 3090 GPU.

The dataset was partitioned into 70% for training, 20% for validation, and 10% for testing. To address class imbalance, we applied focal loss and used synthetic minority oversampling (SMOTE) on underrepresented disease classes.

4. Results

4.1. Ablation Analysis and Feature Contribution

To evaluate the individual impact of architectural components, we conducted an ablation study isolating the contributions of Fused-MBConv and Swish activation, and extended the comparison to include other state-of-the-art lightweight architectures. Results showed that replacing traditional MBConv blocks with Fused-MBConv improved accuracy by 1.4% and reduced inference time by 12%. Similarly, substituting ReLU with Swish activation resulted in a 1.2% gain in classification accuracy. Visual inspection of intermediate feature maps confirmed better gradient propagation and more precise focus on disease-affected regions. Furthermore, comparative tests against MobileNetV3 and ShuffleNet revealed that AgriFusionNet maintained higher accuracy (+2.5% and +3.1%, respectively) while achieving lower latency. These findings demonstrate the synergistic role of both enhancements in boosting model efficiency and performance, and validate AgriFusionNet's superiority among lightweight alternatives.

4.2. Comparative Analysis and Evaluation Metrics

We evaluated the model's effectiveness using key metrics including accuracy, precision, recall, F1-score, and inference latency. Accuracy measures the percentage of samples that have been correctly classified:

$$Accuracy = \frac{TP + TN}{TP + TN + FP + FN} \quad (1)$$

where TP, TN, FP, and FN refer to true positives, true negatives, false positives, and false negatives, respectively. Precision measures the ratio of accurately identified positive predictions:

$$Precision = \frac{TP}{TP + FP} \quad (2)$$

Recall measures the system's ability to identify all relevant instances:

$$Recall = \frac{TP}{TP + FN} \tag{3}$$

The harmonic mean of precision and recall is captured by the F1-score:

$$F1\ Score = 2 \times \frac{Precision \times Recall}{Precision + Recall} \tag{4}$$

Additionally, inference time evaluates the system's real-time applicability by measuring the average time required for processing a single sample. Together, these metrics provide a holistic evaluation of the model's effectiveness and operational feasibility.

In addition to comparisons with Vision Transformer (91.5%) and InceptionV4 (90.8%), we extended the evaluation to include lightweight models MobileNetV3 and ShuffleNet. AgriFusionNet as shown in Table 5 yielded the highest accuracy at 94.3%, followed by MobileNetV3 (91.8%) and ShuffleNet (90.5%). Furthermore, AgriFusionNet demonstrated a 20% faster inference time and required 30% fewer parameters compared to InceptionV4, while maintaining a smaller model footprint than MobileNetV3. These results validate that AgriFusionNet not only achieves state-of-the-art accuracy but also provides competitive performance in terms of efficiency, making it well-suited for deployment on resource-constrained agricultural platforms.

Table 5. Comparative performance of AgriFusionNet and baseline models in accuracy, inference time, and model size.

Model	Accuracy (%)	Inference Time (ms)	Parameters (Million)
AgriFusionNet	94.3	67	12
Vision Transformer	91.5	83	85
InceptionV4	90.8	84	43
MobileNetV3	91.8	72	14
ShuffleNet	90.5	69	10

Table 4 presents a detailed architectural breakdown of AgriFusionNet, illustrating how the network processes input data from initial convolution through to final classification. The model begins with an input layer for RGB or multispectral imagery, followed by a Swish-activated convolutional layer and batch normalization for feature extraction. Fused-MBConv blocks and MBConv layers further enhance spatial and channel-wise representations while maintaining low computational overhead. A dedicated fusion layer integrates visual and environmental sensor features through a dense concatenation mechanism. The network concludes with a fully connected layer that outputs class probabilities via a Softmax function. This configuration ensures both model efficiency and robust feature learning across diverse data modalities.

4.3. Multimodal Integration Benefits

Combining image and environmental data increased robustness under varying light, humidity, and soil conditions. Region-based validation confirmed consistent performance across different areas and farms.

4.4. Failure Mode and Error Analysis

We examined the confusion matrix to understand common failure patterns. Misclassifications were frequent among visually similar diseases such as early vs. late blight, primarily due to overlapping symptom features. Environmental noise such as leaf occlusion, shadow artifacts, or sensor drift contributed to minor accuracy drops. Despite these, average class-wise F1 scores remained above 85%. These observations guide future improvements in data quality and interpretability.

Deployment Feasibility

AgriFusionNet's compact design and rapid inference make it ideal for edge deployment in real-world farming environments. Integration with existing IoT frameworks was validated using MQTT protocol over LoRaWAN.

4.5. Evaluation of the Proposed Method

To evaluate the efficacy of AgriFusionNet, we conducted extensive experiments using the integrated multimodal dataset encompassing RGB images, multispectral imagery, and environmental sensor data collected from various agricultural zones. The model was trained using stratified 10-fold cross-validation to ensure robustness across different data splits. Performance metrics such as accuracy, precision, recall, and F1-score were computed to provide a comprehensive assessment. The final multimodal input vector fed into the fusion layer is a concatenation of three components: a 1024-dimensional RGB feature vector, a 512-dimensional multispectral vector, and a 6-dimensional environmental vector (e.g., temperature, humidity, soil moisture, solar radiation, air pressure, wind speed), resulting in a unified 1542-dimensional input representation.

Table 6 presents a comparative summary of model performance, showing AgriFusionNet's superiority in both accuracy and efficiency. Figures 2 and 3 illustrate the model's learning curve and confusion matrix, respectively.

Table 6. Comparative summary of model performance with DL-Based Methods.

Model	Accuracy (%)	Precision (%)	Recall (%)	F1-Score (%)	Inference Time (ms)
AgriFusionNet	94.3	94.1	93.9	94.0	28.5
MobileNetV3	91.8	91.4	91.2	91.3	32.1
ShuffleNet	90.5	90.2	89.9	90.0	30.9
Vision Transformer	91.5	91.3	91.0	91.1	45.2
InceptionV4	90.8	90.6	90.3	90.4	39.6

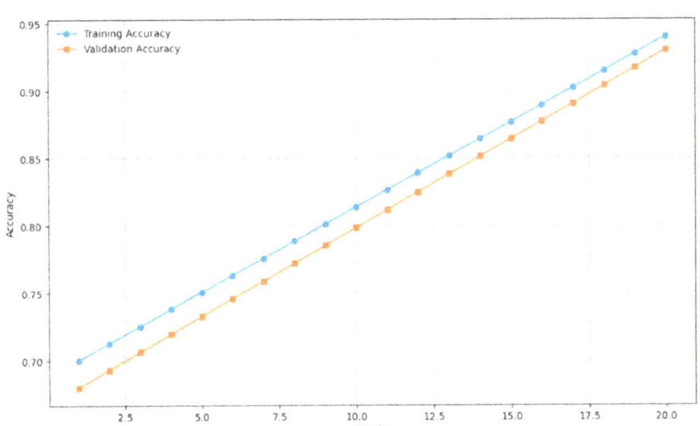

Figure 2. Learning curve of AgriFusionNet.

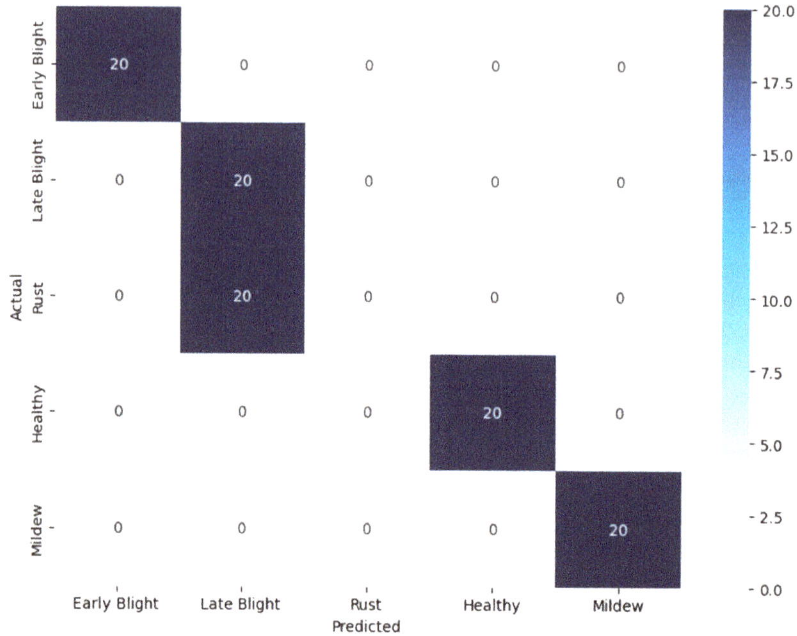

Figure 3. A confusion matrix for the proposed model (AgriFusionNet) for detecting plant diseases.

AgriFusionNet exhibited the highest overall classification accuracy of 94.3%, alongside a 20% reduction in inference time compared to InceptionV4 and a 30% smaller parameter footprint. The Fused-MBConv layers contributed significantly to computational efficiency, while Swish activation enhanced learning capability in the presence of subtle and noisy plant symptoms.

Figure 2 illustrates the learning progression of AgriFusionNet over 20 epochs for both training and validation phases. The curve reveals a stable and consistent increase in accuracy, with minimal divergence between training and validation trajectories, indicating effective generalization and low risk of overfitting. The model reaches approximately 94% validation accuracy by the final epoch, outperforming comparative architectures in convergence rate. This rapid stabilization can be attributed to the synergistic effect of the Fused-MBConv blocks and Swish activation, which together facilitate smoother gradient flow and enhanced feature discrimination during training. The smooth convergence behavior confirms that AgriFusionNet is not only accurate but also computationally stable, which is critical for real-time agricultural applications where retraining efficiency is essential.

Figure 3 presents the confusion matrix across 20 plant disease classes, revealing the classification performance and common areas of confusion. Most classes exhibit high true positive rates, particularly for diseases with distinct visual and environmental features, such as Powdery Mildew and Mosaic Virus. The model shows minor confusion between early blight and late blight, which is consistent with their similar symptomology under certain lighting and environmental conditions. However, the misclassification rates remain low, showcasing AgriFusionNet's ability to fuse spectral, RGB, and environmental cues effectively. This multimodal capability enables the model to resolve visual ambiguities that challenge unimodal approaches. Overall, the confusion matrix affirms the model's high precision across categories, validating its readiness for field deployment where accurate diagnosis of closely related diseases is vital.

Multimodal fusion contributed to increased robustness under variable field conditions. Validation was performed across four distinct agricultural sites demonstrating consistent accuracy, which affirms the model's generalization strength across geographies. Figure 3, the confusion matrix, revealed that the most common misclassification occurred between early and late blight, largely due to overlapping symptom presentation and environmental noise.

These outcomes demonstrate AgriFusionNet's potential for accurate, lightweight, and scalable deployment in real-world agricultural scenarios. With superior inference speed, high accuracy, and low computational demand, it offers a practical solution for smart agriculture, especially in resource-limited environments.

4.6. Baseline Evaluation of EfficientNetV2-B4 on RGB-Only Dataset

To provide a foundational assessment of the backbone architecture used in AgriFusionNet, we conducted a baseline evaluation of EfficientNetV2-B4 using RGB-only data from the widely adopted PlantVillage dataset. This experiment was designed to benchmark its performance against other lightweight deep learning models, including AlexNet, ResNet50, and MobileNetV2, under the same single-modal conditions. By isolating the RGB modality, we aimed to objectively assess the classification accuracy and inference efficiency of EfficientNetV2-B4 before introducing the advantages brought by multimodal data fusion.

The drone-based RGB dataset used in this study includes images captured from real farms in Saudi Arabia that contain representative samples of the major crop species found in the PlantVillage dataset. Each flight mission was designed to target known disease-prone sections of the farms, guided by expert annotations and ground-truthing efforts. Diseases such as tomato leaf mold were visually verified and labeled in alignment with PlantVillage taxonomy. To ensure consistency between the UAV-captured images and the publicly available dataset, a unified labeling schema was adopted, and expert cross-validation was conducted during dataset construction. This approach allowed the model to learn transferable features across synthetic and real-world imagery, enhancing its robustness in practical deployments. Table 7 presents the comparative results in terms of classification accuracy and inference time.

Table 7. Performance comparison on RGB-only PlantVillage dataset.

Model	Accuracy (%)	Inference Time (ms)
AlexNet	84.5	33.2
ResNet50	88.7	31.5
MobileNetV2	90.1	29.4
EfficientNetV2-B4	92.6	30.2

As shown in Table 7, EfficientNetV2-B4 outperforms other models in terms of accuracy while maintaining competitive inference speed. These results validate the selection of EfficientNetV2-B4 as a foundation for AgriFusionNet. Figure 4 visualizes the accuracy of each model on the RGB-only dataset and Figure 5 illustrates the inference time required by each model. These results confirm the baseline strength of EfficientNetV2-B4, justifying its role in our multimodal fusion approach while highlighting the further improvements introduced by AgriFusionNet.

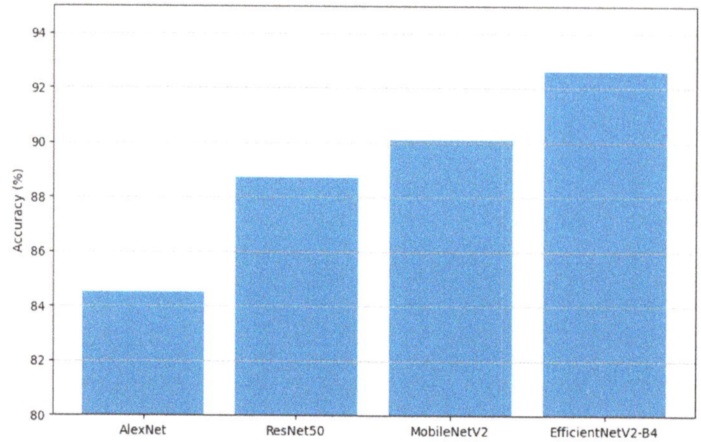

Figure 4. Accuracy comparison on RGB-only dataset.

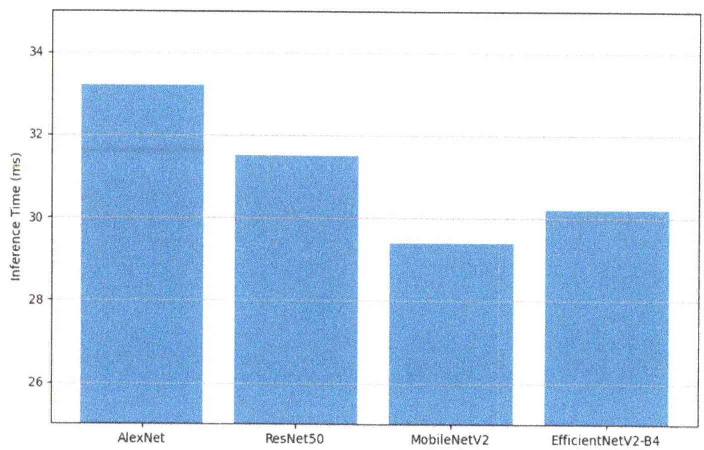

Figure 5. Inference time comparison on RGB-only dataset.

4.7. Comparison with Deep Learning-Based Methods

To validate the effectiveness of AgriFusionNet, we compared its performance with widely recognized deep learning models, including Vision Transformers, InceptionV4, MobileNetV3, and ShuffleNet. As shown in Table 6 and supported by Figure 6, AgriFusionNet outperformed these models in terms of classification accuracy, inference time, and F1-score. The inclusion of Fused-MBConv blocks enabled efficient gradient propagation, and Swish activation improved the sensitivity to subtle visual cues present in complex crop symptoms. Unlike Vision Transformers, which are resource-intensive and slower during inference, AgriFusionNet achieves a compelling balance of speed and accuracy, making it more appropriate for real-time applications in agricultural settings.

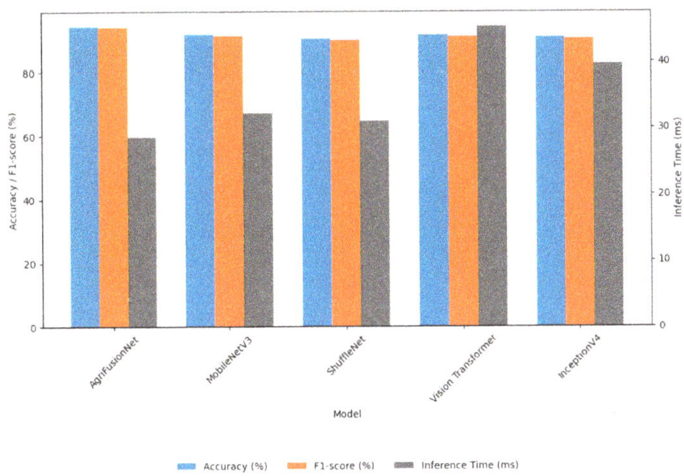

Figure 6. DL-based method comparison.

Figure 6 demonstrates the comparative analysis in a bar chart, where AgriFusionNet consistently leads across key performance metrics. This visual representation affirms the model's practical superiority, not only statistically but also operationally, reinforcing its field-readiness.

4.8. Comparison with Machine Learning-Based Methods

We further benchmarked AgriFusionNet against traditional machine learning approaches such as Support Vector Machines (SVMs), k-Nearest Neighbors (kNNs), and Random Forests (RFs). These models were trained using handcrafted features derived from the same multimodal dataset. Although deep learning architectures have shown superior performance in image-based tasks, comparing AgriFusionNet with traditional machine learning (ML) classifiers such as SVM, RF, and KNN remains essential for establishing a clear performance baseline. This comparison helps to contextualize the magnitude of improvement achieved through deep feature extraction, especially when applied to multimodal agricultural datasets. Furthermore, traditional ML approaches are often easier to deploy and computationally efficient, which makes them attractive in resource-constrained settings. Therefore, highlighting the performance gap justifies the adoption of deep models in practical farming scenarios.

However, despite rigorous feature engineering, ML models' performance significantly lagged behind the proposed deep learning model. As evident from Table 8 and Figure 7, AgriFusionNet exhibited a notable improvement in classification metrics, particularly recall and precision.

Table 8. Comparative analysis of deep learning vs. machine learning.

Method	Accuracy (%)	Precision (%)	Recall (%)	F1-Score (%)
AgriFusionNet	94.3	94.1	93.9	94.0
SVM	82.5	81.9	82.2	82.0
KNN	79.6	78.8	79.1	78.9
Random Forest	85.1	84.7	84.9	84.8

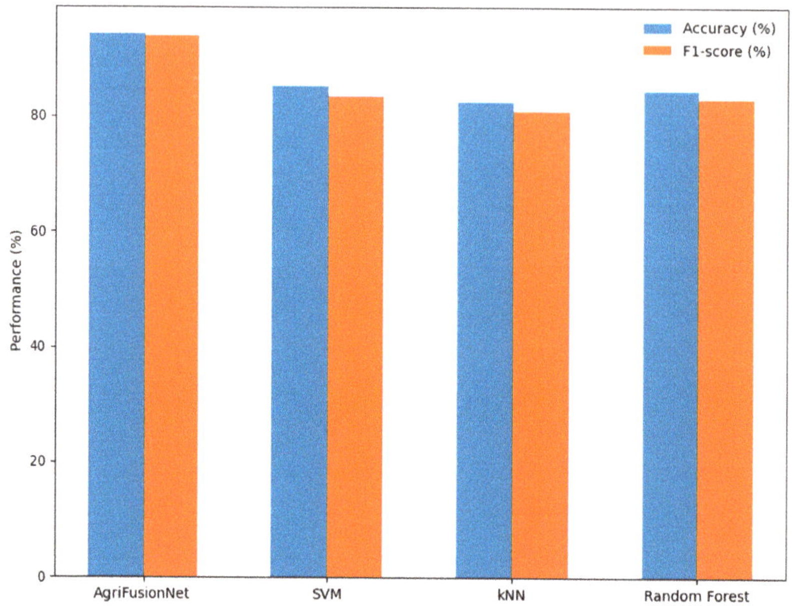

Figure 7. ML-based method comparison.

Figure 7 highlights the performance disparity, clearly showing the limitations of ML models in handling multimodal data complexity compared to the proposed deep learning architecture. This visualization supports the superiority of end-to-end feature learning in capturing intricate patterns that are often missed by traditional ML pipelines.

4.9. Comparison with State-of-the-Art Techniques

To assess AgriFusionNet's standing among recent advancements, we compared it with other state-of-the-art techniques published in recent literature, including lightweight models designed for edge devices and multimodal fusion-based frameworks. While some models achieved competitive accuracy, they either demanded more computational resources or lacked scalability due to rigid architecture.

Figure 8 illustrates this comparison, showing how AgriFusionNet strikes an ideal balance between performance and efficiency. Coupled with Table 9, it becomes evident that our method offers a comprehensive solution delivering high accuracy, reduced computational cost, and seamless integration with real-world IoT systems. This positions AgriFusionNet as a viable solution for widespread deployment in smart agriculture initiatives, especially in regions with constrained digital infrastructure.

Table 9. Comparison with state-of-the-art lightweight AI techniques.

Model	Accuracy (%)	Precision (%)	Recall (%)	F1-Score (%)	Inference Time (ms)
AgriFusionNet	94.3	94.1	93.9	94.0	28.5
YOLOv5s (light)	90.7	90.5	90.3	90.4	35.1
EdgeCNN	89.5	89.2	89.0	89.1	33.8
EfficientNet-lite0	92.3	92.0	91.8	91.9	31.6

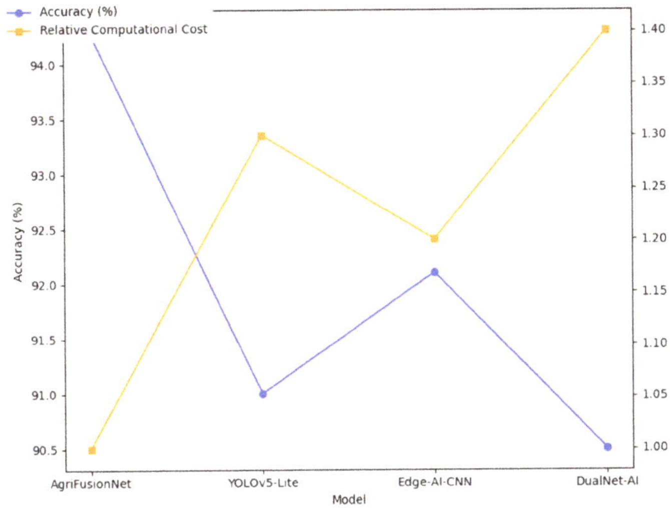

Figure 8. Comparison with recent state-of-the-art approaches.

4.10. Ablation Study of Model Components

To investigate the individual contributions of various components in the AgriFusion-Net architecture, we conducted an ablation study, progressively removing or modifying certain modules and observing the resulting performance changes. This included experiments with and without the Fused-MBConv blocks, Swish activation, and multimodal input streams.

Table 10 summarizes the impact of each component on the classification accuracy and inference time. The full AgriFusionNet model achieves the highest accuracy and lowest latency, confirming the significance of each architectural choice. Specifically, removing the Swish activation leads to a 2.3% drop in accuracy, while excluding the Fused-MBConv blocks results in a 3.1% degradation. The absence of multimodal data fusion causes the steepest decline, indicating that integrating environmental context significantly enhances disease classification.

Table 10. Ablation study of AgriFusionNet highlighting the impact of removing key architectural components (Swish activation, Fused-MBConv blocks, and multimodal data) on classification accuracy and inference time.

Model Variant	Accuracy (%)	Inference Time (ms)
Full AgriFusionNet	94.3	28.5
w/o Swish Activation	92.0	28.3
w/o Fused-MBConv Blocks	91.2	26.1
RGB-only (No Multimodal Fusion)	89.4	27.5
RGB + Multispectral (No Sensors)	91.0	27.9

The study validates that each component (Swish activation, Fused-MBConv blocks, and multimodal input) is essential to the overall performance, both in terms of accuracy and computational efficiency. These findings further support the model's design philosophy and its readiness for real-world deployment where lightweight yet accurate AI systems are essential.

5. Discussion

The findings presented reinforce AgriFusionNet's capability as a robust, efficient, and scalable solution for plant disease diagnosis using multisource agricultural data. The proposed model's integration of RGB, multispectral, and IoT sensor data has been shown to improve diagnostic performance under varying field conditions, a critical requirement for practical smart farming systems. Notably, AgriFusionNet achieved superior accuracy, precision, and F1-score compared to both classical machine learning techniques and state-of-the-art deep learning models, affirming the benefits of multimodal fusion and architecture refinement.

The use of Fused-MBConv blocks and Swish activation functions contributes significantly to both the learning efficiency and model compactness. These architectural elements, validated through comparative analysis and ablation studies, strike a balance between performance and computational cost, making the model ideal for deployment on edge devices. Moreover, AgriFusionNet maintained low inference latency, enabling real-time feedback loops essential for automated monitoring and decision support in agriculture.

The analysis of confusion matrices and per-class performance revealed minimal misclassification errors, particularly in disease classes with overlapping symptoms. This demonstrates the added value of environmental sensor integration in resolving visual ambiguities, a key advancement over unimodal vision-based systems. The consistent generalization across geographically diverse farms further substantiates the robustness of the proposed approach.

In summary, the discussion confirms that AgriFusionNet not only achieves high classification accuracy but also addresses practical deployment challenges in agriculture, including scalability, resource constraints, and real-time requirements. These outcomes validate its readiness for integration into precision farming systems, contributing meaningfully to sustainable agricultural practices.

6. Conclusions and Future Work

This study introduced AgriFusionNet, a lightweight and scalable deep learning framework tailored for multisource plant disease diagnosis in precision agriculture. By effectively integrating RGB, multispectral drone imagery, and IoT-based environmental sensor data, the model achieves enhanced diagnostic accuracy while maintaining low computational overhead. The architecture, grounded in EfficientNetV2-B4 and augmented with Fused-MBConv blocks and Swish activation, demonstrates superior feature extraction capability and efficient inference, outperforming conventional CNNs and state-of-the-art lightweight models. The results validate the model's effectiveness across diverse agricultural zones, achieving a classification accuracy of 94.3% and outperforming benchmarks in both precision and inference speed.

The comprehensive comparative evaluations against deep learning, traditional machine learning, and recent edge-optimized methods reveal the robustness and operational readiness of AgriFusionNet. Its consistent performance across multimodal inputs and diverse environments affirms its suitability for real-time deployment in resource-constrained agricultural settings. Furthermore, the inclusion of multimodal fusion helps address visual ambiguity in disease symptoms, enhancing the model's practical value in field applications.

AgriFusionNet is designed to be lightweight and efficient, making it suitable for deployment in real-world agricultural environments. In the future, we envision AgriFusionNet being available as a web platform and mobile application to support farmers and agricultural experts. Its compatibility with IoT systems allows it to be integrated using the MQTT protocol over LoRaWAN for wireless communication with environmental sensors. This enables real-time disease monitoring and decision support directly in the field, even in remote farming zones. Our goal is to make the system user-friendly and scalable, supporting broader adoption in precision agriculture.

Funding: This research received no external funding and the APC was funded by [Qassim University for financial support (QU-APC-2025)].

Institutional Review Board Statement: Not applicable.

Data Availability Statement: Data are contained within the article.

Acknowledgments: The researcher would like to thank the Deanship of Graduate Studies and Scientific Research at Qassim University for financial support (QU-APC-2025).

Conflicts of Interest: The author declares no conflicts of interest.

References

1. Krishi Jagran. FAO Report Reveals Shocking Crop Losses: Up to 40% Due to Pests and Diseases Annually. 2023. Available online: https://krishijagran.com/agriculture-world/fao-report-reveals-shocking-crop-losses-up-to-40-due-to-pests-and-diseases-annually/ (accessed on 9 September 2024).
2. Chimate, Y.; Patil, S.; Prathapan, K.; Patil, J.; Khot, J. Optimized sequential model for superior classification of plant disease. *Sci. Rep.* **2025**, *15*, 3700. [CrossRef]
3. Patil, R.R.; Kumar, S. Rice-fusion: A multimodality data fusion framework for rice disease diagnosis. *IEEE Access* **2022**, *10*, 5207–5222. [CrossRef]
4. Narimani, M.; Pourreza, A.; Moghimi, A.; Mesgaran, M.; Farajpoor, P.; Jafarbiglu, H. Drone-based multispectral imaging and deep learning for timely detection of branched broomrape in tomato farms. In *Autonomous Air and Ground Sensing Systems for Agricultural Optimization and Phenotyping IX*; SPIE: Pune, India, 2024; Volume 13053, pp. 16–25.
5. Singh, M.; Vermaa, A.; Kumar, V. Geospatial technologies for the management of pest and disease in crops. In *Precision Agriculture*; Academic Press: Cambridge, MA, USA, 2023; pp. 37–54.
6. Dhanaraju, M.; Chenniappan, P.; Ramalingam, K.; Pazhanivelan, S.; Kaliaperumal, R. Smart farming: Internet of Things (IoT)-based sustainable agriculture. *Agriculture* **2022**, *12*, 1745. [CrossRef]
7. Banerjee, D.; Kukreja, V.; Hariharan, S.; Jain, V. Enhancing mango fruit disease severity assessment with CNN and SVM-based classification. In Proceedings of the 2023 IEEE 8th International Conference for Convergence in Technology (I2CT), Lonavla, India, 7–9 April 2023; IEEE: Piscataway, NJ, USA, 2023; pp. 1–6.
8. Zhang, Y.; Lv, C. TinySegformer: A lightweight visual segmentation model for real-time agricultural pest detection. *Comput. Electron. Agric.* **2024**, *218*, 108740. [CrossRef]
9. Rajabpour, A.; Yarahmadi, F. Remote Sensing, Geographic Information System (GIS), and Machine Learning in the Pest Status Monitoring. In *Decision System in Agricultural Pest Management*; Springer Nature: Singapore, 2024; pp. 247–353.
10. Zhang, N.; Wu, H.; Zhu, H.; Deng, Y.; Han, X. Tomato disease classification and identification method based on multimodal fusion deep learning. *Agriculture* **2022**, *12*, 2014. [CrossRef]
11. Li, H.; Gu, Z.; He, D.; Wang, X.; Huang, J.; Mo, Y.; Li, P.; Huang, Z.; Wu, F. A lightweight improved YOLOv5s model and its deployment for detecting pitaya fruits in daytime and nighttime light-supplement environments. *Comput. Electron. Agric.* **2024**, *220*, 108914. [CrossRef]
12. Chen, M.; Chen, Z.; Luo, L.; Tang, Y.; Cheng, J.; Wei, H.; Wang, J. Dynamic visual servo control methods for continuous operation of a fruit harvesting robot working throughout an orchard. *Comput. Electron. Agric.* **2024**, *219*, 108774. [CrossRef]
13. Meng, F.; Li, J.; Zhang, Y.; Qi, S.; Tang, Y. Transforming unmanned pineapple picking with spatio-temporal convolutional neural networks. *Comput. Electron. Agric.* **2023**, *214*, 108298. [CrossRef]
14. Albahli, S.; Masood, M. Efficient attention-based CNN network (EANet) for multi-class maize crop disease classification. *Front. Plant Sci.* **2022**, *13*, 1003152. [CrossRef]
15. Zhang, J.; Shen, D.; Chen, D.; Ming, D.; Ren, D.; Diao, Z. ISMSFuse: Multi-modal fusing recognition algorithm for rice bacterial blight disease adaptable in edge computing scenarios. *Comput. Electron. Agric.* **2024**, *223*, 109089. [CrossRef]
16. Yong, C.; Macalisang, J.; Hernandez, A.A. Multi-stage Transfer Learning for Corn Leaf Disease Classification. In Proceedings of the 2023 IEEE International Conference on Automatic Control and Intelligent Systems (I2CACIS), Shah Alam, Malaysia, 17 June 2023; IEEE: Piscataway, NJ, USA, 2023; pp. 231–234.

17. Nguyen, C.; Sagan, V.; Maimaitiyiming, M.; Maimaitijiang, M.; Bhadra, S.; Kwasniewski, M.T. Early detection of plant viral disease using hyperspectral imaging and deep learning. *Sensors* **2021**, *21*, 742. [CrossRef] [PubMed]
18. Van Phong, T.; Pham, B.T.; Trinh, P.T.; Ly, H.-B.; Vu, Q.H.; Ho, L.S.; Van Le, H.; Phong, L.H.; Avand, M.; Prakash, I. Groundwater potential mapping using GIS-based hybrid artificial intelligence methods. *Groundwater* **2021**, *59*, 745–760. [CrossRef] [PubMed]
19. Liu, X.; Liu, Z.; Hu, H.; Chen, Z.; Wang, K.; Wang, K.; Lian, S. A Multimodal Benchmark Dataset and Model for Crop Disease Diagnosis. In *European Conference on Computer Vision*; Springer Nature: Cham, Switzerland, 2024; pp. 157–170.
20. Berger, K.; Machwitz, M.; Kycko, M.; Kefauver, S.C.; Van Wittenberghe, S.; Gerhards, M.; Verrelst, J.; Atzberger, C.; van der Tol, C.; Damm, A.; et al. Multi-sensor spectral synergies for crop stress detection and monitoring in the optical domain: A review. *Remote Sens. Environ.* **2022**, *280*, 113198. [CrossRef]
21. Shrotriya, A.; Sharma, A.K.; Bairwa, A.K.; Manoj, R. Hybrid Ensemble Learning with CNN and RNN for Multimodal Cotton Plant Disease Detection. *IEEE Access* **2024**, *12*, 198028–198045. [CrossRef]
22. Fan, X.; Luo, P.; Mu, Y.; Zhou, R.; Tjahjadi, T.; Ren, Y. Leaf image based plant disease identification using transfer learning and feature fusion. *Comput. Electron. Agric.* **2022**, *196*, 106892. [CrossRef]
23. Xu, Y.; Zhang, X.; Li, H.; Zheng, H.; Zhang, J.; Olsen, M.S.; Varshney, R.K.; Prasanna, B.M.; Qian, Q. Smart breeding driven by big data, artificial intelligence, and integrated genomic-enviromic prediction. *Mol. Plant* **2022**, *15*, 1664–1695. [CrossRef] [PubMed]
24. Albattah, W.; Nawaz, M.; Javed, A.; Masood, M.; Albahli, S. A novel deep learning method for detection and classification of plant diseases. *Complex Intell. Syst.* **2022**, *8*, 507–524. [CrossRef]
25. Prince, S.J.D. *Understanding Deep Learning*; MIT Press: Cambridge, MA, USA, 2023.
26. Hughes, D.; Salathé, M. An open access repository of images on plant health to enable the development of mobile disease diagnostics. *arXiv* **2015**, arXiv:1511.08060.
27. Rouse, J.W.; Haas, R.H.; Schell, J.A.; Deering, D.W. *Monitoring Vegetation Systems in the Great Plains with ERTS*; NASA Special Publication: Washington, DC, USA, 1974; Volume 351, p. 309.
28. Jacquemoud, S.; Ustin, S.L. Leaf optical properties: A state of the art. In Proceedings of the 8th International Symposium of Physical Measurements & Signatures in Remote Sensing, Aussois, France, 8–12 January 2001; CNES: Aussois, France, 2001; pp. 223–332.
29. Krishnan, R.S.; Julie, E.G. Computer aided detection of leaf disease in agriculture using convolution neural network based squeeze and excitation network. *Automatika Časopis Za Automatiku, Mjerenje, Elektroniku, Računarstvo I Komunikacije* **2023**, *64*, 1038–1053. [CrossRef]
30. Sunkari, S.; Sangam, A.; Suchetha, M.; Raman, R.; Rajalakshmi, R.; Tamilselvi, S. A refined ResNet18 architecture with Swish activation function for Diabetic Retinopathy classification. *Biomed. Signal Process. Control* **2024**, *88*, 105630. [CrossRef]
31. Parez, S.; Dilshad, N.; Alghamdi, N.S.; Alanazi, T.M.; Lee, J.W. Visual intelligence in precision agriculture: Exploring plant disease detection via efficient vision transformers. *Sensors* **2023**, *23*, 6949. [CrossRef] [PubMed]
32. Szegedy, C.; Ioffe, S.; Vanhoucke, V.; Alemi, A. Inception-v4, inception-resnet and the impact of residual connections on learning. In Proceedings of the AAAI Conference on Artificial Intelligence, San Francisco, CA, USA, 4–9 February 2017; Volume 31.
33. Khaki, S.; Wang, L.; Archontoulis, S.V. A CNN-RNN framework for crop yield prediction. *Front. Plant Sci.* **2020**, *10*, 1750. [CrossRef] [PubMed]

agriculture

MDPI

Article

AGRARIAN: A Hybrid AI-Driven Architecture for Smart Agriculture

Michael C. Batistatos [1], Tomaso de Cola [2], Michail Alexandros Kourtis [3,*], Vassiliki Apostolopoulou [4], George K. Xilouris [3] and Nikos C. Sagias [1]

[1] Department of Informatics and Telecommunications, University of Peloponnese, 22100 Tripolis, Greece; mbatist@uop.gr (M.C.B.); nsagias@uop.gr (N.C.S.)
[2] Institute of Communications and Navigation, Deutsches Zentrum für Luft- und Raumfahrt (DLR) Oberpfaffenhofen, 82234 Wessling, Germany; tomaso.decola@dlr.de
[3] Institute of Informatics and Telecommunications, National Centre for Scientific Research "DEMOKRITOS" (NCSRD), 15310 Athens, Greece; xilouris@iit.demokritos.gr
[4] Practin, Kastritsa, 45500 Ioannina, Greece; apostolopoulou@practin.com
* Correspondence: akis.kourtis@iit.demokritos.gr

Abstract: Modern agriculture is increasingly challenged by the need for scalable, sustainable, and connectivity-resilient digital solutions. While existing smart farming platforms offer valuable insights, they often rely heavily on centralized cloud infrastructure, which can be impractical in rural or remote settings. To address this gap, this paper presents AGRARIAN, a hybrid AI-driven architecture that combines IoT sensor networks, UAV-based monitoring, satellite connectivity, and edge-cloud computing to deliver real-time, adaptive agricultural intelligence. AGRARIAN supports a modular and interoperable architecture structured across four layers—Sensor, Network, Data Processing, and Application—enabling flexible deployment in diverse use cases such as precision irrigation, livestock monitoring, and pest forecasting. A key innovation lies in its localized edge processing and federated AI models, which reduce reliance on continuous cloud access while maintaining analytical performance. Pilot scenarios demonstrate the system's ability to provide timely, context-aware decision support, enhancing both operational efficiency and digital inclusion for farmers. AGRARIAN offers a robust and scalable pathway for advancing autonomous, sustainable, and connected farming systems.

Keywords: smart agriculture; AI-driven decision support systems; precision farming

1. Introduction

The global agricultural sector is undergoing a fundamental transformation driven by rapid advancements in digital technologies, artificial intelligence (AI), and the Internet of Things (IoT). These innovations are reshaping traditional farming methods, optimizing resource efficiency, productivity, and sustainability in response to increasing food demand and climate change challenges. The European Commission has emphasized that digitalization in agriculture is a critical component for improving competitiveness, fostering sustainable practices, and ensuring food security in the European Union (EU) and beyond [1].

Modern agricultural systems are increasingly relying on connected infrastructures, including smart sensors, AI-driven decision-making tools, and cloud-based agricultural platforms, to enhance real-time monitoring, automated farm management, and precision agriculture techniques. This transition aligns with the Agriculture 4.0 paradigm, which integrates big data analytics, robotics, and advanced network communication protocols to optimize agricultural operations [2].

The Common Agricultural Policy (CAP) has been a cornerstone of the EU's agricultural strategy, focusing on sustainability, productivity, and the socio-economic well-being of farmers [3]. Recent CAP Strategic Plans for 2023–2027 emphasize the role of digital tools, AI, and IoT in improving decision-making for farmers while promoting climate-resilient agricultural practices [4]. Additionally, EU-backed initiatives on smart farming provide targeted support for the adoption of cloud computing, AI-driven analytics, and blockchain technologies to ensure traceable, transparent, and efficient agricultural supply chains [5]. Despite the potential benefits of precision farming technologies, many farmers still face economic and technical barriers to adoption. According to Eurostat, agricultural labor input in the EU has continued to decline, highlighting the need for automation and smart agricultural solutions to compensate for workforce shortages [6–8].

Several international research initiatives from countries like Israel and China have made substantial progress in smart agriculture. Israel has been a leader in sensor-based drip irrigation systems, leveraging local edge controllers to manage water efficiency in arid environments. Chinese efforts have advanced in areas such as UAV-enabled field monitoring, blockchain-integrated supply chains, and remote sensing for pest management. These technologies are often domain-specific and vertically optimized, addressing critical challenges such as irrigation, crop health, and traceability. However, they tend to rely on centralized architectures, which can be constrained by connectivity gaps or lack of adaptability in remote rural contexts.

AGRARIAN addresses a clear research and deployment gap by offering a modular, horizontally and vertically scalable architecture that unifies edge AI, federated learning, and hybrid connectivity (5G/LEO satellites) within a single framework. Unlike traditional siloed models, AGRARIAN is designed to support simultaneous, real-time operations across layers—from sensor data capture to AI-driven decision support—while optimizing for energy use, latency, and interoperability. This capability is particularly important for rural deployments with intermittent connectivity, making AGRARIAN adaptable where centralized models fall short. Its layered design also supports dynamic resource allocation via slicing, enabling precision agriculture to scale across varied operational environments with different infrastructure constraints.

Driven by the challenges of climate change, resource constraints, and the growing demand for sustainable food production, the AGRARIAN architecture introduces a hybrid AI-driven framework that seamlessly integrates IoT sensors, UAVs, satellite-based remote sensing, edge computing, and AI-powered decision support systems (DSS) to transform precision farming, livestock management, and agricultural sustainability. Unlike conventional systems reliant on centralized cloud processing, AGRARIAN decentralizes data analysis through edge AI and federated learning, significantly reducing latency, bandwidth consumption, and dependency on continuous connectivity. Its four-layered structure—Sensor, Network, Data Processing, and Application Layers—ensures a scalable and modular system where multimodal sensors capture real-time environmental, soil, and livestock data, processed through AI-driven analytics at the edge or in the cloud.

The structure of this paper is organized as follows: Section 2 presents an overview of related technologies, focusing on IoT-based irrigation systems, AI-driven crop protection, decision support systems (DSS), and satellite-enabled precision agriculture. Section 3 introduces the AGRARIAN system architecture, detailing its horizontal and vertical architectural models, along with the integration of edge computing, hybrid networking, and cloud-based AI analytics, an in-depth analysis of the four-layered structure of AGRARIAN, comprising the sensor, network, data processing, and application layers, describing their role, interaction, and impact on smart agriculture. Finally, Section 4 concludes with key findings, potential limitations, and future research directions in hybrid AI-driven agricultural frameworks.

2. Related Technologies

The integration of Internet of Things (IoT) devices and 5G networks is transforming smart irrigation management by enabling real-time monitoring and automated water control. IoT-based sensors, combined with cloud-based decision support systems (DSS), allow precise irrigation scheduling based on soil moisture levels, weather conditions, and crop water demand. The use of 5G connectivity enhances data transmission speed, ensuring low-latency, AI-powered water management solutions that contribute to efficient water resource allocation and climate resilience [9]. Beyond irrigation, DSS also plays a critical role in agrarian policy-making and economic planning, as seen in Ukraine, where data-driven accounting tools are used to align agricultural strategies with European integration frameworks. These systems analyze farm productivity metrics, subsidy allocations, and rural development trends to optimize policy interventions [10].

2.1. AI for Crop Protection and Environmental Monitoring

AI-based decision support technologies are also reshaping crop protection and disease management. Modern deep learning models, when combined with satellite imagery and IoT sensors, enable real-time disease prediction and early detection, reducing the dependency on excessive pesticide use. AI-driven DSS can process historical weather data, pathogen distribution models, and soil health indicators to provide targeted, proactive recommendations for farmers [11]. Satellite technology, particularly CubeSats and GIS-based remote sensing, is revolutionizing precision agriculture by offering high-resolution environmental monitoring. These small satellites provide frequent, real-time imaging, allowing farmers to track crop health, soil moisture levels, and land-use changes, which are then integrated into DSS platforms to facilitate data-driven decision-making for sustainable farming [12].

2.2. DSS for Water Resource and Livestock Management

Water management in agriculture remains a pressing challenge, and DSS frameworks are being developed to guide groundwater resource allocation, especially in drought-prone regions. AI-powered decision tools enable multi-stakeholder collaboration, integrating hydrological models, water demand forecasting, and climate impact assessments to ensure sustainable irrigation practices [13]. In precision livestock farming, AI-driven DSS are being deployed to monitor animal welfare, detect diseases, and optimize feed efficiency. Machine vision, biometric sensors, and predictive analytics allow for real-time tracking of livestock health, reducing operational costs and improving farm productivity [14]. These advancements are not only enhancing individual farm operations but are also influencing renewable energy production in agriculture. DSS frameworks are now used for biogas facility planning, optimizing locations based on geospatial data, waste production metrics, and sustainability indicators. This application supports circular bioeconomy models, where agricultural waste is converted into biofuels and organic fertilizers, reducing carbon footprints and environmental impact [15].

2.3. Circular Bioeconomy and Smart Supply Chains

Decision support technologies also play a crucial role in circular bioeconomy strategies, ensuring efficient resource recycling and sustainable food production systems. AI-powered DSS facilitates agricultural waste management by optimizing the bioconversion of crop residues into bioenergy and minimizing resource wastage [16]. Recent developments in digital agriculture and AI-driven decision systems show that fine-tuned natural language processing (NLP) models outperform traditional chatbot-based farm management solutions. These AI-powered DSS provide context-aware recommendations for crop management, pest control, and supply chain logistics, improving farm productivity and sustainability [17]. Additionally, digital technologies are increasingly being commercialized for nature conservation and ecosystem service provisioning in agriculture. By integrating remote sensing, AI-based environmental modeling, and cloud-based analytics, these tools help farmers balance economic profitability with ecological preservation, ensuring that agriculture remains both productive and environmentally responsible [18].

2.4. Convergence of Digital Agriculture Technologies

As agriculture continues to evolve towards a data-driven, digitally connected ecosystem, the integration of AI, IoT, and DSS technologies is becoming essential. From smart irrigation management and CubeSat-based environmental monitoring to precision livestock farming and biogas facility optimization, DSS is empowering farmers and policymakers with real-time insights for sustainable decision-making. The convergence of edge computing, cloud-based AI analytics, and stakeholder collaboration is paving the way for a resilient, efficient, and sustainable agricultural future.

Enhancing digital infrastructure and hybrid communication technologies is, therefore, a key priority in ensuring widespread accessibility to smart farming solutions across Europe.

2.5. Challenges and Infrastructure Considerations

While digital agriculture offers significant opportunities for growth and sustainability, several challenges must be addressed:

- Connectivity Gaps: Many rural farming areas suffer from limited broadband access, restricting the adoption of real-time IoT monitoring and AI-driven decision systems [19].
- Interoperability Issues: The diverse range of agricultural IoT devices, cloud platforms, and AI models creates integration challenges, requiring standardized data exchange protocols [20].
- Data Privacy and Security: The sensitive nature of farm data necessitates robust cybersecurity frameworks, including secure data transmission protocols like NETCONF and YANG [21,22].

Scalability and Computational Demand: AI-driven edge computing is increasingly being explored to reduce the burden on cloud infrastructure, enabling localized, real-time data processing for smart agriculture [23].

To address these challenges, next-generation networking protocols such as Recursive InterNetwork Architecture (RINA) are being explored to replace traditional IP-based architectures, improving network scalability, data security, and real-time processing capabilities for large-scale agricultural IoT systems [24].

2.6. Edge Computing and Federated AI for Real-Time Farming

A critical technological enabler for precision farming is edge computing, which facilitates real-time data analysis at the farm level without the need for continuous cloud connectivity [23]. By deploying localized AI models on edge devices, UAVs (drones), and satellite nodes, latency is minimized, bandwidth consumption is reduced, and real-time decision-making is enhanced. Recent studies suggest that federated learning and AI-driven edge computing can significantly improve agricultural supply chain efficiency, reducing costs and optimizing farm management strategies [23].

3. AGRARIAN Architecture

The AGRARIAN architecture, Figure 1, is designed to provide a robust, scalable, and intelligent agricultural technology ecosystem, integrating sensor networks, AI-driven decision support, hybrid communication networks, and cloud-based analytics. This system is structured into multiple layers to ensure seamless functionality, interoperability, and efficient data flow between different components. The architecture is conceptualized through two complementary views: the horizontal and vertical architectures, each detailing the organization of system elements and their interactions.

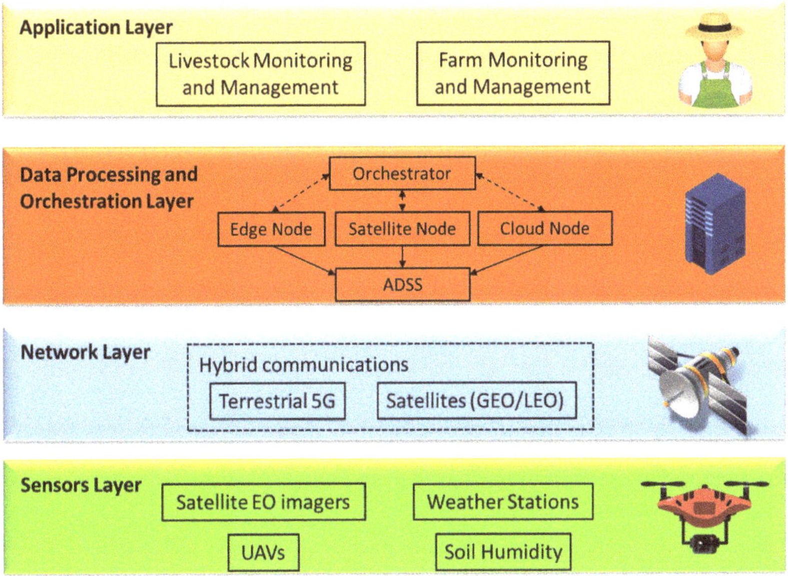

Figure 1. AGRARIAN High-Level Architecture: Integrating IoT, Edge AI, and Hybrid 5G Connectivity for Smart Agriculture.

The AGRARIAN architecture is designed to support precision farming, livestock monitoring, vineyard management, and environmental sustainability. It is composed of multiple interconnected layers that facilitate sensor data acquisition, edge computing, satellite communications, and AI-driven analytics. The horizontal architecture provides an end-to-end view of how different components interact, while the vertical architecture focuses on the service-oriented structure of the system. The horizontal architecture emphasizes the interaction between the customer portal, decision support systems (ADSS), infrastructure, and external data sources. The vertical architecture, on the other hand, categorizes these functionalities into four major layers: Sensor Layer, Network Layer, Data Processing Layer, and Application Layer.

AGRARIAN is currently being evaluated in pilot deployments across vineyard ecosystems and livestock farms, where real-world feedback has helped fine-tune energy management, model accuracy, and connectivity handling. These pilots validate AGRARIAN's layered operation: sensor layer (uRLLC) for fast data capture, network layer for hybrid 5G-satellite backhaul, processing layer (eMBB slicing) for model inference, and application layer for user interaction and alerts. The system also includes API interfaces for future integration with farm machinery, supporting ISOBUS and OPC UA standards. This ensures that AGRARIAN is not only a theoretical model but a deployable, extensible solution for autonomous, sustainable agriculture.

Application Layer

The Application Layer, Figure 2, provides user-facing tools that allow farmers, policymakers, and researchers to interact with the AGRARIAN system, offering AI-powered agricultural insights, decision support, and precision farming applications.

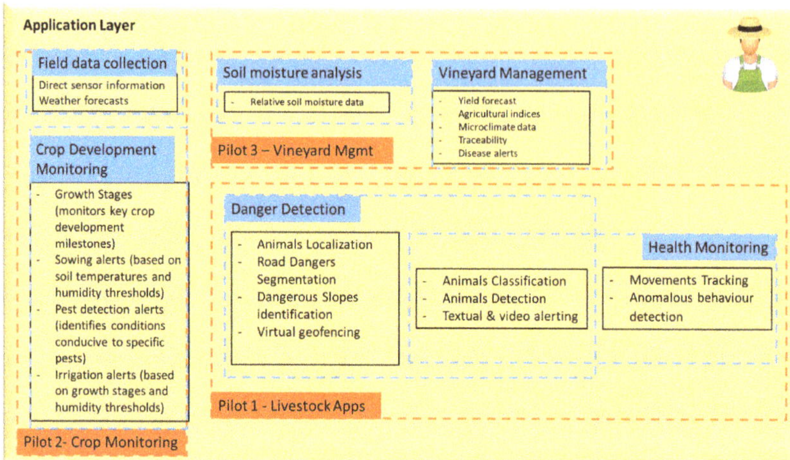

Figure 2. Application layer.

- Agricultural Decision Support System (ADSS): Analyzes sensor, satellite, and UAV data to provide actionable insights on crop health, livestock management, and irrigation scheduling.
- Livestock Monitoring and Anomaly Detection: Uses AI-driven video analytics and GPS tracking to identify anomalous animal behavior, potential health risks, and missing livestock.
- Crop Growth and Yield Forecasting: Predicts crop productivity, pest risks, and optimal harvesting times based on machine learning algorithms and real-time environmental data.
- Smart Irrigation and Water Management: Uses soil moisture analytics, weather forecasts, and AI-based optimization to ensure efficient water usage and minimize waste.
- Disease and Pest Alert Systems: AI models process multispectral and SAR data to predict disease outbreaks and recommend timely interventions.
- Supply Chain Traceability and Food Safety: Blockchain-enabled traceability solutions ensure transparent farm-to-market logistics, improving food safety and regulatory compliance.

By integrating advanced analytics, real-time alerts, and predictive modeling, this layer empowers users with intelligent decision-making tools for sustainable and efficient farming.

Data Processing and Orchestration Layer

The Data Processing Layer, Figure 3, acts as the computational hub of the AGRARIAN system, processing, analyzing, and distributing agricultural data across the network. This layer leverages edge computing, cloud processing, and AI-based analytics to extract meaningful insights from raw sensor data.

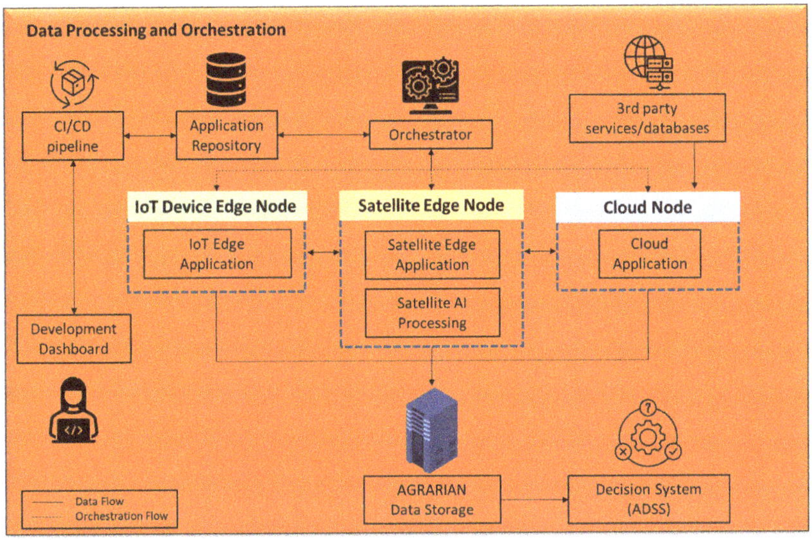

Figure 3. Data Processing and Orchestration layer.

- Edge AI and Federated Learning: Distributed AI models are deployed on satellites, UAVs, and farm-based edge nodes, allowing real-time inference for disease detection, irrigation control, and crop monitoring.
- CI/CD Pipelines for AI Model Deployment: Continuous integration and deployment pipelines ensure real-time AI model updates for improved analytics and decision-making.
- Cloud-Native Orchestration (Kubernetes, KubeEdge, and K3s): Supports scalable, fault-tolerant, and distributed AI computing for precision farming applications.
- Satellite AI Processing: Enables onboard AI inference on CubeSats, reducing latency and bandwidth consumption while providing actionable insights directly from space-based monitoring.
- Data Storage and Integration with External Sources: Ensures secure, efficient storage and retrieval of environmental, livestock, and field data, integrating external climate databases, weather APIs, and agricultural knowledge repositories.

By leveraging advanced AI and edge computing technologies, this layer enhances decision-making efficiency and scalability.

The AGRARIAN architecture incorporates various methodological approaches—such as real-time edge AI, federated learning, and satellite-based inference—not as isolated innovations but as components selected and positioned according to specific agricultural use cases and their corresponding feasibility and problem-solving impact. It is acknowledged that not all agricultural operations require real-time decision-making. For example, strategic planning tasks like yield forecasting or soil nutrient mapping are typically tolerant of batch processing and delayed analytics. However, certain scenarios do benefit from real-time or near-real-time responsiveness. These include livestock anomaly detection (e.g., animal distress or escape), localized irrigation control during heatwaves, and pest outbreak alerts, where immediate data-driven insights can significantly reduce risks or losses. In such contexts, AGRARIAN's edge computing and low-latency satellite links offer value by enabling timely interventions without reliance on centralized cloud infrastructures. The architecture supports hybrid deployment models that allow stakeholders to scale computational resources and analytical intensity according to operational context and economic viability. By not assuming a one-size-fits-all requirement for real-time processing, AGRAR-

IAN is designed with a flexible orchestration layer that balances real-time capability with practical feasibility and cost-effectiveness across scales and farming profiles.

Network Layer

The Network Layer, Figure 4, enables seamless connectivity across all AGRARIAN components, ensuring reliable communication between sensors, computing nodes, and cloud-based systems. It integrates terrestrial and satellite communication networks to provide uninterrupted connectivity in remote agricultural areas.

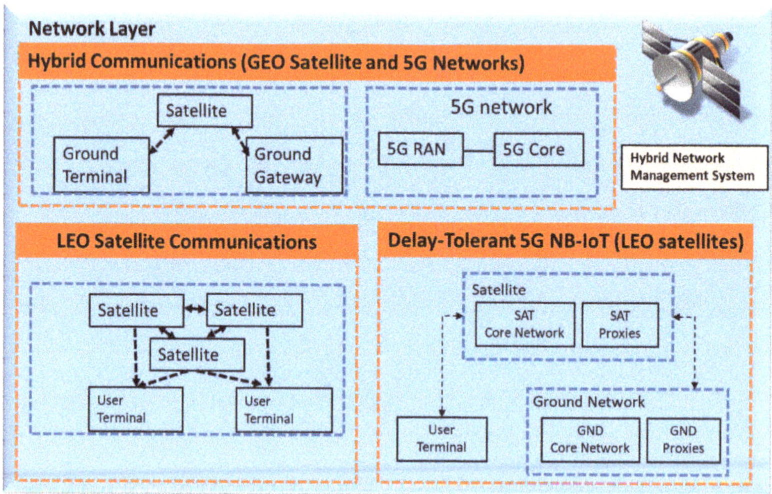

Figure 4. Network layer.

- 5G-Based Communication: Provides high-speed, low-latency connectivity for real-time sensor data transmission and remote farm monitoring.
- Hybrid Satellite Communications (LEO and GEO): LEO satellites facilitate low-latency broadband access, while GEO satellites provide continuous global coverage.
- Edge Network Infrastructure: Supports real-time AI model deployment and inference at the farm level, reducing dependency on centralized cloud computing.
- Delay-Tolerant Networking (DTN) and IoT Protocols: Allow efficient data transmission in rural and disconnected environments, ensuring that time-sensitive agricultural data is not lost.
- Ground Network Infrastructure: Includes 5G base stations, ground terminals, and IoT gateways, allowing seamless integration of AGRARIAN's sensor networks with cloud-based decision support systems.

This layer ensures uninterrupted connectivity, which is essential for real-time agricultural monitoring and automated farming solutions.

Sensor Layer

The Sensor Layer, Figure 5, is the foundation of the AGRARIAN system, comprising various data acquisition technologies that capture environmental, soil, and livestock data. These sensors are deployed in situ, on UAVs, and in satellite-based observation systems, ensuring continuous real-time monitoring of agricultural parameters.

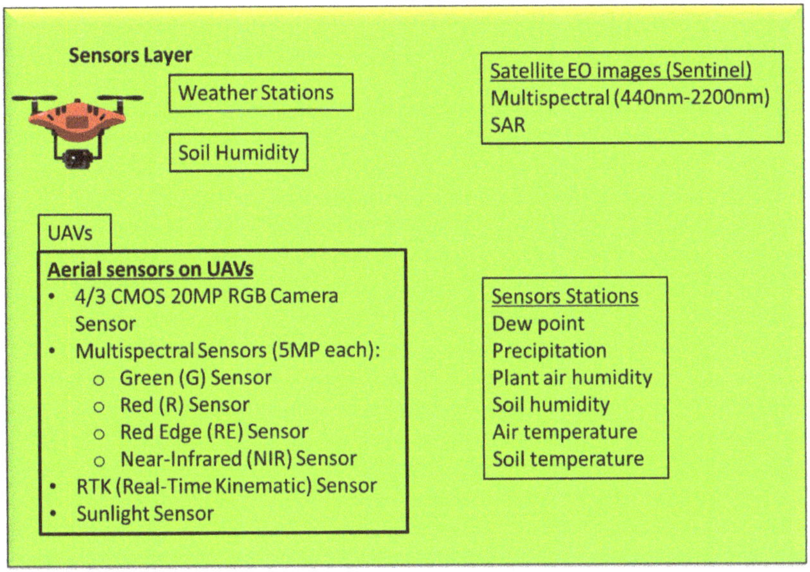

Figure 5. Sensors layer.

- IoT and Ground Sensors: Measure soil moisture, temperature, air humidity, and precipitation, providing critical data for precision irrigation and crop health analysis.
- UAV-Based Sensors: Equipped with multispectral cameras, RGB cameras, and real-time kinematic (RTK) sensors to provide high-resolution field images and topographical mapping.
- Weather and Climate Stations: Monitor meteorological parameters such as wind speed, temperature, solar radiation, and frost prediction, supporting weather-based agricultural decision-making.
- Satellite Earth Observation (EO) Systems: Utilize Sentinel-based multispectral imaging and Synthetic Aperture Radar (SAR) to provide wide-area, high-resolution monitoring for crop health, soil moisture levels, and yield estimation.
- Livestock Tracking Devices: Sensors embedded in wearables and drones to track animal movement, health status, and anomaly detection.

This layer ensures real-time, accurate data collection for informed decision-making within the AGRARIAN ecosystem.

This layer ensures real-time, accurate data collection for informed decision-making within the AGRARIAN ecosystem by integrating a diverse range of sensors deployed across agricultural fields, UAVs, and satellites. Ground-based IoT sensors continuously monitor soil moisture, temperature, humidity, and nutrient levels, providing granular insights into crop health and water needs. UAV-mounted multispectral and thermal sensors capture high-resolution imagery, detecting early signs of crop stress, pest infestations, and irrigation inefficiencies. Weather and climate stations collect atmospheric data, including wind speed, solar radiation, precipitation, and frost risk, enabling microclimate analysis for precision farming. Satellite Earth Observation (EO) systems, leveraging multispectral imaging (Sentinel) and Synthetic Aperture Radar (SAR), offer wide-area crop monitoring, soil moisture assessments, and predictive yield modeling, even under cloud cover and adverse weather conditions. For livestock applications, wearable biometric sensors track movement patterns, body temperature, and feeding behavior, facilitating real-time animal health monitoring and anomaly detection. By ensuring seamless integration of these diverse sensing technologies, the AGRARIAN architecture enables data-driven, AI-enhanced decision-making, fostering sustainable resource management and improved farm productivity.

While the AGRARIAN architecture provides a comprehensive and modular framework for smart agriculture, it is important to critically reflect on its current scope and future integration opportunities. One notable area of expansion involves the direct interfacing of AGRARIAN with agricultural machinery, such as autonomous tractors, robotic sprayers, and harvesters. Although the present architecture primarily focuses on sensor networks, edge/cloud analytics, and decision support, its design is intentionally modular, with API-based interoperability layers that support future integration with smart machinery systems. Technologies such as ISOBUS, MQTT, and OPC UA are being considered for implementing bidirectional communication between the decision support system (ADSS) and farm equipment, enabling automated actuation based on AI-driven recommendations. Furthermore, AGRARIAN's architecture supports future expansion through control-layer protocols and real-time machine interfaces, establishing the foundation for intelligent automation. Implementation-wise, the architecture is already being validated in pilot environments—including livestock farms, vineyards, and field crop settings—where sensor deployment, satellite connectivity, and edge AI models are tested for latency, scalability, and usability. These real-world testbeds are key to evaluating AGRARIAN's readiness for broader deployment and demonstrating its potential to connect directly with autonomous agricultural machinery, enhancing operational efficiency and promoting end-to-end automation in modern farming.

The AGRARIAN architecture leverages a diverse set of AI-driven technologies, IoT connectivity, and hybrid networking to transform modern agricultural practices. The table below highlights how different components of AGRARIAN align with key advancements in smart agriculture, mapping each technology to its impact on various domains such as precision irrigation, crop protection, livestock monitoring, and sustainable resource management. By associating AGRARIAN's sensor, network, data processing, and application layers with emerging decision support systems (DSS), AI models, and satellite-based remote sensing, this comparison demonstrates how AGRARIAN enhances efficiency, sustainability, and productivity across the agricultural sector.

Table 1 illustrates AGRARIAN's role in improving precision farming by integrating IoT and 5G for irrigation, AI-based disease detection, and satellite-powered monitoring. Through real-time sensor data collection, AI-powered decision-making, and seamless connectivity, AGRARIAN enhances resource management, reduces operational costs, and promotes environmental sustainability. Key findings indicate that edge AI and federated learning enable more localized and responsive agricultural intelligence, reducing reliance on centralized cloud computing while improving latency-sensitive applications like livestock health monitoring and irrigation control. By combining AI-enhanced DSS, machine learning-driven crop management, and satellite-based remote sensing, AGRARIAN provides a scalable, modular, and resilient digital agriculture platform that supports data-driven farming, climate adaptation, and food security initiatives.

Table 1. AGRARIAN related works table and how they are mapped to AGRARIAN layers.

AGRARIAN Component	Author, Year, Ref. No.	How AGRARIAN Benefits the Field	Mapped AGRARIAN Layer(s)	Benefit to Agriculture
IoT and 5G for Smart Irrigation	Oppong, R.A. (2025) [9]	Enhances precision irrigation by leveraging real-time sensor data.	Sensor Layer, Network Layer	Enhances irrigation efficiency, prevents overwatering, and improves water conservation.

Table 1. *Cont.*

AGRARIAN Component	Author, Year, Ref. No.	How AGRARIAN Benefits the Field	Mapped AGRARIAN Layer(s)	Benefit to Agriculture
Decision Support Systems (DSS) for Agrarian Policy	Vasylishyn, S. (2025) [10]	Provides AI-driven policy recommendations based on real-time agricultural data.	Application Layer, Data Processing Layer	Optimizes agricultural resource allocation, policy effectiveness, and economic sustainability.
AI-Based Crop Protection and DSS	Jensen, A. et al. (2025) [11]	Enables early disease detection and pest management through AI-powered analytics.	Data Processing Layer, Sensor Layer	Reduces pesticide use, increases farm productivity, and enhances sustainability.
CubeSats for Agricultural Monitoring	Calka, B.; Szostak, M. (2025) [12]	Offers high-resolution environmental monitoring for precision farming.	Sensor Layer, Network Layer	Provides real-time insights into soil health, crop growth, and environmental conditions.
Smart Agriculture and DSS for Water Resource Management	Firoozzare, A. et al. (2025) [13]	Improves sustainable water resource management using AI-driven climate data.	Application Layer, Data Processing Layer	Ensures sustainable water allocation, mitigates drought impacts, and supports climate resilience.
AI for Precision Livestock Farming	Distante, D. et al. (2025) [14]	Enhances livestock welfare via real-time biometric monitoring and disease detection.	Sensor Layer, Data Processing Layer	Reduces livestock mortality, increases efficiency, and improves farm profitability.
Sustainable Agricultural Planning using DSS	Kaynak, T.; Gümüş, M.G. (2025) [15]	Supports energy-efficient agriculture through AI-driven biogas plant planning.	Application Layer, Network Layer	Supports renewable energy integration and reduces the carbon footprint in agriculture.
Circular Bioeconomy and DSS in Agriculture	Nguyen, T.H. et al. (2025) [16]	Facilitates circular agriculture by optimizing waste recycling.	Application Layer, Data Processing Layer	Promotes waste reduction, circular economy strategies, and resource-efficient food production.
Digital Agriculture and AI Decision Systems	De, S.; Sanyal, D.K.; Mukherjee, I. (2025) [17]	Improves real-time farm management with AI-enhanced automation tools.	Application Layer, Data Processing Layer	Enhances farm decision-making with AI-driven insights and real-time analytics.
Digital Technologies for Sustainable Agriculture	Krachunova, T. et al. (2025) [18]	Enables sustainable farming through AI-integrated remote sensing and DSS tools.	Application Layer, Sensor Layer, Network Layer	Encourages climate-smart farming through AI, IoT, and sustainable land management practices.

4. Preliminary Validation of AGRARIAN over 5G Network Slicing for Data Processing and Sensor Layers

To validate the AGRARIAN architecture in terms of its integration with modern communication technologies, we conducted a series of experiments focusing on 5G network slicing and its implications on energy consumption and latency performance across system layers. These trials were designed to explore how eMBB (enhanced Mobile Broadband) slices align with AGRARIAN's Data Processing Layer, while uRLLC (ultra-Reliable Low Latency Communication) slices support the responsiveness required by the Sensor Layer. Experiments were executed on an experimental 5G system that was deployed in Standalone Mode (SA) using Amarisoft 5GC with Huawei P40 Pro UEs and OpenStack-based virtualized services running on a Dell R730xd. The slicing mechanism leveraged RAN numerology manipulation to enforce low-latency configurations via tuning of srPeriod and slot duration.

In the context of 5G network slicing, particularly at the RAN (Radio Access Network) level, srPeriod and slot are two key parameters that directly influence latency and scheduling responsiveness. The srPeriod (Scheduling Request Period) defines how frequently a user equipment (UE) can request uplink transmission resources, with lower values enabling faster response times, Table 2 provides a representative analysis of different configurations and corresponding latency metrics. The slot parameter represents the transmission duration within a time frame, affecting how quickly data can be scheduled and processed. Reducing both parameters leads to lower end-to-end latency, a critical factor for uRLLC (ultra-Reliable Low Latency Communication) performance in applications such as real-time sensor feedback and UAV coordination in smart agriculture. These parameters were varied across several configurations to examine their impact on latency and energy consumption under different slicing conditions.

Table 2. 5G network slicing configuration information and corresponding expected latency.

Configuration	srPeriod	Slot Duration	Expected Latency
Config 1 (Min Latency)	1	2.5	~10 ms
Config 2	1	5.0	~15–18 ms
Config 3	10	2.5	~20–25 ms
Config 4	10	5.0	~30 ms
Config 5	40	2.5	~35 ms
Config 6 (Max Latency)	40	5.0	~40 ms

The first set of experiments evaluated energy consumption under varying eMBB slice bitrates (100 to 300 Mbps) and latency configurations (10 ms and 40 ms). Results (Figure 6) revealed that smaller packet sizes significantly increase energy usage due to higher transmission frequency, and lower latency configurations consistently consume more energy, confirming a trade-off between latency and energy performance at the data processing level. To evaluate uRLLC slicing for the sensor layer, latency measurements were collected under diverse configurations. With an eMBB slice allocated for backhaul, multiple uRLLC slices were imposed (srPeriod = {1, 10, 40}, slot = {2.5, 5}). As illustrated in Figures 6 and 7, these settings significantly affected system latency, with the best result (~12 ms) observed under the most aggressive uRLLC configuration (sr = 1, slot = 2.5). These findings support AGRARIAN's capacity to provide low-latency edge responsiveness for sensor-triggered alerts and UAV communications.

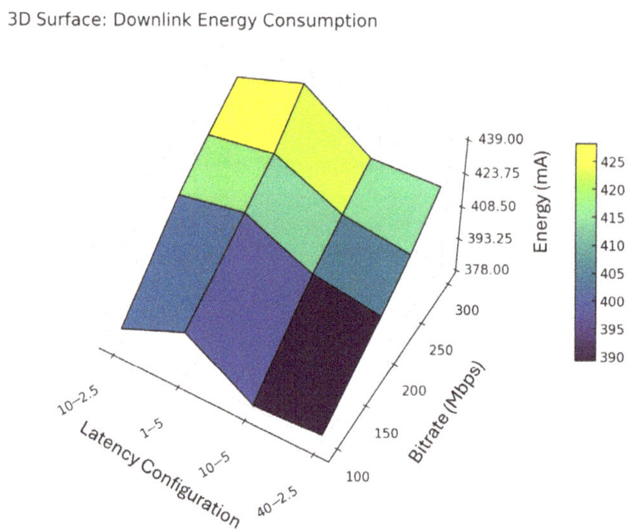

Figure 6. A 3D Surface Plot of Downlink Energy Consumption across Bitrate and Latency Configurations.

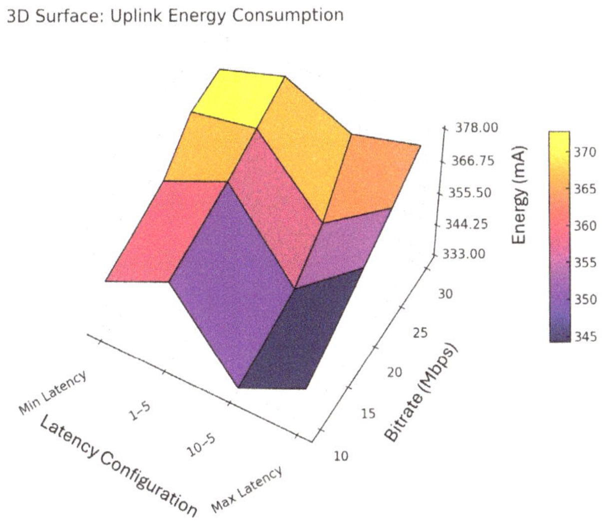

Figure 7. A 3D Surface Plot of Uplink Energy Consumption across Bitrate and Latency Configurations.

To understand the impact of 5G slicing on energy efficiency in AGRARIAN's Data Processing Layer, we analyzed the energy consumption of downlink transmissions across different latency configurations and bitrate levels. These configurations, defined by the slicing parameters srPeriod and slot, were mapped to typical eMBB profiles. Figure 6 presents a 3D surface plot illustrating energy usage (in mA) as a function of both bitrate and latency configuration, offering a visual overview of the trade-offs involved.

From the surface plot, it becomes evident that energy consumption increases with both higher bitrates and more aggressive low-latency configurations. This validates the architectural decision to employ edge computing in AGRARIAN's data processing layer, where compute-intensive tasks can be selectively handled based on energy budgets. The visualization confirms that while high throughput improves data processing responsiveness, it should be balanced against energy constraints—especially for deployments in energy-limited environments such as remote or sensor-heavy agricultural fields.

Complementing the downlink analysis, Figure 7 visualizes uplink energy consumption, which is particularly relevant to the Sensor Layer in AGRARIAN. This layer frequently transmits real-time data from IoT sensors and UAVs back to the system. The 3D surface plot represents energy consumption across the same set of latency configurations and bitrates, this time focusing on the energy demand of uplink transmissions under different uRLLC slice conditions.

The plot reveals a similar trend to the downlink case—energy usage escalates under low-latency configurations and higher data rates, reinforcing the known latency-energy trade-off. However, because the sensor layer often operates with small, frequent packets rather than continuous streams, the relative energy cost becomes a critical design parameter. These findings support AGRARIAN's conservative approach to using uRLLC slices only where real-time sensing is essential while deferring less critical transmissions to energy-optimized configurations. This balance allows for both performance and energy efficiency in field deployments.

5. Conclusions

The AGRARIAN architecture offers a modular and intelligent framework for modern agriculture, combining IoT sensors, UAVs, satellite connectivity, edge computing, and

AI-based analytics to enhance precision farming and sustainability. By decentralizing data processing and enabling real-time insights through edge AI, AGRARIAN supports efficient, scalable, and resilient agricultural operations. Its layered design ensures adaptability across various farming environments and levels of digital maturity.

In contrast to conventional cloud-centric systems, AGRARIAN supports hybrid deployment models that reduce connectivity constraints and promote real-time decision-making where it is contextually valuable. Pilot implementations demonstrate its relevance for applications such as irrigation control, crop disease alerts, and livestock monitoring. Future work will focus on expanding AGRARIAN's integration with agricultural machinery, enhancing interoperability, and validating its performance across broader agro-ecological zones.

Author Contributions: Conceptualization, M.C.B.; methodology, T.d.C.; formal analysis, G.K.X.; writing—original draft preparation, M.C.B. and M.A.K.; writing—review and editing, N.C.S.; visualization, V.A.; project administration, M.C.B.; funding acquisition, M.A.K. and V.A. All authors have read and agreed to the published version of the manuscript.

Funding: This research was funded by the European Commission Programme, Horizon Europe AGRARIAN project, grant number 101134128.

Institutional Review Board Statement: Not applicable.

Data Availability Statement: No new data were created or analyzed in this study. Data sharing is not applicable to this article.

Conflicts of Interest: Author Vassiliki Apostolopoulou was employed by the company Practin. The remaining authors declare that the research was conducted in the absence of any commercial or financial relationships that could be construed as a potential conflict of interest.

Abbreviations

The following abbreviations are used in this manuscript:

AI	Artificial Intelligence
DSS	Decision Support System
IoT	Internet of Things
UAV	Unmanned Aerial Vehicle
LEO	Low Earth Orbit
GEO	Geostationary Earth Orbit
SAR	Synthetic Aperture Radar
5G NTN	5G Non-Terrestrial Network
RINA	Recursive InterNetwork Architecture
NETCONF	Network Configuration Protocol
YANG	Yet Another Next Generation
DTN	Delay-Tolerant Networking
MCDA	Multi-Criteria Decision Analysis
GIS	Geographic Information System
EO	Earth Observation

References

1. European Commission. The Digitalisation of the European Agricultural Sector. Available online: https://digital-strategy.ec.europa.eu/en/policies/digitalisation-agriculture (accessed on 6 September 2024).
2. De Clercq, M.; Vats, A.; Biel, A. Agriculture 4.0: The Future of Farming Technology. In Proceedings of the World Government Summit, Dubai, United Arab Emirates, 11–13 February 2018.
3. European Union. Common Agricultural Policy. Available online: https://agriculture.ec.europa.eu/common-agricultural-policy_en (accessed on 6 September 2024).

4. European Commission. *Explore Farm Incomes in the EU, Farm Economics Overview Based on 2021 FADN Data*; European Commission: Brussels, Belgium, 2023.
5. European Commission. Report from the Commission to the European Parliament and the Council—Summary of CAP Strategic Plans for 2023–2027: Joint Effort and Collective Ambition; Brussels, Belgium, November 2023. Available online: https://agriculture.ec.europa.eu/document/download/6b1c933f-84ef-4b45-9171-debb88f1f757_en?filename=com-2023-707-report_en.pdf (accessed on 6 September 2024).
6. Latruffe, L. *Competitiveness, Productivity and Efficiency in the Agricultural and Agri-Food Sectors*; OECD Food, Agriculture and Fisheries Papers, No. 30; OECD Publishing: Paris, France, 2010. [CrossRef]
7. Eurostat. Agricultural Labour Input (Annual Rate of Change, 2022–2023). Available online: https://ec.europa.eu/eurostat/statistics-explained/index.php?title=File:F4rv_Agricultural_labour_input_(%25,_annual_rate_of_change,_2022-2023).png (accessed on 6 September 2024).
8. Directorate-General for Agriculture and Rural Development. Catalogue of CAP Interventions. Available online: https://agridata.ec.europa.eu/extensions/DashboardCapPlan/catalogue_interventions.html (accessed on 6 September 2024).
9. Oppong, R.A. Integration of IoT-Based Sprinklers, Embedded Systems, Data, and Cloud Computing for Smart Irrigation Management. *World J. Adv. Res. Rev.* **2025**, *25*, 126–151. [CrossRef]
10. Vasylishyn, S. FADNFSDN as a Tool for Accounting Information and Statistical Support of Ukraine's State Agrarian Policy in the Context of European Integration. *J. Agric. Econ.* **2025**.
11. Jensen, A.; Brown, P.; Groves, K.; Morshed, A. Next Generation Crop Protection: A Systematic Review of Trends in Modelling Approaches for Disease Prediction. *Comput. Electron. Agric.* **2025**, *234*, 110245. [CrossRef]
12. Calka, B.; Szostak, M. GIS-Based Environmental Monitoring and Analysis. *Appl. Sci.* **2025**, *15*, 3155. [CrossRef]
13. Firoozzare, A.; Dourandish, A.; Sabouhi, M. Group Decision-Making of Agricultural Stakeholders Towards Sustainable Groundwater Resources Management: A Case Study in North Khorasan. *J. Environ. Agric. Dev.* **2025**, *38*, 443–460.
14. Distante, D.; Albanello, C.; Zaffar, H.; Faralli, S. Artificial Intelligence Applied to Precision Livestock Farming: A Tertiary Study. *Smart Agric. Syst.* **2025**, *11*, 100889. [CrossRef]
15. Kaynak, T.; Gümüş, M.G. Biogas Potential Estimation and Geospatial Decision-Making: A Hybrid MCDA Approach to Optimizing Biogas Facility Locations. *Environ. Dev. Sustain.* **2025**. [CrossRef]
16. Nguyen, T.H.; Wang, X.; Utomo, D.; Gage, E.; Xu, B. Circular Bioeconomy and Sustainable Food Systems: What Are the Possible Mechanisms? *Clean. Circ. Econ.* **2025**, *11*, 100145. [CrossRef]
17. De, S.; Sanyal, D.K.; Mukherjee, I. Fine-Tuned Encoder Models with Data Augmentation Beat ChatGPT in Agricultural Named Entity Recognition and Relation Extraction. *Expert Syst. Appl.* **2025**, *277*, 127126. [CrossRef]
18. Krachunova, T.; Geppert, F.; Lemke, N. Digital Technologies Commercially Available in Germany in the Context of Nature Conservation and Ecosystem Service Provisioning in Agriculture. *Front. Sustain. Food Syst.* **2025**, *9*, 1464020. [CrossRef]
19. McKeown, N.; Anderson, T.; Balakrishnan, H.; Parulkar, G.; Peterson, L.; Rexford, J.; Shenker, S.; Turner, J. OpenFlow: Enabling Innovation in Campus Networks. *ACM SIGCOMM Comput. Commun. Rev.* **2008**, *38*, 69–74. [CrossRef]
20. Enns, R.; Björklund, M.; Bierman, A.; Schönwälder, J. *Network Configuration Protocol (NETCONF)*; Internet Engineering Task Force (IETF): Wilmington, DE, USA, 2011. [CrossRef]
21. Björklund, M. *YANG—A Data Modeling Language for the Network Configuration Protocol (NETCONF)*; Internet Engineering Task Force (IETF): Wilmington, DE, USA, 2010. [CrossRef]
22. *ETSI-GR NGP 009*; Next Generation Protocols (NGP); An Example of a Non-IP Network Protocol Architecture Based on RINA Design Principles. European Telecommunications Standards Institute (ETSI): Sophia Antipolis, France, 2019.
23. Maffione, V.; Salvestrini, F.; Grasa, E.; Bergesio, L.; Tarzan, M. A Software Development Kit to Exploit RINA Programmability. In Proceedings of the 2016 IEEE International Conference on Communications (ICC), Kuala Lumpur, Malaysia, 22–27 May 2016; IEEE: Piscataway, NJ, USA, 2016; pp. 1–7. [CrossRef]
24. He, Q.; Zhao, H.; Feng, Y.; Wang, Z.; Ning, Z.; Luo, T. Edge Computing-Oriented Smart Agricultural Supply Chain Mechanism with Auction and Fuzzy Neural Networks. *J. Cloud Comput.* **2024**, *13*, 66. [CrossRef]

agriculture

MDPI

Article

The Extraction of *Torreya grandis* Growing Areas Using a Spatial–Spectral Fused Attention Network and Multitemporal Sentinel-2 Images: A Case Study of the Kuaiji Mountain Region

Yanyan Lyu [1,2], Yong Wang [1,*] and Xiaoling Shen [3]

[1] School of Ecology and Applied Meteorology, Nanjing University of Information Science and Technology, Nanjing 210044, China; yylv@nuist.edu.cn
[2] Reading Academy, Nanjing University of Information Science and Technology, Nanjing 210044, China
[3] Shaoxing Meteorological Bureau, Shaoxing 312000, China; sxqxsxl@126.com
* Correspondence: ywang@nuist.edu.cn

Abstract: Global climate change poses a serious threat to *Torreya grandis*, a rare and economically important tree species, making the accurate mapping of its spatial distribution essential for forest resource management. However, extracting forest-growing areas remains challenging due to the limited spatial and temporal resolution of remote sensing data and the insufficient classification capability of traditional algorithms for complex land cover types. This study utilized monthly Sentinel-2 imagery from 2023 to extract multitemporal spectral bands, vegetation indices, and texture features. Following minimum redundancy maximum relevance (mRMR) feature selection, a spatial–spectral fused attention network (SSFAN) was developed to extract the distribution of *T. grandis* in the Kuaiji Mountain area and to analyze the influence of topographic factors. Compared with traditional deep learning models such as 2D-CNN, 3D-CNN, and HybridSN, the SSFAN model achieved superior performance, with an overall accuracy of 99.1% and a Kappa coefficient of 0.961. The results indicate that *T. grandis* is primarily distributed on the western, southern, and southwestern slopes, with higher occurrence at elevations above 500–600 m and on slopes steeper than 20°. The SSFAN model effectively integrates spectral–spatial information and leverages a self-attention mechanism to enhance classification accuracy. Furthermore, this study highlights the joint influence of natural factors and human land-use decisions on the distribution pattern of *T. grandis*. These findings aid precision planting and resource management while advancing methods for identifying tree species.

Keywords: *Torreya grandis* tree; Sentinel-2 image; Kuaiji Mountain; spatial–spectrum fused attention network; growth region extraction

1. Introduction

Global climate change has led to increases in temperature and changes in precipitation patterns, significantly impacting the distribution and ecosystems of forest resources [1,2]. As a rare tree species with important economic value endemic to China [3,4], *Torreya grandis* has also been affected by climate change. The increased frequency of extreme weather events, such as low temperatures, droughts [5], soil loss [6], and pest infestations [7], poses a serious threat to the survival of *T. grandis* trees. Additionally, ecological and environmental problems have become increasingly prominent, leading to a reduction in the ecological and economic benefits associated with *T. grandis*. The lack of comprehensive distribution data for *T. grandis* has hindered the government's ability to implement effective forestry policies and management strategies. Furthermore, the extraction accuracy of *T. grandis* distributions based solely on object-oriented or machine learning methods still needs improvement [8–11]. Therefore, the accurate identification of the spatial distribution of *T. grandis* has become an urgent problem.

Traditional tree species identification methods rely mainly on ground surveys, which usually require considerable manpower and material resources. With advancements in remote sensing technology and machine learning methods, researchers have carried out numerous studies on tree species classification using remote sensing data. Various machine learning algorithms, such as the maximum likelihood method [12], decision trees [13], random forests [14–16] and support vector machines [17–20], have been applied. While these methods offer advantages such as fast training speeds and a high degree of automation, they struggle to fully utilize the complex feature structures of high-dimensional remote sensing data. Deep learning has emerged as a powerful approach, capable of automatically extracting more abstract and deeper feature information [21] and are well-suited for processing large-scale remote sensing datasets. Among deep learning models, convolutional neural networks (CNNs) are most commonly used for classifying tree species in remote sensing images [22–24]. Based on the dimensionality of the input remote sensing data, deep learning methods for tree species classification can be categorized into one-dimensional CNNs (1D-CNNs), two-dimensional CNNs (2D-CNNs), and three-dimensional CNNs (3D-CNNs). In remote sensing tree species classification, 1D-CNNs mainly focus on spectral features and offer certain advantages [25,26]. However, they only consider spectral information and do not consider spatial information. In contrast, 2D-CNNs have demonstrated strong performance in tree species classification by reducing the dimensionality of the original 3D remote sensing image data and extracting the spatial information around each pixel [27,28]. Meanwhile, 3D-CNNs utilize convolution kernels to process input data in spatial and spectral dimensions simultaneously to extract spectral and spatial features. Despite their ability to capture richer joint features, 3D-CNNs often come with high computational costs and a large number of parameters, making them prone to overfitting. Researchers frequently construct self-designed models for tree species classification based on hyperspectral data and light detection and ranging (LiDAR) data [29–33]. Furthermore, some researchers have used different networks to separately extract spectral and spatial features and then combined them, providing an approach that takes both dimensions of spatial and spectral information into account for remote sensing data [34–36]. The dual-branch network can independently process spectral and spatial features, effectively preserving the integrity of each feature and fully leveraging their respective advantages [37]. In this study, drawing inspiration from the dual-attention network (DANet) with an adaptive self-attention mechanism [38] and a dual-branch dual-attention (DBDA) mechanism network [39] for hyperspectral image classification, this paper introduces a spatial–spectral fused attention network (SSFAN) to mine spectral and spatial information more efficiently for accurate identification of tree species such as *T. grandis*.

As remote sensing technology continues to mature, the spectral resolution and spatiotemporal resolution of satellite images have also greatly improved [40]. This progress has facilitated the acquisition of multitemporal and multiscale remote sensing data, providing a more reliable source. Various types of remote sensing data have been used for tree species identification, including multispectral data, hyperspectral data, and LiDAR data [12,16–19,33,41–43]. However, traditional multispectral imagery typically has limited spatial resolution and relatively few spectral bands, which restricts its capacity to capture subtle spectral differences among tree species. As a result, challenges such as "same spectrum for different objects" and "different spectrum for the same object" frequently arise. Moreover, although hyperspectral and LiDAR data offer greater detail, their high cost and limited accessibility constrain their use in large-scale applications. The multispectral instruments onboard the Sentinel-2 series satellites, launched by the European Space Agency (ESA), offer high spatial, spectral, and temporal resolution, providing detailed information at a relatively small scale. Importantly, Sentinel-2 imagery is freely and publicly available through the ESA Copernicus program, making it highly accessible for research use. In addition, multitemporal remote sensing images enable the extraction and analysis of vegetation phenology information, which can accurately reflect seasonal periodic changes in the vegetation growth process [15,44]. Therefore, the application of high-resolution Sentinel-2 multitemporal remote sensing imagery presents significant potential for advancing tree species identification methods.

At present, despite advancements in deep learning networks for tree species classification, several challenges remain. On the one hand, existing CNN-based methods are applied primarily to uncrewed aerial vehicle (UAV) hyperspectral remote sensing images, which contain hundreds of spectral bands. However, when applied to satellite multispectral images with fewer bands, classification accuracy decreases because of insufficient band information. On the other hand, many studies are based only on single-temporal imagery and fail to make full use of the differences in different surface objects in the temporal dimension. To address these problems, in this study, based on the spectral information, vegetation index, and texture features of 12 multitemporal Sentinel-2 images that were synthesized monthly in 2023, a structure similar to that of hyperspectral data was constructed, and after applying minimum redundancy maximum relevance (mRMR) feature optimization, the spatial–spectral fused attention network (SSFAN) was constructed. The SSFAN was used to extract the *T. grandis* growing areas in the Kuaiji Mountain region. Then, the distribution characteristics of *T. grandis* were analyzed under different topographic factors. The SSFAN was compared with other deep learning models to explore its advantages, and factors affecting the distribution of *T. grandis* were analyzed. The proposed model provides a new extraction method for *T. grandis*, has high research and application value, and provides a scientific basis for the management and protection of *T. grandis* resources in the Kuaiji Mountain region.

2. Data and Methods

2.1. Study Area

Kuaiji Mountain (Figure 1) is located in Shaoxing City, Zhejiang Province, China (120°44′–120°50′ E, 29°18′–29°59′ N), at the junction of the hilly areas of Eastern Zhejiang and the plains of Northern Zhejiang. The region's average elevation is approximately 500 m, with Dongbai Mountain reaching the highest point at 1187 m above sea level. The Kuaiji Mountain area experiences a typical subtropical monsoon climate, with an annual average sunshine duration of approximately 1900 h, an annual average temperature of 16 °C, and an annual average precipitation of approximately 1200 mm. These climatic conditions are highly favorable for the growth of *T. grandis*. The area also holds the best-preserved group of ancient *T. grandis* trees in China, with more than 70,000 trees over 100 years old and several thousand over 1000 years old, making it a valuable site for research on *T. grandis* cultivation and conservation. Given its typical southern hilly terrain and representative climate, Kuaiji Mountain serves as a suitable case study area for *T. grandis*, reflecting the characteristics of other major *T. grandis* distribution regions in Southern China.

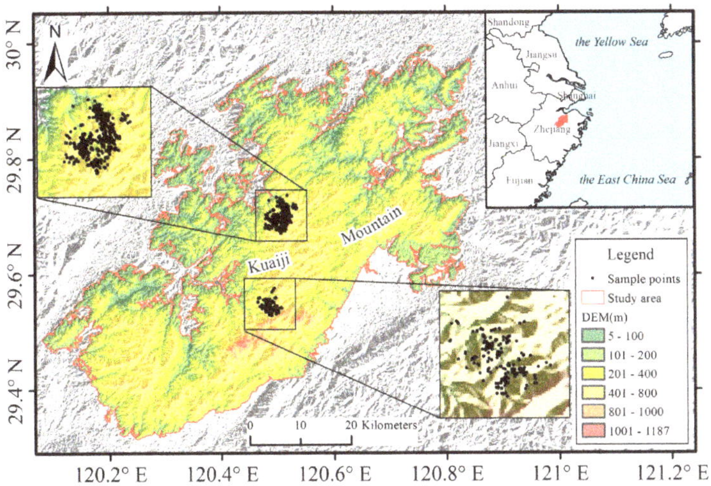

Figure 1. Overview of the Kuaiji Mountain study area in Shaoxing City, Zhejiang Province, China.

2.2. Data

2.2.1. Sentinel-2 Data

The remote sensing data used in this study were obtained from Sentinel-2 imagery available from the ESA Copernicus Data Centre. The Sentinel-2 satellites consist of two multispectral imaging satellites, Sentinel-2A and Sentinel-2B, launched by the ESA. These satellites have slight differences in their launch dates and orbital positions, but both are equipped with the same multispectral imager (MSI). With a revisit period of up to 5 days, these two satellites capture images of vegetation, soil and water cover, inland waterways and coastal areas. They are widely utilized for applications such as land cover detection, forestry and water monitoring, and natural disaster assessment [45–48]. In addition, the freely available Sentinel-2 data offer a favorable balance between cost-effectiveness and classification accuracy, making them highly suitable for large-scale vegetation mapping applications. Each Sentinel-2 image comprises 13 spectral bands, covering visible, near-infrared, and shortwave infrared wavelengths, with widths of up to 290 km and spatial resolutions of 10, 20, and 60 m, respectively, as shown in Table 1.

Table 1. Specifications of spectral bands for Sentinel-2 remote sensing images.

Band Name	Central Wavelength (μm)	Spatial Resolution (m)
B1—Coastal aerosol	0.443	60
B2—Blue	0.490	10
B3—Green	0.560	10
B4—Red	0.665	10
B5—Vegetation Red Edge 1	0.705	20
B6—Vegetation Red Edge 2	0.740	20
B7—Vegetation Red Edge 3	0.783	20
B8—NIR	0.842	10
B8A—Vegetation Red Edge 4	0.865	20
B9—Water vapor	0.945	60
B10—SWIR Cirrus	1.375	60
B11—SWIR	1.610	20
B12—SWIR	2.190	20

2.2.2. *T. grandis* Sample Data

Field survey data on *T. grandis* were collected from two representative zones with dense *T. grandis* populations in the Kuaiji Mountain area. These zones were selected to capture spatial heterogeneity, encompassing diverse topographic conditions such as variations in elevation, slope gradients, and slope aspects. In addition, the sampling areas contained various non-target land cover types, including buildings, farmland, shrubs, and broadleaf forests. The recorded attributes included geographic coordinates (latitude and longitude), elevation, tree height, east–west crown width, and north–south crown width, which were all obtained using a handheld GPS device with a typical position accuracy of approximately less than 1 m. After data cleaning and abnormal sample removal, a total of 1348 valid sample points for *T. grandis* were retained. The spatial distribution of the sample points is shown in Figure 1.

2.2.3. Supplementary Data

Additional data used for analyzing the distribution characteristics of *T. grandis* included a digital elevation model (DEM) provided by the National Aeronautics and Space Administration (NASA) Shuttle Radar Topography Mission (SRTM), with a spatial resolution of 30 m. Other supplementary data included vector diagrams of the study area and high-resolution Google Earth images. High-resolution Google Earth images were used to train labels for the deep learning model, and the prediction results of different models were compared to ensure the accuracy and reliability of the study results.

2.3. Methods

2.3.1. Data Preprocessing

In this work, a total of 159 Sentinel-2 Level-1C images acquired throughout 2023 were selected, each with less than 30% cloud cover, covering the entire Kuaiji Mountain study area using four scenes. These images were downloaded from the ESA Copernicus Data Centre and underwent orthorectification and geometric refinement. Atmospheric correction and band resampling to a 10 m resolution were performed using Sen2Cor [49] and SNAP software [50], respectively, for the L1C-level multispectral data. Sen2Cor is an open-source atmospheric correction processor developed by the ESA, while SNAP is an open-source platform provided by ESA for remote sensing image preprocessing tasks such as resampling. Both tools were freely downloaded from the ESA website. During resampling, the 20 m and 60 m bands of Sentinel-2 were unified to a 10 m resolution using the nearest neighbor interpolation method in SNAP. This method preserves the original spectral information, thereby minimizing the introduction of additional spectral noise during downscaling, and offers a simple and efficient solution [51]. Subsequently, the ENVI software (version 5.3) was employed to mosaic the Sentinel-2 imagery into a seamless dataset covering the entire study area. ENVI is a commercial remote sensing image processing and analysis software used here for image mosaicking, cropping, and other related operations. The vector files defining the study area boundaries were generated by calculating the slope from the DEM and extracting areas with a slope greater than 3 degrees [52] to delineate the mountainous region of Kuaiji Mountain. These vector files were then used to clip the mosaicked imagery. Following the mosaicking and clipping processes, a monthly synthesis was conducted using ENVI to generate a multitemporal remote sensing image set consisting of 12 synthesized monthly average images, where all available Sentinel-2 images within each calendar month were combined through pixel-wise averaging to produce one representative image per month. The clipped DEM was further used to derive aspect and slope datasets, and all resulting datasets, including elevation, aspect, and slope, were resampled to a 10 m resolution.

For each synthesized monthly average Sentinel-2 image, all 13 spectral bands were selected as the original input spectral features. On this basis, six vegetation indices were calculated pixel-wise using the corresponding spectral bands according to their standard formulas, including the normalized difference vegetation index (NDVI), green normalized difference vegetation index (GNDVI), soil-adjusted vegetation index (SAVI), enhanced vegetation index (EVI), Sentinel-2 red-edge position index (S2REP), and normalized difference red-edge index (NDREI), with all computations performed in ENVI. The calculation formulas for these vegetation indices are provided in Table 2.

Table 2. Vegetation index and calculation formulas.

Name	Calculation Formulas
NDVI [53]	NDVI = (B8 − B4)/(B8 + B4)
GNDVI [54]	GNDVI = (B4 − B2)/(B4 + B2)
SAVI [55]	SAVI = (1 + L) × (B8 − B4)/(B8 + B4 + L)
EVI [56]	EVI = 2.5 × (B8 − B4)/(B8 + 6 × B4 − 7.5 × B2 + 1)
S2REP [57]	S2REP = 705 + 35 × ((B7 + B4)/2 − B5) / (B6 − B5)
NDREI [58]	NDREI = (B8 − B5)/(B8 + B5)

Texture features play an important role in tree species identification. They effectively capture the spatial heterogeneity caused by variations in crown morphology and stand structure, which cannot be fully represented by spectral features alone [59]. As a valuable complement to spectral information, texture features not only enrich the description of tree species but also significantly enhance the separability between different species, thereby improving the accuracy and stability of classification results. In particular, *T. grandis* plantations typically exhibit large crown sizes and are planted at high densities with regular spacing, resulting in distinctive and consistent spatial structures compared to the surrounding vegetation. To capture these spatial patterns, texture features were extracted using the gray-level co-occurrence matrix (GLCM) method with a 3 × 3 moving window in ENVI. GLCM characterizes the spatial distribution and structural relationships of pixel intensity values by analyzing gray-level spatial correlations [60]. Based on the texture features of *T. grandis* and a comprehensive consideration of the correlations, differences and redundancies among texture parameters, six key texture features were selected in this study: information entropy (ENT), correlation (CORR), contrast (CON), an-gular second moment (ASM), dissimilarity (DISS) and variance (VAR).

Ultimately, the 13 spectral bands, six vegetation indices, and six texture features for each synthesized monthly average Sentinel-2 image were combined. The bands of all 12 monthly images were merged to construct a raw feature set consisting of multitemporal spectral features, vegetation indices, texture features, and temporal information, resulting in a dataset with a structure similar to that of a hyperspectral image and comprising a total of 300 bands (Figure 2).

Based on existing sample point data for *T. grandis*, buffer zones were created using ArcGIS 10.8 software by incorporating east–west and north–south crown width information obtained from field surveys. These were then overlaid on contemporaneous high-resolution Google Earth imagery for visual interpretation, allowing for the correction of *T. grandis* distribution boundaries. The resulting vector data were subsequently rasterized into pixel-level labels for training the deep learning model (Figure 3). Non-*T. grandis* samples were similarly generated through visual interpretation of the same imagery to obtain vector labels, which were then converted into pixel-level classification labels, resulting in a complete labeled dataset.

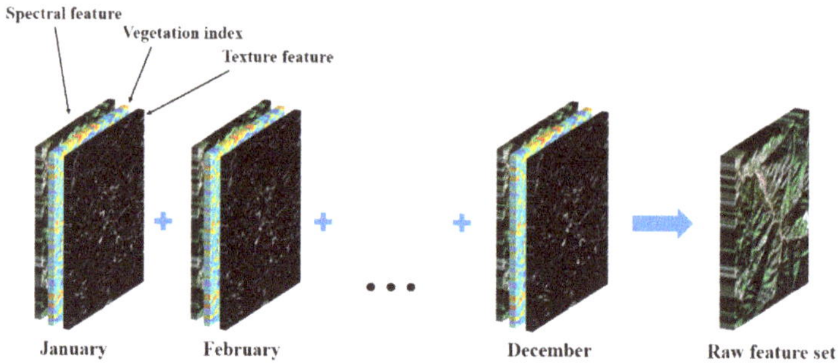

Figure 2. The construction process of the raw feature set.

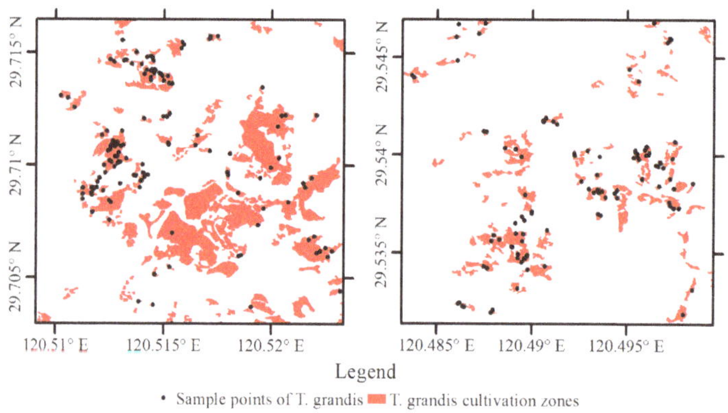

Figure 3. Example of partial *T. grandis* labels generated from sample points.

A total of 215,640 pixels from the study areas were selected as the sample set, including 29,104 *T. grandis* pixels and 186,536 non-*T. grandis* pixels. After feature optimization, the input band data were normalized, and a 3D patch was generated for each pixel. The label of each patch was determined by the class of the central pixel. The sample data were then randomly divided into training, validation, and test sets in a 60%, 20%, and 20% ratio, respectively. The training set was used to train the deep learning model, the validation set was used to select the optimal model parameters, and the test set was used to evaluate the classification accuracy of the model [61].

2.3.2. Feature Selection via the mRMR Algorithm

The comprehensive use of various features for tree species identification can effectively increase the amount of information, thereby improving classification accuracy. However, as the number of features increases, weakly correlated or highly redundant features are often introduced [62]. Feature optimization helps eliminate redundant bands and prevent classification errors caused by poor separation between target and background objects. The mRMR algorithm is a widely used feature selection method originally proposed by Peng et al. [63]. Compared with PCA and random forest-based methods, mRMR retains the interpretability of original features and improves model generalization by reducing redundancy with lower computational cost. Its principle is based on information theory, which measures feature relevance through mutual information with the target variable and quantifies redundancy as the average mutual information among features. The goal of mRMR is to select a subset of features that are highly relevant to the target variable while minimizing redundancy among them.

In the mRMR algorithm, the maximum correlation criterion is used to select the features with high correlation with class c, whereas the minimum redundancy criterion is used to remove redundant features in the feature set. The formulas are as follows:

$$maxD(S,c), D = \frac{1}{|S|}\sum_{x_i \in s} I(x_i;c) \tag{1}$$

$$minR(S), R = \frac{1}{|S|^2}\sum_{x_i,x_j \in S} I(x_i,x_j) \tag{2}$$

where $|S|$ is the number of features selected in feature subset S, $I(x_i;c)$ is the mutual information between characteristic x_i and class c, and $I(x_i;j)$ the mutual information between characteristic x_i and characteristic x_j. The selection of the final feature subset is based on the following comprehensive evaluation indicator:

$$max\Phi(D,R), \Phi = D - R \tag{3}$$

In this study, the mRMR algorithm was used to select 300 bands from the raw feature set, the redundant bands were excluded, and several optimal bands were selected. These bands were input into the deep learning network for subsequent analysis.

2.3.3. SSFAN

To fully utilize the spectral and spatial features of multispectral imagery, the SSFAN was introduced in this paper. The overall network architecture is shown in Figure 4 and consists of two main branches, namely, the spectral branch and the spatial branch. The spectral branch employs a 3D-CNN to extract spectral features from pixels, and the spatial branch uses a 2D-CNN to extract spatial features from pixels. After the original data are input into the two branches, the spectral feature map and spatial feature map can be obtained. To enhance the characterization ability of the model and improve the performance of object detection, the spectral attention module and spatial attention module are embedded in the spectral branch and spatial branch, respectively. Finally, after the feature maps of the two branches are fused, the prediction result is obtained through the pooling and fully connected layers.

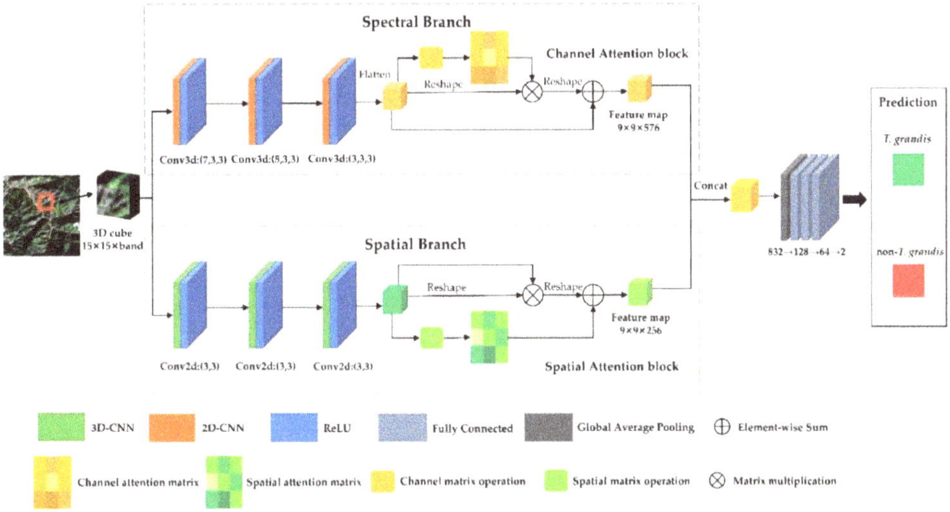

Figure 4. Structure of the SSFAN model.

1. Spectral branch and spectral attention

The 3D-CNN can simultaneously extract the spatial and spectral features of remote sensing imagery. Therefore, in this study, a 3D-CNN is employed as the basic structure of the spectral branch. The 3D convolution kernels across the height, width, and channel dimensions to compute the activation value for each pixel at the spatial position (x, y, z) in the j-th feature map of the i-th layer, which is denoted as $v_{i,j}^{x,y,z}$. The specific calculation formula is as follows:

$$v_{i,j}^{x,y,z} = \varphi\left(b_{i,j} + \sum_{m}\sum_{p=0}^{P_i-1}\sum_{q=0}^{Q_i-1}\sum_{r=0}^{R_i-1} w_{i,j,m}^{p,q,r} \times v_{i-1,m}^{x+p,y+q,z+r}\right) \tag{4}$$

where φ is the activation function, $b_{i,j}$ is the bias term, P_i, Q_i, and R_i are the sizes of the convolution kernels, $w_{i,j,m}^{p,q,r}$ is the weight of the m-th feature map connected to the position of (p, q, r), $v_{i-1,m}^{x+p,y+q,z+r}$ is the activation value of the pixel in the m-th feature map of the $i-1$ layer at $(x+p, y+q, z+r)$.

The spectral branch employs multiple 3D convolution layers in conjunction with a channel attention mechanism to fully extract spectral features. The input data is formatted as (1, 15, 15, band), representing a 15 × 15 spatial neighborhood centered on the target pixel. Three consecutive 3D convolutional layers are applied with kernel sizes of 3 × 3 × 7, 3 × 3 × 5, and 3 × 3 × 3, respectively, and the number of filters in these layers is set to 8, 16, and 32. All layers use a stride of (1, 1, 1) and are followed by the rectified linear unit (ReLU) activation function to introduce nonlinearity. Following the convolution operations, the channel and spectral dimensions of the output are fused, and the channel attention mechanism is applied to obtain the spectral feature map.

The channel attention module is designed to capture the dependency between different channels through a self-attention mechanism. The main process is as follows. First, the input feature map $A \in R^{C\times H\times W}$ is reconstructed as $A \in R^{C\times N}$, and the channel attention map $X \in R^{C\times N}$ is calculated. The formula is as follows:

$$x_{ji} = \frac{exp(A_i \cdot A_j)}{\sum_{i=1}^{C} exp(A_i \cdot A_j)} \tag{5}$$

where x_{ij} denotes the effect of the i-th channel on the j-th channel. Then, the channel attention map X is used to weight the original feature map A, and the spectral feature map E is obtained by combining with a learnable scaling factor α, as shown in Equation (6):

$$E_j = \alpha\sum_{i=1}^{N}(x_{ji}A_i) + A_j \tag{6}$$

2. Spatial branch and spatial attention

In the spatial branch, the model uses a 2D-CNN as the fundamental structure to extract the spatial features from remote sensing imagery. The 2D convolution kernel slides along the height and width dimensions, applying nonlinear transformations through an activation function. The activation value of the pixel at the spatial position (x, y) in the j-th feature map of the i-th layer is denoted as $v_{i,j}^{x,y}$, and the calculation formula is as follows:

$$v_{i,j}^{x,y} = \varphi\left(b_{i,j} + \sum_{m}\sum_{p=0}^{P_i-1}\sum_{q=0}^{Q_i-1} w_{i,j,m}^{p,q} \times v_{i-1,m}^{x+p,y+q}\right) \tag{7}$$

The parameters in Equation (7) are the same as those in Equation (4).

The spatial branch incorporates multiple 2D convolution layers combined with a spatial attention mechanism to capture spatial semantic information. The input data patch has the shape (15, 15, band), centered on the target pixel. Specifically, the spatial branch consists of three sequential 2D convolutional layers, each with a kernel size of 3 × 3 and a stride of 1. The number of filters in these layers is set to 64, 128, and 256, respectively. Each layer is followed by a ReLU activation function. After feature extraction, the spatial attention mechanism is applied to generate the final spatial feature map.

The spatial attention module incorporates a self-attention mechanism to capture the long-range dependencies between spatial locations, and the specific process is as follows. The original feature map $A \in R^{C\times H\times W}$ is processed by two convolutional layers and then reshaped to obtain two new feature maps $B, C \in R^{C\times N}$. Then, the spatial attention feature map $S \in R^{N\times N}$ is computed, as shown in Equation (8):

$$s_{ij} = \frac{exp(B_i + C_j)}{\sum_{i=1}^{N} exp(B_i + C_j)} \quad (8)$$

where s_{ij} denotes the effect of the i-th position on the j-th position. Next, the initial feature map A passes through a convolutional layer, is reshaped to obtain a new feature map $D \in R^{C\times N}$ and combined with the scaling factor β. Then, the spatial feature map E is obtained, as shown in Equation (9).

$$E_j = \beta\sum_{i=1}^{N} (s_{ji}D_i) + A_j \quad (9)$$

3. Feature fusion

After the input data pass through the spectral and spatial branches, the spectral and spatial feature maps are obtained, respectively. These two feature maps are concatenated along the channel dimension to preserve the independence and integrity of both spectral and spatial information. The fused feature map is then passed through an adaptive average pooling layer with an output size of 1 × 1, which reduces the spatial dimensions while retaining global contextual information. The resulting feature vector is fed into a fully connected classifier consisting of three sequential linear layers. The first layer projects the feature vector from 832 to 128 dimensions, followed by a ReLU activation and a dropout layer with a dropout rate of 0.4. The second layer further reduces the dimensionality from 128 to 64, also followed by a ReLU activation and dropout (rate = 0.4). The final linear layer maps the 64-dimensional vector to two output neurons corresponding to the classification categories (*T. grandis* and non-*T. grandis*). The output is directly used for pixel-wise classification.

2.3.4. Evaluation Methods

The extraction of *T. grandis* growing areas is essentially a binary classification task. To evaluate the prediction performance of the modem in determining the spatial distribution of *T. grandis* in the Kuaiji Mountain area, a confusion matrix was introduced based on the test set to evaluate the accuracy of the classification results of the different methods. The confusion matrix is the most basic and intuitive method for measuring the accuracy of a classification model. For each class, the number of samples predicted correctly and incorrectly by the classification model can be calculated. On this basis, the user accuracy (UA), producer accuracy (PA), overall accuracy (OA), and Kappa coefficient are chosen as the evaluation indicators. UA represents the proportion of correctly classified pixels for a specific class. PA measures the proportion of actual class pixels that were correctly identified. OA denotes the proportion of correctly classified pixels relative to the total number of pixels. The Kappa coefficient is a statistical measure used to assess the agreement between classified data and reference data, quantifying the reduction in classification errors compared to what would be expected by chance. The specific formulas are as follows:

$$UA = \frac{TP}{TP + FP} \tag{10}$$

$$PA = \frac{TP}{TP + FN} \tag{11}$$

$$OA = \frac{TP + TN}{TP + FP + FN + TN} \tag{12}$$

$$Kappa = \frac{OA - p_e}{1 - p_e} \tag{13}$$

$$p_e = \frac{(TP + FN)(TP + EP) + (FP + FN)(TN + FN)}{(TP + FN + FP + TN)^2} \tag{14}$$

In Equations (10)–(14), where TP represents the number of correctly classified pixels of *T. grandis*, FP represents the number of incorrectly classified pixels of *T. grandis*, TN represents the number of correctly classified pixels of other surface objects, and FN represents the number of incorrectly classified pixels of other surface objects.

To better analyze the distribution characteristics of *T. grandis* in the study area, a distribution index [64] was introduced in this study. The relationships between the spatial distribution of *T. grandis* and topographic factors (including aspect, slope and elevation) are described below:

$$D = \left(\frac{S}{S_k}\right) \times \left(\frac{S_{ik}}{S_i}\right) \tag{15}$$

In Equation (15), D is the distribution index, S is the total area of the study area, S_k is the area of a specific grade of topographic factor k in the study area, S_{ik} is the area of a topographic factor k within class i at that specific grade, and S_i is the total area of class i in the study area.

3. Result Analysis

3.1. Extraction of *T. grandis* Growing Areas Using the SSFAN

3.1.1. Feature Optimization

The original band features used as input for the deep learning model were first selected through the mRMR algorithm. The number of optimized features plays a critical role in the model performance. To evaluate the effects of different numbers of features, while keeping other experimental conditions unchanged, the top 10 to 50 features (with a step of 10) were selected as model input for classification, and the results are shown in Figure 4.

As shown in Figure 5, an increase in the number of selected features leads to continuous improvements in the model's classification performance. However, when the feature count exceeds 30, the rate of improvement becomes more gradual. This trend occurs because too few features may reduce the model's ability to distinguish between different classes. Conversely, too many features can introduce redundancy and noise, which may increase the training cost and slow convergence. Based on a comprehensive consideration of the speed and performance of the model, the top 30 optimal features were selected as the model input. Table 3 provides a detailed ranking and description of these 30 optimal features.

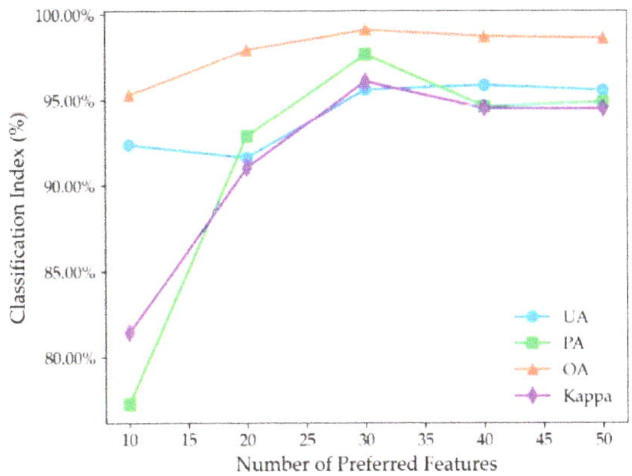

Figure 5. Classification indices under different numbers of features.

Table 3. Bands obtained from feature optimization.

Rank	Category	Name	Rank	Category	Name
1	Vegetation index (May)	GNDVI	16	Spectral feature (Mar)	B4
2	Texture feature (Apr)	CORR	17	Spectral feature (May)	B11
3	Vegetation index (Nov)	S2REP	18	Spectral feature (May)	B3
4	Spectral feature (May)	B5	19	Vegetation index (Mar)	EVI
5	Vegetation index (Mar)	GNDVI	20	Vegetation index (Nov)	NDREI
6	Spectral feature (Jul)	B9	21	Spectral feature (Apr)	B11
7	Vegetation index (Dec)	S2REP	22	Spectral feature (Mar)	B3
8	Vegetation index (May)	S2REP	23	Spectral feature (May)	B12
9	Vegetation index (Jul)	GNDVI	24	Vegetation index (Jan)	GNDVI
10	Vegetation index (Apr)	NDVI	25	Spectral feature (Mar)	B2
11	Spectral feature (Jul)	B12	26	Spectral feature (Mar)	B12
12	Vegetation index (Feb)	GNDVI	27	Spectral feature (Apr)	B5
13	Texture feature (Nov)	CORR	28	Texture feature (Mar)	CORR
14	Texture feature (Jun)	ENT	29	Vegetation index (Oct)	S2REP
15	Spectral feature (Jul)	B11	30	Spectral feature (May)	B4

After feature selection, we selected sample points of *T. grandis* and other typical non-*T. grandis* land cover types (coniferous forest, broadleaf forest, buildings, and shrubs) to compare their spectral features, vegetation indices, and texture features. As shown in Figure 6, *T. grandis* exhibits significant differences from other categories in the selected features—specifically, it shows generally lower spectral reflectance and relatively higher vegetation index values. Due to its large canopy and concentrated distribution, *T. grandis* also presents distinct texture patterns compared to natural vegetation. In addition, the selected temporal features are mainly concentrated in March and May. March marks a critical transition period from dormancy to active growth for *T. grandis*, during which the canopy structure changes significantly, making it more distinguishable from other vegetation still in early spring conditions. May represents the peak of canopy development, with increased spectral reflectance and vegetation indices, and *T. grandis* exhibits notable spectral differences from coniferous forests, shrubs, and other land cover types during this period. These differences indicate that the selected features have strong discriminative power for identifying *T. grandis*.

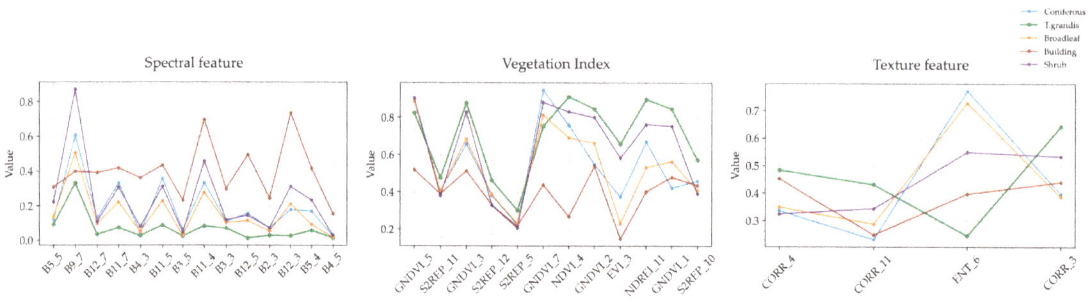

Figure 6. Comparison of spectral, vegetation index, and texture features among *Torreya grandis* and other land cover types.

3.1.2. Deep Learning Environment Settings

In this study, a cloud-based deep learning server was utilized for model training. Specifically, the system ran Ubuntu 20.04, equipped with 80 GB of RAM. The computational resources included a 16 vCPU Intel(R) Xeon(R) Platinum 8481C CPU and an NVIDIA RTX 4090D GPU with 24 GB of memory. The deep learning model was implemented using the PyTorch 2.0.0 framework with Python 3.8 as the programming language. During model training, the image bands were first normalized and then segmented into 15 $\times$ 15 patches before being fed into the deep learning model. The initial learning rate was set to 0.01, the batch size to 64, and the number of epochs to 200. The cross-entropy loss function was employed, and the Adam optimizer was used to adaptively adjust the learning rate based on historical gradient information. Using a larger learning rate in the early stages of training facilitates rapid convergence, while a smaller learning rate in later stages helps achieve more precise identification of the minimum loss function value. The specific hyperparameter settings before training are shown in Table 4.

Table 4. Hyperparameter settings for the experiments.

Patch Size	Initial Learning Rate	Batch Size	Epoch	Loss Function
15	0.01	64	200	cross entropy loss

3.1.3. Model Validation and Extraction Results for *T. grandis*

The test set was used to assess the model's performance. The OA of the proposed model was 99.1%, the Kappa coefficient was 0.961, the UA was 95.2%, and the PA was 97.67% (Table 5). The trained model was applied to identify the spatial distribution of *T. grandis* in the Kuaiji Mountain area. Owing to the large size of the study area, the system memory cannot process the entire image at one time. Therefore, the image of the study area was divided into several small blocks, which were then input to the model block by block for prediction. The prediction results were stitched based on geographic location information to obtain a distribution map of *T. grandis* in the Kuaiji Mountain area (Figure 7).

Table 5. Comparison of classification accuracy of different models across train, validation, and test sets.

Method	Data Set	Class	UA (%)	PA (%)	OA (%)	Kappa
2D-CNN	Train	*T. grandis*	94.56	98.32	98.40	0.955
		non-*T. grandis*	99.60	98.66		
	Validation	*T. grandis*	91.10	82.45	96.18	0.843
		non-*T. grandis s*	96.97	98.59		
	Test	*T. grandis*	81.11	88.14	95.62	0.818
		non-*T. grandis*	98.10	96.80		
3D-CNN	Train	*T. grandis*	97.16	99.09	99.11	0.975
		non-*T. grandis*	99.72	99.72		
	Validation	*T. grandis*	91.22	88.12	96.92	0.867
		non-*T. grandis*	97.83	98.52		
	Test	*T. grandis*	91.08	85.90	96.78	0.865
		non-*T. grandis*	97.67	98.60		
HybridSN	Train	*T. grandis*	99.16	99.42	99.66	0.990
		non-*T. grandis*	99.82	99.74		
	Validation	*T. grandis*	93.68	84.71	96.87	0.872
		non-*T. grandis*	97.36	99.01		
	Test	*T. grandis*	94.26	84.29	96.85	0.871
		non-*T. grandis*	97.26	99.09		
SSFN	Train	*T. grandis*	98.51	99.04	99.42	0.983
		non-*T. grandis*	99.70	99.54		
	Validation	*T. grandis*	92.21	92.50	97.92	0.911
		non-*T. grandis*	98.89	98.85		
	Test	*T. grandis*	92.61	91.71	97.87	0.909
		non-*T. grandis*	98.69	98.85		
SSFAN	Train	*T. grandis*	98.61	99.12	99.46	0.985
		non-*T. grandis*	99.73	99.57		
	Validation	*T. grandis*	94.85	98.71	99.14	0.962
		non-*T. grandis*	99.81	99.20		
	Test	*T. grandis*	95.62	97.67	99.11	0.961
		non-*T. grandis*	99.64	99.32		

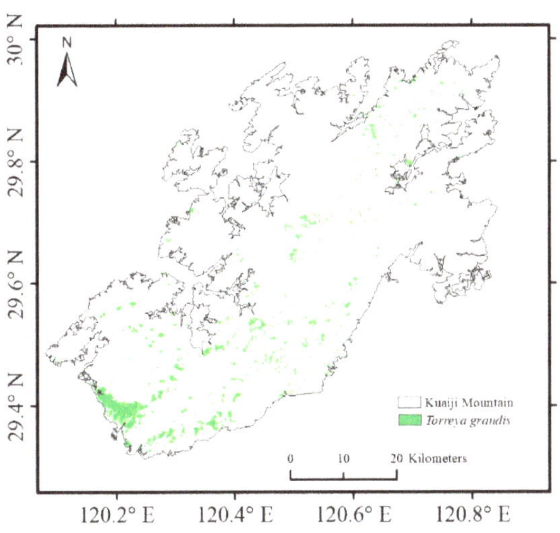

Figure 7. Distribution of *T. grandis* predicted by the SSFAN model.

3.2. Analysis of the Distribution Characteristics of *T. grandis*

To analyze the distribution characteristics of *T. grandis* in the Kuaiji Mountain area, DEM data were used to analyze the extracted *T. grandis* data. The distribution characteristics of *T. grandis* were investigated from three aspects: elevation, aspect and slope. The distribution index could be used to describe the distribution characteristics of *T. grandis* under different slopes, aspects and elevations. The aspect was divided into eight directions, and the elevation and slope were graded at steps of 100 m and 5°, respectively. Figure 8a shows that the distribution indices of *T. grandis* are different at different aspects, with the highest distribution index occurring on the western slope and high distribution indices occurring on the southern and southwestern slopes, which may be attributed to more suitable sunshine and climatic conditions on these slopes. As shown in Figure 8b, as the slope increases, the distribution index of *T. grandis* generally tends to increase, and the distribution index is high on steep slopes > 20°, indicating that a steep slope environment is more conducive to the growth of *T. grandis*. Figure 8c shows that the distribution index at 500–600 m is the highest, and the distribution indices in other high-elevation areas are also relatively high, indicating that *T. grandis* has a clear distribution advantage at relatively high elevations.

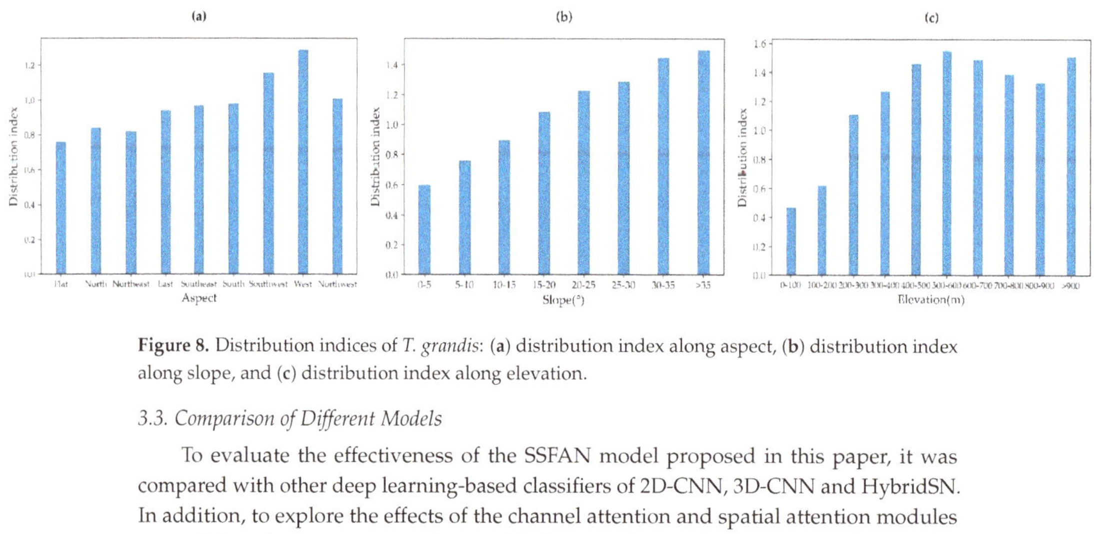

Figure 8. Distribution indices of *T. grandis*: (**a**) distribution index along aspect, (**b**) distribution index along slope, and (**c**) distribution index along elevation.

3.3. Comparison of Different Models

To evaluate the effectiveness of the SSFAN model proposed in this paper, it was compared with other deep learning-based classifiers of 2D-CNN, 3D-CNN and HybridSN. In addition, to explore the effects of the channel attention and spatial attention modules on model performance, comparisons with the attention mechanism removed were also performed. All the experiments were implemented in PyTorch on the same platform with the same data. These methods are briefly described below:

1. 2D-CNN: Makantasis et al. described the specific network architecture [65]. This model is mainly based on the 2D-CNN.
2. 3D-CNN: Zhang et al. described the specific network architecture [33]. This model is mainly based on the 3D-CNN.
3. HybridSN: This hybrid model combines the 3D-CNN and 2D-CNN in a series. Roy et al. described the specific network architecture [66].
4. SSFN: The attention mechanism is removed from the original SSFAN model, while the remainder of the structure remains unchanged.

A confusion matrix was introduced to the test set, and various evaluation indicators were calculated accordingly. The confusion matrix of each model is shown in Figure 9, and the corresponding evaluation indicators are shown in Table 5.

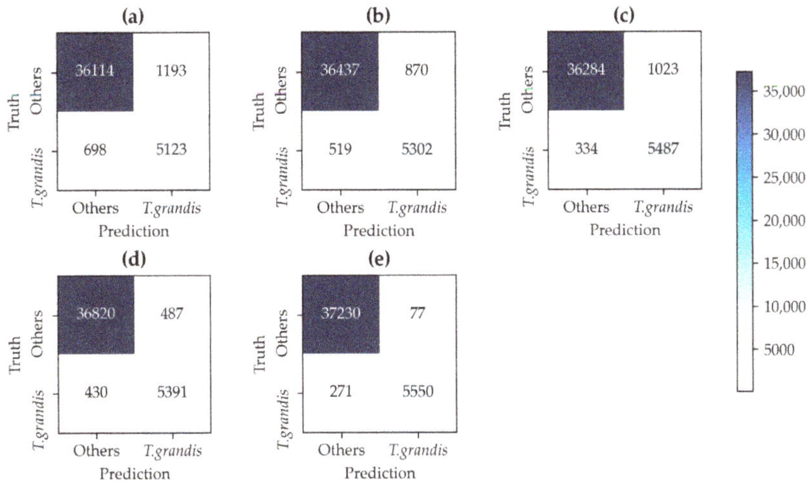

Figure 9. Confusion matrices of different models on the test set: (**a**) 2D-CNN, (**b**) 3D-CNN, (**c**) HybridSN, (**d**) SSFN, (**e**) SSFAN.

Table 5 presents the classification performance of different models across the training, validation, and testing datasets. All models adopted a consistent data split strategy and training configuration, with sufficient training epochs to ensure model convergence. Except for the 2D-CNN, the other four models (3D-CNN, HybridSN, SSFN, and SSFAN) achieved over 99% OA on the training set, with less than 1% fluctuation across performance metrics, indicating a stable and fully converged training process. This setup provides a reliable foundation for evaluating the models' generalization ability on testing sets. The proposed SSFAN model achieved the best classification performance, with an OA of 99.11% and a Kappa coefficient of 0.961 on the testing set. This indicates that SSFAN not only has strong fitting ability but also excellent generalization capacity, effectively identifying key features in unseen samples. In contrast, the 2D-CNN model exhibited lower performance due to its relatively simple architecture and limited feature extraction capability. The 3D-CNN outperformed 2D-CNN on the testing set as it can extract both spectral and spatial features simultaneously rather than relying solely on spatial convolution, thus enhancing its representational power. The HybridSN model further integrated the 2D-CNN and 3D-CNN architectures in series, aiming to capture both spectral and spatial abstractions. However, its testing performance did not surpass that of SSFN or SSFAN. The SSFN model maintained good consistency between the training and validation phases, showing strong stability and generalizability. Nevertheless, its UA and PA for both classes were slightly lower than those of SSFAN, especially in edge cases or small fragmented patches. Compared with SSFN, which lacks attention mechanisms, SSFAN incorporates both channel and spatial attention modules into its architecture. These allow the model to focus more effectively on critical bands and spatially informative regions, resulting in an additional improvement in classification accuracy.

In addition, to more intuitively represent the prediction effect of each model, the prediction results of different models can be visualized. Since the resolution of Sentinel-2 images is 10 m, it is difficult to distinguish *T. grandis* ground objects from non-*T. grandis* ground objects directly with the naked eye. Therefore, in this study, a high-resolution Google Earth image of the corresponding area was selected as a control to aid in evaluation.

As shown in Figure 10, the extraction results of the *T. grandis* growing areas by the 2D-CNN and 3D-CNN models exhibit some degree of misclassification and omission. In addition, the distribution areas of *T. grandis* extracted by these two networks and HybridSN are continuous; however, there is still a serious lack of edge information. In general, the extraction results of the SSFAN model are the most consistent with the ground truth. To enhance the model's ability to capture feature information from *T. grandis* blocks and reduce the interference of complex background information, an attention mechanism was introduced in the network to highlight the key features of *T. grandis* identification and reduce the influence of the complex environment on the extraction accuracy of the *T. grandis* blocks. However, all the model results inevitably differ from actual conditions. These errors are mainly concentrated in small or more scattered *T. grandis* blocks, which is partly due to the insufficient resolution of the remote sensing images and partly due to the more severe mixing between *T. grandis* and other ground objects.

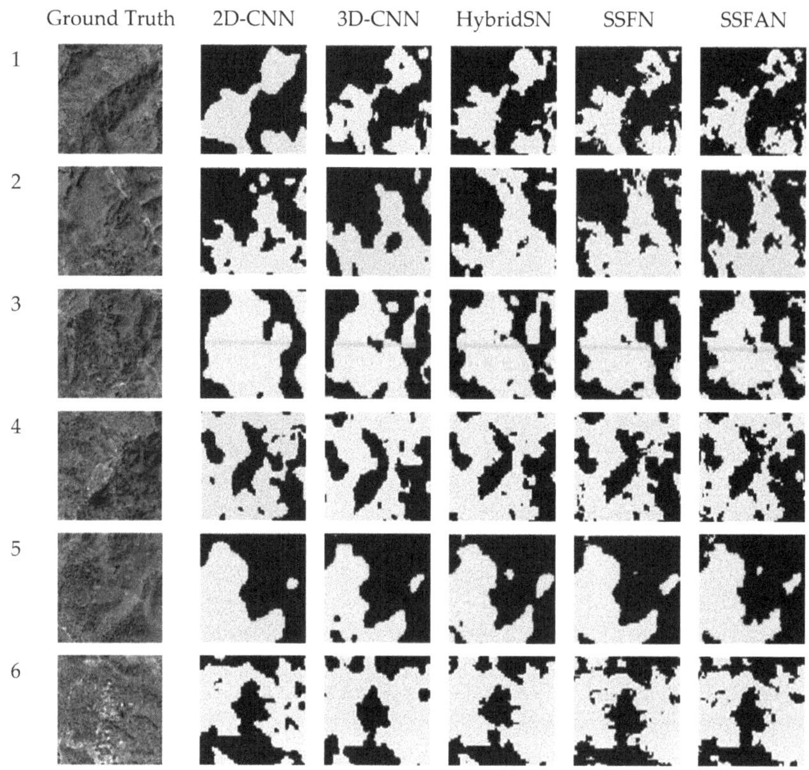

Figure 10. Prediction maps of different models and ground truths of high-resolution remote sensing images.

4. Discussion

The extraction results of *T. grandis* growing areas (Figure 10) and the evaluation indicators based on the test set (Table 5) between the SSFAN constructed in this paper and other deep learning networks, such as 2D-CNN, 3D-CNN, and HybridSN, were compared. The results showed the superiority of the SSFAN. Studies have shown that methods using only 2D-CNNs or 3D-CNNs, which are based solely on spatial or spectral features, may result in the loss of critical information, thereby limiting the ability to fully exploit the comprehensive characteristics of remote sensing images [37]. In contrast, approaches that simultaneously consider both spectral and spatial information tend to be more robust and yield improved classification performance [34]. The dual-branch network architecture allows the design of different spectral and spatial networks for feature extraction, offering greater flexibility and advantages. Therefore, in this study, 3D-CNN and 2D-CNN models were selected as the main structures of the spectral and spatial branches, respectively. Considering that the spectral and spatial features are in uncorrelated domains, compared with the 3D-2D tandem structure used in HybridSN, the dual-branch structure that first extracts spectral and spatial features in parallel, followed by series fusion in the later stage, can better retain the independence between the two branches [39]. Moreover, this study introduced a self-attention mechanism in both the spectral and spatial dimensions, and the effectiveness of this mechanism was verified by comparing the results of models with and without the spectral and spatial attention mechanisms. The self-attention mechanism can adaptively integrate local semantic features and global dependencies [38], thus effectively capturing long-range contextual information. The spectral attention module focuses on information-rich bands, whereas the spatial attention module emphasizes pixels with significant spatial features. In summary, the SSFAN introduced in this study can effectively extract *T. grandis* growing areas and fully demonstrate the effectiveness and superiority of fused space-spectral features and a self-attention mechanism.

This study also presented an analysis of the distribution characteristics of *T. grandis* (Figure 8). The results revealed that topographic factors were the key factors affecting the distribution of *T. grandis* in the Kuaiji Mountain area. The higher distribution indices on the west and southwest slopes are related to the warm growth habit of *T. grandis* [67]; these two slopes have relatively warm climatic conditions and long sunshine hours. Slope also has a significant effect on the distribution of *T. grandis*. Large slopes provide sufficient sunlight, which makes the distribution of *T. grandis* more concentrated on steep slopes. *T. grandis* is also suitable for planting on mountainous or hilly terrain [68]. Due to human land-use preferences, residential areas are generally concentrated on flat land or gentle slopes on the west side of Kuaiji Mountain. Low-elevation and relatively flat areas near residential zones are therefore often prioritized for cultivating staple crops such as rice and corn. To avoid occupying arable land, *T. grandis* is typically planted on steep slopes in the west and southwest near residential areas. Elevation is also an important factor affecting the planting of *T. grandis*, which exhibits strong adaptability to high-elevation environments. In contrast, land-use decisions in low-elevation areas have limited the distribution of *T. grandis*, which has a long growth cycle, bearing fruit in five to six years. Therefore, residents often choose to plant *T. grandis* on relatively barren mountains at relatively high elevations. In addition, elevation has a significant effect on the quality of *T. grandis* fruit, which is best in high-elevation areas [69] and result in greater economic benefits. Therefore, the distribution of *T. grandis* is the result of the combined influence of natural and human factors.

In this study, a spatial–spectral fused attention network (SSFAN) was constructed to extract *Torreya grandis* growing areas based on multitemporal Sentinel-2 imagery. Compared with traditional machine learning methods previously used for *T. grandis* extraction [11], the overall accuracy was significantly improved from 78% to 99.1%. Although deep learning methods have primarily been applied in tree species identification based on hyperspectral imagery, this study successfully adapted SSFAN to multispectral data by constructing a multitemporal band combination that mimics a hyperspectral structure. However, this study still has certain limitations. The sample data were collected from two typical *T. grandis* distribution areas in the Kuaiji Mountain region. While these areas are representative within the study region and the model performed well, its generalizability to regions with different geographical and climatic conditions requires further validation. In addition, although most *T. grandis* individuals have large canopies and are densely distributed, the 10 m spatial resolution of Sentinel-2 imagery still presents challenges in edge areas or scattered distributions, where mixed pixels may lead to local misclassification. To address this issue, future studies could consider incorporating higher spatial resolution remote sensing data to further improve extraction accuracy and model performance.

5. Conclusions

In this study, on the basis of multitemporal Sentinel-2 images and the SSFAN model, the *T. grandis* growing areas in the Kuaiji Mountain area were identified, and the distribution characteristics of *T. grandis* under different aspects, slopes, and elevations were analyzed. In addition, the extraction results obtained using SSFAN were compared with those of other deep learning models, and the superiority of the proposed model and the factors affecting the distribution of *T. grandis* were discussed. The following conclusions were reached:

1. On the basis of the spectral bands, vegetation indices, and texture features of 12 synthesized monthly average Sentinel-2 images from 2023, a hyperspectral image-like band structure was formed. After the mRMR feature selection, the SSFAN was constructed. The *T. grandis* growing areas in the Kuaiji Mountain area were extracted. The model's accuracy was validated. The model achieved an OA of 99.1% and a Kappa coefficient of 0.961, with the UA and PA of *T. grandis* reaching 95.2% and 97.67%, respectively, indicating good extraction results.
2. Elevation, aspect, and slope have significant impacts on the distribution of *T. grandis*. The distribution index of *T. grandis* is the most concentrated on the western, southern, and southwestern slopes, where light and heat conditions are more suitable. As slope steepness increased, the *T. grandis* distribution index exhibited an upward trend, with a particular preference for steep slopes (>20°). Additionally, regions at 500–600 m or greater elevation exhibited a relatively high distribution index.
3. The test set was used to assess the evaluation indicators for each model. The performance ranking from highest to lowest was as follows: SSFAN > SSFN > HybridSN > 3D-CNN > 2D-CNN. The proposed model performed best and had an advantage in the extraction of *T. grandis* growing areas.
4. In the extraction of *T. grandis* growing areas, the SSFAN, which fuses spectral and spatial features and introduces a self-attention mechanism, exhibited notable effectiveness and superiority. Additionally, the distribution of *T. grandis* is influenced by a combination of natural and human factors, providing a valuable scientific basis for the planting and management of *T. grandis*.

In future research, this method could be extended to other economic tree species or different regions and integrated with higher-resolution remote sensing data and multi-source environmental factors to further improve the model's generalization ability and classification accuracy.

Author Contributions: Conceptualization, Y.L. and Y.W.; methodology, Y.L.; software, Y.L.; validation, Y.L., Y.W. and X.S.; formal analysis, Y.L.; investigation, Y.W.; resources, Y.W. and X.S.; data curation, Y.L.; writing—original draft preparation, Y.L.; writing—review and editing, Y.W.; visualization, Y.L.; supervision, Y.W.; project administration, Y.W. and X.S.; funding acquisition, Y.W. All authors have read and agreed to the published version of the manuscript.

Funding: This research was funded by the NUIST Students' Platform for Innovation and Entrepreneurship Training Program, grant number XJDC202410300172; the Jiangsu Students' Platform for Innovation and Entrepreneurship Training Program, grant number 202410300169Y; and the National Key R&D Program of China, grant number 2019YFB2102003.

Institutional Review Board Statement: Not applicable.

Data Availability Statement: The data presented in this study are available upon request from the corresponding author. The data are not publicly available due to the privacy policy of the authors' institution.

Acknowledgments: We would like to thank the NUIST Students' Platform for Innovation and Entrepreneurship Training Program, the Jiangsu Students' Platform for Innovation and Entrepreneurship Training Program, and the National Key R&D Program of China for their financial support of this research.

Conflicts of Interest: The authors declare no conflicts of interest.

References

1. Milad, M.; Schaich, H.; Bürgi, M.; Konold, W. Climate change and nature conservation in Central European forests: A review of consequences, concepts and challenges. *For. Ecol. Manag.* **2011**, *261*, 829–843. [CrossRef]
2. Nunes, L.J.R.; Meireles, C.I.R.; Gomes, C.J.P.; Ribeiro, N.M.C.A. The Impact of Climate Change on Forest Development: A Sustainable Approach to Management Models Applied to Mediterranean-Type Climate Regions. *Plants* **2021**, *11*, 69. [CrossRef] [PubMed]
3. Chen, X.; Jin, H. A Case Study of Enhancing Sustainable Intensification of Chinese Torreya Forest in Zhuji of China. *Environ. Nat. Resour. Res.* **2019**, *9*, 53–60. [CrossRef]
4. Chen, X.; Jin, H. Review of cultivation and development of Chinese torreya in China. *For. Trees Livelihoods* **2018**, *28*, 68–78. [CrossRef]
5. Cook, B.I.; Mankin, J.S.; Anchukaitis, K.J. Climate Change and Drought: From Past to Future. *Curr. Clim. Chang. Rep.* **2018**, *4*, 164–179. [CrossRef]
6. Ramos, M.C.; Martínez-Casasnovas, J.A. Climate change influence on runoff and soil losses in a rainfed basin with Mediterranean climate. *Nat. Hazards* **2015**, *78*, 1065–1089. [CrossRef]
7. Skendzic, S.; Zovko, M.; Zivkovic, I.P.; Lesic, V.; Lemic, D. The Impact of Climate Change on Agricultural Insect Pests. *Insects* **2021**, *12*, 440. [CrossRef]
8. Han, N.; Zhang, X.; Wang, X.; Wang, K. Extraction of *Torreya grandis* Merrillii based on object-oriented method from IKONOS imagery. *J. Zhejiang Univ. (Agric. Life Sci.)* **2009**, *35*, 670–676. [CrossRef]
9. Han, N.; Zhang, X.; Wang, X.; Chen, L.; Wang, K. Identification of distributional information Torreya Grandis Merrlllii using high resolution imagery. *J. Zhejiang Univ. (Eng. Sci.)* **2010**, *44*, 420–425.
10. Wang, Y. Distribution Information Extraction and Dynamic Change Monitoring of *Torreya Grandis* Based on Multi-Source Data. Master's Thesis, Zhejiang A&F University, Hangzhou, China, 2017.
11. Ji, J.; Yu, X.; Yu, C.; Zhang, Y.; Li, D. The Spatial Distribution Extraction of Torreya grandis Forest Based on Remote Sensing and Random Forest. *South China Agric.* **2018**, *12*, 93–95. [CrossRef]
12. Axelsson, A.; Lindberg, E.; Reese, H.; Olsson, H. Tree species classification using Sentinel-2 imagery and Bayesian inference. *Int. J. Appl. Earth Obs. Geoinf.* **2021**, *100*, 102318. [CrossRef]
13. Franklin, S.E.; Ahmed, O.S. Deciduous tree species classification using object-based analysis and machine learning with unmanned aerial vehicle multispectral data. *Int. J. Remote Sens.* **2017**, *39*, 5236–5245. [CrossRef]
14. Bjerreskov, K.S.; Nord-Larsen, T.; Fensholt, R. Forest type and tree species classification of nemoral forests with Sentinel-1 and 2 Time Series data. *Preprints* **2021**. [CrossRef]
15. Ghorbanian, A.; Zaghian, S.; Asiyabi, R.M.; Amani, M.; Mohammadzadeh, A.; Jamali, S. Mangrove Ecosystem Mapping Using Sentinel-1 and Sentinel-2 Satellite Images and Random Forest Algorithm in Google Earth Engine. *Remote Sens.* **2021**, *13*, 2565. [CrossRef]
16. Guo, Q.; Zhang, J.; Guo, S.; Ye, Z.; Deng, H.; Hou, X.; Zhang, H. Urban Tree Classification Based on Object-Oriented Approach and Random Forest Algorithm Using Unmanned Aerial Vehicle (UAV) Multispectral Imagery. *Remote Sens.* **2022**, *14*, 3885. [CrossRef]
17. Modzelewska, A.; Fassnacht, F.E.; Sterenczak, K. Tree species identification within an extensive forest area with diverse management regimes using airborne hyperspectral data. *Int. J. Appl. Earth Obs. Geoinf.* **2020**, *84*, 101960. [CrossRef]
18. Nasiri, V.; Beloiu, M.; Darvishsefat, A.A.; Griess, V.C.; Maftei, C.; Waser, L.T. Mapping tree species composition in a Caspian temperate mixed forest based on spectral-temporal metrics and machine learning. *Int. J. Appl. Earth Obs. Geoinf.* **2023**, *116*, 103154. [CrossRef]
19. Nguyen, H.M.; Demir, B.; Dalponte, M. Weighted support vector machines for tree species classification using lidar data. In Proceedings of the IGARSS 2019-2019 IEEE International Geoscience and Remote Sensing Symposium, Yokohama, Japan, 28 July–2 August 2019; pp. 6740–6743.
20. Yang, G.; Zhao, Y.; Li, B.; Ma, Y.; Li, R.; Jing, J.; Dian, Y. Tree Species Classification by Employing Multiple Features Acquired from Integrated Sensors. *J. Sens.* **2019**, *2019*, 3247946. [CrossRef]

21. Alzubaidi, L.; Zhang, J.; Humaidi, A.J.; Al-Dujaili, A.; Duan, Y.; Al-Shamma, O.; Santamaría, J.; Fadhel, M.A.; Al-Amidie, M.; Farhan, L. Review of deep learning: Concepts, CNN architectures, challenges, applications, future directions. *J. Big Data* **2021**, *8*, 53. [CrossRef]
22. La Rosa, L.E.C.; Sothe, C.; Feitosa, R.Q.; de Almeida, C.M.; Schimalski, M.B.; Oliveira, D.A.B. Multi-task fully convolutional network for tree species mapping in dense forests using small training hyperspectral data. *ISPRS J. Photogramm. Remote Sens.* **2021**, *179*, 35–49. [CrossRef]
23. Li, Q.; Wong, F.K.K.; Fung, T. Mapping multi-layered mangroves from multispectral, hyperspectral, and LiDAR data. *Remote Sens. Environ.* **2021**, *258*, 112403. [CrossRef]
24. Zhao, H.; Zhong, Y.; Wang, X.; Hu, X.; Luo, C.; Boitt, M.; Piiroinen, R.; Zhang, L.; Heiskanen, J.; Pellikka, P. Mapping the distribution of invasive tree species using deep one-class classification in the tropical montane landscape of Kenya. *ISPRS J. Photogramm. Remote Sens.* **2022**, *187*, 328–344. [CrossRef]
25. Wei, X.; Yu, X.; Liu, B.; Zhi, L. Convolutional neural networks and local binary patterns for hyperspectral image classification. *Eur. J. Remote Sens.* **2019**, *52*, 448–462. [CrossRef]
26. Xi, Y.; Ren, C.; Wang, Z.; Wei, S.; Bai, J.; Zhang, B.; Xiang, H.; Chen, L. Mapping Tree Species Composition Using OHS-1 Hyperspectral Data and Deep Learning Algorithms in Changbai Mountains, Northeast China. *Forests* **2019**, *10*, 818. [CrossRef]
27. Fricker, G.A.; Ventura, J.D.; Wolf, J.A.; North, M.P.; Davis, F.W.; Franklin, J. A Convolutional Neural Network Classifier Identifies Tree Species in Mixed-Conifer Forest from Hyperspectral Imagery. *Remote Sens.* **2019**, *11*, 2326. [CrossRef]
28. Sothe, C.; De Almeida, C.M.; Schimalski, M.B.; La Rosa, L.E.C.; Castro, J.D.B.; Feitosa, R.Q.; Dalponte, M.; Lima, C.L.; Liesenberg, V.; Miyoshi, G.T.; et al. Comparative performance of convolutional neural network, weighted and conventional support vector machine and random forest for classifying tree species using hyperspectral and photogrammetric data. *GIScience Remote Sens.* **2020**, *57*, 369–394. [CrossRef]
29. Guan, H.; Yu, Y.; Yan, W.; Li, D.; Li, J. 3D-Cnn Based Tree species classification using mobile lidar data. *Int. Arch. Photogramm. Remote Sens. Spat. Inf. Sci.* **2019**, *42*, 989–993. [CrossRef]
30. Liu, M.H.; Han, Z.W.; Chen, Y.M.; Liu, Z.; Han, Y.S. Tree species classification of LiDAR data based on 3D deep learning. *Measurement* **2021**, *177*, 109301. [CrossRef]
31. Mäyrä, J.; Keski-Saari, S.; Kivinen, S.; Tanhuanpää, T.; Hurskainen, P.; Kullberg, P.; Poikolainen, L.; Viinikka, A.; Tuominen, S.; Kumpula, T.; et al. Tree species classification from airborne hyperspectral and LiDAR data using 3D convolutional neural networks. *Remote Sens. Environ.* **2021**, *256*, 112322. [CrossRef]
32. Nezami, S.; Khoramshahi, E.; Nevalainen, O.; Pölönen, I.; Honkavaara, E. Tree species classification of drone hyperspectral and RGB imagery with deep learning convolutional neural networks. *Remote Sens.* **2020**, *12*, 1070. [CrossRef]
33. Zhang, B.; Zhao, L.; Zhang, X. Three-dimensional convolutional neural network model for tree species classification using airborne hyperspectral images. *Remote Sens. Environ.* **2020**, *247*, 111938. [CrossRef]
34. Liang, J.; Li, P.; Zhao, H.; Han, L.; Qu, M. Forest species classification of UAV hyperspectral image using deep learning. In Proceedings of the 2020 Chinese Automation Congress (CAC), Shanghai, China, 6–8 November 2020; pp. 7126–7130.
35. Shi, C.; Liao, D.; Xiong, Y.; Zhang, T.; Wang, L. Hyperspectral Image Classification Based on Dual-Branch Spectral Multiscale Attention Network. *IEEE J. Sel. Top. Appl. Earth Obs. Remote Sens.* **2021**, *14*, 10450–10467. [CrossRef]
36. Xu, J.; Li, K.; Li, Z.; Chong, Q.; Xing, H.; Xing, Q.; Ni, M. Fuzzy graph convolutional network for hyperspectral image classification. *Eng. Appl. Artif. Intell.* **2024**, *127*, 107280. [CrossRef]
37. Hou, C.; Liu, Z.; Chen, Y.; Wang, S.; Liu, A. Tree Species Classification from Airborne Hyperspectral Images Using Spatial–Spectral Network. *Remote Sens.* **2023**, *15*, 5679. [CrossRef]
38. Fu, J.; Liu, J.; Tian, H.; Li, Y.; Bao, Y.; Fang, Z.; Lu, H. Dual attention network for scene segmentation. In Proceedings of the IEEE/CVF Conference on Computer Vision and Pattern Recognition, Long Beach, CA, USA, 15–20 June 2019; pp. 3146–3154.
39. Li, R.; Zheng, S.; Duan, C.; Yang, Y.; Wang, X. Classification of Hyperspectral Image Based on Double-Branch Dual-Attention Mechanism Network. *Remote Sens.* **2020**, *12*, 582. [CrossRef]
40. Mura, M.; Bottalico, F.; Giannetti, F.; Bertani, R.; Giannini, R.; Mancini, M.; Orlandini, S.; Travaglini, D.; Chirici, G. Exploiting the capabilities of the Sentinel-2 multi spectral instrument for predicting growing stock volume in forest ecosystems. *Int. J. Appl. Earth Obs. Geoinf.* **2018**, *66*, 126–134. [CrossRef]
41. Ferreira, M.P.; Wagner, F.H.; Aragao, L.; Shimabukuro, Y.E.; de Souza, C.R. Tree species classification in tropical forests using visible to shortwave infrared WorldView-3 images and texture analysis. *Isprs J. Photogramm. Remote Sens.* **2019**, *149*, 119–131. [CrossRef]
42. Liu, L.X.; Coops, N.C.; Aven, N.W.; Pang, Y. Mapping urban tree species using integrated airborne hyperspectral and LiDAR remote sensing data. *Remote Sens. Environ.* **2017**, *200*, 170–182. [CrossRef]
43. Valderrama-Landeros, L.; Flores-de-Santiago, F.; Kovacs, J.M.; Flores-Verdugo, F. An assessment of commonly employed satellite-based remote sensors for mapping mangrove species in Mexico using an NDVI-based classification scheme. *Environ. Monit. Assess.* **2017**, *190*, 23. [CrossRef]

44. Persson, M.; Lindberg, E.; Reese, H. Tree species classification with multi-temporal Sentinel-2 data. *Remote Sens.* **2018**, *10*, 1794. [CrossRef]
45. Sefrin, O.; Riese, F.M.; Keller, S. Deep Learning for Land Cover Change Detection. *Remote Sens.* **2020**, *13*, 78. [CrossRef]
46. Russo, L.; Mauro, F.; Memar, B.; Sebastianelli, A.; Gamba, P.; Ullo, S.L. Using Multi-Temporal Sentinel-1 and Sentinel-2 data for water bodies mapping. *arXiv* **2024**, arXiv:2402.00023. [CrossRef]
47. Sukmono, A.; Hadi, F.; Widayanti, E.; Nugraha, A.L.; Bashit, N. Identifying Burnt Areas in Forests and Land Fire Using Multitemporal Normalized Burn Ratio (NBR) Index on Sentinel-2 Satellite Imagery. *Int. J. Saf. Secur. Eng.* **2023**, *13*, 469–477. [CrossRef]
48. Tavus, B.; Kocaman, S.; Gokceoglu, C. Flood damage assessment with Sentinel-1 and Sentinel-2 data after Sardoba dam break with GLCM features and Random Forest method. *Sci. Total Environ.* **2022**, *816*, 151585. [CrossRef]
49. Louis, J.; Debaecker, V.; Pflug, B.; Main-Knorn, M.; Bieniarz, J.; Mueller-Wilm, U.; Cadau, E.; Gascon, F. Sentinel-2 Sen2Cor: L2A processor for users. In Proceedings of the Living Planet Symposium 2016, Prague, Czech Republic, 9–13 May 2016; pp. 1–8.
50. Sentinel Application Platform (SNAP). Available online: https://step.esa.int/main/toolboxes/snap/ (accessed on 20 December 2024).
51. Zheng, H.; Du, P.; Chen, J.; Xia, J.; Li, E.; Xu, Z.; Li, X.; Yokoya, N. Performance evaluation of downscaling Sentinel-2 imagery for land use and land cover classification by spectral-spatial features. *Remote Sens.* **2017**, *9*, 1274. [CrossRef]
52. Qi, H.; Tian, W.; Zhang, X. Index System of Landform Regionalization for Highway in China. *J. Chang. Univ. Nat. Sci. Ed.* **2011**, *31*, 33–38.
53. Rouse Jr, J.W.; Haas, R.H.; Deering, D.; Schell, J.; Harlan, J.C. *Monitoring the Vernal Advancement and Retrogradation (Green Wave Effect) of Natural Vegetation*; NASA: Washington, DC, USA, 1974.
54. Gitelson, A.A.; Kaufman, Y.J.; Merzlyak, M.N. Use of a green channel in remote sensing of global vegetation from EOS-MODIS. *Remote Sens. Environ.* **1996**, *58*, 289–298. [CrossRef]
55. Huete, A.R. A soil-adjusted vegetation index (SAVI). *Remote Sens. Environ.* **1988**, *25*, 295–309. [CrossRef]
56. Huete, A.; Didan, K.; Miura, T.; Rodriguez, E.P.; Gao, X.; Ferreira, L.G. Overview of the radiometric and biophysical performance of the MODIS vegetation indices. *Remote Sens. Environ.* **2002**, *83*, 195–213. [CrossRef]
57. Frampton, W.J.; Dash, J.; Watmough, G.; Milton, E.J. Evaluating the capabilities of Sentinel-2 for quantitative estimation of biophysical variables in vegetation. *ISPRS J. Photogramm. Remote Sens.* **2013**, *82*, 83–92. [CrossRef]
58. Fitzgerald, G.; Rodriguez, D.; Christensen, L.; Belford, R.; Sadras, V.; Clarke, T. Spectral and thermal sensing for nitrogen and water status in rainfed and irrigated wheat environments. *Precis. Agric.* **2006**, *7*, 233–248. [CrossRef]
59. Jiang, K.; Zhao, Q.; Wang, X.; Sheng, Y.; Tian, W. Tree Species Classification for Shelterbelt Forest Based on Multi-Source Remote Sensing Data Fusion from Unmanned Aerial Vehicles. *Forests* **2024**, *15*, 2200. [CrossRef]
60. Tassi, A.; Vizzari, M. Object-Oriented LULC Classification in Google Earth Engine Combining SNIC, GLCM, and Machine Learning Algorithms. *Remote Sens.* **2020**, *12*, 3776. [CrossRef]
61. Yang, S.; Gu, L.; Li, X.; Jiang, T.; Ren, R. Crop Classification Method Based on Optimal Feature Selection and Hybrid CNN-RF Networks for Multi-Temporal Remote Sensing Imagery. *Remote Sens.* **2020**, *12*, 3119. [CrossRef]
62. Liu, G.; Jiang, X.; Tang, B. Application of feature optimization and convolutional neural network in crop classification. *J. Geoinf. Sci.* **2021**, *23*, 1071–1081.
63. Peng, H.; Long, F.; Ding, C. Feature selection based on mutual information: Criteria of max-dependency, max-relevance, and min-redundancy. *IEEE Trans. Pattern Anal. Mach. Intell.* **2005**, *27*, 1226–1238. [CrossRef]
64. Sun, L.; Chen, H.; Pan, J. Analysis of the land use spatiotemporal variation based on DEM—Beijing Yanqing County as an example. *J. Mt. Res.* **2004**, *22*, 762–766.
65. Makantasis, K.; Karantzalos, K.; Doulamis, A.; Doulamis, N. Deep supervised learning for hyperspectral data classification through convolutional neural networks. In Proceedings of the 2015 IEEE International Geoscience and Remote Sensing Symposium (IGARSS), Milan, Italy, 26–31 July 2015; pp. 4959–4962.
66. Roy, S.K.; Krishna, G.; Dubey, S.R.; Chaudhuri, B.B. HybridSN: Exploring 3D–2D CNN Feature Hierarchy for Hyperspectral Image Classification. *IEEE Geosci. Remote Sens. Lett.* **2020**, *17*, 277–281. [CrossRef]
67. Zhong, Z.; Wang, J.; Wen, Y.; Zhang, Q. Research on the Resource Development and Industrialization Strategies of Chinese Torreya grandis. *J. Zhejiang Agric. Sci.* **2023**, *64*, 3020–3025.
68. Wu, L.; Yang, L.; Li, Y.; Shi, J.; Zhu, X.; Zeng, Y. Evaluation of the Habitat Suitability for Zhuji Torreya Based on Machine Learning Algorithms. *Agriculture* **2024**, *14*, 1077. [CrossRef]
69. Yang, X.; Yao, H.; Fan, F.; Yan, F.; Li, K.; Chen, C.; Wu, L. Effects of different altitudes on Torreya grandis 'Merrillii' fruit quality and its comprehensive evaluation. *J. Anhui Agric. Univ.* **2023**, *50*, 792–797. [CrossRef]

agriculture

MDPI

Article

Research on Ginger Price Prediction Model Based on Deep Learning

Fengyu Li [1,2,3], Xianyong Meng [1,2,3,*], Ke Zhu [1,2,3], Jun Yan [1,2,3], Lining Liu [1,2,3] and Pingzeng Liu [1,2,3,*]

1 School of Information Science and Engineering, Shandong Agricultural University, Tai'an 271018, China; 2023121117@sdau.edu.cn (F.L.); kezhu@sdau.edu.cn (K.Z.); yanj2016@sdau.edu.cn (J.Y.); 2021010104@sdau.edu.cn (L.L.)
2 Key Laboratory of Huanghuaihai Smart Agricultural Technology, Ministry of Agriculture and Rural Affairs, Tai'an 271018, China
3 Agricultural Big Data Research Center, Shandong Agricultural University, Tai'an 271018, China
* Correspondence: xym@sdau.edu.cn (X.M.); pzliu@sdau.edu.cn (P.L.)

Abstract: In order to ensure the price stability of niche agricultural products and enhance farmers' income, the study delves into the pattern of the ginger price fluctuation rule and its main influencing factors. By combining seasonal decomposition STL, long and short-term memory network LSTM, attention mechanism ATT and Kolmogorov-Arnold network, a combined STL-LSTM-ATT-KAN prediction model is developed, and the model parameters are finely tuned by using multi-population adaptive particle swarm optimisation algorithm (AMP-PSO). Based on an in-depth analysis of actual data on ginger prices over the past decade, the STL-LSTM-ATT-KAN model demonstrated excellent performance in terms of prediction accuracy: its mean absolute error (MAE) was 0.111, mean squared error (MSE) was 0.021, root mean squared error (RMSE) was 0.146, and the coefficient of determination (R^2) was 0.998. This study provides the Ginger Industry, agricultural trade, farmers and policymakers with digitalised and intelligent aids, which are important for improving market monitoring, risk control, competitiveness and guaranteeing the stability of supply and price.

Keywords: ginger price; price prediction; LSTM network; KAN network; AMP-PSO algorithm

1. Introduction

As a large agricultural country, China's agriculture is the foundation of the country's economy. The stability of agricultural prices has a direct impact on the rural economy, market regulation and the interests of farmers and consumers. Ginger price fluctuations are influenced by a variety of factors and have a significant impact on the downstream industry chain [1]. Therefore, accurate prediction of ginger price is crucial for farmers' production planning, consumers' consumption arrangement and government market regulation.

For the price fluctuation of agricultural products, researchers have widely used the ARIMA model in the field of agricultural price forecasting and verified its effectiveness. And successfully predicted the price trend of agricultural products such as cabbage, corn, mango, cucumber and potato [2–5]. The model is used to forecast the price trend of agricultural products such as cabbage, corn, mango, cucumber and potato. Meanwhile, the researchers improved the accuracy of the model by applying machine learning techniques such as time series analysis, support vector machine and neural network [6–11]. Nevertheless, researchers still need to face the challenges of insufficient model generalisation

 https://doi.org/10.3390/agriculture15060596

capability and incomplete consideration of complex influencing factors. In recent years, significant progress has been made in research on agricultural price forecasting, covering from traditional time series analysis to advanced machine learning and deep learning techniques. Sun et al. provide a review of existing methods, pointing out that combinatorial models and intelligent forecasting methods have a promising application in the future. However, the complexity and nonlinear nature of time series makes it still challenging to accurately forecast price fluctuations [12].

Deep learning techniques are pushing traditional boundaries, and their potential in the field of prediction is becoming increasingly significant. At the same time, elements such as online public opinion and government support are being integrated into the prediction model to improve the accuracy of prediction. Jinling Teng et al. used Prophet model to predict ginger price [13]; Wang Yan improved ginger price prediction accuracy using STCN-LSTM model and improved loss function [14]; Xingchen Lv showed that online public opinion can improve the price prediction accuracy of small agricultural products [15]; Chaudhary et al. used value chain analysis to propose ginger market optimisation strategies, pointing out the importance of government and technical support [16]; Manjubala et al. used complex exponential smoothing to accurately predict weekly prices of garlic and ginger [17]; Hongfeng Li et al. investigated the price factors of ginger [18]; Fu et al. revealed the mechanism of multiple and complex factors affecting the prices of small-scale agricultural commodities [19]; Anamisa et al. used a normalised LSTM model to improve the accuracy of ginger price prediction [20]; Lin et al.'s GCRNN model improves time series classification accuracy and model interpretability [21]; CH-LSTM model developed by Linmei Hu et al. to predict future sub-events [22]; ATTAIN network proposed by Zhang et al. which improves the prediction accuracy and model interpretability through the attention mechanism [23]; Zhang Dabin et al. proposed a VMD-ELM-based integrated prediction model for agricultural futures price decomposition, which significantly improved the prediction accuracy by optimising the VMD parameters and combining with the learning capability of ELM, providing a new method for agricultural futures price prediction [24]; Wang Runzhou et al. proposed a prediction model for agricultural commodity prices based on signal decomposition and deep learning, which effectively improves the prediction accuracy and stability of agricultural commodity prices by complementary ensemble empirical modal decomposition (CEEMD) and deep learning network to extract features, combined with multidimensional data, effectively improved the prediction accuracy and stability of the price of agricultural products such as white pork, spinach, apples, eggs, etc. [25]; Xueqing Fang et al. proposed EEMD-LSTM model for short-term price prediction of agricultural products, which significantly improved the prediction performance through feature extraction and LSTM network training [26]; Gu Yeong Hyeon et al. proposed a DIA-LSTM model for cabbage and radish price forecasting in Korea, which significantly reduces the forecasting error by dynamically selecting meteorological and trading volume data from the main production areas and provides a new method to stabilise the supply and demand of agricultural products [27]; Sun Changxia et al. proposed a hybrid VMD-EEMD-LSTM forecasting model, which effectively improves the pork in China through quadratic decomposition and LSTM training, leek, mushroom and cauliflower price prediction accuracy [28].

Deep learning techniques have made remarkable achievements in the field of agricultural price prediction, but the research for multi-factor models still needs to be further deepened. Sun et al. suggested that future research directions should include the development of more effective combinatorial models in order to integrate the advantages of different models so as to improve the accuracy and stability of the prediction [12]. Price volatility is the result of a combination of factors, so it is particularly critical to analyse these factors and develop integrated forecasting models. Integrated models can effectively improve the accuracy of forecasts by combining the advantages of different models. Building an integrated model requires integrating multiple data sources and optimising the model structure. Future research needs to explore the development and optimisation of multi-factor models further to capture the factors that influence price volatility in a more comprehensive way. Sun et al. also highlighted this point and suggested that structured and unstructured data should be integrated into the model to improve the accuracy of the prediction [12]. Based on this, this paper constructs a combined STL-LSTM-ATT-KAN (Seasonal Trend Decomposition-Long and Short-Term Memory Network-Attention Mechanism-Kolmogorov-Arnold Network) model for the purpose of predicting ginger prices with the expectation that the accuracy of the prediction will be further improved by this model and its practical application value will be enhanced. In order to further improve the accuracy of ginger price prediction, an improved particle swarm optimisation algorithm, the Adaptive Multi-Population Particle Swarm Optimisation (AMP-PSO) algorithm, is introduced in this study [29]. AMP-PSO effectively improves the efficiency and accuracy of the model parameter optimisation through the multi-population structure and adaptive parameter tuning mechanism and provides a powerful model for ginger price prediction. Ginger price prediction provides powerful technical support.

2. Materials and Methods

2.1. Analysis of Factors Affecting the Price of Ginger

This study aims to investigate the fluctuation pattern of ginger prices, analyse the price history and fluctuation pattern, and identify the key influencing factors through literature review, expert interviews and fieldwork. The specific influencing factors are shown in Table 1.

Table 1. Key influences on ginger prices.

Level 1 Impact Factors	Specific Influencing Factors
Agricultural price factors	The daily wholesale price of garlic The daily wholesale price of scallions
Agricultural production factors	Average daily temperature in the main ginger-producing areas Area under crops Production Production volume index
Economic indicator factors	Fresh Vegetables Consumer Price Index (CPI) Broad money supply M2 Exchange rate (USD/CNY)
International market factors	Ginger export amount Ginger export volume International crude oil prices
Market Opinion Factors	Ginger public opinion

Ginger price fluctuations reflect regional and inter-variety trends, and market linkage effects indicate shifts in supply and demand, improving forecasting accuracy. Supply and demand indicators map supply conditions and market dynamics to predict price volatility. Agricultural valuation indicators reflect production conditions and price changes that affect the ginger market. Economic indicators show the macroeconomic impact on prices, including money supply and exchange rate movements. A comprehensive analysis of these factors allows for a more comprehensive forecast of ginger prices and improves the model's accuracy and generalizability.

2.2. Theoretical Approach and Model Construction

This paper innovatively constructs a model consisting of a combination of STL-LSTM-ATT-KAN aimed at forecasting ginger prices. The model first decomposes the time series using STL (Seasonal Trend Decomposition), then captures the long-term dependencies through LSTM (Long Short-Term Memory Network), followed by ATT (Attention Technique) to strengthen the identification of key information, and finally optimises the integration of information through KAN network [30]. In addition, the AMP-PSO (Adaptive Multi-Particle Swarm Optimisation) algorithm was used to optimise the model parameters to improve the accuracy and performance of the prediction [29]. AMP-PSO provides an efficient and accurate solution for ginger price prediction by effectively balancing global and local search capabilities through a multi-population structure and adaptive parameter adjustment mechanism.

As shown in Figure 1, the STL-ATT-LSTM-KAN model structure contains three main steps:

(i) The STL (Seasonal Trend Decomposition) module serves to decompose the ginger price time series into a trend component, a seasonal component and a stochastic component.

(ii) In the LSTM-ATT-KAN process, LSTM processes time series data to capture long-term dependent information; the attention mechanism highlights key time point information, and the KAN module optimally integrates the information to improve prediction accuracy.

(iii) The AMP-PSO optimisation step ensures optimal model prediction performance by adaptively adjusting the particle swarm search space and optimising the model parameters.

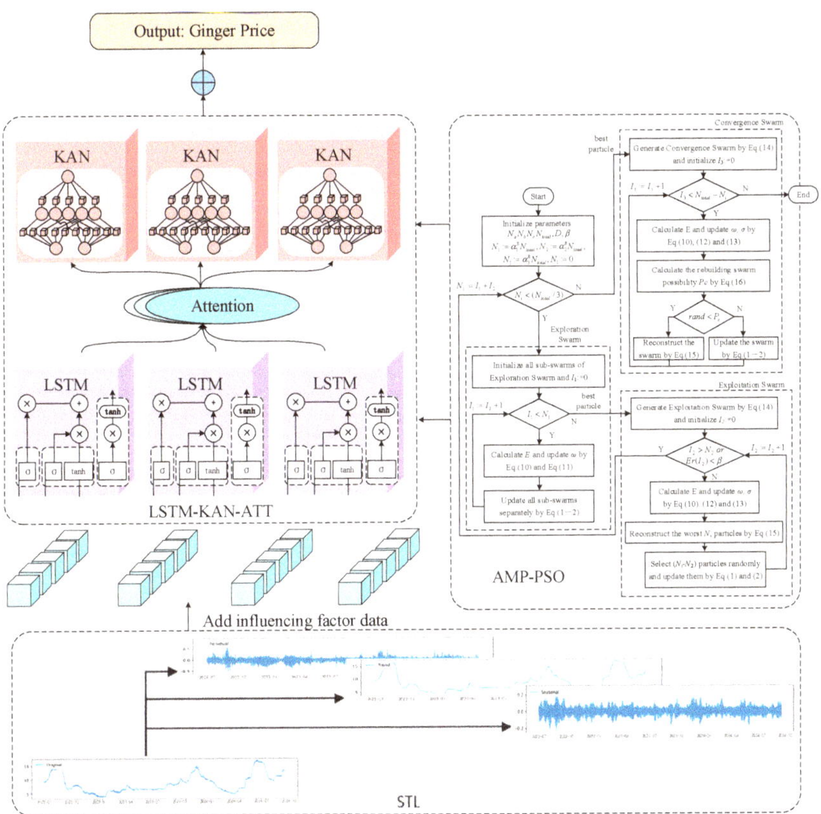

Figure 1. STL-ATT-LSTM-KAN model structure.

The final output module displays the model's predictions, revealing the trend changes in the future price of ginger.

2.2.1. Seasonal-Trend Decomposition Using STL for Ginger Price Analysis

The Seasonal-Trend Decomposition using LOESS (STL) is a robust time series decomposition technique that decomposes ginger price data into three independent components:trend (T_t), seasonality (S_t), and residual (R_t), estimated separately using the Locally Weighted Scatterplot Smoothing (LOESS) method. The calculation formula for this method, as shown in Equation (1):

$$y_t = T_t + S_t + R_t \tag{1}$$

where y_t represents the time series data, T_t denotes the trend component, which reflects the overall upward or downward trend of prices; S_t denotes the seasonal component, which reveals the regularity of price fluctuations over a one-year cycle; R_t denotes the residual component, which represents stochastic fluctuations in prices that cannot be explained by the model and may be caused by factors such as weather changes, natural disasters, or sudden market events.

The trend component T_t and the seasonal component S_t are calculated as shown in Equations (2) and (3).

$$T_t = \text{LOESS}(y_t, \lambda_T) \tag{2}$$

$$S_t = \text{LOESS}(y_t - T_t, \lambda_S) \tag{3}$$

Of these, the λ_T and λ_S denote the smoothing parameters for the trend and seasonal components, respectively.

In this study, STL enhances the forecasting accuracy and stability of our STL-LSTM-ATT-KAN model by isolating seasonal and trend components, enabling the subsequent LSTM layer to better capture temporal dependencies. Later experimental results demonstrate the effectiveness of STL in coping with the high volatility of ginger prices, especially during the peak harvest season.

2.2.2. LSTM-ATT-KAN Model Construction

Step 1: LSTM in STL-LSTM-ATT-KAN Model

LSTM (Long Short-Term Memory Network) is an improved recurrent neural network (RNN) that efficiently captures long-term and short-term dependencies in time-series data by means of memory units and gating mechanisms (input gates, forgetting gates, and output gates). In this study, LSTM is combined with seasonal decomposition (STL) to process the trend and seasonal components of the decomposition and extract time-dependent features. The first layer of LSTM focuses on serial pattern modelling, while the second layer of LSTM further refines the advanced features to improve the accuracy of ginger price prediction. The basic unit structure of LSTM is shown in Figure 2.

The basic unit of the LSTM can be expressed as the following equation:

Oblivion Gate:

$$f_t = \sigma(W_f \cdot [h_{t-1}, x_t] + b_f) \tag{4}$$

Input Gate:

$$i_t = \sigma(W_i \cdot [h_{t-1}, x_t] + b_i) \tag{5}$$

$$\widetilde{C}_t = \tanh(W_C \cdot [h_{t-1}, x_t] + b_C) \tag{6}$$

Memory Updates:

$$C_t = f_t \cdot C_{t-1} + i_t \cdot \widetilde{C}_t \tag{7}$$

Output Gate:

$$o_t = \sigma(W_o \cdot [h_{t-1}, x_t] + b_o) \tag{8}$$

$$h_t = o_t \cdot \tanh(C_t) \tag{9}$$

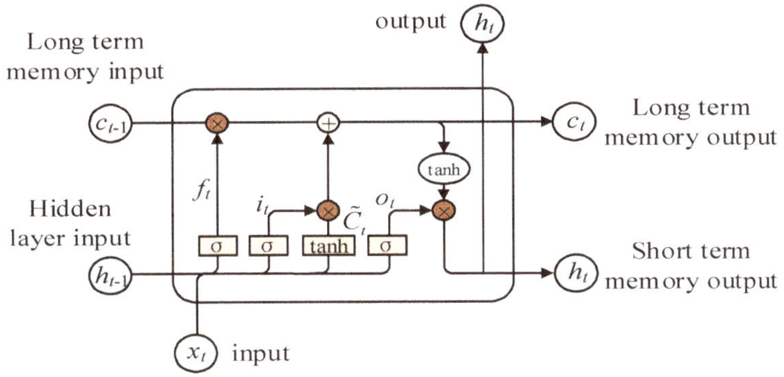

Figure 2. LSTM Basic Unit.

Step 2: Attention Mechanism in STL-LSTM-ATT-KAN Model

ATT (Attention Mechanism) is a dynamic feature selection technique that focuses on critical information in time series data while de-emphasising less relevant details, mimicking human attention allocation by calculating weights of different feature vectors to achieve high-quality feature extraction. In this study, ATT is integrated into the STL-LSTM-ATT-KAN model to enhance the focus on key time points in ginger price sequences, improving the prediction of price fluctuations. The basic principle of the attention mechanism is illustrated in Figure 3, where it dynamically prioritises influential time steps based on their relevance. Figure 3 illustrates the attention mechanism's workflow, highlighting the dynamic weighting of time steps to prioritise relevant features in ginger price time series data.

The weight calculation is formalised in Equations (10) and (11):

$$e_t = tanh(W_e \cdot h_t + b_e) \tag{10}$$

$$\alpha_t = \frac{exp(e_t)}{\sum_{i=1}^{T} exp(e_i)} \tag{11}$$

where e_t represents the energy score for time t, W_e is a weight matrix, h_t is the hidden state vector, b_e is a bias term, tanh is the hyperbolic tangent function compressing values between -1 and 1, and exp with the denominator ensures α_t forms a probability distribution over all time steps T.

This mechanism acts as a dynamic feature selector, allowing the model to focus on key price patterns in the ginger price data. In the experiments, ATT significantly improves the performance of the STL-LSTM-ATT-KAN model by improving the capture of key temporal dependencies.

Step 3: Kolmogorov-Arnold Network (KAN) in STL-LSTM-ATT-KAN Model

KAN (Kolmogorov-Arnold Network) is a neural network architecture based on the Kolmogorov-Arnold theorem, which decomposes complex multivariate functions into combinations of simple one-dimensional functions [30]. This enables KAN to effectively extract key features from high-dimensional data, enhancing model performance. In this study, KAN is integrated into the STL-LSTM-ATT-KAN model to process the high-dimensional features extracted by the LSTM and Attention Mechanism, further improving the expressive power for ginger price prediction.

The structure of KAN is depicted in Figure 4, comprising:

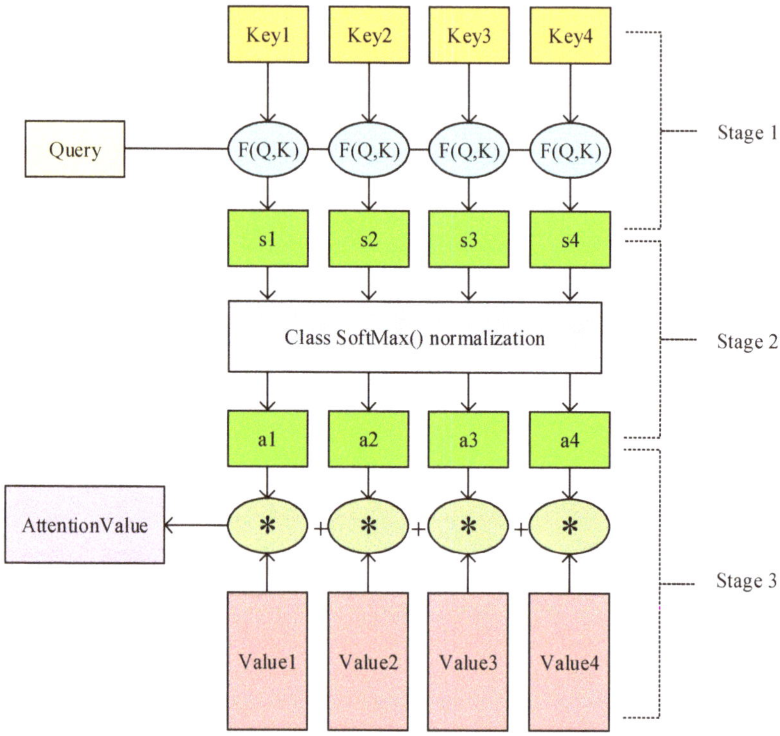

Figure 3. Structure of the Attention Mechanism.

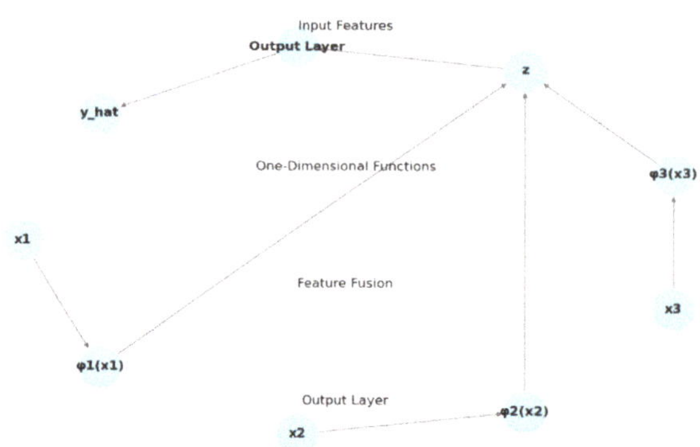

Figure 4. KAN network structure.

Input Features: Represented as x1, x2, x3, these are the initial features (e.g., decomposed trends and seasonal components from ginger price data).

One-Dimensional Functions: Each input feature is transformed by a corresponding one-dimensional function ($\phi_{ij}(x_j)$).

Feature Fusion: The outputs are aggregated into a fused feature y, as shown in Equation (12):

$$y = \sum_{i=1}^{n} g_i \left(\sum_{j=1}^{m} \phi_{ij}(x_j) \right) \tag{12}$$

where y is the fused feature, g_i are transformation functions, ϕ_{ij} are one-dimensional functions, and x_j are input variables, with n representing the number of transformation functions and m the number of input features.

Output Layer: The fused feature y is mapped through a fully connected layer to generate the final prediction $\hat{y}$, as described in Equation (13):

$$\hat{y} = ReLU\left(W_{fc} \cdot y + b_{fc}\right) \tag{13}$$

where W_{fc} is the weight matrix mapping high-dimensional features to the target dimension, b_{fc} is the bias adjusting the feature space centre, and ReLU is the rectified linear unit activation function, enabling nonlinear modelling of future ginger price trends.

This architecture enhances the STL-LSTM-ATT-KAN model's ability to capture complex nonlinear patterns in ginger prices (e.g., market shocks and seasonal dynamics), achieving superior prediction performance for agricultural market monitoring.

2.2.3. Adaptive Multi-Population Particle Swarm Optimisation (AMP-PSO) Algorithm

The Adaptive Multi-Population Particle Swarm Optimisation (AMP-PSO) algorithm is an enhanced version of the Particle Swarm Optimisation (PSO) algorithm, incorporating a multi-population strategy and adaptive parameter adjustments to significantly improve global search capability and convergence speed. In this study, AMP-PSO is applied to optimise the parameters of the STL-LSTM-ATT-KAN model, ensuring optimal prediction performance for complex ginger price time series data.

(1) Multi-Population Strategy

AMP-PSO divides the particle swarm into multiple sub-populations, each performing independent optimisation to balance global exploration and local exploitation. Particles are categorised into:

- Leader Population: Conducts global searches by interacting with other sub-populations to expand the solution space.
- Follower Population: Focuses on local refinement, leveraging local optima to enhance precision.

Through periodic particle exchanges, AMP-PSO maintains population diversity and accelerates convergence, effectively preventing traditional PSO from getting trapped in local optima.

(2) Adaptive Parameter Adjustment

AMP-PSO dynamically adjusts the inertia weight ($\omega(t)$) and learning factors (c_1, c_2) to balance global and local search capabilities. The adaptive inertia weight is updated as:

$$\omega(t) = \omega_{max} - \frac{(\omega_{max} - \omega_{min}) \cdot t}{T_{max}} \tag{14}$$

where ω_{max} and ω_{min} are the maximum and minimum inertia weights, t is the current iteration, and T_{max} is the maximum iteration number. This mechanism enhances the search efficiency and convergence speed for optimising the STL-LSTM-ATT-KAN model's parameters in ginger price prediction.

(3) Particle Update Mechanism

Particle positions and velocities are updated for each dimension j of particle i using the following Equation (15):

- Velocity Update:

$$v_{i,j}(t+1) = \omega(t) \cdot v_{i,j}(t) + c_1 \cdot r_1 \cdot (p_{i,j}(t) - x_{i,j}(t)) + c_2 \cdot r_2 \cdot \left(g_{best,j}(t) - x_{i,j}(t)\right) \quad (15)$$

where $v_{i,j}(t)$ and $x_{i,j}(t)$ are the velocity and position of particle i at dimension j and iteration t, $p_{i,j}(t)$ is the individual best position, $g_{best,j}(t)$ is the global best position, $\omega(t)$ is the adaptive inertia weight (updated in Equation (14)), c_1, c_2 are the learning factors, and $r_1, r_2 \sim U(0,1)$ are random numbers in [0, 1].

- Position Update:

$$x_{ij}^{t+1} = x_{ij}^{t} + v_{ij}^{t+1} \quad (16)$$

where x_{ij}^{t+1} and v_{ij}^{t+1} are the position and velocity of particle i at dimension j and iteration $t+1$, respectively, and x_{ij}^{t} is the position at iteration t.

To constrain particle states within reasonable bounds, AMP-PSO applies a *sigmoid* function to limit velocity and position elements to [0, 1]:

$$sigmoid(x) = \frac{1}{1+e^{-x}} \quad (17)$$

During each iteration, the algorithm evaluates the objective function F for each particle. If the current position yields a better fitness than the individual best, $p_{best,ij}$ is updated for each dimension j, as described in Equation (18). Suppose the fitness surpasses the global best, $g_{best,j}$ is updated for each dimension j, as described in Equation (19). Here, F is the objective function, defined as minimising the prediction error of ginger prices in the STL-LSTM-ATT-KAN model, $x_{ij}(t+1)$ denotes the current position for dimension j of particle i at iteration $(t+1)$, and $p_{best,ij}(t)$ and $g_{best,j}(t)$ denote the individual and global best positions, respectively, for dimension j at iteration t.

$$p_{best,ij}(t+1) = x_{ij}(t+1) \ if \ F(x_{ij}(t+1)) > F(p_{best,ij}(t)) \quad (18)$$

If the fitness surpasses the $g_{best,j}$ is updated for each dimension j:

$$g_{best,j}(t+1) = x_{ij}(t+1) \ if \ F(x_{ij}(t+1)) > F(g_{best,j}(t)) \quad (19)$$

where F is the objective function, defined as minimising the prediction error of ginger prices in the STL-LSTM-ATT-KAN model, $p_{i,j}(t)$ denotes the current position, and $p_{best,ij}(t)$ denotes the individual best position, both for dimension j of particle i at iteration t, while $g_{best,j}(t)$ denotes the global best position for dimension j at iteration updated across all particles i to optimise the model parameters. Note that $p_{best,ij}(t+1)$ and $g_{best,j}(t+1)$ are updated in the next iteration if the fitness improves, as described in Equations (18) and (19).

(4) **Particle Exchange Mechanism**

To enhance diversity and prevent premature convergence, AMP-PSO periodically exchanges particles between the leader and follower populations after a predefined number of iterations, introducing new solution space information to accelerate convergence.

(5) **Determination of Global Optimal Solution**

At each iteration, AMP-PSO updates individual and global optima based on fitness values. The algorithm terminates when reaching the maximum iteration T_{max}, achieving convergence of the objective function, or finding a solution meeting performance criteria (e.g., minimising prediction error). The global optimal position determines the optimal parameters for the STL-LSTM-ATT-KAN model, balancing multiple objectives to yield a Pareto-optimal solution set.

Compared to traditional PSO, AMP-PSO avoids local optima through its multi-population strategy and adaptive adjustments. The follower population's local search capability improves parameter precision, while particle exchanges and adaptive inertia weights enhance diversity and adaptability. In our experiments, AMP-PSO significantly improved the STL-LSTM-ATT-KAN model's prediction accuracy and stability for ginger prices, maintaining high computational efficiency for agricultural applications.

2.3. Experimental Design

2.3.1. Experimental Setup

Experimental environment

The experiments were conducted using a Suttai ST-SN260 server with an AMD EPYC 9754 CPU (128 cores, 2.25 GHz), NVIDIA RTX 4060 GPU (8 GB), Samsung 384 GB DDR5 RAM, Samsung 2 TB SSD and 16 TB HDD. The operating system was Ubuntu 20.04 LTS version, with the installation of Python 3.8.10 and libraries PyTorch 1.9.0, NumPy 1.21.0, and Pandas 1.3.0.

Data sources

The agricultural product price data used in the study comes from the National Key Agricultural Product Market Information Platform of the Ministry of Agriculture and Rural Affairs (PRC). The platform compiles and publishes price information from major wholesale markets for agricultural products across the country. The time span chosen for the experiment is nearly ten years, from 1 January 2014 to 29 September 2024, for the daily wholesale price data of ginger. In order to deeply explore the formation mechanism of ginger price and its fluctuation pattern, 13 dimensions of key influencing factor indicators, including daily wholesale price of garlic, daily wholesale price of green onion, temperature of main production area, planting area, production, production volume index, consumer price index (CPI) of fresh vegetables, export volume, export volume, M2 money supply, exchange rate, crude oil price, and public opinion of ginger, are selected in a comprehensive manner and in combination with the actual situation. The data for the selected indicators come from the Ministry of Agriculture and Rural Affairs (PRC), Ministry of Commerce of the People's Republic of China (MOFCOM), National Bureau of Statistics of China (NBS), General Administration of Customs of the People's Republic of China (GACC), The People's Bank Of China (PBC), China Meteorological Administration (CMA), Brick Agricultural Database, Baidu Index, as well as authoritative organisations and platforms such as the Food and Agriculture Organisation of the United Nations (FAO), and the Energy Information Administration (EIA). These data sources are rich and varied and of high reference value, providing solid support for the study. Specific platform website information is detailed in the subsequent description of data sources (Abbreviations).

This study collects public data from websites through crawling techniques. Static web pages are mainly crawled using Python's BeautifulSoup and Request libraries, while dynamically loaded content (e.g., sections requiring login authorisation) is efficiently processed using Selenium and XPath tools.

Data processing

With regard to the data preprocessing stage, the collected ginger dataset was systematically organised and cleaned to ensure data quality. The specific steps involved are as follows:

① Initially, the data underwent a thorough cleansing process, encompassing the removal of duplicate records and the management of outliers.

② For instances where values were missing, the interpolation of similar mean values was employed to ensure the integrity and consistency of the data.

③ Conversion of the data format was conducted to facilitate the seamless progression of subsequent analyses.

The efficacy of these preprocessing steps in enhancing the reliability and applicability of the dataset is evident.

The aggregated dataset (3925 data points from 1 January 2014 to 29 September 2024) was divided into a training set (65%), a validation set (20%), and a test set (15%) according to the time series. The division is as follows: The training set contained 2551 entries (1 January 2014 to 26 December 2020), the validation set contained 785 entries (27 December 2020 to 23 February 2023), and the test set contained 589 entries (24 February 2023 to 29 September 2024). This division increases the proportion of training data to reduce the risk of overfitting while ensuring that the validation and test sets have sufficient sample sizes on unseen data to improve the generalisation ability of the model.

In order to enhance the interpretability and predictive ability of the model, the price volatility analysis was simplified by dividing the ginger price series data using the STL method. The STL decomposition separates the raw data into three components: trend, seasonal and residual. The results are displayed in Figure 5.

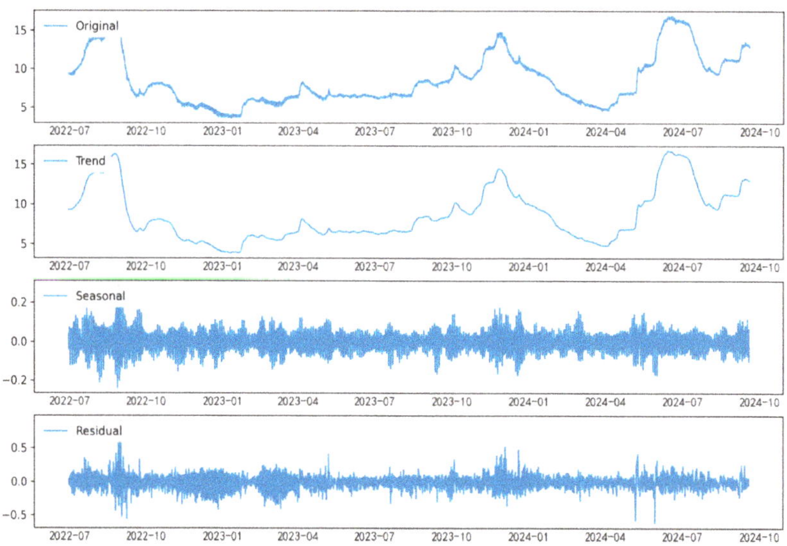

Figure 5. Exploded view of STL time series.

2.3.2. Forecast Evaluation Indicators

When forecasting ginger prices, the assessment of the model's predictive performance is crucial. Metrics such as mean square error (MSE), mean absolute error (MAE), root mean square error (RMSE), and coefficient of determination (R^2) are usually used for the assessment.

The mean square error (MSE) is the average of the squares of the differences between the predicted and true values:

$$MSE = \frac{1}{n}\sum_{i=1}^{n}(y_i - \hat{y}_i)^2 \tag{20}$$

where y_i is the true value and $\hat{y}_i$ is the predicted value, and n is the sample size.

The root mean square error (RMSE) is the difference between the observed value and the true value. The square root of the ratio of the square of the deviation of an observation from the true value to the number n of observations:

$$RMSE = \sqrt{\frac{1}{n}\sum_{i=1}^{n}\cdot(y_i - \hat{y}_i)^2} \tag{21}$$

The Mean Absolute Error (MAE) is the average of the absolute values of the difference between the predicted and true values:

$$MAE = \frac{1}{n}\sum_{i=1}^{n}|y_i - \hat{y}_i| \tag{22}$$

The coefficient of determination (R^2) is used to measure the proportion of total variance in the explanatory variables of the model and indicates the goodness of fit of the model:

$$R^2 = 1 - \frac{\sum_{i=1}^{n}(y_i - \hat{y}_i)^2}{\sum_{i=1}^{n}(y_i - \overline{y})^2} \tag{23}$$

where $\overline{y}$ is the mean of the true values. The value range of R^2 is between 0 and 1, which can intuitively reflect the goodness of fit of the model, and the closer the value of R^2 is to 1, the stronger the explanatory power of the model.

2.4. Model Optimisation Parameter Setting

To ensure that all parts of the model are fully functional, the parameters of the model were finely tuned in this study using the Adaptive Particle Swarm Optimisation algorithm for Multiple Populations (AMP-PSO). The main parameters of the model were set as follows: kan_units (number of KAN network units) to 10, lstm_units (number of LSTM network units) to 10, learning_rate (learning rate) to 0.005, and batch_size (batch size) to 26.908, as shown in Table 2 below.

Table 2. AMP-PSO parameter selection.

Model Parameter	Parameter Value
kan_units	22
lstm_units	50
learning_rate	0.004
batch_size	36.451

3. Results

3.1. Experimental Results

To better understand and optimise the training process of the model, in order to further validate the effect of model training, this paper plots the loss map of the STL-LSTM-ATT-KAN model during the training process, as shown in Figure 6 below.

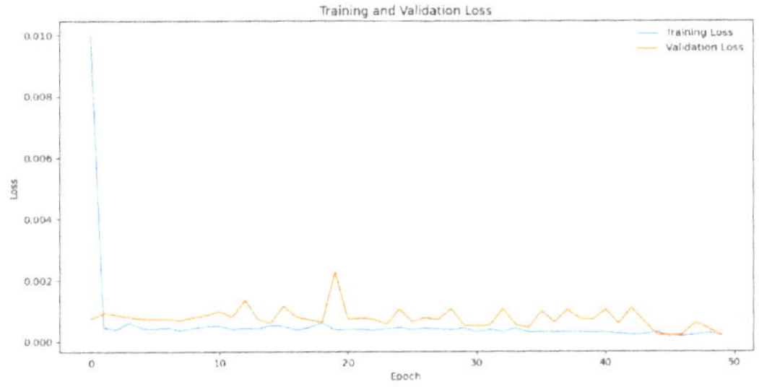

Figure 6. Training loss curve.

As shown in Figure 6, the model training loss (blue line) and validation loss (orange line) vary with epoch. The initial training loss decreases rapidly, indicating that the model converges quickly and captures the training data patterns. The training loss is low and stable for most of the training cycles, indicating that the model fits well and the training process is stable.

The residual plot shows the distribution of the difference between the true value and the predicted value (i.e., the residual). By looking at the distribution of the residuals, it is possible to visually assess the model's prediction bias and error patterns. This is shown in Figure 7.

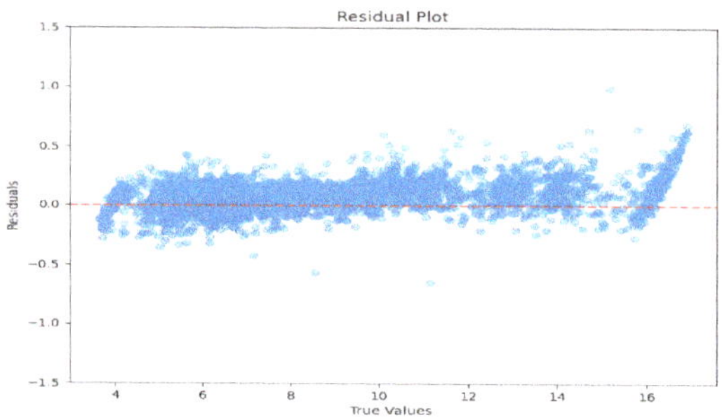

Figure 7. Model residual plot.

Figure 7 shows the distribution of the residuals between the model's predicted and true values, where the residuals are defined as the difference between the true value and the predicted value, which is used to assess prediction accuracy. In the figure, the red dashed line indicates the baseline, where the residuals are zero, and the blue dots show the residuals for each sample. Most of the blue points are distributed around the zero line (residuals range from about −0.5 to 0.5), indicating that the model predicts most of the data more accurately and close to the true value. However, when the true value is large (especially above 10), the distribution of blue points becomes more spread out and extends to a range of −1.0 to 1.5, suggesting that the model's error in predicting high values increases. In particular, the residuals are skewed towards positive values (above the zero line) in regions where the true value is greater than 14, reflecting the fact that the model may be underestimating the true data for these high values.

In addition, the model performs better in regions with low (less than 8) or moderate (around 8–12) true values, with less fluctuation in the residuals, whereas the residuals in regions with high values fluctuate more, suggesting a limitation of the model in dealing with large or extreme values. If the distribution of residuals shows a clear pattern (e.g., trend or periodicity), this may indicate that the model is not fully capturing certain features in the time series data. In the future, the prediction accuracy and stability of the model can be improved by increasing the data set, optimising the data processing method, and improving the prediction algorithm to reduce errors in the high-value range.

Through the training and prediction process of the model, it is possible to obtain the predicted values of ginger prices. In order to visualise these predictions more, a trend comparison graph between the real and predicted values and a predicted trend graph for the next seven days were produced, respectively, as shown in Figures 8 and 9.

This study demonstrates the effectiveness of AMP-PSO in optimising the parameters of the ginger price prediction model. The multi-population structure and adaptive parameters of AMP-PSO improve the searchability and convergence speed of the model, providing an efficient and accurate solution for ginger price prediction. Experiments show that the AMP-PSO optimised STL-LSTM-ATT-KAN model significantly outperforms other algorithms in terms of prediction accuracy, with specific metrics of MAE: 0.111, MSE: 0.021, RMSE: 0.146, and R^2: 0.998. This demonstrates that the model has high accuracy and stability when dealing with complex time series data. Therefore, the optimised model of AMP-PSO performs well in terms of prediction accuracy, stability, and adaptability, and it provides powerful technical support for ginger price prediction. Traditional PSO is easy to fall into a local optimum, has limited accuracy, and performs poorly when dealing with complex data. Genetic algorithm GA has strong global search capability but is computationally complex and slow. Multi-population PSO (MP-PSO) improves the optimisation ability but lacks adaptive tuning and has limited effect. AMP-PSO combines multi-population and adaptive parameters and is superior in optimisation accuracy and speed. Experiments show that AMP-PSO performs well on all indicators with high computational efficiency, provides technical support for ginger price prediction, and verifies its effectiveness in complex time series prediction.

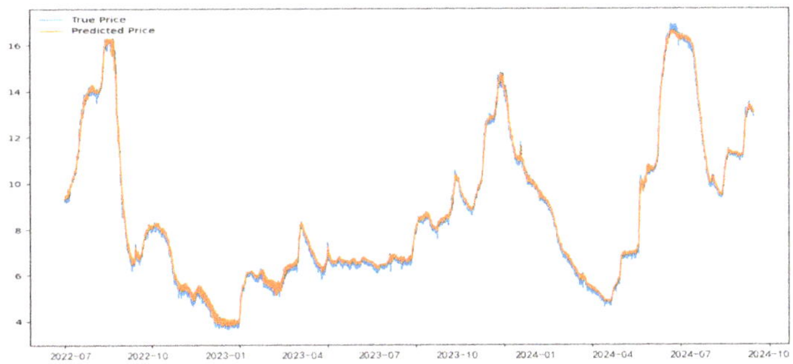

Figure 8. Comparison of true value prediction trends.

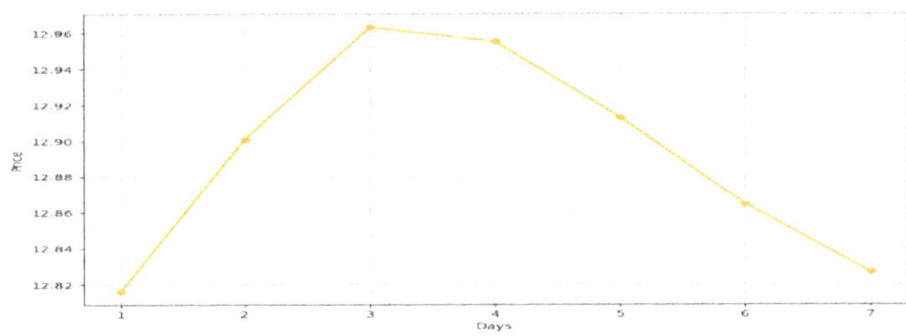

Figure 9. Forecast trend for the future seven days.

3.2. *Comparative Experimental Results*

In order to verify the prediction advantages of this model, it is compared with RF, SVM, ARIMA, XGBoost, CNN, RNN, LSTM, GRU, etc. RF and XGBoost are integrated models based on decision trees, which are good at dealing with nonlinear relationships. RF reduces overfitting, and XGBoost reduces errors. SVM is a linear classifier that is used by the kernel method to deal with nonlinear problems but has limited flexibility in time series prediction tasks. ARIMA is suitable for linear time series data. CNN is good at extracting local features, RNN is the basic model for time series modelling, LSTM deals with long-term dependencies, and GRU has fewer parameters and is computationally efficient. By comparing these eight models, the prediction results of different models are shown in Figure 10.

Comparative experiments and quantitative analysis were conducted to evaluate the performance of the model in ginger price prediction. The evaluation metrics and error comparisons are shown in Table 3 and Figure 11, respectively.

The MAE measures the mean absolute error between the predicted and true values. From Figure 11, the STL-LSTM-ATT-KAN model and the RF model have the smallest Mean Absolute Error (MAE) in prediction, showing that they have the least deviation from the true value. XGBoost also shows a low MAE, demonstrating the ability of these integrated learning models to capture trends and patterns in the data. In contrast, the ARIMA model has a high MAE, indicating that it performs poorly in capturing nonlinear features, leading to larger prediction errors.

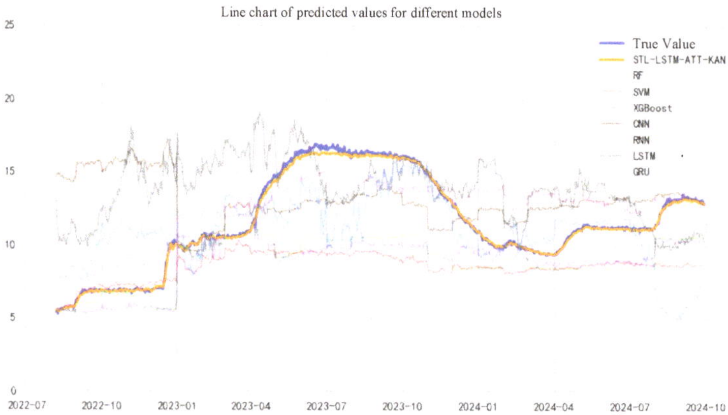

Figure 10. Comparison of prediction results.

Table 3. Comparative experimental results.

Evaluation Metrics	AMP-PSO Optimised STL-LSTM-ATT-KAN	RF	SVM	XGBoost	CNN	RNN	LSTM	GRU
MAE	0.111	0.113	0.224	0.123	0.398	0.295	0.643	0.707
MSE	0.021	0.032	0.143	0.0409	0.268	0.1737	0.782	1.195
RMSE	0.146	0.181	0.378	0.202	0.518	0.416	0.884	1.093
R^2	0.998	0.996	0.986	0.996	0.975	0.984	0.928	0.890

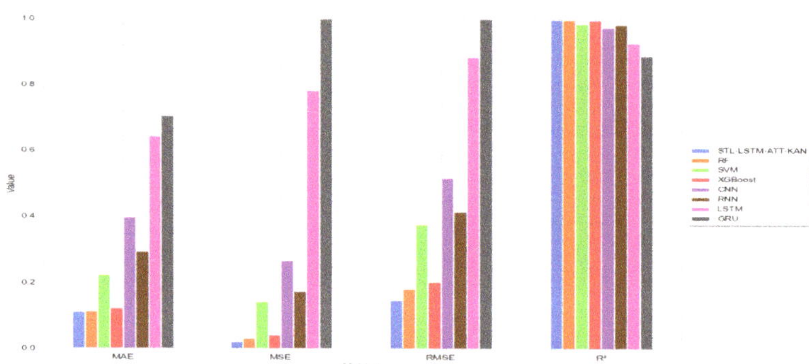

Figure 11. Comparison error bar chart.

The MSE not only reflects the magnitude of the prediction error but also weights the larger errors. From Figure 11, the STL-LSTM-ATT-KAN model has the lowest MSE, implying that the model not only has small errors but also exhibits better robustness and immunity to interference in the case of extreme values or outliers. This is due to its integrated model architecture of STL decomposition and attention mechanism, which can finely regulate the weights of different features to reduce large errors; the model effectively captures trends and seasonal variations in the time series and accurately identifies the key time points through the attention mechanism. The MSEs of RF and XGBoost are slightly higher, but the difference is not that big compared with STL-LSTM-ATT-KAN, which is not significant, indicating that they also have better performance in extreme error control.

RMSE, root mean square error, is the square root of MSE (mean square error), which provides a measure of error that is consistent with the scale of the original data. From Figure 11, the STL-LSTM-ATT-KAN model exhibits the lowest RMSE value in terms of processing complex time series data, which further confirms the high accuracy and consistency of the model. This suggests that the LSTM structure incorporating the attention mechanism is able to effectively capture the temporal and feature dependencies in time series data, thereby significantly reducing the overall error. Following closely behind, the RF (Random Forest) and XGBoost models also exhibit good performance, showing that their errors are relatively small and evenly distributed.

The R^2 metric is used to measure the model's ability to explain changes in the data. From Figure 11, all models except the ARIMA model have high R^2 values, which indicates that they are able to effectively explain the variation in ginger prices. However, the STL-LSTM-ATT-KAN model not only maintains a high R^2 value but also significantly reduces the MAE, MSE, and RMSE, which indicates that the model demonstrates a significant advantage in terms of all-round performance when dealing with complex time series data.

3.3. Results of Ablation Experiments

The results in Table 4 show that the number of ATT-KAN has a significant impact on the performance of the STL-LSTM model: Too few (e.g., 20) or too many (e.g., 75, 100) ATT-KAN lead to performance degradation, while the baseline model (which may use the optimal number of ATT-KAN) performs best, suggesting that the number of ATT-KAN needs to be experimentally optimised to achieve optimal performance. The ablation analysis in Table 5 also shows that STL and ATT are crucial for performance improvement, while KAN contributes little, but the complete model (STL-LSTM-ATT-KAN) has the best performance. In contrast, traditional LSTM performance deteriorates significantly when used alone, demonstrating that advanced features such as STL and ATT are essential for the task at hand. The evaluation indices show that the smaller the MAE, MSE and RMSE, the more accurate the model prediction. The closer R^2 is to 1, the better the model fit, and these metrics consistently validate the superior performance of the complete model on the task at hand.

Table 4. Effect of the number of ATT-KANs on STL-LSTM performance.

Evaluation Metrics	STL-LSTM-ATT-KAN	STL-LSTM(20)-ATT-KAN	STL-LSTM(75)-ATT-KAN	STL-LSTM(100)-ATT-KAN
MAE	0.111	0.129	0.181	0.139
MSE	0.021	0.026	0.044	0.030
RMSE	0.146	0.160	0.211	0.174
R^2	0.998	0.997	0.995	0.997

Table 5. STL-LSTM ablation analysis.

Evaluation Metrics	STL-LSTM-ATT-KAN	STL-LSTM-ATT	STL-LSTM	LSTM
MAE	0.111	0.113	0.124	0.643
MSE	0.021	0.019	0.023	0.782
RMSE	0.146	0.139	0.151	0.884
R^2	0.998	0.998	0.997	0.928

Figure 12, the significant decrease in MAE, suggests that the STL-LSTM-ATT-KAN model significantly improves the prediction accuracy by incorporating the attention mechanism and the KAN network. In contrast, the STL-LSTM model that lacks the attention mechanism and KAN network leads to a significant increase in MAE, while the LSTM model that lacks the advantage of time series decomposition further exacerbates the prediction error.

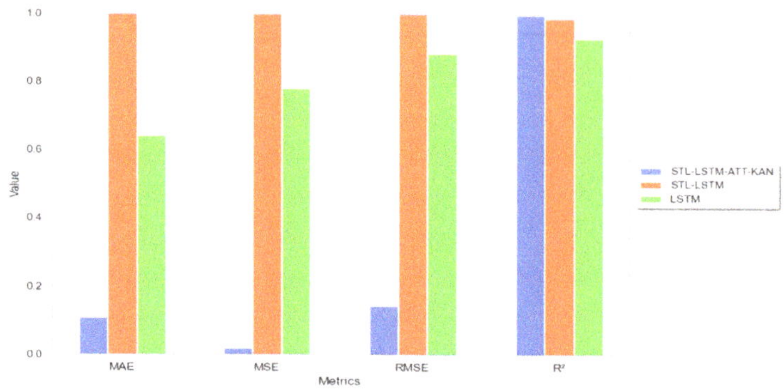

Figure 12. Comparison bar chart of ablation errors.

The significant decrease in MSE reveals that the combination of the ATT and KAN networks not only improves forecasting accuracy but also strengthens the stability of the model, making it perform better in the face of abnormal volatility. Although the STL-LSTM model lacking ATT and KAN networks still retains the advantages of STL decomposition, it is still inferior to the full STL-LSTM model integrating ATT and KAN networks in terms of error control in complex scenarios. The LSTM model lacking STL decomposition, on the other hand, shows a higher error rate, which further proves the key role of STL in time series feature decomposition.

The significant reduction in RMSE further confirms the overall superiority of the STL-LSTM-ATT-KAN model in time series forecasting. The lower RMSE values indicate that the model is able to provide more accurate forecasts in most cases. Although the RMSE of the STL-LSTM model is lower than that of the standard LSTM model, its prediction results still suffer from a large error due to the lack of enhancement by the ATT and KAN networks. In contrast, the standard LSTM model is unable to decompose the time series efficiently, which leads to the poor stability of its prediction results.

The R^2 values indicate that both the STL-LSTM-ATT-KAN and STL-LSTM models can excellently elucidate the variance of ginger price fluctuations, which is mainly due to the STL decomposition technique and the LSTM network's accurate capture of the long-term dependence of the time series. However, the R^2 value of the LSTM model is almost zero, which indicates that it is extremely weak in explaining the ginger price fluctuations without applying the time series decomposition, and the prediction effect is significantly reduced.

The STL-LSTM-ATT-KAN model performs optimally on the evaluation metrics with low MAE, MSE and RMSE values, showing its ability to reduce prediction errors. The achievement is attributed to the combination of STL decomposition and the attention mechanism, which improves the prediction accuracy. The increase in error after removing the attention mechanism emphasises its role in error reduction. The LSTM model, although fitting the data well, does not have as good a prediction accuracy as the STL-LSTM-ATT-KAN, confirming the enhancement of the model performance by the STL decomposition and the attention mechanism.

By applying the STL (Seasonal Trend Decomposition) technique to decompose the time series data, the model is able to handle the trend, seasonality, and residual components independently, which significantly reduces the forecast error. The Long Short-Term Memory Network (LSTM), with its superior ability to capture long-term dependencies in time-series data, enables the model to more accurately handle the long-term trend of ginger prices.

Incorporating the attention mechanism in the model allows the model to focus more on key time points during the forecasting process, which improves the accuracy of the forecasts and further reduces errors. The integration of the KAN network integrates external knowledge sources, improves the generalisation ability of the model, and ensures the stability and reliability of the model when dealing with new data.

4. Discussion

The STL-LSTM-ATT-KAN ginger price prediction model proposed in this study shows significant advantages in prediction accuracy and stability. This result is due to the optimal design of the model structure and the efficient adjustment of parameters by the adaptive multi-population particle swarm optimisation (AMP-PSO) algorithm. AMP-PSO balances the global and local search ability through the multi-population strategy and the adaptive adjustment mechanism, which improves the efficiency of parameter optimisation and enhances the accuracy and stability of ginger price prediction.

However, the model has some limitations. The study covers the last ten years, and the data are more obviously affected by seasonal fluctuations, economic situations and policies. The current analysis is based on data availability and the typical cycle of ginger prices, and the performance of the model in other time horizons has not been adequately tested, and the generalisability has not been verified in a wider range of scenarios.

Future research proposes to improve the model in the following directions. The model will be applied to the price forecasting of other agricultural products, such as garlic, shallots, chillies and peppers, to assess its applicability and robustness. The dataset will be extended to more agricultural product types by introducing macroeconomic indicators (e.g., inflation rate, gross domestic product, trade-weighted exchange rate index, agricultural production costs, domestic diesel price) and climatic variables (e.g., rainfall, temperature, relative humidity, sunshine hours and frequency of extreme weather events) to capture the drivers of price fluctuations. The optimisation of data sources will improve accuracy and completeness and provide high-quality support to the model. The forecasting horizon will be extended from short-term fluctuations to long-term trends to test its adaptability over different business cycles. Improvements in algorithm design and parameter tuning methods are expected to enhance forecasting performance.

The practical value of the model will need to be validated through collaboration with agribusiness and government agencies, and it is expected to play a role in agricultural market monitoring, risk management, and policy formulation.

5. Conclusions

In the context of agricultural markets, price volatility carries significant ramifications for a range of stakeholders, including farmers, traders, consumers and policymakers. This is particularly evident in the case of agricultural products such as ginger, which are produced and consumed on a seasonal basis. Price uncertainty can result in market imbalances, fluctuations in farmers' incomes and unpredictability in consumer spending. Consequently, the development of a model capable of accurately predicting ginger price fluctuations is imperative for enhancing market efficiency, stabilising farmers' incomes, guiding consumer behaviour and facilitating policy formulation.

The objective of the present study is to enhance the precision and reliability of predictions by integrating seasonal decomposition, long- and short-term memory networks, attention mechanisms, and Kolmogorov-Arnold networks. The proposed STL-LSTM-ATT-KAN prediction model has been developed to address this need, with a focus on capturing seasonal fluctuations and long-term trends in ginger prices. The incorporation of an attention mechanism is a key innovation as it enables the identification of key factors affecting prices, thereby providing more reliable decision support for relevant stakeholders.

The model was constructed and validated in this study, and it was found to significantly improve the accuracy and stability of forecasting by accurately capturing the linkage effects among markets and the interactions between ginger production and macroeconomics. The introduction of the AMP-PSO algorithm further optimises the model parameters and enhances the performance of the model in complex time series forecasting. The experimental results demonstrate that the proposed model significantly outperforms alternative algorithms in terms of forecasting accuracy (MAE: 0.111, MSE: 0.021, RMSE: 0.146, R^2: 0.998), thereby substantiating its efficacy in complex time series forecasting.

In addition, the model demonstrates considerable potential in the domains of agricultural market monitoring and risk early warning control, thereby providing substantial support for the stable operation of agricultural markets and the prosperous development of agricultural economies. Notwithstanding the positive outcomes achieved by this study, there is still scope for further enhancement and refinement. Future research directions include the following: the application of the model to a wider range of agricultural price forecasts to assess its generalisation ability; the addition of further influencing factors, such as climate change and policy adjustments, to capture the influences on price fluctuations more comprehensively; and the exploration of more efficient optimisation algorithms to reduce the computational complexity of the model. Expanding the model's temporal range and incorporating macroeconomic indicators and seasonal data to assess its performance in long-term trend forecasting, and collaborating with agribusinesses and government departments to apply the model to real market monitoring and policy formulation to validate its application value in real-world scenarios.

Through these future endeavours, it is expected that the application scope of the model will be further validated and expanded, thereby providing more powerful support for agricultural market monitoring, risk early warning and control, and policy formulation.

Author Contributions: Conceptualisation, F.L. and X.M.; methodology, F.L.; software, F.L.; validation, L.L. and K.Z.; formal analysis, P.L.; investigation, F.L. and X.M.; resources, P.L.; data curation, J.Y.; writing—original draft preparation, F.L.; writing—review and editing, F.L. and X.M.; visualisation, F.L.; supervision, P.L.; project administration, K.Z.; funding acquisition, P.L. All authors have read and agreed to the published version of the manuscript.

Funding: This research was funded by the Shandong Province Higher Education Program for the Introduction and Cultivation of Young Innovative Talents (2021) and the Shandong Province Key Research and Development Program under the Rural Revitalisation Science and Technology Innovation Enhancement Action Plan (grant number 2022TZXD0030).

Data Availability Statement: The data in this study are available upon request from the authors and corresponding author. This study analyses ginger price information as well as a variety of indicators across multiple industries in China, and the aggregated dataset cannot be publicly disclosed due to confidentiality restrictions. The aggregated data is not publicly available due to the confidentiality requirements of the project team. Daily wholesale prices for ginger, garlic and shallots were obtained from the State Key Agricultural Products Market Information Platform of the Ministry of Agriculture and Rural Development of China. Data on key indicators affecting the price of ginger (daily wholesale prices of garlic and onion, temperature in main production areas, planting area, production, production index, consumer price index (CPI) for fresh vegetables, export volume, import volume, M2 money supply, exchange rate, crude oil price, ginger public opinion, etc.) were obtained from the Ministry of Agriculture and Rural Affairs (MArA), Ministry of Commerce (MOFCOM), the National Bureau of Statistics (NBS), the General Administration of Customs (GAC), the People's Bank of China (PBOC), the China Meteorological Administration (CMA), the Bricker Agricultural Database, Baidu Index, Food and Agriculture Organisation of the United Nations, U.S. Energy

Information Administration and other authoritative institutions. Specific data sources are listed below: Daily wholesale prices of ginger, garlic and green onion from the Ministry of Agriculture and Rural Development (PRC) National Key Agricultural Products Market Information Platform (https://ncpscxx.moa.gov.cn (accessed 24 September 2024)); Temperature data for major ginger-producing areas from the China Meteorological Administration (CMA) (https://weather.cma.cn/ (accessed 24 September 2024)); Ginger planting area and consumer price index (CPI) for fresh vegetables from the National Bureau of Statistics of China (NBS). (https://data.stats.gov.cn (accessed 24 September 2024)); M2 money supply and exchange rate data from the Ministry of Commerce of China (https://cif.mofcom.gov.cn (accessed 24 September 2024)) and the People's Bank of China (http://www.pbc.gov.cn (accessed 24 September 2024)); Ginger public opinion data from Baidu Index (https://index.baidu.com (accessed 24 September 2024)); Ginger export data from China's General Administration of Customs (http://stats.customs.gov.cn (accessed 24 September 2024)); International crude oil prices from the U.S. Energy Information Administration (https://atlas.eia.gov (accessed 24 September 2024)); Production index and volume data from the Food and Agriculture Organisation of the United Nations (https://www.fao.org (accessed 24 September 2024)), China Statistical Yearbook (https://www.stats.gov.cn/sj/ndsj/ (accessed 24 September 2024)), Brick Database (http://www.chinabric.com/service.html (accessed 24 September 2024)), China Ginger Website (https://www.jiang7.com (accessed 24 September 2024)). All data used were accessed and retrieved on 24 September 2024 and some data required authorised access to retrieve. This study was supported by the Key Laboratory of Smart Agricultural Technology in the Huanghuaihai Region of the Ministry of Agriculture and Rural Affairs (MARD), which provided important contributions to the data collection, financial and technical resources for this study. We would also like to thank Prof Pingzeng Liu and Prof Xianyong Meng of Shandong Agricultural University for their valuable guidance on this study.

Conflicts of Interest: The authors declare no conflicts of interest.

Abbreviations

Abbreviation	Full Name
ATT	Attention Mechanism
ATTAIN	Attention-based Time-Aware LSTM Network
AMP-PSO	Adaptive Multi-Population Particle Swarm Optimization
ARIMA	Autoregressive Integrated Moving Average
CEEMD	Complementary Ensemble Empirical Mode Decomposition
CES	Complex Exponential Smoothing
CNN	Convolutional Neural Network
CMA	China Meteorological Administration
CPI	Consumer Price Index
CH-LSTM	Contextualized Hierarchical Long Short-Term Memory
DIA	Dual Input Attention
EEMD	Ensemble Empirical Mode Decomposition
EIA	U.S. Energy Information Administration
FAO	Food and Agriculture Organization of the United Nations
GACC	General Administration of Customs, PRC
GCRNN	Group-Constrained Convolutional Recurrent Neural Network
GRU	Gated Recurrent Unit
KAN	Kolmogorov-Arnold Network
LOESS	Locally Estimated Scatterplot Smoothing
LSTM	Long Short-Term Memory Networks
MAE	Mean Absolute Error
M2	Broad Money Supply
MARA	Ministry of Agriculture and Rural Affairs, PRC
MSE	Mean Squared Error

MOFCOM	Ministry of Commerce, PRC
NBS	National Bureau of Statistics
PBC	People's Bank of China
PROPHET	Prophet
PRC	People's Republic of China
PSO	Particle Swarm Optimization
R^2	Coefficient of Determination
RNN	Recurrent Neural Network
RMSE	Root Mean Squared Error
RF	Random Forest
STL	Seasonal-Trend Decomposition using LOESS
STL-LSTM-ATT-KAN	Seasonal-Trend Decomposition-Long Short-Term Memory-Attention Mechanism-Kolmogorov-Arnold Network Combination Model
SVM	Support Vector Machine
STCN-LSTM	Spatial-Temporal Convolutional Network with Long Short-Term Memory
USD/CNY	United States Dollar to Chinese Yuan
VMD	Variational Mode Decomposition
XGBoost	Extreme Gradient Boosting

Term	**Synonyms**
AMP-PSO Algorithm	Adaptive Multi-Population Particle Swarm Optimization
Attention Mechanism	Attention Model, Attention Network
CES	Complex Exponential Smoothing
Coefficient of Determination	R2, Goodness of Fit
Data Preprocessing	Data Cleaning, Data Processing
Feature Extraction	Feature Selection, Feature Processing
Ginger Price Prediction	Ginger Price Forecasting, Ginger Market Prediction
KAN Network	Kolmogorov-Arnold Network
LOESS Method	Locally Estimated Scatterplot Smoothing
LSTM Networks	Long Short-Term Memory Networks, Long Short-Term Memory Units
LSTM Unit	Long Short-Term Memory Unit
MAE	Mean Absolute Error, Average Error
MSE	Mean Squared Error, Squared Error
Market Regulation	Market Adjustment, Market Intervention
Model Ensemble	Model Fusion
Model Fitting	Model Training, Model Learning
Model Optimization	Parameter Optimization, Model Tuning
Model Testing	Model Validation, Model Evaluation
Pareto Optimality	Pareto Optimal Solution, Non-Dominated Solution
Prediction Accuracy	Prediction Precision, Prediction Performance
Price Forecasting	Time Series Forecasting
RMSE	Root Mean Squared Error, Root Mean Square Error
STL Decomposition	Seasonal-Trend Decomposition
Sequence Decomposition	Data Decomposition, Time Series Decomposition

References

1. Xu, K. Short-Term Price Forecasting Model for Fresh and Live Agricultural Products Based on Price Decomposition. Ph.D. Thesis, Chinese Academy of Agricultural Sciences, Beijing, China, 2016.
2. Liu, F.; Wang, R.; Li, C. Application of ARIMA model in agricultural products price forecasting. *Comput. Eng. Appl.* **2009**, *45*, 238–239+248.
3. Ding, W.; He, Y.; Liu, C. China's corn price fluctuation trend and short-term forecast. *Price Theory Pract.* **2014**, 83–84. [CrossRef]

4. Ravishankar, P.; Rakesh, S.; Ranjit, P.K. Price Forecasting of Mango in Varanasi Market of Uttar Pradesh. *Price Theory Pract.* **2018**, *6*, 218–224.
5. Wang, B.; Song, Y.; Zong, Y. Cucumber price trend analysis and market forecast based on ARIMA model. *Stat. Manag.* **2019**, 87–90. [CrossRef]
6. Van Gestel, T.; Suykens, J.; Baestaens, D.-E.; Lambrechts, A.; Lanckriet, G.; Vandaele, B.; De Moor, B.; Vandewalle, J. Financial time series prediction using least squares support vector machines within the evidence framework. *IEEE Trans. Neural Networks* **2001**, *12*, 809–821. [CrossRef]
7. Das, M.; Ghosh, S.K. A probabilistic approach for weather forecast using spatio-temporal inter-relationships among climate variables. In Proceedings of the 2014 9th International Conference on Industrial and Information Systems (ICIIS), Gwalior, India, 15–17 December 2014; IEEE: Piscataway, NJ, USA, 2014; pp. 1–6.
8. Duan, Q.; Zhang, L.; Wei, F.; Xiao, X.; Wang, L. Forecasting model and validation of aquatic product price based on time series GA-SVR. *J. Agric. Eng.* **2017**, *33*, 308–314.
9. Wang, P.; Zhang, H.; Qin, Z.; Zhang, G. A novel hybrid-Garch model based on ARIMA and SVM for $PM_{2.5}$ concentrations forecasting. *Atmos. Pollut. Res.* **2017**, *8*, 850–860. [CrossRef]
10. Polyiam, K.; Boonrawd, P. A hybrid forecasting model of cassava price based on artificial neural network with support vector machine technique. In Proceedings of the 2017 3rd International Conference on Information Management (ICIM), Chengdu, China, 21–23 April 2017; IEEE: Piscataway, NJ, USA; pp. 123–127. [CrossRef]
11. Zhao, B.-H.; Huo, Q.; Lu, Y.-H. Research on the characteristics of ginger price fluctuation and its law—Analysis based on ARCH class model. *Price Theory Pract.* **2022**, *5*, 130–133.
12. Sun, F.; Meng, X.; Zhang, Y.; Wang, Y.; Jiang, H.; Liu, P. Agricultural Product Price Forecasting Methods: A Review. *Agriculture* **2023**, *13*, 1671. [CrossRef]
13. Teng, J.; Liu, P.; Zhang, Y.; Xu, S.; Xu, G.; Han, W. A study on ginger price prediction based on Prophet. *Chin. J. Agric. Mech. Chem.* **2020**, *41*, 211–216.
14. Wang, Y. *Research on the Combination Prediction Model of Ginger Price Based on Multiple Influencing Factors*; Shandong Agricultural University: Tai'an, China, 2023.
15. Lv, X. Research on Price Prediction of Small Agricultural Products Based on the Influence of Network Public Opinion. Ph.D. Thesis, Northeast Agricultural University, Harbin, China, 2023.
16. Chaudhary, R.; Giri, H.N.; Bhusal, N.R.; Dahal, S.; Timalsina, P.; Bhandari, N. An analysis of the ginger value chain in palpa, Nepal. *Cogent Food Agric.* **2023**, *9*, 2270799. [CrossRef]
17. Manjubala, M. Weekly price prediction of garlic and ginger using complex exponential smoothing. *Int. J. Stat. Appl. Math.* **2023**, *8*, 214–220.
18. Li, H.; Jia, Y.; Zhang, Y. Research on Pricing Strategies Based on Time Series Analysis Model and Greedy Algorithm Model. *Highlights Bus. Econ. Manag.* **2024**, *25*, 127–133. [CrossRef]
19. Fu, L.; Zhang, H. Analysis of Price Influencing Factors of Small-Scale Agricultural Products Based on Network Perspective. *Front. Sustain. Food Syst.* **2023**, *8*, 1355853.
20. Anamisa, D.R.; Mufarroha, F.A.; Jauhari, A.; Yusuf, M.; Khotimah, B.K.; Haq, A.F. Long Short-Term Memory Method Based on Normalisation Data for Forecasting Analysis of Madura Ginger Selling Price. In Proceedings of the 4th International Conference on Informatics, Technology and Engineering 2023 (InCITE 2023), Yogyakarta, Indonesia, 14–15 September 2023.
21. Lin, S.; Runger, C.G. GCRNN: Group-Constrained Convolutional Recurrent Neural Network. *IEEE Trans. Neural Netw. Learn. Syst.* **2018**, *29*, 4709–4718. [CrossRef]
22. Hu, L.; Li, J.; Nie, L.; Li, X.-L.; Shao, C. What Happens Next? Future Subevent Prediction Using Contextual Hierarchical LSTM. In Proceedings of the Thirty-First AAAI Conference on Artificial Intelligence (AAAI'17), San Francisco, CA, USA, 4–9 February 2017; AAAI Press: Palo Alto, CA, USA; pp. 3450–3456. [CrossRef]
23. Zhang, Y.; Yang, X.; Ivy, J. ATTAIN: Attention-based Time-Aware LSTM Networks for Disease Progression Modeling. In Proceedings of the Twenty-Eighth International Joint Conference on Artificial Intelligence {IJCAI-19}, Macao, China, 10–16 August 2019. [CrossRef]
24. Zhang, D.B.; Zeng, L.L.; Ling, L.W. An integrated prediction model based on VMD-ELM for agricultural futures price decomposition. *Oper. Res. Manag.* **2023**, *32*, 127–133.
25. Wang, R.; Zhang, X.; Wang, M. Agricultural product price prediction based on signal decomposition and deep learning. *J. Agric. Eng.* **2022**, *38*, 256–267.
26. Fang, X.; Wu, C.; Yu, S.; Zhang, D.; Ouyang, Q. Research on short-term prediction model of agricultural products price based on EEMD-LSTM. *China Manag. Sci.* **2021**, *29*, 68–77. [CrossRef]
27. Gu, Y.H.; Jin, D.; Yin, H.; Zheng, R.; Piao, X.; Yoo, S.J. Forecasting Agricultural Commodity Prices Using Dual Input Attention LSTM. *Agriculture* **2022**, *12*, 256. [CrossRef]

28. Sun, C.; Pei, M.; Cao, B.; Chang, S.; Si, H. A Study on Agricultural Commodity Price Prediction Model Based on Secondary Decomposition and Long Short-Term Memory Network. *Agriculture* **2023**, *14*, 60. [CrossRef]
29. Zhi, L.; Zuo, Y. Collaborative Path Planning of Multiple AUVs Based on Adaptive MultiPopulation PSO. *J. Mar. Sci. Eng.* **2024**, *12*, 223. [CrossRef]
30. Liu, Z.; Wang, Y.; Vaidya, S.; Ruehle, F.; Halverson, J.; Soljačić, M.; Hou, T.Y.; Tegmark, M. Kan: Kolmogorov-arnold networks. *arXiv* **2024**, arXiv:2404.19756.

agriculture

MDPI

Article

Plant Disease Segmentation Networks for Fast Automatic Severity Estimation Under Natural Field Scenarios

Chenyi Zhao [1], Changchun Li [1,*], Xin Wang [1], Xifang Wu [1], Yongquan Du [1], Huabin Chai [1], Taiyi Cai [1], Hengmao Xiang [2,*] and Yinghua Jiao [2]

1 School of Surveying and Land Information Engineering, Henan Polytechnic University, Jiaozuo 454000, China; 212204010026@home.hpu.edu.cn (C.Z.); 212204010013@home.hpu.edu.cn (X.W.); wxf2020@hpu.edu.cn (X.W.); dyq@home.hpu.edu.cn (Y.D.); chaihb@hpu.edu.cn (H.C.); caity2008@hpu.edu.cn (T.C.)

2 Shandong Provincial Institute of Land Surveying and Mapping, Jinan 250102, China; jiaoyinghua@shandong.cn

* Correspondence: lcc@hpu.edu.cn (C.L.); xianghengmao@shandong.cn (H.X.)

Abstract: The segmentation of plant disease images enables researchers to quantify the proportion of disease spots on leaves, known as disease severity. Current deep learning methods predominantly focus on single diseases, simple lesions, or laboratory-controlled environments. In this study, we established and publicly released image datasets of field scenarios for three diseases: soybean bacterial blight (SBB), wheat stripe rust (WSR), and cedar apple rust (CAR). We developed Plant Disease Segmentation Networks (PDSNets) based on LinkNet with ResNet-18 as the encoder, including three versions: $\times 1.0$, $\times 0.75$, and $\times 0.5$. The $\times 1.0$ version incorporates a 4×4 embedding layer to enhance prediction speed, while versions $\times 0.75$ and $\times 0.5$ are lightweight variants with reduced channel numbers within the same architecture. Their parameter counts are 11.53 M, 6.50 M, and 2.90 M, respectively. PDSNetx0.5 achieved an overall F1 score of 91.96%, an Intersection over Union (IoU) of 85.85% for segmentation, and a coefficient of determination (R^2) of 0.908 for severity estimation. On a local central processing unit (CPU), PDSNetx0.5 demonstrated a prediction speed of 34.18 images (640×640 pixels) per second, which is 2.66 times faster than LinkNet. Our work provides an efficient and automated approach for assessing plant disease severity in field scenarios.

Keywords: deep learning; lightweight; plant disease; crop disease; semantic segmentation; severity estimation

1. Introduction

Plant diseases represent a substantial threat to global agricultural productivity, leading to estimated annual crop yield losses of 20–40%, which severely compromise food security and economic stability [1]. Among these, foliar diseases, the most prevalent type, significantly reduce yields by impairing photosynthetic efficiency. Traditionally, researchers have relied on manual visual estimation to assess disease severity in the field. However, this method is not only labor-intensive and time-consuming but also prone to human error [2,3]. Disease severity is quantified as the proportion of leaf area affected by disease spots. While image segmentation techniques, such as threshold segmentation, have been effective in isolating disease spots from healthy leaf tissue [4–6], they struggle with complex backgrounds and variable lighting conditions. Moreover, these methods necessitate the destructive collection of diseased leaves for laboratory imaging, which risks altering the disease's natural characteristics and is highly impractical. Additionally, detached leaves often rapidly dehydrate and deform, complicating the application of these techniques. In contrast, deep learning approaches, particularly Convolutional Neural Networks (CNNs), offer an end-to-end solution. Once trained, CNNs can directly predict outcomes from input images without manual intervention.

With the rapid advancement of deep learning technologies, numerous classical network models [7–9] have been extensively employed in plant disease image recognition tasks. For instance, Long et al. [10] employed Convolutional Neural Networks to classify six types of wheat diseases in field environments, achieving an accuracy rate of 97.05%. Similarly, Bever et al. [11] applied the DenseNet model to classify eight types of soybean diseases, attaining a classification accuracy of 96.8%. Furthermore, Lin et al. [12] introduced a dual-branch network architecture that integrates attention mechanisms for simultaneous prediction, achieving an accuracy of 88.74% when classifying 10 distinct diseases.

In the field of disease severity assessment, several researchers have explored classification-based approaches to evaluate disease severity. For instance, Mi et al. [13] manually categorized wheat stripe rust images into six severity levels and employed a DenseNet model integrated with attention mechanisms, achieving a classification accuracy of 97.99%. Similarly, Joshi et al. [14] classified Vigna mungo disease images into three levels—healthy, moderate, and severe—using their custom VirLeafNet, attaining an accuracy of 97.4%. In a comparable study, Yin et al. [15] manually divided field maize disease images into five severity levels, achieving a classification accuracy of 97.12%. Despite the high classification accuracy of these methods, their dependence on manual severity level categorization introduces significant subjectivity, thereby limiting their reliability and broader applicability.

Another group of researchers has focused on the semantic segmentation of disease images with the goal of extracting lesion and leaf pixels from the images. For instance, Lin et al. [16] employed an enhanced U-Net [17] model to segment cucumber powdery mildew images, achieving an Intersection over Union (IoU) of 72.11%. Similarly, Li et al. [18] enhanced DeepLabv3+ [19] by integrating attention mechanisms and transfer learning, attaining an average IoU of 81.23% for cucumber disease image segmentation, and further assessed disease severity, achieving a coefficient of determination (R^2) of 0.958. Furthermore, Wang et al. [20] introduced a two-stage segmentation approach: the first stage utilizes DeepLabv3+ to segment diseased leaves in complex backgrounds, while the second stage applies U-Net for lesion segmentation. Despite the notable success of these methods in segmentation tasks, their models are characterized by substantial parameter sizes and high computational demands.

Model lightweight design has emerged as a prominent research focus in the field of deep learning. For instance, Feng et al. [21] introduced a U-shaped DFFANet model integrated with attention mechanisms for segmenting rice blast images, achieving an Intersection over Union (IoU) of 96.15% with a remarkably low parameter count of only 1.4 M. Similarly, Shi et al. [22] developed a U-Net model utilizing MobileNetV3 [23] as the encoder, incorporating attention mechanisms to segment wheat scab, achieving an IoU of 88.87% with a parameter count of 4.52 M. Furthermore, Xu et al. [24] implemented the DeepLabv3+network, leveraging channel scaling and knowledge distillation techniques, to perform two-stage segmentation of cotton diseases. This approach achieved IoU values of 87.66% and 85.18% for the two stages, respectively, with a parameter count of 5.02 M. Despite the notable success of these methods in minimizing network parameters and computational demands, their validation is limited to single-disease scenarios, and the lack of publicly available data poses significant challenges for reproducibility and broader application.

To address the limitations identified in current research, this study introduces improvements across three key areas: First, a comprehensive multi-disease dataset encompassing field scenarios was developed and publicly released. This dataset features high-resolution images of soybean bacterial blight (SBB), wheat stripe rust (WSR), and cedar apple rust (CAR). Second, leveraging the LinkNet network architecture, the PDSNets network family was proposed, which employs channel scaling techniques to achieve lightweight model design. Lastly, a quantitative evaluation of disease severity was performed based on the segmentation outcomes.

2. Materials and Methods

2.1. Data Acquisition and Preprocessing

This study utilized three types of disease images (Table 1), including SBB and CAR obtained from public resources, as well as WSR collected in the field by our team at the Jiaozuo Academy of Agriculture and Forestry Sciences (Jiaozuo, China) in May 2023. As illustrated in Figure 1, each image was padded from a rectangular format to a square format using black pixels and subsequently downsampled to 640 × 640 pixels using Lanczos interpolation [25]. A total of 80 images for each disease were processed using a custom Python 3.12.7 program. The integrated dataset is well suited for field plant disease segmentation due to its inclusion of the following characteristics: (1) high quality and resolution; (2) complex backgrounds and lighting conditions; (3) variable camera angles; (4) diseased leaves attached to the plant; and (5) diverse shooting equipment and locations. The dataset is publicly accessible at https://doi.org/10.34740/kaggle/ds/6620728, accessed on 9 March 2025.

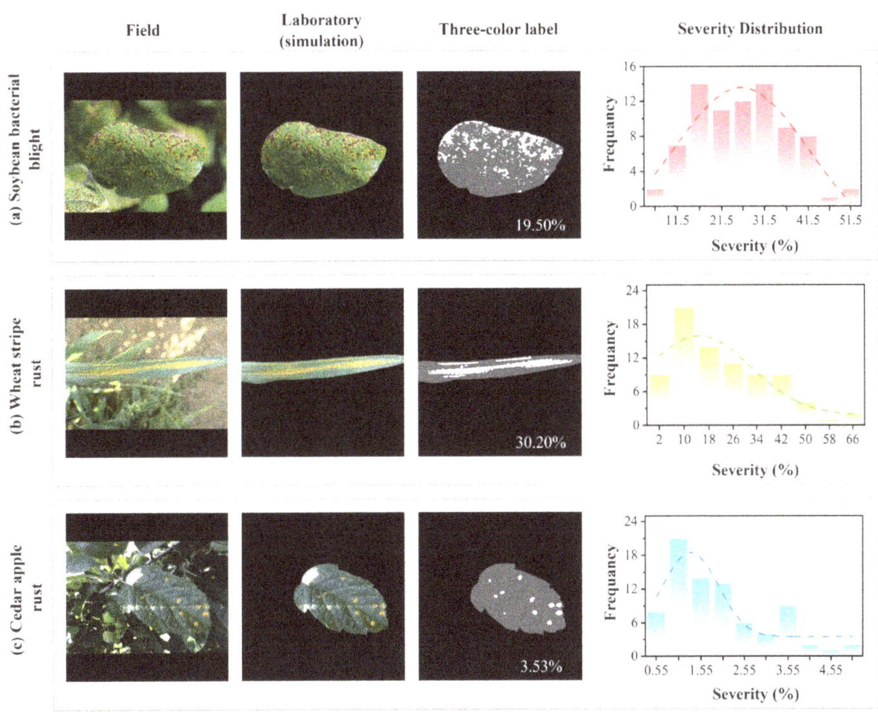

Figure 1. Three-color labels and reference severity distribution for disease images. Black, gray, and white in labels represent background (B), healthy leaf (H), and disease spots (S), respectively. Severity is calculated according to white × 100/(white + gray).

Table 1. Basic information of the original disease images.

Disease	Pixels	Year	Equipment	Location	Reference
Soybean bacterial blight	5472 × 3648	2020, 2021	Canon EOS 7D Mark II	Tallassee, AL, USA	[11]
Wheat stripe rust	4272 × 2848	2023	Canon EOS 450D	Jiaozuo, China	Our study
Cedar apple rust	2048 × 1365	2019	Canon Rebel T5i DSLR	Geneva, NY, USA	[26]

2.2. Pixel-Level Annotation

Before conducting supervised learning, each pixel of each image must be labeled as one of the pixel classes: background (B), healthy leaf (H), and disease spots (S). Severity is the proportion of a leaf occupied by disease spots.

$$Severity = \frac{S}{S+H} \times 100 \tag{1}$$

Due to the complexity of the data, we have different labeling schemes for each type of disease, as shown below. These schemes were inspired by reference [4] and were adapted during the annotation process. The implementation of all schemes was based on self-developed Python 3.12.7 programs.

SBB images contain leaves with regular margins and spots with an extremely complex distribution (Figure 1a). We used LabelMe v5.7.0 software [27] to manually label the polygonal boundary of a major leaf, thus setting the B pixels to black. Then, we converted the black background image from RGB color space to LAB color space and took its a-channel (LAB-A) because the a-channel can highlight S pixels very well. After normalizing the a-channel to the range of 0 to 255, we used Otsu threshold [28] to set the H pixel to gray and the S pixel to white. Finally, a three-color pixel label for training was completed.

The WSR images contain long and thin leaves with regular margins and spots with complex distribution (Figure 1b). We tried to continue using the soybean bacterial blight scheme but found that the LAB-A method labeled many S pixels as H pixels. Instead of the LAB-A-highlighted diseased spots, we used the healthy leaves highlighted by the modified green–red vegetation index (MGRVI) [29].

$$MGRVI = \frac{G^2 - R^2}{G^2 + R^2} \tag{2}$$

where R and G are the green and red channels of the RGB color space, respectively. Their value range is from 0 to 1. Before performing threshold segmentation, the MGRVI also needs to be normalized to the range of 0 to 255. In this way, we successfully labeled all the S pixels, but at the same time, the white edges and leaf stalks of wheat leaves in some images were incorrectly labeled as S pixels. Therefore, we used morphological opening operation to remove these errors.

The CAR images contain leaves and spots with regular margins, but reflections and shadows caused by direct sunlight render the above scheme ineffective (Figure 1c). So, we used LabelMe to manually annotate the edge polygon of a main leaf and the spots on it and then converted it into the required three-color pixel label.

2.3. PDSNets

Figure 2 presents the Plant Disease Segmentation Networks (PDSNets). They are based on LinkNet, using a new 4 × 4 embedding layer and the GELU activation function [30]. To represent feature size simply, we used the fraction between its height (width) and the height (width) of the input image. For example, a feature with a size of {64, 1/4} means that its channel size is 64 and its height (width) is 1/4 of the height (width) of the input image.

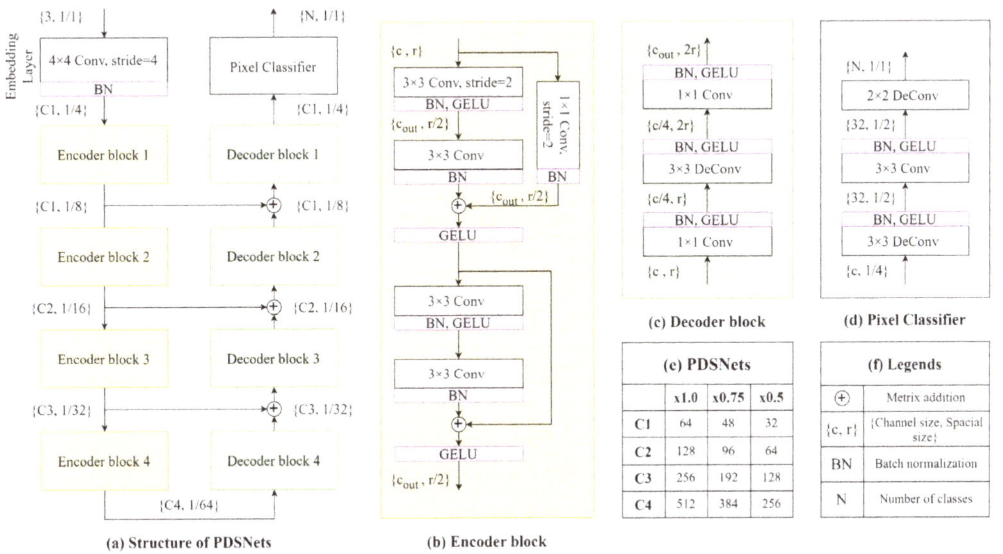

Figure 2. The structure of Plant Disease Segmentation Networks (PDSNets). The spatial size is defined as the fraction between the height (width) of the feature and the height (width) of the input image.

The encoder of PDSNets consists of a 4 × 4 embedding layer and 4 encoder units. It can extract image features step by step as the channel size of the features increases and the spatial size decreases. The 4 × 4 embedding layer, which can effectively increase the processing speed, is responsible for mapping the RGB image into the function domain defined by the network. Each encoder unit consists of two residual blocks from ResNet [8], with the first changing the size of the feature and the second keeping the feature size unchanged (Figure 2b). The size of the highest-level feature in the final output of the encoder is {512, 1/64}.

The decoder of PDSNets is made of 4 decoder units and a pixel-level classifier. Each decoder unit uses two 1 × 1 convolutional layers to reduce and increase the channel size of the features, making the 3 × 3 deconvolution operation [31] in the middle cheaper to compute (Figure 2c). Each decoder unit accepts not only the output features of the previous decoder but also the output features of the encoder of the same level, also known as skip connection. The pixel-level classifier uses two deconvolutions to restore the spatial size of the features to the image size (Figure 2d). In the network, due to the usage of batch normalization after each convolutional layer, we removed the bias from all convolutional and deconvolutional layers except the last deconvolutional layer of the pixel-level classifier.

As shown in Figure 2e, we scaled the channel size of PDSNetx1.0 to obtain its lightweight versions. The channel sizes of PDSNetx0.75 are 0.75 times the baseline, meaning its 4 encoders have channel sizes of 48, 96, 192, and 384, respectively. Similarly, the channel sizes of PDSNetx0.5 are 0.5 times the baseline, meaning its 4 encoders have channel sizes of 32, 64, 128, and 256, respectively. During the scaling process, the number of input and output channels of the intermediate 3 × 3 convolutional layer in the pixel-level classifier is fixed at 32.

2.4. Evaluation Indices

*F*1 and *IoU* were used to evaluate the pixel classification performance of the network. For pixels labeled as *i*, *F*1 can evaluate the accuracy of pixel classification, and *IoU* can assess the similarity between true label i pixels and the predicted label i pixels. They are defined as follows:

$$Precision = \frac{TP}{TP + FP} \quad (3)$$

$$Recall = \frac{TP}{TP + FN} \quad (4)$$

$$F1 = \frac{2 \times Precision \times Recall}{Precision + Recall} \quad (5)$$

$$IoU = \frac{TP}{TP + FP + FN} \quad (6)$$

where *true positive* (TP) is the number of pixels where both the true label and the predicted label are i, *false positive* (FP) is the number of pixels where the predicted label is i but the true label is not i, and *false negative* (FN) is the number of pixels where the true label is i but the predicted label is not i. Therefore, *precision* represents the proportion of pixels with the predicted label i that were correctly predicted, and *recall* represents the proportion of pixels with the true label i that were correctly predicted. $F1$ is the harmonic mean of precision and recall.

The coefficient of determination (R^2) and root mean square error (RMSE) between the reference severity and the predicted severity were used to evaluate the network's performance in estimating severity. In addition, scatter plots and linear regression lines were drawn for auxiliary evaluation. FPS is used to evaluate the prediction speed of the network, indicating the number of 640 × 640 pixel images the network can predict per second.

$$FPS = \frac{1}{T} \quad (7)$$

where T is the average time taken for the network to make 300 predictions on the local i5-12400 CPU.

2.5. Experiment Settings

All experiments in this work were conducted on the same computer, which is equipped with Windows 11 (10.0.26100, Microsoft Corporation, Redmond, WA, USA), Pytorch 12.4, Intel i5-12400 CPU (Intel Corporation, Santa Clara, CA, USA), and NVIDIA GeForce 4060ti GPU (Nvidia Corporation, Santa Clara, CA, USA). The maximum number of epochs and the batch size for network training are 200 and 8, respectively. The learning rate (LR) is 0.0003, and it is reduced every epoch.

$$LR = \left(1 - \frac{epoch}{epoch_{max}}\right)^{0.9} \quad (8)$$

The optimizer is AdamW [32] with a default decay rate of 0.01. The weighted cross-entropy loss (WCE-Loss) was used to balance the naturally occurring pixel count imbalance between B, H and S. The WCE-Loss for a single pixel is defined as follows:

$$WCE\text{-}Loss = -\sum_{i=1}^{N} \omega_i \cdot p_i \cdot log\left(\frac{\exp(q_i)}{\sum_{j=1}^{N} \exp(q_j)}\right) \quad (9)$$

where p and q are the one-hot true label and the predicted logits for the pixel, respectively. N is the number of pixel classes. We empirically set the weight as B: $\omega_1 = 1$; H: $\omega_2 = 3$; and S: $\omega_3 = 6$.

For each disease, we randomly selected 60% of the data for training and the remaining 40% for testing. During the training process, we used random image transformations [33] to make the training data different for each epoch, which increased the diversity of the training data. These image transformation methods included random rotations within the range of -25° to 25° and random flips up, down, left, and right. The probability of applying these image transformation methods to each input image was 0.5. The training data were shuffled and combined into new mini-batches at the start of each epoch. In addition, when training the U-Net, we set the batch size to 4 to prevent running out of GPU memory. For the backbone of the DeepLabv3+, we used the dilated ResNet-18 with the dilation method described in [19].

3. Results

3.1. Comparison of Networks

As shown in Table 2, we compared the segmentation results of the PDSNets with the common U-Net, DeepLabv3+ with the ResNet-18 backbone, and LinkNet under the same experimental conditions. The *F*1 scores and *IoU* values in the table indicate the cross-disease segmentation accuracy of these networks for *B*, *H*, and *S* classes. Params represent the numbers of learnable parameters, and FPS_{CPU} indicates their prediction speed on the local CPU.

Table 2. The comparison of networks under field scenarios. *B*, *H*, and *S* represent the background, healthy leaf, and disease spots, respectively. FPS_{CPU} indicates how many 640 × 640 pixel images can be processed per second on the local central processing unit (CPU). The F1 score and IoU values are shown as the means across diseases.

Networks	*F*1 (%)			*IoU* (%)			Params (M)	FPS_{CPU} (F/s)
	B	*H*	*S*	*B*	*H*	*S*		
U-Net	98.24	91.56	89.14	96.59	85.41	80.83	31.04	0.549
DeepLabV3+(ResNet-18)	98.66	93.05	87.15	97.40	87.53	77.52	16.60	2.673
LinkNet	98.80	93.05	86.19	97.66	87.54	76.07	11.54	12.86
PDSNetx1.0	98.55	93.30	86.38	97.19	87.71	76.34	11.53	19.88
PDSNetx0.75	98.46	92.74	85.37	97.03	86.91	74.88	6.497	25.78
PDSNetx0.5	98.49	92.52	84.87	97.07	86.37	74.11	2.898	34.18

To evaluate the segmentation performance of these networks, we averaged the *F*1 score and *IoU* values across pixel classes in the table. U-Net achieved the highest *F*1 and *IoU* values at 92.98% and 87.61%, respectively, but its prediction speed is only half an image per second. Compared to U-Net, DeepLabv3+(ResNet-18) has fewer parameters, but its prediction speed is still very low. As the benchmark for our work, the prediction speed of LinkNet is 23.4 times that of U-Net, with only a 0.3% *F*1 and 0.52% *IoU* cost. The PDSNetx1.0 we proposed has a prediction speed 1.55 times that of LinkNet with almost no cost to accuracy. The prediction speed of the lightest PDSNetx0.5 is 2.66 times that of LinkNet with costs of 0.72% for *F*1 and 1.24% for *IoU*.

To evaluate the segmentation performance of CNNs on the pixel classes *B*, *H*, and *S*, we averaged the *F*1 and *IoU* values across networks in the table. The *F*1 and *IoU* values for class *B* are 98.53% and 97.16%, respectively; for class *H*, they are 92.7% and 86.91%, respectively; and for class *S*, they are 86.52% and 76.63%, respectively.

In summary, our PDSNets achieved the best balance between prediction speed and accuracy. CNNs have the highest segmentation ability for the background, followed by healthy leaf, and the lowest for disease spots [34].

Figure 3 visually demonstrates the prediction results of the networks, showing their excellent ability to extract target diseased leaves from field background images. However, in complex scenarios with multiple leaves, reflections, and shadows, they exhibited lower segmentation accuracy (Figure 4). Since most images in our experiments are single subject leaves and avoid direct sunlight, we cannot yet evaluate the networks' true ability to handle such complex scenarios.

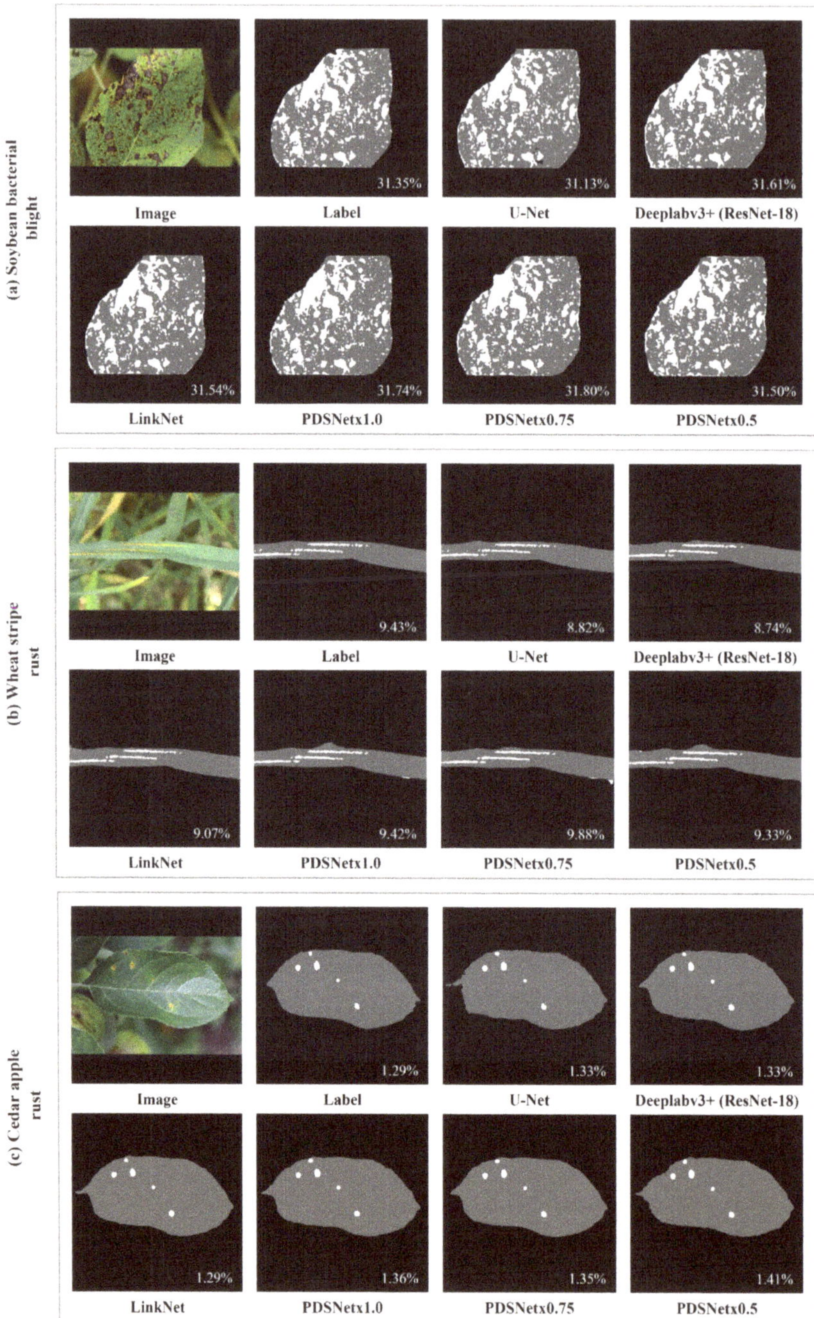

Figure 3. Examples of segmentation in conventional field scenarios. The percentage is the severity defined as white × 100/(white + gray).

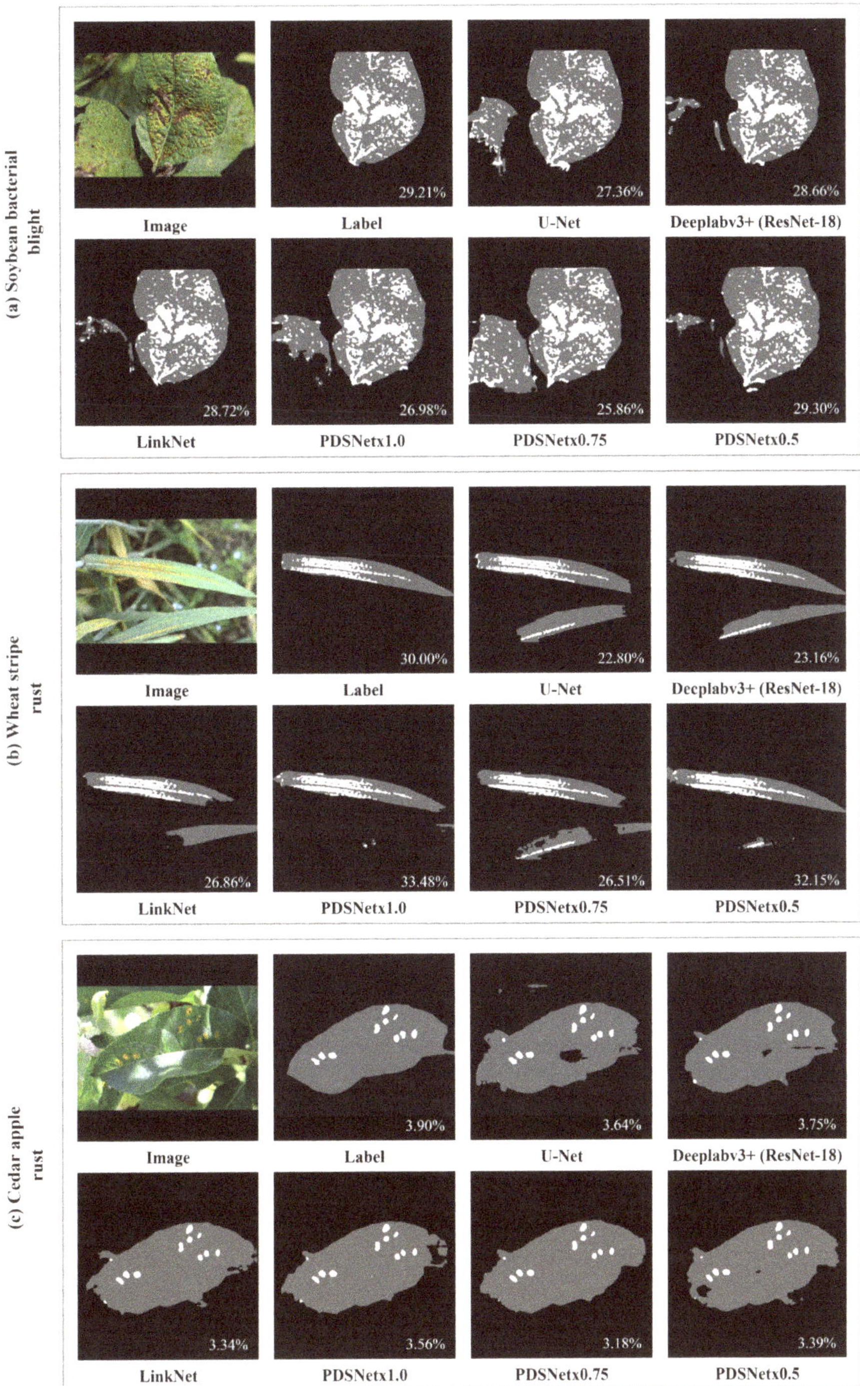

Figure 4. Examples of segmentation of complex field scenarios with multiple leaves, reflections, and shadows. The percentage is the severity defined as white × 100/(white + gray).

3.2. Ablation Experiment of PDSNetx1.0

As shown in Table 3, we conducted ablation experiments under field scenarios to verify the effects of the 4 × 4 embedding layer and the GELU activation function. The *F1* and *IoU* in the table indicate the cross-disease segmentation accuracy of these networks for *B*, *H*, and *S*. FPS_{CPU} indicates their prediction speed on the local CPU. We took the average of *F1* and *IoU* values across pixel classes. After applying the 4 × 4 embedding layer, the network's prediction speed was 1.58 times that of LinkNet with only a 0.17% cost for *F1* and 0.38% cost for *IoU*. After simultaneously applying the 4 × 4 embedding layer and the GELU activation function, the model's prediction speed slightly decreased but was still 1.55 times that of LinkNet, while the model's segmentation accuracy slightly improved.

Table 3. The ablation results of PDSNetx1.0 on field-acquired images. *B*, *H*, and *S* represent the background, healthy leaf, and disease spots, respectively. FPS_{CPU} indicates how many 640 × 640 pixel images can be processed per second on the local central processing unit (CPU). The F1 score and IoU values are shown as means across diseases. "−" indicates below the baseline and "+" indicates above the baseline.

Networks	F1 (%)			IoU (%)			FPS_{CPU} (F/s)
	B	*H*	*S*	*B*	*H*	*S*	
LinkNet + ReLU (Baseline)	98.80	93.05	86.19	97.66	87.54	76.07	12.86
LinkNet + 4 × 4 embedding + ReLU	98.56 (−)	92.77 (−)	86.20 (+)	97.20 (−)	86.94 (−)	76.00 (−)	20.37 (+)
LinkNet + 4 × 4 embedding + GELU (PDSNetx1.0)	98.55 (−)	93.30 (+)	86.38 (+)	97.19 (−)	87.71 (+)	76.34 (+)	19.88 (+)

In summary, the 4 × 4 embedding layer can effectively improve the model's prediction speed, while the GELU activation function effectively compensates for the segmentation accuracy loss brought by the 4 × 4 embedding layer.

3.3. CE-Loss vs. WCE-Loss of PDSNetx1.0

As shown in Figure 5, we recorded the training information of PDSNetx1.0 using cross-entropy loss (CE-Loss) and weighted cross-entropy loss (WCE-Loss), including the training loss and test *F1* score for each disease in each training epoch. We also calculated the proportion of *B*, *H*, and *S* pixel classes in each disease training set. The results for SBB are *B*:69.55%, *H*:22.77%, and *S*:7.68%; for WSR, they are *B*:90.65%, *H*:7.30%, and *S*:2.05%; and for CAR, they are *B*:75.90%, *H*:23.66%, and *S*:0.44%. The improvement in WCE-Loss on PDSNetx1.0 in SBB and WSR is not significant, with the cross-class average *F1* score increasing by 0.66% and 0.92%, respectively, and the *F1* score of *S* pixels increasing by 2.03% and 1.80%. However, the improvement in WCE-Loss on CAR is significant, especially in the case of disease spots. The cross-pixel class average *F1* score of CAR improved by 2.74%, with the *F1* score of *S* pixels improving by 7.52%.

In summary, a very small number of samples will significantly increase the difficulty of the network's learning. WCE-Loss can effectively enhance PDSNetx1.0's classification performance of *S* pixels in disease images, particularly for diseases like CAR with spot-like distributions that result in too few *S* pixels, where the improvement is especially noticeable.

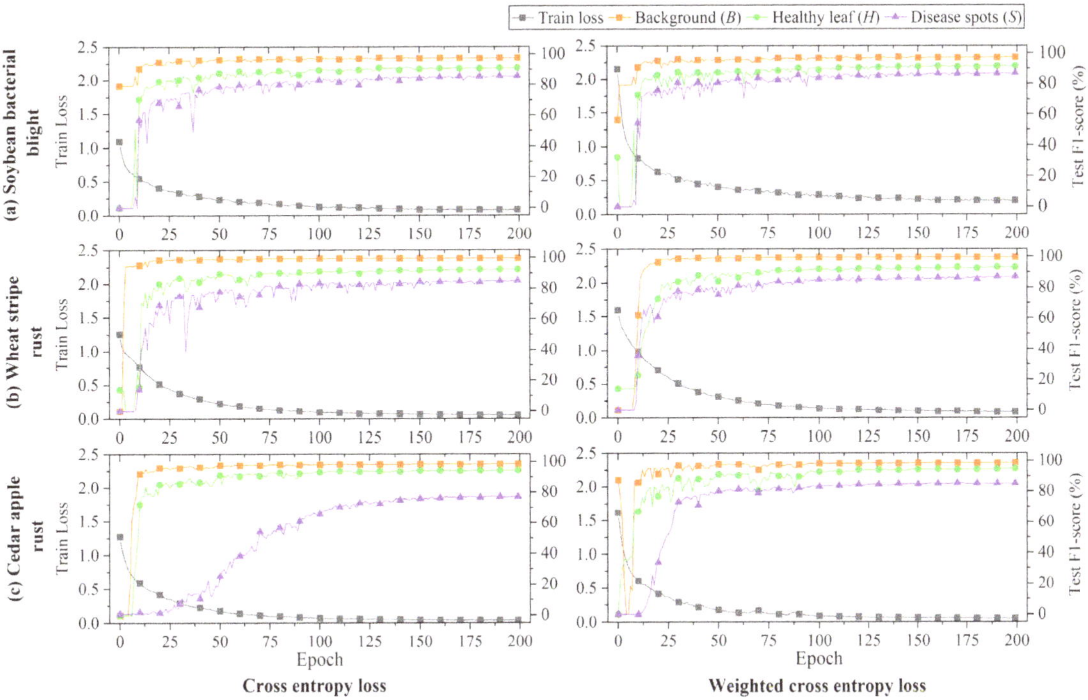

Figure 5. Comparison of PDSNetx1.0 using cross-entropy loss and weighted cross-entropy loss.

3.4. Generalization Test on Field, Laboratory, and Mixed Images

As shown in Figure 6, we used training sets of field, laboratory, and mixed images to train U-Net, DeepLabv3+(ResNet-18), LinkNet, and PDSNetx1.0, respectively, and then tested them on the test sets of field images and laboratory images. The mixed images are half field images and half laboratory images. Such an experimental setup can test the model's generalization ability under the field and laboratory scenarios.

To conduct a quantitative evaluation, we took the cross-network average of the $F1$ for each disease. For SBB, the networks trained with field images performed similarly in lab image tests, but the networks trained with lab images decreased by 32.65% in field image tests. For WSR, the networks trained with field images decreased by 26.76% in lab image tests, and the networks trained with lab images decreased by 60.97% in field image tests. For CAR, the networks trained with field images decreased by 29.36% in lab image tests, and the networks trained with lab images decreased by 40.87% in field image tests. The networks trained with mixed images performed well in the tests of two background images for each disease.

In summary, CNNs cannot accurately predict data that are not present in the training set, even though the lab images in this work are merely simulated results of field images with the background removed. It can be reasonably inferred that if images are collected from real laboratory scenarios with greater differences, the CNNs will be even less capable of carrying out cross-background prediction.

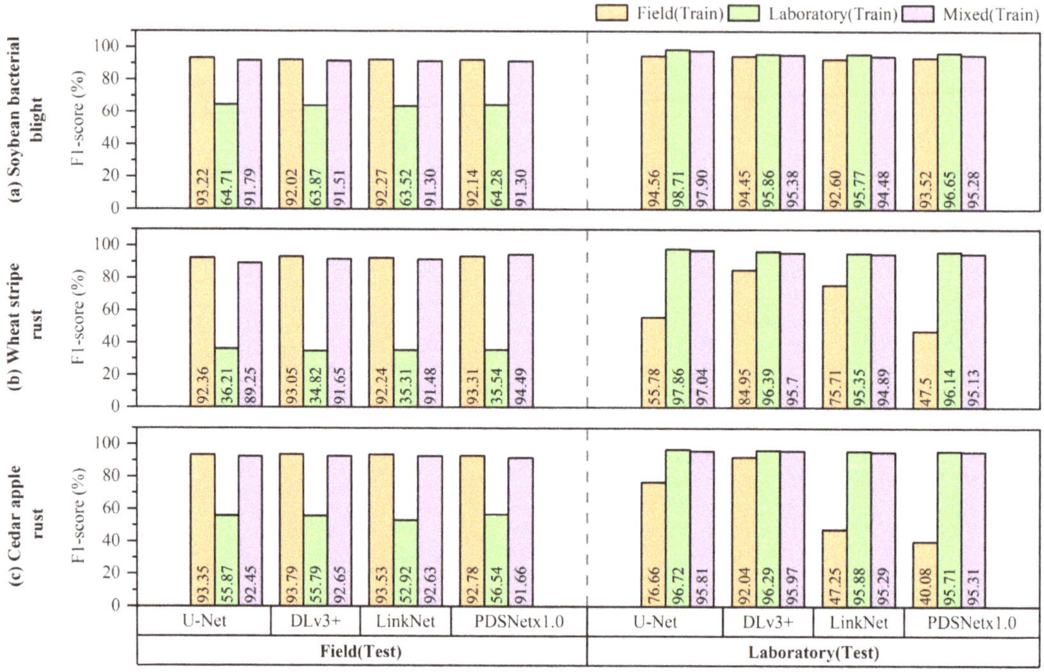

Figure 6. Test results of the networks' generalization ability under field and laboratory scenarios. DLv3+ is the abbreviation for DeepLabv3+(ResNet-18).

3.5. Severity Estimation

To evaluate the severity estimation capability of the networks under field scenarios, we calculated the reference severity of the labels and the predicted severity of the networks. Severity refers to the proportion of a leaf occupied by disease spots, calculated using Equation (1).

As shown in Figure 7, the average R^2 and RMSE values of SBB are 0.888 and 2.942, respectively. DeepLabV3+ achieved the best estimation results, but the differences among these networks were not significant. According to the regression lines, most networks overestimated the severity. As shown in Figure 8, the average R^2 and RMSE values of WSR are 0.966 and 2.711, respectively. Like SBB, DeepLabV3+ achieved the highest estimation performance, but the differences among the networks were not significant. Most models overestimated severity, while U-Net underestimated it. As shown in Figure 9, the average R^2 and RMSE values of CAR are 0.921 and 0.331, respectively. LinkNet achieved the best estimation results with little difference among the networks. Most networks underestimated the severity.

In summary, without considering the number of parameters and prediction speed, DeepLabv3+(ResNet-18) demonstrated the best estimation performance. The proposed PDSNetx0.5, with its parameter count and prediction speed being 0.175 times and 12.79 times that of DeepLabv3+, respectively, only saw a reduction in R^2 for the three diseases by 0.069 for SBB, 0.026 for WSR, and 0.029 for CAR. Therefore, PDSNetx0.5 achieved a good balance between prediction accuracy and speed in severity estimation.

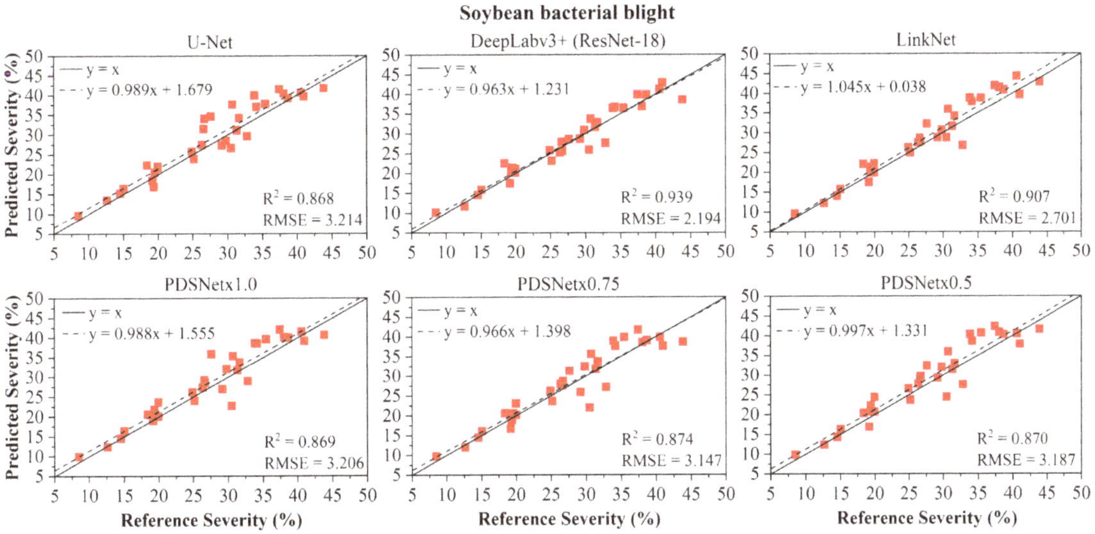

Figure 7. The severity estimation results of soybean bacterial blight. The dashed line is the linear regression line, and the solid line is the best fit line.

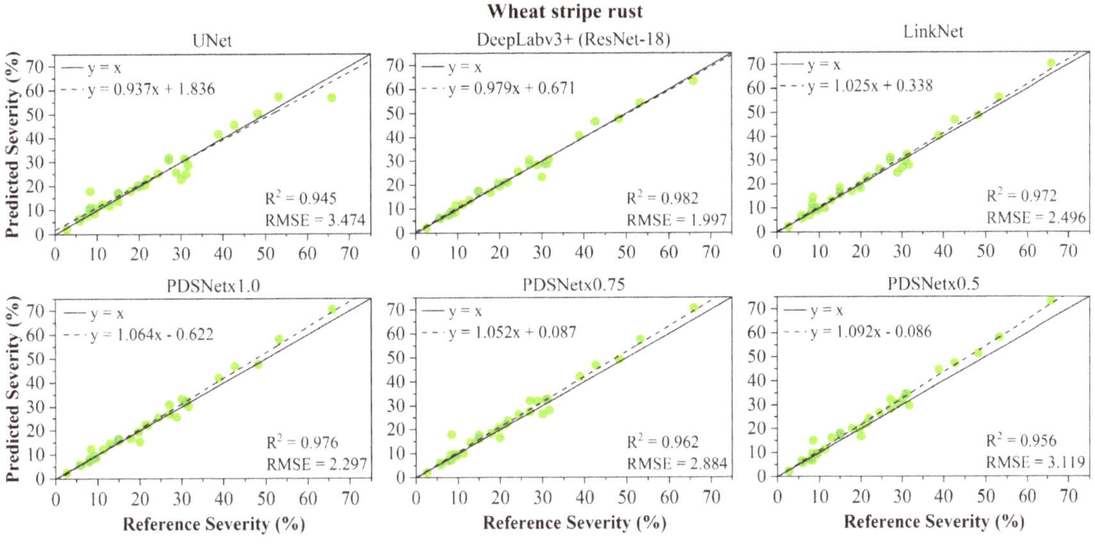

Figure 8. The severity estimation results of wheat stripe rust. The dashed line is the linear regression line, and the solid line is the best fit line.

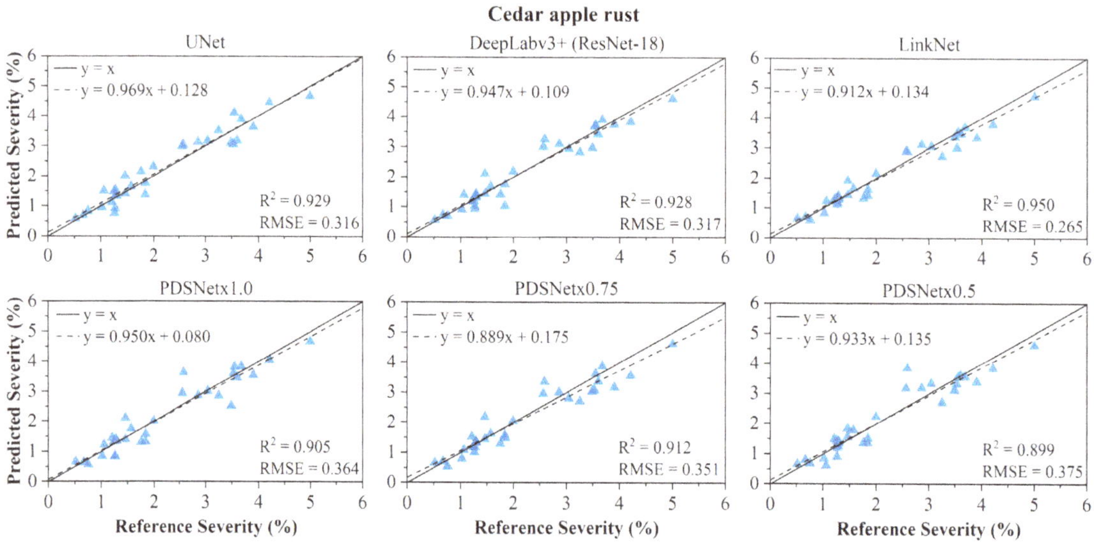

Figure 9. The severity estimation results of cedar apple rust. The dashed line is the linear regression line, and the solid line is the best fit line.

4. Discussion

4.1. Lightweight Networks

In this study, we introduced the PDSNets family, built upon the LinkNet architecture, which employs ResNet-18 as the encoder and incorporates a 4 × 4 embedding layer to enhance the prediction speed by 50%. As a flagship model of the PDSNets family, PDSNetx0.5 demonstrates exceptional performance in segmenting multiple diseases, achieving an overall *F*1 score of 91.96% and an Intersection over Union (*IoU*) of 85.85%, alongside a coefficient of determination (R^2) of 0.908 for disease severity evaluation. With a compact parameter count of only 2.9 M, PDSNetx0.5 represents just 25.1% and 9.33% of the parameter sizes of LinkNet and U-Net, respectively. In contrast to existing lightweight network research [21,22,24], PDSNetx0.5 has been rigorously tested on diverse disease images, showcasing superior generalization capabilities. Furthermore, the implementation of the weighted cross-entropy loss function has been shown to significantly improve the network's ability to capture lesion characteristics [18].

4.2. Limitations

The primary limitation of PDSNetx0.5 stems from the reduction in network parameters, resulting in a decline in prediction accuracy. To address this, it is essential to incorporate advanced deep learning techniques, such as attention mechanisms [22] and knowledge distillation [35], which can significantly enhance the accuracy of lightweight networks. Additionally, the limited sample size, particularly the absence of data from multi-leaf and complex lighting scenarios, constrains the network's generalization capabilities. Moreover, the model has not been deployed on mobile platforms for real-world testing, further limiting its practical applicability.

The experimental findings of this study demonstrate that networks trained in laboratory scenarios cannot be directly utilized for field scenario predictions, and the opposite is also true. This observation holds across multiple mainstream architectures, including U-Net, DeepLabV3+, and LinkNet. These results underscore the critical importance of developing field scenario datasets to advance research in field applications.

4.3. Future Work

In future studies, we will prioritize the following research directions: (1) integrating advanced deep learning techniques, including attention mechanisms and knowledge distillation, to further enhance the accuracy of PDSNetx0.5; (2) expanding the scale and diversity of the field scenario dataset by increasing the sample size and incorporating additional disease types, thereby improving the network's generalization capabilities; and (3) deploying PDSNetx0.5 on mobile platforms for real-world testing to evaluate its practical performance. Overall, the end-to-end design enables farmers to utilize the tool effectively without specialized technical expertise. Furthermore, the automated disease severity assessment approach not only significantly boosts the efficiency of field disease surveys conducted by farmers and experts but also offers robust support for optimizing agricultural production and disease management strategies.

5. Conclusions

In conclusion, this study introduced PDSNets, a family of lightweight semantic segmentation networks, with PDSNetx0.5 achieving remarkable performance in disease segmentation tasks. With only 2.9 M parameters, PDSNetx0.5 surpasses mainstream networks like LinkNet and U-Net in efficiency. The experimental findings reveal that networks trained in laboratory settings cannot be directly utilized in field scenarios, and the opposite is also true, underscoring the critical need for high-quality field datasets.

Author Contributions: Conceptualization C.Z. and C.L.; methodology, C.Z.; software, C.Z.; validation, Xin Wang.; investigation, C.Z., X.W. (Xin Wang), X.W. (Xifang Wu) and Y.D.; resources, C.Z. and C.L.; data curation, C.Z.; writing—original draft preparation, C.Z.; writing—review and editing, C.L.; visualization, C.Z. and Xin Wang; supervision, C.L., H.X. and Y.J.; funding acquisition, C.L., T.C. and H.C. All authors have read and agreed to the published version of the manuscript.

Funding: This research was funded by the Fundamental Research Funds for the Universities of Henan Province, grant number 242300420221; the Fundamental Research Funds for the Universities of Henan Province, grant number 242300421365; and the National Major Scientific Research Achievement Cultivation Fund, grant number NSFRF240101.

Institutional Review Board Statement: Not applicable.

Data Availability Statement: The original data presented in this study are openly available in Kaggle at https://doi.org/10.34740/kaggle/ds/6620728, accessed on 9 March 2025.

Acknowledgments: The authors are thankful for the support provided by the Jiaozuo Academy of Agriculture and Forestry Sciences (http://www.jznlkxy.com/, accessed on 9 March 2025).

Conflicts of Interest: The authors declare no conflicts of interest.

Abbreviations

The following abbreviations are used in this manuscript:

LAB-A	A-channel of LAB color space
MGRVI	Modified green–red vegetation index
CNN(s)	Conventional neural network(s)
SBB	Soybean bacterial blight
WSR	Wheat stripe rust
CAR	Cedar apple rust
B	Background
H	Healthy leaf
S	Disease spots

CPU	Central processing unit
GPU	Graph processing unit
FPS	Frame per second
GELU	Gaussian error linear unit
CE-Loss	Cross-entropy loss
TP	True positive
FP	False positive
FN	False negative
IoU	Interaction over union
LR	Learning rate

References

1. Savary, S.; Willocquet, L.; Pethybridge, S.J.; Esker, P.; McRoberts, N.; Nelson, A. The Global Burden of Pathogens and Pests on Major Food Crops. *Nat. Ecol. Evol.* **2019**, *3*, 430–439. [CrossRef] [PubMed]
2. Bock, C.H.; Chiang KuoSzu, C.K.; Del Ponte, E.M. Accuracy of Plant Specimen Disease Severity Estimates: Concepts, History, Methods, Ramifications and Challenges for the Future. *CABI Rev.* **2016**, 1–21. [CrossRef]
3. Shi, T.; Liu, Y.; Zheng, X.; Hu, K.; Huang, H.; Liu, H.; Huang, H. Recent Advances in Plant Disease Severity Assessment Using Convolutional Neural Networks. *Sci. Rep.* **2023**, *13*, 2336. [CrossRef]
4. Barbedo, J.G.A. An Automatic Method to Detect and Measure Leaf Disease Symptoms Using Digital Image Processing. *Plant Dis.* **2014**, *98*, 1709–1716. [CrossRef] [PubMed]
5. Bai, X.; Li, X.; Fu, Z.; Lv, X.; Zhang, L. A Fuzzy Clustering Segmentation Method Based on Neighborhood Grayscale Information for Defining Cucumber Leaf Spot Disease Images. *Comput. Electron. Agric.* **2017**, *136*, 157–165. [CrossRef]
6. Sibiya, M.; Sumbwanyambe, M. An Algorithm for Severity Estimation of Plant Leaf Diseases by the Use of Colour Threshold Image Segmentation and Fuzzy Logic Inference: A Proposed Algorithm to Update a "Leaf Doctor" Application. *AgriEngineering* **2019**, *1*, 205–219. [CrossRef]
7. Simonyan, K.; Zisserman, A. Very Deep Convolutional Networks for Large-Scale Image Recognition. *arXiv* **2015**, arXiv:1409.1556. [CrossRef]
8. He, K.; Zhang, X.; Ren, S.; Sun, J. Deep Residual Learning for Image Recognition. In Proceedings of the 2016 IEEE Conference on Computer Vision and Pattern Recognition (CVPR), Las Vegas, NV, USA, 27–30 June 2016; IEEE: Piscataway, NJ, USA; pp. 770–778. [CrossRef]
9. Huang, G.; Liu, Z.; van der Maaten, L.; Weinberger, K.Q. Densely Connected Convolutional Networks. In Proceedings of the IEEE Conference on Computer Vision and Pattern Recognition, Salt Lake City, UT, USA, 18–22 June 2018.
10. Long, M.; Hartley, M.; Morris, R.J.; Brown, J.K.M. Classification of Wheat Diseases Using Deep Learning Networks with Field and Glasshouse Images. *Plant Pathol.* **2023**, *72*, 536–547. [CrossRef]
11. Bevers, N.; Sikora, E.J.; Hardy, N.B. Soybean Disease Identification Using Original Field Images and Transfer Learning with Convolutional Neural Networks. *Comput. Electron. Agric.* **2022**, *203*, 107449. [CrossRef]
12. Lin, J.; Zhang, X.; Qin, Y.; Yang, S.; Wen, X.; Cernava, T.; Migheli, Q.; Chen, X. Local and Global Feature-Aware Dual-Branch Networks for Plant Disease Recognition. *Plant Phenomics* **2024**, *6*, 0208. [CrossRef]
13. Mi, Z.; Zhang, X.; Su, J.; Han, D.; Su, B. Wheat Stripe Rust Grading by Deep Learning With Attention Mechanism and Images From Mobile Devices. *Front. Plant Sci.* **2020**, *11*, 558126. [CrossRef] [PubMed]
14. Joshi, R.C.; Kaushik, M.; Dutta, M.K.; Srivastava, A.; Choudhary, N. VirLeafNet: Automatic Analysis and Viral Disease Diagnosis Using Deep-Learning in Vigna Mungo Plant. *Ecol. Inform.* **2021**, *61*, 101197. [CrossRef]
15. Yin, C.; Zeng, T.; Zhang, H.; Fu, W.; Wang, L.; Yao, S. Maize Small Leaf Spot Classification Based on Improved Deep Convolutional Neural Networks with a Multi-Scale Attention Mechanism. *Agronomy* **2022**, *12*, 906. [CrossRef]
16. Lin, K.; Gong, L.; Huang, Y.; Liu, C.; Pan, J. Deep Learning-Based Segmentation and Quantification of Cucumber Powdery Mildew Using Convolutional Neural Network. *Front. Plant Sci.* **2019**, *10*, 155. [CrossRef]
17. Ronneberger, O.; Fischer, P.; Brox, T. U-Net: Convolutional Networks for Biomedical Image Segmentation. In Proceedings of the Medical Image Computing and Computer-Assisted Intervention—MICCAI 2015, Munich, Germany, 5–9 October 2015; Navab, N., Hornegger, J., Wells, W.M., Frangi, A.F., Eds.; Springer International Publishing: Cham, Switzerland, 2015; pp. 234–241. [CrossRef]
18. Li, K.; Zhang, L.; Li, B.; Li, S.; Ma, J. Attention-Optimized DeepLab V3 + for Automatic Estimation of Cucumber Disease Severity. *Plant Methods* **2022**, *18*, 109. [CrossRef]

19. Chen, L.-C.; Zhu, Y.; Papandreou, G.; Schroff, F.; Adam, H. Encoder-Decoder with Atrous Separable Convolution for Semantic Image Segmentation. In Proceedings of the Computer Vision—ECCV 2018, Munich, Germany, 8–14 September 2018; Ferrari, V., Hebert, M., Sminchisescu, C., Weiss, Y., Eds.; Springer International Publishing: Cham, Switzerland, 2018; pp. 833–851. [CrossRef]
20. Wang, C.; Du, P.; Wu, H.; Li, J.; Zhao, C.; Zhu, H. A Cucumber Leaf Disease Severity Classification Method Based on the Fusion of DeepLabV3+ and U-Net. *Comput. Electron. Agric.* **2021**, *189*, 106373. [CrossRef]
21. Feng, C.; Jiang, M.; Huang, Q.; Zeng, L.; Zhang, C.; Fan, Y. A Lightweight Real-Time Rice Blast Disease Segmentation Method Based on DFFANet. *Agriculture* **2022**, *12*, 1543. [CrossRef]
22. Shi, L.; Liu, Z.; Yang, C.; Lei, J.; Wang, Q.; Yin, F.; Wang, J. Lightweight U-Net-Based Method for Estimating the Severity of Wheat Fusarium Head Blight. *Agriculture* **2024**, *14*, 938. [CrossRef]
23. Howard, A.; Sandler, M.; Chen, B.; Wang, W.; Chen, L.-C.; Tan, M.; Chu, G.; Vasudevan, V.; Zhu, Y.; Pang, R.; et al. Searching for MobileNetV3. In Proceedings of the 2019 IEEE/CVF International Conference on Computer Vision (ICCV), Seoul, Republic of Korea, 27 October–2 November 2019; pp. 1314–1324. [CrossRef]
24. Xu, Y.; Ma, B.; Yu, G.; Zhang, R.; Tan, H.; Dong, F.; Bian, H. Accurate Cotton Verticillium Wilt Segmentation in Field Background Based on the Two-Stage Lightweight DeepLabV3+model. *Comput. Electron. Agric.* **2025**, *229*, 109814. [CrossRef]
25. Komzsik, L. *The Lanczos Method: Evolution and Application*; Society for Industrial and Applied Mathematics: Philadelphia, PA, USA, 2003; ISBN 978-0-89871-537-8.
26. Thapa, R.; Zhang, K.; Snavely, N.; Belongie, S.; Khan, A. The Plant Pathology Challenge 2020 Data Set to Classify Foliar Disease of Apples. *Appl. Plant Sci.* **2020**, *8*, e11390. [CrossRef]
27. Wada, K. Labelme: Image Polygonal Annotation with Python. Available online: https://github.com/wkentaro/labelme (accessed on 11 February 2025).
28. Otsu, N. A Threshold Selection Method from Gray-Level Histograms. *IEEE Trans. Syst. Man Cybern.* **1979**, *9*, 62–66. [CrossRef]
29. Bendig, J.; Yu, K.; Aasen, H.; Bolten, A.; Bennertz, S.; Broscheit, J.; Gnyp, M.L.; Bareth, G. Combining UAV-Based Plant Height from Crop Surface Models, Visible, and near Infrared Vegetation Indices for Biomass Monitoring in Barley. *Int. J. Appl. Earth Obs. Geoinf.* **2015**, *39*, 79–87. [CrossRef]
30. Hendrycks, D.; Gimpel, K. Gaussian Error Linear Units (GELUs). *arXiv* **2023**. [CrossRef]
31. Zeiler, M.D.; Krishnan, D.; Taylor, G.W.; Fergus, R. Deconvolutional Networks. In Proceedings of the 2010 IEEE Computer Society Conference on Computer Vision and Pattern Recognition, San Francisco, CA, USA, 13–18 June 2010; IEEE: Piscataway, NJ, USA; pp. 2528–2535. [CrossRef]
32. Loshchilov, I.; Hutter, F. Decoupled Weight Decay Regularization. *arXiv* **2019**, arXiv:1711.05101. [CrossRef]
33. Cubuk, E.D.; Zoph, B.; Mane, D.; Vasudevan, V.; Le, Q.V. AutoAugment: Learning Augmentation Strategies from Data. In Proceedings of the 2019 IEEE/CVF Conference on Computer Vision and Pattern Recognition (CVPR), Long Beach, CA, USA, 15–20 June 2019; IEEE: Piscataway, NJ, USA; pp. 113–123. [CrossRef]
34. Gonçalves, J.P.; Pinto, F.A.C.; Queiroz, D.M.; Villar, F.M.M.; Barbedo, J.G.A.; Del Ponte, E.M. Deep Learning Architectures for Semantic Segmentation and Automatic Estimation of Severity of Foliar Symptoms Caused by Diseases or Pests. *Biosyst. Eng.* **2021**, *210*, 129–142. [CrossRef]
35. Huang, Q.; Wu, X.; Wang, Q.; Dong, X.; Qin, Y.; Wu, X.; Gao, Y.; Hao, G. Knowledge Distillation Facilitates the Lightweight and Efficient Plant Diseases Detection Model. *Plant Phenomics* **2023**, *5*, 62. [CrossRef]

agriculture

MDPI

Article

DHS-YOLO: Enhanced Detection of Slender Wheat Seedlings Under Dynamic Illumination Conditions

Xuhua Dong [1,2,*] and Jingbang Pan [3]

1 Key Laboratory of Industrial Control Technology, Institute of Cyber-Systems and Control, Zhejiang University, Hangzhou 310058, China
2 Huzhou Institute, Zhejiang University, Huzhou 313000, China
3 Department of Architecture and Civil Engineering, University of Bath, Bath BA2 7AY, UK; jp2259@bath.ac.uk
* Correspondence: xuhuadong@zju.edu.cn

Abstract: The precise identification of wheat seedlings in unmanned aerial vehicle (UAV) imagery is fundamental for implementing precision agricultural practices such as targeted pesticide application and irrigation management. This detection task presents significant technical challenges due to two inherent complexities: (1) environmental interference from variable illumination conditions and (2) morphological characteristics of wheat seedlings characterized by slender leaf structures and flexible posture variations. To address these challenges, we propose DHS-YOLO, a novel deep learning framework optimized for robust wheat seedling detection under diverse illumination intensities. Our methodology builds upon the YOLOv11 architecture with three principal enhancements: First, the Dynamic Slender Convolution (DSC) module employs deformable convolutions to adaptively capture the elongated morphological features of wheat leaves. Second, the Histogram Transformer (HT) module integrates a dynamic-range spatial attention mechanism to mitigate illumination-induced image degradation. Third, we implement the ShapeIoU loss function that prioritizes geometric consistency between predicted and ground truth bounding boxes, particularly optimizing for slender plant structures. The experimental validation was conducted using a custom UAV-captured dataset containing wheat seedling images under varying illumination conditions. Compared to the existing models, the proposed model achieved the best performance with precision, recall, mAP50, and mAP50-95 values of 94.1%, 91.0%, 95.2%, and 81.9%, respectively. These results demonstrate our model's effectiveness in overcoming illumination variations while maintaining high sensitivity to fine plant structures. This research contributes an optimized computer vision solution for precision agriculture applications, particularly enabling automated field management systems through reliable crop detection in challenging environmental conditions.

Keywords: dynamic slender convolution; histogram transformer; ShapeIoU; illumination-induced degeneration

1. Introduction

The rapid and precise identification of wheat seedlings in the field is crucial for achieving precise pesticide spraying or irrigation [1]. As a key data source for precision agriculture, unmanned aerial vehicle (UAV) remote sensing imagery plays a vital role in crop protection, quality assessment, and plant seedling identification [2–5]. Nowadays, acquiring RGB image data from wheat seedling fields using UAV remote sensing and developing automated identification models are common technological approaches. However, the task of identifying wheat seedlings faces challenges due to complex adverse factors, such as varying illumination conditions and the slender and flexible structure of the seedlings themselves.

Illumination intensity is a crucial factor that affects UAV remote sensing imaging. Both excessively low and high illumination can severely degrade visual quality, thereby impacting target detection performance. Periodic monitoring of wheat seedlings is often hindered by poor image quality due to these varying illumination conditions. Therefore, processing images affected by illumination variations is essential to improve the accuracy of wheat seedling identification.

Currently, many effective object detection methods have been developed to handle low-illumination environments. Traditional techniques, such as histogram equalization and filtering, enhance the contrast and brightness of low-light images, thereby improving detection performance [6,7]. However, owing to the relatively fixed algorithm framework, these traditional techniques often fail to achieve the expected enhancement effect in certain aspects. During the process of image enhancement, they cannot take into account the detailed and color information of the image itself. Meanwhile, when dealing with complex images under different lighting conditions, manual parameter adjustment is required, which leads to fewer practical applications in multiple scenarios. oHHowAlternatively, deep learning-based approaches that combine image enhancement and object detection networks [8] have gained traction. These can be broadly categorized into CNN-based [9] and transformer-based models [10]. CNN-based methods typically first enhance the image quality and then detect objects. The CNN model offers strong feature extraction capabilities to some extent. However, due to the difficulty in capturing global and contextual information, the relationship between colors and details cannot be coordinated well, which makes them less adaptive to varying illumination conditions. In contrast, the transformer-based method has been investigated for adverse weather removal tasks, exceeding the effectiveness of CNNs, such as TransWeather [11], Retinexformer [12], MSATr [13], and so on. Outdoor detection in agricultural environments of varying illumination (including low illumination, high illumination, and non-uniform illumination) may need a more refined transformer framework, improving the ability to reduce color deviation and noise interference and better balancing the differentiated illumination. Meanwhile, these transformer-based approaches normally make concessions regarding efficient memory utilization by limiting self-attention operations to a predetermined spatial range or only within the channel dimension. This compromise impedes the intrinsic capability of transformers, initially intended for enhanced global feature modeling, hence resulting in a decline in restoration performance and a limitation of feature extraction in varying illumination-induced degradation. This paper aims to develop a transformer-based module to overcome feature extraction limitations and achieve adaptive feature extraction for image degradation caused by varying lighting intensities, which would help to eliminate degradation effects and noise and enhance wheat seedling identification accuracy.

Another challenge is the feature extraction of the slender and flexible structure of wheat seedlings themselves. In UAV remote sensing images, the flexible and slender structural characteristics of the wheat seedlings are amplified, making them appear even more elongated. Therefore, there is a critical need to enhance the perception of this kind of structure, but perception is difficult owing to the following challenges. First, these slender and thin local structures account for only a relatively small part of the overall image, with limited pixel composition, making it sensitive to interference from complex backgrounds, which complicates precise detection. Second, the flexible global morphology causes targets to intertwine with each other, making the identification of individual targets more complex and challenging.

Recently, numerous studies have advocated for the integration of domain-specific knowledge, such as geometric topology and tree structures, to enhance the model's ability to accurately perceive the unique features of slender and flexible structures. This approach aims to improve the precision of local perception while preserving the continuity of global morphology. The previous research can be classified into three directions. First, network-based methods develop specific network architectures based on the characteristics of the slender structures that guide the model to focus on key features [14,15]. However, the network may miss the perception of the slender structures due to the small proportion of structures. Second, the feature-based methods can provide more feature representations to the model to improve understanding of the specific geometric characteristics of slender structures [16]. Nevertheless, certain redundant feature representations increase the computational load without improving the model. Third, the loss-based methods use the typical loss function to supplement constraints during the training process [17]. For the wheat seedling detection task in the remote sensing images, we should consider the slender morphology of leaf structures more and supplement the free learning process with constraints that allow for targeted enhancement of the perception of wheat seedling structures.

To tackle the above obstacles, including the interference of different illumination intensities and the slender structure of wheat seedlings simultaneously, we proposed a novel object detection model, DHS-YOLO, involving a Dynamic Slender Convolution (DSC), Histogram Transformer (HT), and ShapeIoU loss function. To address the challenge of different image degradation in a complete illumination period, the HT module is proposed to extract intricate features and capture dynamically distributed degradation factors, which improve the accuracy of wheat seedling detection in illumination-induced adverse environments. To address the challenge of the flexible and slender structure of wheat seedlings, which is difficult for the model to focus on, the DSC module is developed to enhance the perception of the geometric structure by adaptively focusing on the slender and flexible local features of the leaf-shaped structure of wheat seedlings. The main contributions of this paper are as follows:

1. The remote sensing image dataset of wheat seedlings is collected by a UAV under different illumination intensities.
2. A Dynamic Slender Convolution (DSC) module is developed to enhance the perception of the geometric structure by adaptively focusing on the slender and flexible local features of the leaf-shaped structure of wheat seedlings.
3. A novel Histogram Transformer (HT) module is introduced for unified adverse illumination removal, utilizing dynamic-range spatial attention to address illumination-induced degradation.
4. A ShapeIoU loss function is proposed to emphasize the shape and scale of the bounding box itself and make it more precise.
5. The DHS-YOLO model is developed to detect wheat seedlings under different illumination intensities, which adopts the YOLOv11 as the basics and consists of a DSC module, HT module, and ShapIoU loss.
6. DHS-YOLO is trained by the collected dataset and validated using an ablation experiment and comparison experiment.

2. Materials and Methods

2.1. Data Acquisition

The RGB image dataset acquisition of wheat seedlings was performed by a consumer-grade DJI Mavic 3E UAV (DJI, Shenzhen, China) with an RGB digital zoomable camera with a resolution of 4000 × 3000 pixels. The UAV flight tasks were conducted in the two-leaf period of wheat seedlings and under different illumination conditions to capture the wheat seedling images with illumination intensities (II) at five levels, from weak to strong, plus the effect of shadows, as shown in Figure 1. The image acquisition based on UAVs was carried out from November to December 2023, generating a dataset of 270 images covering six types of different illumination scenarios.

Figure 1. Wheat seedling dataset at the six different illumination intensities (II), i.e., (**a**) II level #1 (II $\leq$ 0.2k Lux), (**b**) II level #2 (0.2k Lux < II $\leq$ 0.5k Lux), (**c**) II level #3 (0.5k Lux < II $\leq$ 10k Lux), (**d**) II level #4 (10k Lux < II $\leq$ 40k Lux), (**e**) II level #5 (40Lux < II $\leq$ 100k Lux), and (**f**) II level #6-Shadow (10k < II $\leq$ 100k, with over 50% shadow area), where Lux is the universal unit of II and k means 1000. The image data in (**a**,**b**) were collected when the sky was overcast, while (**c**) was collected when it was cloudy. The image data in (**d–f**) were collected on a sunny day, with a lot of shadows in (**f**).

The dataset labels were generated in a semi-automatic manner using the LabelImg tool [18]. Initially, the processing in LabelImg was conducted on 10% of all images with the manual labeling of object polygons. The resulting real class and position of the wheat seedlings in each image were stored in a TXT file in the format of [class ID, x coordinate of center, y coordinate of center, width, height]. Subsequently, a YOLOv11 [19] model was trained by these manually labeled images to predict and generate approximate annotations for the remaining images in the dataset. Finally, the generated annotations were corrected manually using the LabelImg tool to ensure accuracy.

Due to the large size of the original UAV images, the images could not be directly imported into the deep learning model for training. Therefore, after labeling the UAV images, both the UAV images and their corresponding label files were cropped to improve training efficiency. By keeping an overlap of 20%, each original UAV image was cropped along the x and y directions, resulting in 12 sub-images with a resolution of 1200 $\times$ 1200 pixels generated from the original image with a resolution of 4000 $\times$ 3000 pixels. Then, the resize operation was conducted to reduce the image size from 1200 × 1200 to 640 × 640. The final dataset only contained one class named "Seedlings", randomly divided into training, validation, and prediction sets, comprising 70%, 10%, and 20% of images from each II level, respectively. The specific dataset configuration is shown in Table 1.

Table 1. Dataset configuration.

Number of Images Collected by UAV		Number of Cropped Images	Number of Instances	Resizing	Training	Validation	Prediction
II level #1	46	552	7728	640 × 640	386	55	111
II level #2	65	780	8580		546	78	156
II level #3	35	420	6720		294	42	84
II level #4	48	576	4968		403	58	115
II level #5	53	636	6996		445	64	127
II level #6-Shadow	23	276	1656		193	28	55
Total	270	3240	36,648	-	2267	325	648

2.2. Proposed Method

YOLOv11 is the latest version of the YOLO family of real-time high-performance object detectors. Compared to the previous version, YOLOv11 offers several advances in feature extraction, accuracy, speed, and other aspects so that it can be better applied in complex scenes. Therefore, we adopt YOLOv11 as the base model of DHS-YOLO. In this section, we introduce the architecture and components of the developed DHS-YOLO module, which includes two modules, i.e., Dynamic Slender Convolution (DSC) and Histogram Transformer (HT), and one loss function, i.e., ShapeIoU loss. The DSC module enhances the extraction of the characteristics of slender morphology. The HT module is a type of histogram self-attention for the features in an adverse environment, which focuses on spatial features of dynamic range and processes similar degraded pixels of the long-range together. ShapeIoU loss pays more attention to the shape and scale of the bounding box itself, which makes the bounding box regression more accurate.

2.2.1. Overview of DHS-YOLO

DHS-YOLO is a detector for the remote sensing of slender objects in adverse environments. As shown in Figure 2, the model consists of three parts: backbone, neck, and head. The backbone acts as the core component of the model, especially designed to extract multi-scale features from input images. The backbone is mainly structured by the CBS module (including 2D Convolution, Batch Normalization, and SiLU activation function), the DSC block, the SPPF module, and the HT block. The CBS module, as the initial convolutional layer, processes input images to extract fundamental low-level features, laying the foundation for subsequent high-level feature extraction. The DSC blocks emphasize and extract the flexible and slender features of wheat seedlings. The SPPF module utilizes pooling operations at multiple scales to improve the model's ability to capture global context through multi-scale feature representations. The HT block effectively filters out irrelevant information from the adverse environment and enhances detection capabilities in complex wheat seedling fields.

The neck serves as the intermediate layer connecting the backbone and head, enhancing wheat seedling detection by merging multi-scale features, which is particularly advantageous for detecting wheat seedlings of varying sizes. The neck includes Upsample operation, Concat (Feature Concatenation) operation, CBS modules, and DSC blocks. The Upsample module enhances the resolution of low-resolution feature maps, similar to zooming in on a blurry section of an image, which improves the detection of smaller objects. The Concat module merges feature maps with different resolutions, similar to overlaying transparent layers with varying details, allowing the model to utilize multi-level information and improve detection performance. The DSC blocks reused in the neck part integrate features across different scales through repeated Dynamic Slender Convolutional operations and shortcut connections, which enhances feature representations and improves detection accuracy.

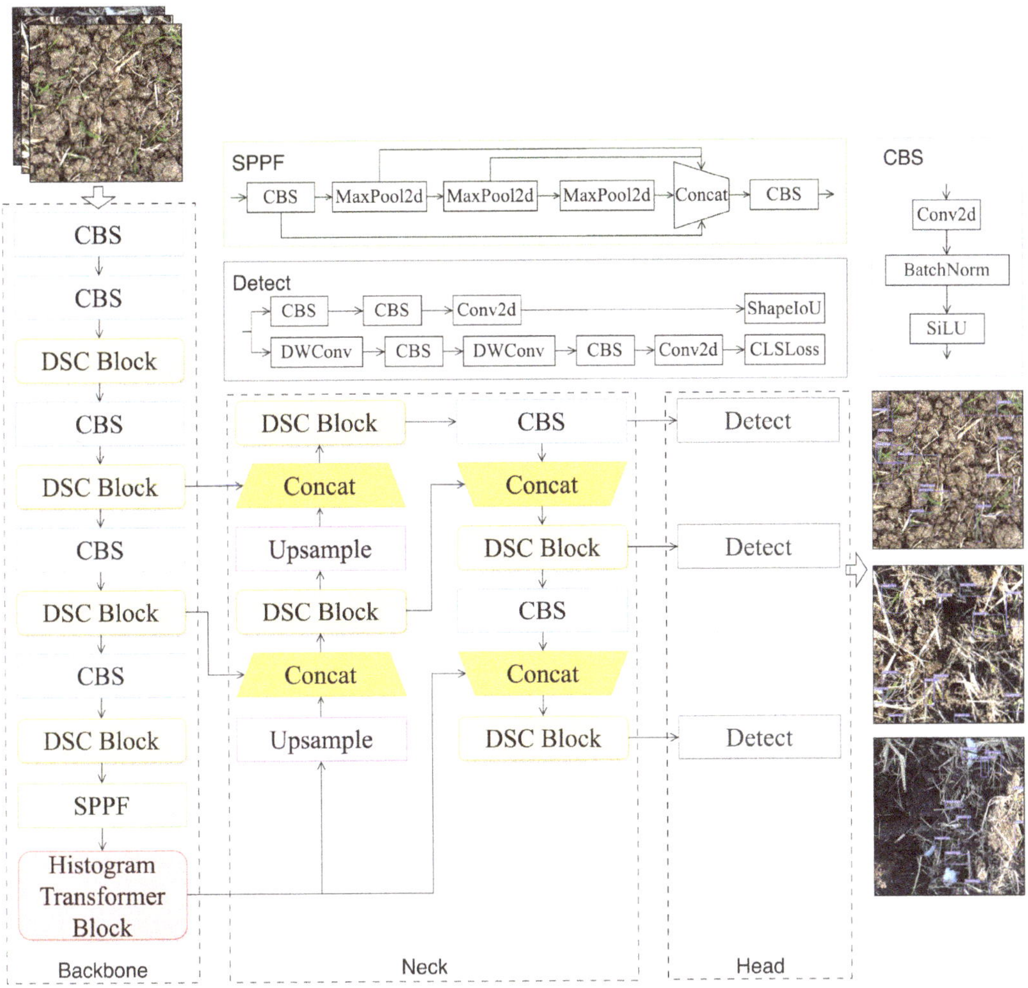

Figure 2. Overall structure of DHS-YOLO.

The head, as the final stage of DHS-YOLO, converts the processed feature maps into accurate object detection outputs. To precisely recognize both large and small objects, the detection module processes feature maps across different sizes, much like zooming in and out of an image. To ensure consistent performance in a variety of situations, this multi-scale capability is essential for recognizing objects of different sizes, from massive machines to smaller items. The head is inspired by YOLOv11. Except for the CBS modules, the detection head contains two depth-wise separable convolution (DWConv) modules, which significantly lower the model's parameters and computing needs while preserving excellent accuracy and accelerating processing.

2.2.2. Dynamic Slender Convolution (DSC) Module

Due to the flexible and slender morphology of the structure of wheat seedlings in the complex field, the extracted features by conventional convolution are insufficient and inaccurate. Therefore, it is crucial to develop a special convolution to fully focus on and utilize the key features. In this section, we discuss how to perform the Dynamic Slender Convolution (DSC) module to extract the local features of flexible and slender structures.

Given the standard 2D Convolution kernel coordinates as K, the central coordinate is defined as $K_i = (x_i, y_i)$. A conventional kernel K_{ij} of 3×3 with dilation 1 is described as:

$$K = \{(x-1, y-1), (x-1, y), \ldots, (x+1, y+1)\}, \tag{1}$$

To provide the convolution kernel with greater flexibility in focusing on complex geometric features of the wheat seedlings, the deformation offsets Δ are crucial. However, if the model is allowed to freely learn the deformation offsets, the perceptual field may shift outside the objects, particularly in the case of thin and slender structures. Therefore, an iterative strategy is employed (Figure 3), sequentially selecting the next position to observe for each target. This approach ensures continuity of attention and prevents the perceptual field from expanding too far due to large deformation offsets.

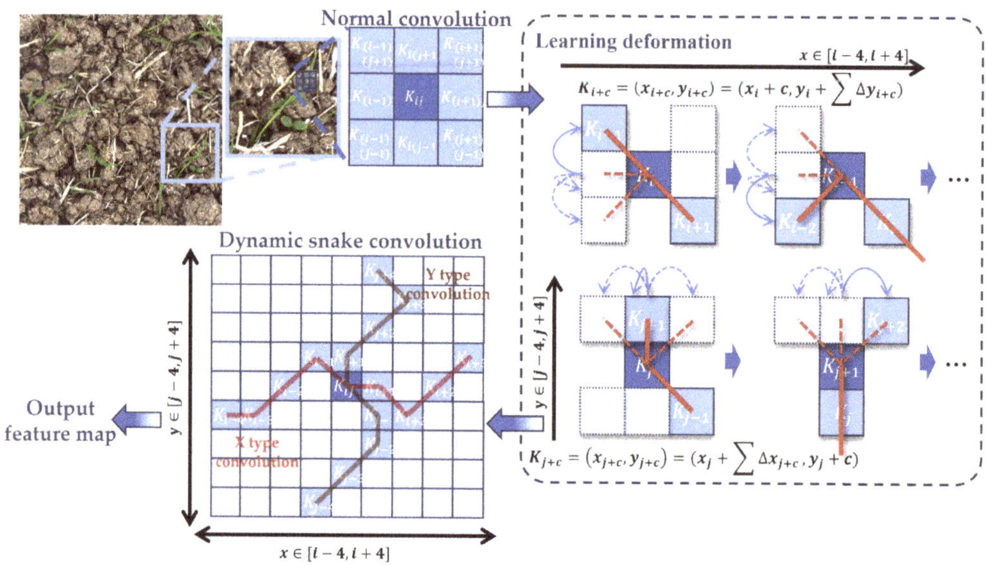

Figure 3. Dynamic Slender Convolution (DSC) module, which learns the deformation according to the input feature map, adaptively focuses on the slender local features of the wheat leaves under the knowledge of the slender structure morphology.

In the DSC module, the standard convolution kernel is straightened in the direction of both the x-axis and y-axis. A convolution kernel of size 9 is designed. For example, in the x-axis direction, the specific position of each grid in K is described as $K_{i\pm c} = (x_{i\pm c}, y_{i\pm c})$, where $c = \{0, 1, 2, 3, 4\}$ represents the horizontal distance from the central grid. The decision of each grid position $K_{i\pm c}$ in the kernel K of convolution is an iterative and cumulative process. Beginning from the center grid K_i, the position of each subsequent grid is determined by the location of the previous grid. K_{i+1} is deduced through an offset $\Delta = \{\delta | \delta \in [-1, 1]\}$ compared to K_i. Then, the offsets need to be accumulated to ensure that the convolution kernel adheres to a slender morphological structure. The kernel in the direction of the x-axis is described as:

$$K_{i\pm c} = \begin{cases} (x_{i+c}, y_{i+c}) = (x_i + c, y_i + \sum_i^{i+c} \Delta y) \\ (x_{i-c}, y_{i-c}) = (x_i - c, y_i + \sum_{i-c}^{i} \Delta y) \end{cases}, \tag{2}$$

In the direction of the y-axis, it is described as:

$$K_{j\pm c} = \begin{cases} (x_{j+c}, y_{j+c}) = (x_i + \sum_j^{j+c} \Delta x, y_j + c) \\ (x_{j-c}, y_{j-c}) = (x_j + \sum_{j-c}^{j} \Delta x, y_j - c) \end{cases}, \tag{3}$$

As the offset Δ is typically fractional, bilinear interpolation is implemented as:

$$K = \sum_{K'} B(K', K) \cdot K', \tag{4}$$

where K is a fractional location for Equations (2) and (3) and B represents the bilinear interpolation kernel, separated into two 1D kernels as:

$$B(K, K') = b(K_x, K'_x) \cdot b(K_y, K'_y). \tag{5}$$

As shown in Figure 3, the output of the DSC module covers a perceptual field with a 9×9 range after the deformation process due to the adjustment in two directions (x-axis and y-axis). The DSC module is developed to better adapt to the dynamic and slender structure to better perceive the key features.

The DSC blocks in DHS-YOLO are the feature extraction blocks, which handle the input feature graph by splitting it into two parts and using the DSC modules introduced above for multi-scale feature extraction in one part. Finally, the feature maps from the two parts are fused to obtain the final feature map, as shown in Figure 4.

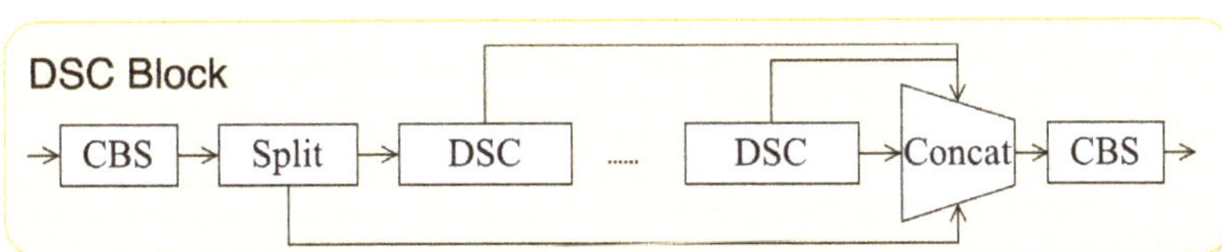

Figure 4. DSC block.

2.2.3. Histogram Transformer (HT) Module

The image quality of UAVs is usually affected by adverse environments, such as low illumination, high illumination, shadows, and so on. Therefore, the HT module is designed to extract intricate features and capture dynamically distributed degradation factors. The HT module includes the Dynamic-range Histogram Self-Attention (DHSA) and Dual-scale Gated Feed-Forward (DGFF) modules. DHSA extracts spatially dynamic environment degradation and enhances both global and local feature aggregation. DGFF enhances the representation of multi-range features, playing a significant role in the process of image restoration. These two components are operated to interact with layer normalization (Figure 5) and can be formulated as follows:

$$F_l = F_{l-1} + \text{DHSA}(\text{LN}(F_{l-1})), \tag{6}$$

$$F_l = F_l + \mathrm{DGFF}(\mathrm{LN}(F_l)), \tag{7}$$

where LN represents layer normalization and F_l denotes the feature at the l-th stage. The details of DHSA and DGFF are described below.

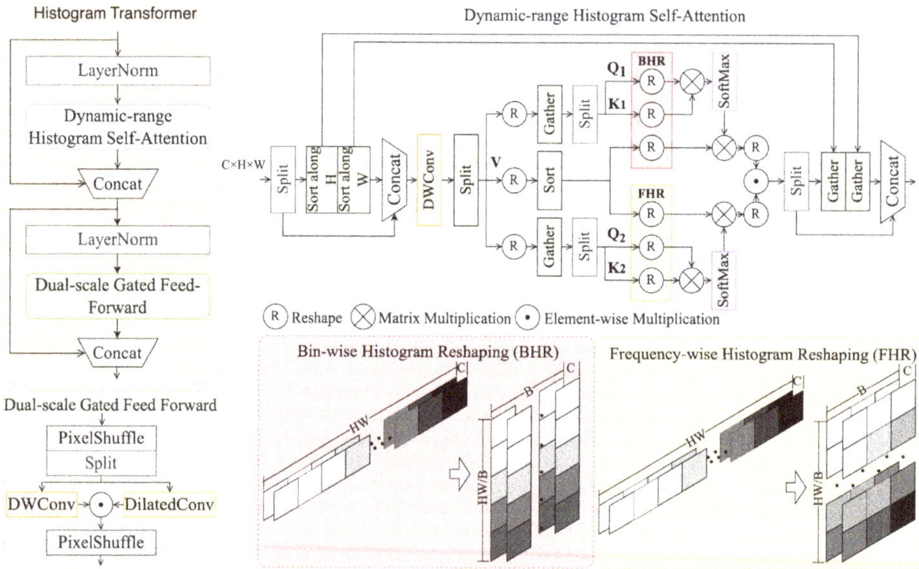

Figure 5. Architecture of our Histogram Transformer module for illumination removal. The main components are the Dynamic-range Histogram Self-Attention (DHSA) part and the Dual-scale Gated Feed-Forward (DGFF) part. There are two types of reshaping mechanisms in DHSA, such as Bin-wise Histogram Reshaping (BHR) and Frequency-wise Histogram Reshaping (FHR).

To better capture dynamically distributed illumination-induced degradation, the DHSA module is designed as a method of dynamic-range convolution that reorganizes the spatial distribution of fractional features, alongside a dual-path histogram self-attention mechanism that integrates global and local dynamic feature aggregation.

Given an input feature $F \in \mathbb{R}^{C\times H\times W}$, it can be divided into two branches along the channel dimension, named as F_1 and F_2. For the features in the first branch, the sorting operations are performed both horizontally and vertically, subsequently concatenating the sorted features with the features in the second branch. The resulting reassembled features are then passed through the DWConv module. The entire process is formulated as follows:

$$F_1, F_2 = \mathrm{Split}(F),\ F_1 = \mathrm{Sort}_v(\mathrm{Sort}_h(F_1)), \tag{8}$$

$$F = \mathrm{Conv}^{\mathrm{d}}_{3\times 3}(\mathrm{Conv}_{1\times 1}(\mathrm{Concat}(F_1, F_2))), \tag{9}$$

where $Conv^d_{3\times 3}$ is the 3×3 depth-wise convolution, $Conv_{1\times 1}$ is the 1×1 point-wise convolution, Split is the operation of splitting features along the channel dimension, Concat represents the concatenation operation along the channel, and Sort_v and Sort_h denote the horizontal and vertical sorting operations, respectively. This method arranges pixels of varying intensities into structured patterns at the diagonal corners of the matrices, facilitating convolution to execute calculations across dynamic ranges. Considering that illumination-induced degradation generally displays tightly correlated patterns, deteriorated pixels tend to cluster in adjacent areas, isolated from the clean pixels. This operation allows convolution kernels to partially concentrate on maintaining clean information while independently recovering degraded features.

The output of dynamic-range convolution can be separated into the value feature $V \in \mathbb{R}^{C\times H\times W}$ and two pairs of Query-key $F_{QK,1}, F_{QK,2} \in \mathbb{R}^{2C\times H\times W}$, subsequently passing to two branches, expressed as follows:

$$V, d = \text{Sort}(R_{C\times H\times W}^{C\times HW}(V)), \quad (10)$$

$$Q_1, K_1 = \text{Split}(\text{Gather}(R_{C\times H\times W}^{C\times HW}(F_{QK,1}), d)), \quad (11)$$

$$Q_2, K_2 = \text{Split}(\text{Gather}(R_{C\times H\times W}^{C\times HW}(F_{QK,2}), d)), \quad (12)$$

where $R_{C\times H\times W}^{C\times HW}$ denotes the operation of reshaping features from $\mathbb{R}^{C\times H\times W}$ to $\mathbb{R}^{C\times HW}$, Gather represents the operation of retrieving the elements' tensor based on a given index, and d is the index of the sorted value. Then, given the number of bins B, the sorted features are reshaped from $C \times HW$ to $C \times HW/B$. Therefore, two types of reshaping are defined to extract both local and global information, such as bin-wise histogram reshaping (BHR) and frequency-wise histogram reshaping (FHR). The first is to assign the number of bins equal to B so that each bin includes HW/B elements, while the second is to set the frequency of each bin equal to B so that the number of bins is HW/B. By this method, large-scale information can be extracted by BHR, where each bin includes a large number of dynamically located pixels, and fine-grained information can be extracted by FHB, where each bin includes modicum pixels neighboring based on intensity. The two pairs of Query-key characteristics undergo two distinct reshaping processes and corresponding self-attention mechanisms, after which their outputs are multiplied element-wise to produce the final result. The process can be expressed as follows:

$$A_B = \text{Softmax}(\frac{R_B(Q_1)R_B(K_1)^T)}{\sqrt{k}})R_B(V), \quad (13)$$

$$A_F = \text{Softmax}(\frac{R_F(Q_2)R_F(K_2)^T)}{\sqrt{k}})R_F(V), \quad (14)$$

$$A = A_B \odot A_F, \quad (15)$$

where R_B and R_F represent the reshaping operation of BHR and FHR, respectively, k is the number of heads, and A_B and A_F denote the obtained attention map.

Previous research [10,20] has typically employed single-range or single-scale convolution in the traditional feed-forward network to bolster the local context. However, these techniques frequently overlook the linkages among dynamically distributed illumination-induced deterioration. In practice, multi-scale information can be obtained by both increasing the kernel size and utilizing the dilation method [21]. Therefore, a Dual-scale Gated Feed-Forward (DGFF) module is designed, which integrates two different multi-range and multi-scale depth-wise convolution pathways into the transmission process. Given an input tensor $F_l \in \mathbb{R}^{C\times H\times W}$ after the DHSA module and Concat and LayerNorm operation, a point-wise convolution operation is initially employed to augment the channel dimension by a factor of r. After this augmentation, the enlarged tensor is channeled into two parallel pathways. During the feature transformation process, 5×5 and dilated 3×3 depth-wise convolutions are utilized to improve the extraction of multi-range and multi-scale information. By the gating mechanism, the output of the second branch, after passing an activation function, serves as a gating map for the other branch. Therefore, the comprehensive feature fusion procedure within the DGFF module is delineated as follows:

$$F_{l,1}, F_{l,2} = \text{Split}(\text{Shuffle}(\text{Conv}_{1\times1}(F_l))), \tag{16}$$

$$F_{l,1} = \text{Conv}^{\text{d}}_{5\times5}(F_{l,1}), F_{l,2} = \text{Conv}^{\text{d,dilated}}_{3\times3}(F_{l,2}), \tag{17}$$

$$F_{l+1} = \text{Conv}_{1\times1}(\text{Unshuffle}(\text{Mish}(F_{l,2}) \odot F_{l,1})), \tag{18}$$

where Shuffle and Unshuffle denote the operations of pixel shuffling and unshuffling, $\text{Conv}^{\text{d}}_{5\times5}$ is 5×5 depth-wise convolution, $\text{Conv}^{\text{d,dilated}}_{3\times3}$ represents 3×3 dilated depth-wise convolution, and Mish represents Mish activation.

The HT blocks in the DHS-YOLO are designed to help the model focus on the key feature while avoiding interference from the adverse environment, which handles the input feature graph by splitting it into two parts and using the Histogram Transformer introduced above for multi-range and multi-scale feature extraction in one part. Finally, the feature maps from the two parts are fused to obtain the final feature map, as shown in Figure 6.

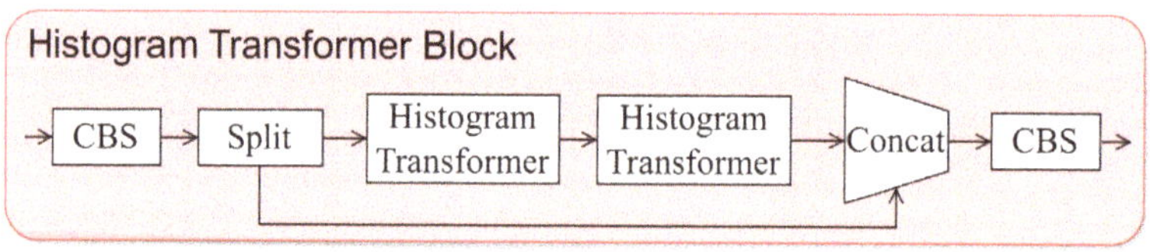

Figure 6. HT block.

2.2.4. ShapeIoU Loss Function

As an essential component of the detector's localization branch, bounding box regression loss plays a crucial role in object detection tasks. Existing bounding box regression methods typically focus on the geometric relationship between ground truth (GT) boxes and predicted boxes, calculating the loss based on their relative positions and shapes while neglecting the inherent attributes of the bounding boxes, such as shape and scale, which can influence regression performance. For example, the bounding box of a slender object is normally a slender box whose accuracy is sensitive to the accuracy of the long edge. Traditional losses like CIoU, DIoU, or GIoU treat all edge errors equally, making them less sensitive to the critical long-edge alignment required for slender objects. To make up for these shortcomings, this paper proposes a bounding box regression method that emphasizes the shape and scale of the bounding boxes themselves. First, we analyze the regression characteristics of bounding boxes and discover that their shape and scale factors significantly impact the regression results. Based on this observation, we propose the Shape IoU method, which calculates the loss by focusing on the shape and scale of the bounding boxes, thereby improving the accuracy of bounding box regression.

The function for ShapeIoU can be derived from Figure 7:

$$IoU = \frac{|B \cap B^{gt}|}{|B \cup B^{gt}|}, \tag{19}$$

$$ww = \frac{2 \times (w^{gt})^{scale}}{(w^{gt})^{scale} + (h^{gt})^{scale}}, \tag{20}$$

$$hh = \frac{2 \times (h^{gt})^{scale}}{(w^{gt})^{scale} + (h^{gt})^{scale}}, \tag{21}$$

$$distance^{shape} = hh \times \frac{(x_c - x_c^{gt})^2}{c^2} + ww \times \frac{(y_c - y_c^{gt})^2}{c^2}, \tag{22}$$

$$\Omega^{shape} = \sum_{t=w,h} (1 - e^{-\omega t})^{\theta}, \theta - 4, \tag{23}$$

$$\begin{cases} \omega_w = hh \times \frac{|w - w^{gt}|}{\max(w, w^{gt})} \\ \omega_h = ww \times \frac{|h - h^{gt}|}{\max(h, h^{gt})} \end{cases}, \tag{24}$$

where scale represents the scale factor, which is related to the scale of the wheat seedlings in the image dataset and ww and hh represent the weight coefficients in the horizontal and vertical orientations, respectively, with their values contingent upon the configuration of the GT box. The corresponding bounding box regression loss function of the designed ShapeIou is as follows:

$$L_{ShapeIoU} = 1 - IoU + distance^{shape} + 0.5 \times \Omega^{shape}. \tag{25}$$

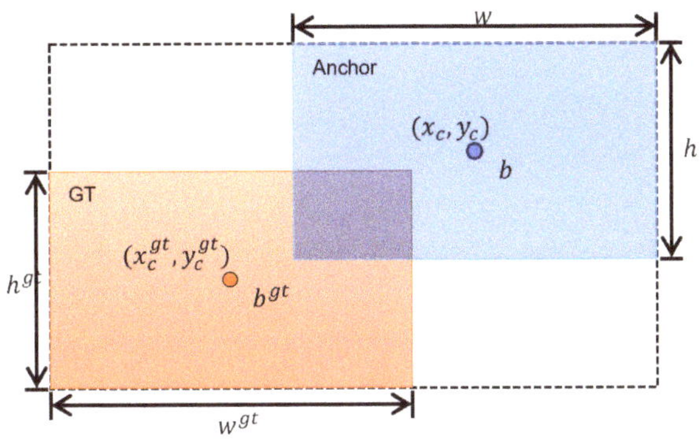

Figure 7. Graphical representation for the introduction of ShapeIoU loss.

2.3. Metric Methods

In order to comprehensively estimate the performance of the model in the detection task of wheat seedlings, this paper adopted commonly used evaluation metrics in the object detection domain, such as precision (P), recall (R), mAP50, and mAP50-95.

Precision (P) evaluates the accuracy of the model's identification results, specifically indicating the proportion of samples classified by the model as real wheat seedlings. In wheat seedling detection, high accuracy means that the model is able to identify wheat seedlings more accurately, reducing the number of cases where weeds or other objects are misjudged as wheat seedlings.

$$P = \frac{TP}{TP + FP}, \tag{26}$$

where TP means True Positive, which refers to situations in which the model correctly classifies objects in images as actual wheat seedlings, and FP represents False Positive and refers to situations in which the model incorrectly classifies weeds or other objects as actual wheat seedlings.

The recall rate (R) is the proportion of all images that contained wheat seedlings that were correctly detected by the model. The recall rate estimates the comprehensiveness of the model, for example, whether wheat seedlings are missed. In wheat seedling detection, a high recall rate indicates that the model is able to classify as many wheat seedlings in the image as possible and avoid omissions. It can be calculated as follows:

$$R = \frac{TP}{TP + FN}, \tag{27}$$

where *FN* means True Negative, which refers to situations in which the model incorrectly classifies a wheat seedling as weed or another object.

The mean Average Precision (mAP) is a metric that incorporates both model accuracy and recall rate. The Average Precision (AP) value ranges from 0 to 1, where higher values signify better performance of the model. The mAP is calculated as the average of the AP values across all categories. In wheat seedling detection, we should pay attention to mAP50 (the mAP value when the IoU threshold is set to 0.5) and mAP50-95 (the average mAP value when the IoU threshold changes from 0.5 to 0.95 with a step of 0.05). The mAP50 can reflect the accuracy of matching the detected box with the real box, and the mAP50-95 can comprehensively provide performance under different IoU thresholds. These two indicators can be calculated as follows:

$$AP = \int_0^1 p(r)dr, \tag{28}$$

$$mAP = \frac{\sum_{i=1}^{N} AP_i}{N}. \tag{29}$$

3. Experiments

3.1. Experiment Environment and Hyperparameter Setting

This experiment was built on a Pytorch-based deep learning framework and executed in the Anaconda environment. Table 2 shows the specific experiment devices and experiment configuration, and Table 3 shows the important hyperparameter settings.

Table 2. Experimental environment configuration.

Environment Configuration	Parameter
Operating system	Ubuntu 22.04
CPU	13th Gen Intel(R) Core(TM) i9-13900KF @ 3.0 GHz
GPU	2 × NVIDIA GeForce RTX3090 (48 GB)
Development environment	PyCharm 2023.3
Programming language	Python 3.9
Deep learning frame	PyTorch 2.0.1
Acceleration platform	CUDA 11.8

Table 3. Hyperparameter setting.

Hyperparameter	Parameter
Epochs	500
Batch	32
Patience	100
Adam learning rate	0.001
Momentum	0.937
Weight decay	0.0005

3.2. Implementation Details

During the training process, the experiment was conducted on a workstation with RTX3090 GPU. The batch size was set to 32, and the Adam algorithm was used to optimize the model with an initial learning rate of 0.001. The models used image preprocessing methods, such as mosaic enhancement, shearing, and flipping. These preprocessing methods can partially replicate the impacts of motion blur, defocusing, pixel occlusion, and jitter on object detection in remote sensing imaging processes and can be utilized to enhance the robustness of the proposed model. The epochs were set to 500, while the patience was set to 100 epochs, which was used to wait for no observable improvement in the early stopping of the training. In this way, during the model training process, overfitting could be prevented. Additionally, the same training epochs and early stopping strategy were adopted when comparing it to other algorithms to ensure fairness and validity in the comparison.

4. Results

4.1. Ablation Analysis

To assess the effectiveness of the developed modules, five sets of ablation experiments were conducted using the same wheat seedling dataset and equipment for training and testing to ensure comparability. Utilizing the classic YOLOv11 as the baseline, these five experiments consisted of the original YOLOv11, YOLOv11 with DSC, YOLOv11 with HT, the proposed method without ShapeIoU loss, and the proposed comprehensive method. The experimental results are presented in Table 4. Compared to YOLOv11 as the baseline method, adding the DSC and HT modules separately increases the accuracy of detection. Adding the DSC module increases the precision by reducing False Positives in slender regions so that the P is improved. The HT module prevents the influence of illumination-induced degradation, which increases the detection ratio of actual wheat seedlings that are under high illumination or shadow so as to improve the R. When combining two modules simultaneously, there is further improvement (Table 4). ShapeIoU loss considers the edge-length-dependent weighting that disproportionately rewards the accurate alignment of longer edges. It is through the configuration of ShapeIoU loss that the highest indicators are achieved by combining the designed DSC and HT modules and ShapeIoU loss function, which are the key factors determining the scope and ability of algorithm applications in the actual wheat seedling detection application.

Table 4. Ablation experiment results.

Model	P (%)	R (%)	mAP50 (%)	mAP50-95 (%)
YOLOv11	87.5	82.2	88.8	72.5
YOLOv11 + DSC	91.3	85.6	91.7	76.9
YOLOv11 + HT	90.9	86.2	92.2	78.2
YOLOv11 + DSC + HT	93.6	89.4	94.0	80.1
Our model: YOLOv11 + DSC + HT + ShapeIoU	94.1	91.0	95.2	81.9

To visualize the functional features of each module in enhancing model performance, heatmaps of the network layers for each module were generated using Grad-CAM [22] (Figure 8). Each pixel in the heatmap represents the active degree of the corresponding model. A higher degree value correlates with an increased likelihood of the target's presence in the position, as depicted in the heatmap by enhanced brightness and significance. By fusing the DSC module, the highlighted region in the heatmap is more refined and precise for the position of wheat seedlings compared to those in the heatmap of the original YOLOv11. Therefore, the DSC module can focus on and capture more features of flexible and slender structures. In addition, this paper also considers the detection of performance in conditions with different illumination intensities. The heatmap of the original YOLOv11 shows several noises at low illumination (II level #1) and high illumination (II levels #4 and #5) intensities, which focus on the incorrect objects, such as straw and soil. The HT module can remove these kinds of noise caused by extreme illumination environments.

Therefore, the DSC and HT modules can significantly enhance the model's ability to recognize slender-shaped wheat seedlings under various lighting conditions.

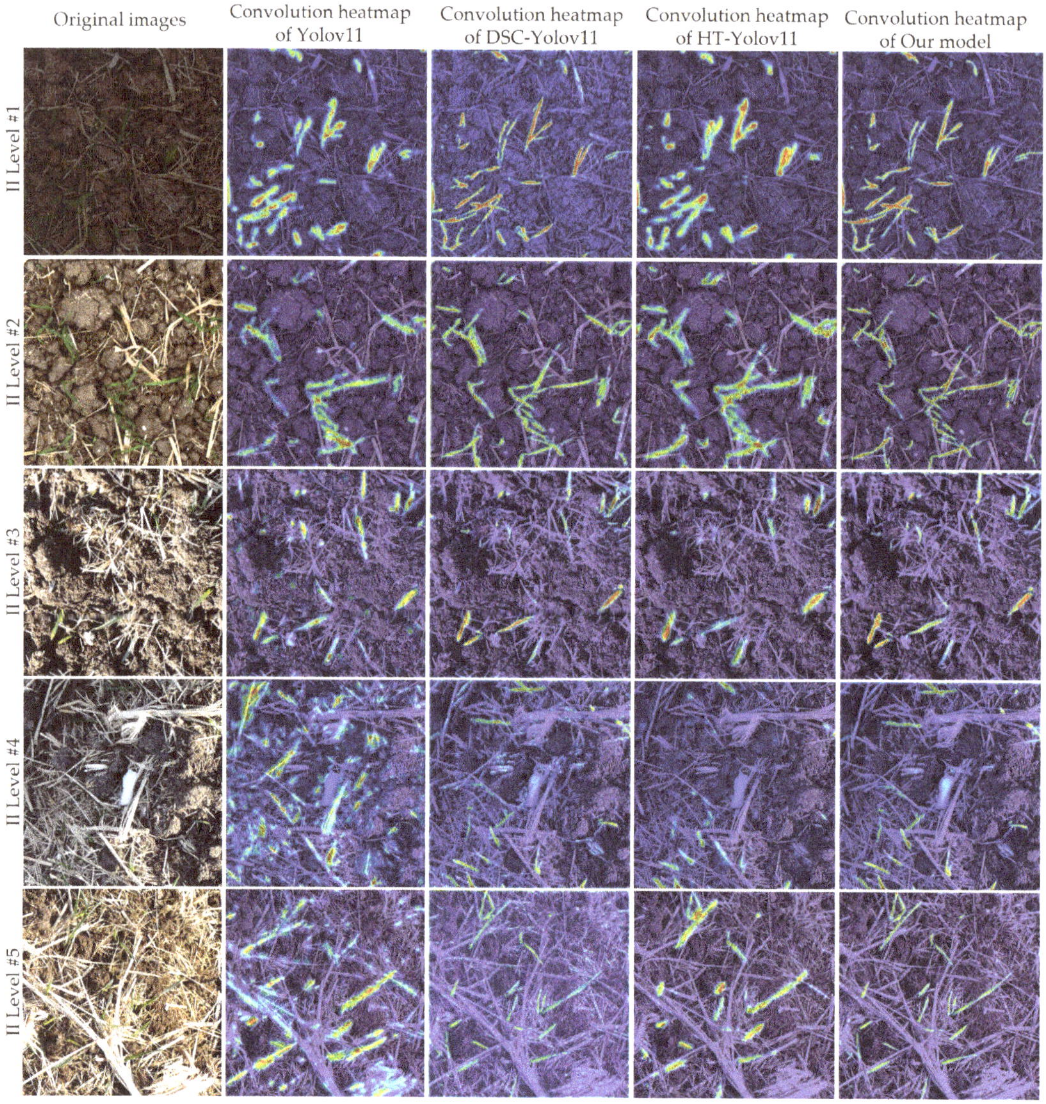

Figure 8. Heatmap of YOLOv11 with different modules in different illumination intensities. From left to right, the columns represent the original images, heatmap of original YOLOv11, heatmap of YOLOv11 with DSC modules, heatmap of YOLOv11 with HT modules, and heatmap of our model (YOLOv11 with DSC and HT modules), respectively. From top to bottom, the rows correspond to illumination intensities of II level #1, #2, #3, #4, and #5, respectively. The heatmap uses a rainbow color coding scheme, with a gradient of color bands from blue (low active degree value) to red (high active degree value).

4.2. Algorithm Performance Comparison

To verify the performance of the proposed algorithm, we conducted comparative experiments with several advanced models from two perspectives. First, the proposed model was compared to the original YOLOv11 and WSA-YOLO [23] using the wheat seedling prediction dataset under different illumination intensities (Table 1). Second, the proposed model was verified by the wheat seedling dataset with different density distributions. Based on the number of wheat seedlings in each image, the predicted dataset was divided into four categories: density levels #1 (wheat seedling counts < 24), #2 ($24 \leq$ wheat seedling counts < 28), #3 ($28 \leq$ wheat seedling counts < 32), and #4 (wheat seedling counts ≥ 32). The model was compared to the original YOLOv11 and Deformable DERT [24].

In order to maintain the fairness and rationality of the comparison, the same hyperparameters were adopted for the models, as shown in Table 3. However, the specific parameters in Deformable DERT still used the parameters from the original paper. The same data augmentation processing was also employed, including translation, scaling, and inversion.

4.2.1. Comparison of Different Illumination Intensities

Currently, mainstream research on object detection algorithms under varying lighting conditions primarily focuses on low-light or nighttime object detection, with representative models such as WSA-YOLO. In general, object detection in UAV imagery is susceptible to interference from different lighting intensities, such as low and high illumination. In this experiment, the proposed model is validated using prediction data under six different lighting intensities and compared to mainstream models. The detection results, as shown in Table 5, demonstrate that our proposed model achieves higher performance metrics.

Table 5. Comparison results using the image dataset in different illumination intensities.

Model in Different II		P (%)	R (%)	mAP50 (%)	mAP50-95 (%)
II Level #1	YOLOv11	83.8	77.4	85.1	64.8
	WAS-YOLO	94.1	89.7	94.3	84.3
	Our model	95.6	93.0	97.7	82.7
II Level #2	YOLOv11	90.3	84.9	89.7	74.8
	WAS-YOLO	88.8	85.1	91.5	80.5
	Our model	94.2	90.9	95.6	83.0
II Level #3	YOLOv11	92.5	86.1	93.8	78.9
	WAS-YOLO	87.7	84.8	88.1	76.2
	Our model	93.8	89.4	94.9	81.8
II Level #4	YOLOv11	89.2	83.0	91.4	76.3
	WAS-YOLO	84.8	81.4	84.7	74.1
	Our model	91.6	88.7	96.4	79.6
II Level #5	YOLOv11	85.2	81.3	87.8	71.4
	WAS-YOLO	83.1	79.5	85.4	72.5
	Our model	95.9	93.8	94.0	81.2
II Level #6-Shadow	YOLOv11	84.0	80.5	85.0	68.8
	WAS-YOLO	90.7	86.5	92.4	82.2
	Our model	93.5	90.2	92.6	83.1

To visually compare the detection performance of the proposed method with representative detection models, we selected several typical scenes under six different illumination conditions for display. The detection results of these scenes using different algorithms are shown in Figure 9. Notably, the blue boxes denote the detection results, the detection count is displayed in the bottom left corner of the image, and the red regions with markers (A, B, and C) indicate incorrect detection areas. Marker A means over-detection, marker B represents missed detection, and marker C represents the low-quality detection box with low IoU.

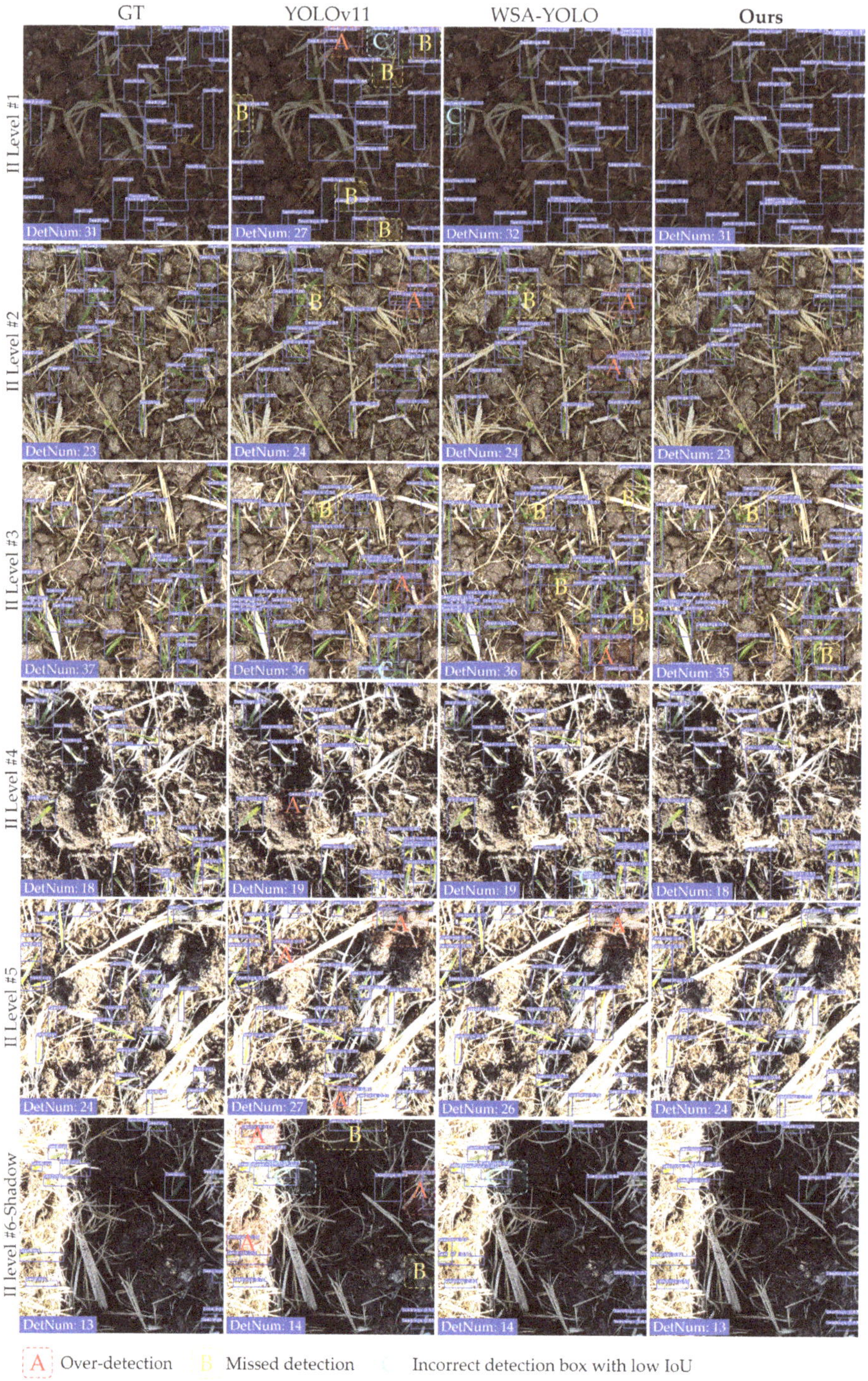

Figure 9. Comparison of wheat seedling detection results of different models using the prediction image dataset under six different illumination conditions. From left to right, the columns represent

the GT (real bounding box of wheat seedlings), the detection results of YOLOv11, WAS-YOLO, and our proposed algorithm, respectively. From top to bottom, the rows represent the six different illumination conditions in Table 1. It is worth noting that the blue boxes denote the detection results, the detection count is displayed in the bottom left corner of the image, e.g., DetNum: 24, and regions with markers (A, B, and C) indicate incorrect detection areas. Marker A means over-detection, marker B represents missed detection, and marker C means the incorrect detection box with low IoU.

For the original YOLOv11 model, extreme illumination conditions, including low illumination, high illumination, and shadow coverage, significantly interfere with the model's ability to detect wheat seedlings. In low-illumination areas, missed detections are predominant, while in high-illumination areas, over-detections are more common. Additionally, low-quality detection boxes are also present. The WSA-YOLO model significantly enhances wheat seedling detection under low-illumination conditions. WSA-YOLO mainly adopts an Adaptive Residual Feature Block (ARFB), which enhances the image quality in low-illumination conditions adaptively through weak supervision and uses a consistency loss function to impose brightness enhancement constraints on the images. However, this ARFB is unable to comprehensively analyze pixel degradation caused by lighting variations and fails to extract effective features from degraded pixels for accurate wheat seedling detection, which results in a lack of ability to process high-illumination images. As the illumination intensity increases, its detection performance declines, posing certain limitations for wheat seedling detection in full-lighting periods and different weather environments. As shown in Table 5, from II levels #3 to #5, WSA-YOLO has poorer performance compared to the original YOLOv11 and our model. Under II level #1 and II level #6-Shadow, WSA-YOLO can achieve similar performance with the proposed method because of the low illumination and large shadow in the image dataset. Compared to these two algorithms, the proposed model can address the wheat seedling recognition problem under varying illumination intensities. The DHSA in the HT module can extract spatially dynamic illumination degradation and enhance both global and local feature aggregation. Moreover, the integrated DGFF module can enrich the representation of multi-range and multi-scale features, contributing to the processing of degraded images. The HT module is capable of utilizing both the adaptive elements of illumination-irrelevant backgrounds and the intrinsic features of illumination-degraded patterns, enabling the distinction of unwanted degradation from the underlying clean background. In Figure 9, compared to the original YOLOv11 under low illumination (II levels #1, #2, and #3), the proposed model can reduce the missed detection. Meanwhile, the proposed model can reduce over-detection under high illumination (II levels #4 and #5). Therefore, in full-lighting periods, including extreme illumination conditions, the proposed model can detect the wheat seedlings well.

4.2.2. Comparison of Different Distribution Densities

The conventional convolutional kernel is designed to extract localized characteristics. On this basis, the deformable kernels have been developed to enrich their application and adaptation to geometric deformations of various objects, such as the Deformable DERT. The wheat seedlings have flexible and slender structures, which makes it difficult to perceive the thin features. In this experiment, the proposed model is validated and compared to representative models (the original YOLOv11 and Deformable DERT) using the image data with different distribution density levels of wheat seedlings. As shown in Table 6, the detection performance of these three algorithms decreases with an increase in wheat seedling density. When the wheat seedling density is too high, such as density levels #3 and #4, the deformable kernels in Deformable DERT almost fail, such that the detection results are similar to those of the original YOLOv11. The main reason is that the wheat seedlings' leaves are intertwined and crossed, making the structure very complex, which greatly affects the extraction of features for individual wheat seedlings. By contrast, the proposed model is more robust, with the density of wheat seedlings having a relatively small impact on seedling identification, achieving the highest detection performance.

Table 6. Comparison results using image datasets in different distribution densities.

Model in Different Distribution Densities		P (%)	R (%)	mAP50 (%)	mAP50-95 (%)
Density level #1	YOLOv11	93.1	88.2	92.2	78.9
	Deformable DERT	96.1	90.9	95.3	82.3
	Our model	97.3	93.2	98.5	87.3
Density level #2	YOLOv11	90.2	86.1	90.4	73.4
	Deformable DERT	94.4	87.5	94.7	74.1
	Our model	94.9	91.5	96.3	83.9
Density level #3	YOLOv11	84.2	78.7	87.8	70.3
	Deformable DERT	86.1	81.6	90.3	71.6
	Our model	92.8	90.2	93.4	80.4
Density level #4	YOLOv11	82.5	75.9	84.7	67.5
	Deformable DERT	83.7	78.2	85.4	69.1
	Our model	91.4	87.2	92.7	75.8

To visually compare the detection performance of the proposed method to those of representative detection models, we selected several typical scenes under four different density conditions for display. The detection results of these scenes using different algorithms are shown in Figure 10. Notably, the blue boxes denote the detection results, the detection count is displayed in the bottom left corner of the image, and the red regions with markers (A, B, and C) indicate incorrect detection areas. Marker A means over-detection, marker B represents missed detection, and marker C represents the low-quality detection box with low IoU.

It can be observed from Figure 10 that the original YOLOv11 and Deformable DERT can detect the individual wheat seedling precisely in low-density conditions, such as density levels #1 and #2, but there are some missed detections. In those situations, the wheat seedlings are relatively dispersed, with almost no overlap between individual seedlings, allowing the model to achieve high detection performance. Only some incomplete wheat seedlings distributed at the edges of the image are not detected by the model. However, as the distribution density increases, its detection performance declines, posing certain limitations for wheat seedling detection in high-density scenes of wheat fields. Moreover, with YOLOv11, there are over-detections at density level #3 (33 detections) and at density level #4 (38 detections) compared to the actual number of seedlings present (30 at level #3 and 35 at level #4), and these are correctly detected by our model. With Deformable DERT, there is over-detection at level #3 (35 detections) compared to the actual 30 seedlings present. Meanwhile, there is missed detection at level #4 (34 detections) compared to the actual 35 seedlings present. The fundamental reason lies in its inability to comprehensively analyze complex features caused by the interlacing of wheat leaves and its failure to extract effective and complete features from the slender wheat seedling for accurate detection. The interlaced wheat leaves would be recognized as a new wrong wheat seedling. Standard convolution cannot focus on processing slender feature targets. If wheat seedlings are too dense, it is easy to cause a large number of duplicate detections. Deformable convolution in Deformable DERT permits unrestricted learning of geometrical alterations and results in the offset of perceptual regions, particularly in flexible and slender structures. As the density of seedlings increases, both over-detection and missed detection can occur. Compared to these two algorithms, the proposed model can address the wheat seedling detection problem under different density scenes of the field. The DSC modules can enhance the perception of the geometric structure by adaptively focusing on the slender and flexible local features of the leaf-shaped structure of wheat seedlings. In contrast to deformable convolution in Deformable DERT, our DSC module takes into account the slender morphology of these structures and incorporates constraints that facilitate the focused enhancement of their perception. Therefore, in high-density conditions, such as density levels #3 and #4, the proposed model can detect the wheat seedlings well.

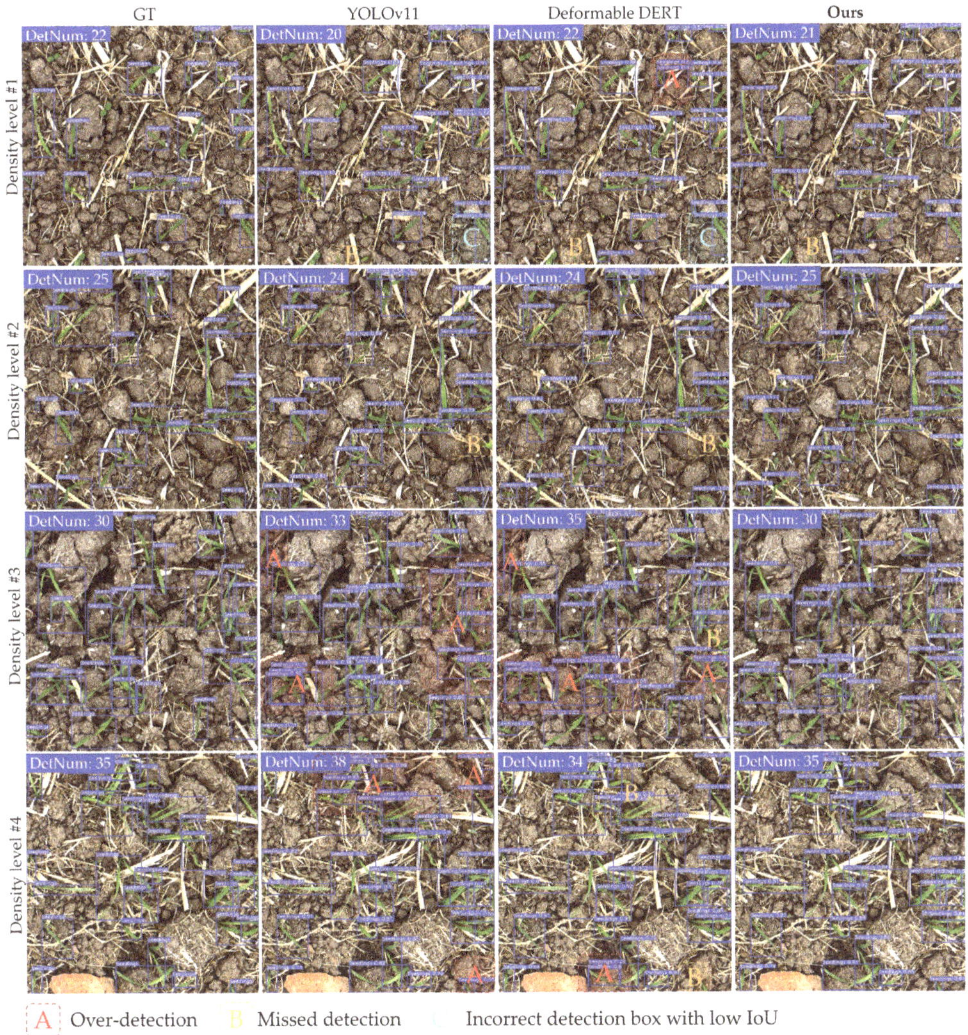

Figure 10. Comparison of wheat seedling detection results of different models under four different density conditions. From left to right, the columns represent the GT (real bounding box of wheat seedlings), the detection results of YOLOv11, Deformable DERT, and our proposed algorithm, respectively. From top to bottom, the rows represent the four different density conditions of wheat seedling distribution (density levels #1, #2, #3, and #4). It is worth noting that the blue boxes denote the detection results, the detection count is displayed in the bottom left corner of the image, e.g., DetNum: 38, and regions with markers (A, B, and C) indicate incorrect detection areas. Marker A means over-detection, marker B represents missed detection, and marker C means the incorrect detection box with low IoU.

5. Discussion

In this research, we used the RGB images acquired by UAV-based remote sensing to detect wheat seedlings whose growth conditions should be closely monitored. Our investigation focused on the importance of the morphological structure of objects and image quality (due to the illumination) to develop an accurate model, with an overarching target of practical deployment in future works. The results demonstrated that our DHS-YOLO model can detect wheat seedlings with an accuracy (mAP50 is referred to as accuracy) exceeding 95%.

Feature extraction of flexible and slender objects has always been a hot topic in computer vision. The proposed DSC module outperformed the common standard convolution and deformable convolution in this task. While the existing deformable convolution permits unrestricted learning of geometrical alterations, it can result in the offset of perceptual regions, particularly in flexible and slender structures. On the contrary, the DSC module takes into account the slender morphology of these structures and incorporates constraints that facilitate the focused enhancement of their perception. In agricultural production, there are many applications for the detection of slender objects, such as seedlings of various crops, some slender fruits, and so on. However, agriculture presents a certain level of complexity, making algorithm performance highly susceptible to limitations. For instance, in this study on wheat seedling detection, the experiment was designed to detect two-leaf seedlings. If the number of leaves increases to three, the entire wheat field becomes significantly more complex, with overlapping seedling leaves, which may render the algorithm ineffective. Additionally, labeling the dataset under such conditions would become considerably more challenging. Perhaps pure 2D detection is insufficient, and 3D reconstruction, along with 3D object detection, may be required.

To address the problem of the different image degradation in a complete illumination period, the HT module is proposed to extract intricate features and capture dynamically distributed degradation factors, which improve the accuracy of object detection in illumination-induced adverse environments. Our histogram self-attention module can focus on illumination-related information with a dynamic range by segmenting spatial features into numerous bins and distributing varied attention along the bin or frequency dimension. The DGFF module is introduced to make learning multi-range and multi-scale information easier. We show the superiority and efficacy of our method through experimentation. Furthermore, this algorithm can be applied to detect objects in similar agricultural scenes. However, since this study is conducted based on real challenges of projects, the primary focus is on the accuracy of wheat seedling detection rather than real-time data processing. Therefore, factors such as computational cost, model size, and processing speed were not heavily considered. Under some actual field conditions, real-time performance becomes crucial. Future research should explore model pruning or hardware acceleration to ensure viability for onboard UAV processing.

6. Conclusions

This paper proposed DHS-YOLO for effective slender and flexible wheat seedling detection under different illumination intensities. DHS-YOLO adopts YOLOv11 as the basics and incorporates the DSC module, HT module, and ShapeIoU loss function. The DSC module is developed to enhance the perception of the geometric structure by adaptively focusing on the slender and flexible local features of the leaf-shaped structure of wheat seedlings. The HT module is proposed for unified adverse illumination removal, which can effectively and universally remove deterioration since it has dynamic-range spatial attention to illumination-induced degeneration. The ShapeIoU loss function is proposed to emphasize the shape and scale of the bounding box itself and make the bounding box more precise. The DHS-YOLO model is trained and verified by the wheat seedling dataset collected by a UAV under different illumination intensities. Meanwhile, the model is compared to typical models, achieving the best performance. This study developed an effective model to detect wheat seedlings, which is of great significance for achieving precise pesticide spraying or irrigation in wheat fields.

Author Contributions: Conceptualization, X.D.; methodology, X.D.; software, X.D.; validation, X.D. and J.P.; formal analysis, X.D.; investigation, X.D.; resources, X.D.; data curation, J.P.; writing—original draft preparation, X.D.; writing—review and editing, J.P.; visualization, J.P.; supervision, X.D.; project administration, X.D.; funding acquisition, X.D. All authors have read and agreed to the published version of the manuscript.

Funding: This research was supported by the Zhejiang Provincial Department of Science and Technology Project (2024C04031).

Institutional Review Board Statement: Not applicable.

Data Availability Statement: The dataset cannot be made fully public due to a non-disclosure agreement, but limited samples can be provided upon reasonable request.

Acknowledgments: We sincerely thank the Huzhou Institute of Zhejiang University.

Conflicts of Interest: The authors declare no conflicts of interest.

References

1. Srivastava, K.; Pandey, P.C.; Sharma, J.K. An Approach for Route Optimization in Applications of Precision Agriculture Using UAVs. *Drones* **2020**, *4*, 58. [CrossRef]
2. Nguyen, L.H.; Robinson, S.; Galpern, P. Medium-resolution multispectral satellite imagery in precision agriculture: Mapping precision canola (*Brassica napus* L.) yield using Sentinel-2 time series. *Precis. Agric.* **2021**, *23*, 1051–1071. [CrossRef]
3. Sethy, P.K.; Pandey, C.; Sahu, Y.K.; Behera, S.K. Hyperspectral imagery applications for precision agriculture—A systemic survey. *Multimed. Tools Appl.* **2022**, *81*, 3005–3038. [CrossRef]
4. Dong, X.; Kim, W.-Y.; Zheng, Y.; Oh, J.-Y.; Ehsani, R.; Lee, K.-H. Three-Dimensional Quantification of Apple Phenotypic Traits Based on Deep Learning Instance Segmentation. *Comput. Electron. Agric.* **2023**, *212*, 108156. [CrossRef]
5. Zhang, Y.; Liu, X.; Wa, S.; Chen, S.; Ma, Q. GANsformer: A Detection Network for Aerial Images with High Performance Combining Convolutional Network and Transformer. *Remote Sens.* **2022**, *14*, 923. [CrossRef]
6. Wu, Y.; Guo, H.; Chakraborty, C.; Khosravi, M.R.; Berretti, S.; Wan, S. Edge Computing Driven Low-Light Image Dynamic Enhancement for Object Detection. *IEEE Trans. Netw. Sci. Eng.* **2023**, *10*, 3086–3098. [CrossRef]
7. Xu, X.; Wang, S.; Wang, Z.; Zhang, X.; Hu, R. Exploring Image Enhancement for Salient Object Detection in Low Light Images. *arXiv* **2020**, arXiv:2007.16124. [CrossRef]
8. Zhang, G.; Wang, L.; Chen, Z. A Step-Wise Domain Adaptation Detection Transformer for Object Detection under Poor Visibility Conditions. *Remote Sens.* **2024**, *16*, 2722. [CrossRef]
9. Wang, Y.; Wan, R.; Yang, W.; Li, H.; Chau, L.P.; Kot, A. Low-light image enhancement with normalizing flow. In Proceedings of the AAAI Conference on Artificial Intelligence, Virtual, 22 February–1 March 2022; Volume 36, pp. 2604–2612.
10. Chen, X.; Li, H.; Li, M.; Pan, J. Learning A Sparse Transformer Network for Effective Image Deraining. In Proceedings of the 2023 IEEE/CVF Conference on Computer Vision and Pattern Recognition (CVPR), Vancouver, BC, Canada, 18–22 June 2023; IEEE: Piscataway, NJ, USA, 2023; pp. 5896–5905.
11. Jose Valanarasu, J.M.; Yasarla, R.; Patel, V.M. TransWeather: Transformer-Based Restoration of Images Degraded by Adverse Weather Conditions. In Proceedings of the 2022 IEEE/CVF Conference on Computer Vision and Pattern Recognition (CVPR); IEEE: New Orleans, LA, USA, 19–24 June, 2022; pp. 2343–2353.
12. Cai, Y.; Bian, H.; Lin, J.; Wang, H.; Timofte, R.; Zhang, Y. Retinexformer: One-Stage Retinex-Based Transformer for Low-Light Image Enhancement. In Proceedings of the 2023 IEEE/CVF International Conference on Computer Vision (ICCV), Paris, France, 1 October 2023; IEEE: Piscataway, NJ, USA, 2023; pp. 12470–12479.
13. Fang, X.; Gao, X.; Li, B.; Zhai, F.; Qin, Y.; Meng, Z.; Lu, J.; Xiao, C. A Non-Uniform Low-Light Image Enhancement Method with Multi-Scale Attention Transformer and Luminance Consistency Loss. *arXiv* **2023**, arXiv:2312.16498. [CrossRef]
14. Yang, X.; Li, Z.; Guo, Y.; Zhou, D. DCU-Net: A Deformable Convolutional Neural Network Based on Cascade U-Net for Retinal Vessel Segmentation. *Multimed. Tools Appl.* **2022**, *81*, 15593–15607. [CrossRef]

15. He, Y.; Ge, R.; Wu, J.; Coatrieux, J.-L.; Shu, H.; Chen, Y.; Yang, G.; Li, S. Thin Semantics Enhancement via High-Frequency Priori Rule for Thin Structures Segmentation. In Proceedings of the 2021 IEEE International Conference on Data Mining (ICDM), Auckland, New Zealand, 7–10 December 2021; IEEE: Piscataway, NJ, USA, 2022; pp. 1096–1101.
16. Zhao, G.; Liang, K.; Pan, C.; Zhang, F.; Wu, X.; Hu, X.; Yu, Y. Graph Convolution Based Cross-Network Multiscale Feature Fusion for Deep Vessel Segmentation. *IEEE Trans. Med. Imaging* **2023**, *42*, 183–195. [CrossRef] [PubMed]
17. Shit, S.; Paetzold, J.C.; Sekuboyina, A.; Ezhov, I.; Unger, A.; Zhylka, A.; Pluim, J.P.W.; Bauer, U.; Menze, B.H. clDice—a Novel Topology-Preserving Loss Function for Tubular Structure Segmentation. In Proceedings of the 2021 IEEE/CVF Conference on Computer Vision and Pattern Recognition (CVPR), Nashville, TN, USA, 19–25 June; IEEE: Piscataway, NJ, USA, 2021; pp. 16555–16564.
18. LabelImg. Available online: https://github.com/HumanSignal/labelImg (accessed on 15 January 2025).
19. YOLOv11. Available online: https://github.com/ultralytics/ultralytics/tree/main?tab=readme-ov-file (accessed on 15 January 2025).
20. Wang, Z.; Cun, X.; Bao, J.; Zhou, W.; Liu, J.; Li, H. Uformer: A General U-Shaped Transformer for Image Restoration. In Proceedings of the 2022 IEEE/CVF Conference on Computer Vision and Pattern Recognition (CVPR), New Orleans, LA, USA, 19–24 June 2022; IEEE: Piscataway, NJ, USA, 2022; pp. 17662–17672.
21. Li, Y.; Lu, J.; Chen, H.; Wu, X.; Chen, X. Dilated Convolutional Transformer for High-Quality Image Deraining. In Proceedings of the 2023 IEEE/CVF Conference on Computer Vision and Pattern Recognition Workshops (CVPRW), Vancouver, BC, Canada, 18–23 June 2023; IEEE: Piscataway, NJ, USA, 2023; pp. 4199–4207.
22. Selvaraju, R.R.; Das, A.; Vedantam, R.; Cogswell, M.; Parikh, D.; Batra, D. Grad-CAM: Why Did You Say That? *arXiv* **2017**, arXiv:1611.07450.
23. Hui, Y.; Wang, J.; Li, B. WSA-YOLO: Weak-supervised and Adaptive object detection in the low-light environment for YOLOV7. IEEE Trans. *Instrum. Meas.* **2024**, *73*, 2507012. [CrossRef]
24. Wang, D.; Li, Z.; Du, X.; Ma, Z.; Liu, X. Farmland Obstacle Detection from the Perspective of UAVs Based on Non-Local Deformable DETR. *Agriculture* **2022**, *12*, 1983. [CrossRef]

agriculture

MDPI

Article

Potato Disease and Pest Question Classification Based on Prompt Engineering and Gated Convolution

Wentao Tang and Zelin Hu *

School of Physics and Electronic Information, Gannan Normal University, Ganzhou 341000, China; tang764873104@163.com
* Correspondence: huzelin@gnnu.edu.cn

Abstract: Currently, there is no publicly available dataset for the classification of potato pest and disease-related queries. Moreover, traditional query classification models generally adopt a single maximum-pooling strategy when performing down-sampling operations. This mechanism only extracts the extreme value responses within the local receptive field, which leads to the degradation of fine-grained feature representation and significantly amplifies text noise. To address these issues, a dataset construction method based on prompt engineering is proposed, along with a question classification method utilizing a gated fusion–convolutional neural network (GF-CNN). By interacting with large language models, prompt words are used to generate potato disease and pest question templates and efficiently construct the Potato Pest and Disease Question Classification Dataset (PDPQCD) by batch importing named entities. The GF-CNN combines outputs from convolutional kernels of varying sizes, and after processing with max-pooling and average-pooling, a gating mechanism is employed to regulate the flow of information, thereby optimizing the text feature extraction process. Experiments using GF-CNN on the PDPQCD, Subj, and THUCNews datasets show $F1$ scores of 100.00%, 96.70%, and 93.55%, respectively, outperforming other models. The prompt engineering-based method provides a new paradigm for constructing question classification datasets, and the GF-CNN can also be extended for application in other domains.

Keywords: potato pests and diseases; question classification; prompt engineering; GF-CNN; large language model

1. Introduction

Potatoes, as an important crop with multiple functions, are widely cultivated around the world and are one of the four major staple crops globally [1]. However, in agricultural practice, potatoes face a variety of disease threats, among which early blight and late blight, caused by the pathogens *Alternaria solani* and *Phytophthora infestans*, respectively, are considered the most destructive fungal diseases due to their rapid spread and severe impact [2,3]. In addition to diseases, pest issues are also significant. Pests such as the *Epilachna vigintioctomaculata* and the *Acanthoscelides obtectus* cause compound damage by feeding on leaves and spreading pathogens, leading to yield losses of 15–30%. In severe cases, entire fields may be lost, significantly impacting the economic benefits of the industry. Therefore, establishing a comprehensive potato pest and disease control system is crucial for ensuring stable potato production [4].

With the development of agricultural informatization, intelligent monitoring and decision-making systems based on agricultural data services are gradually replacing traditional experience-based agricultural production models [5]. In this context, question-answering systems, due to their portability and efficiency, have become ideal tools for querying. Users interact with the system using natural language, and the system analyzes these inputs to generate useful answers, helping users obtain timely and accurate agricultural advice. Previous studies have shown that question-answering systems have demonstrated significant effectiveness in disease and pest identification, improving crop yields, and reducing pesticide use [6]. In this regard, developing a question-answering system integrated with potato pest and disease knowledge is of great practical significance for reducing the threat of potato pests and diseases and minimizing economic losses for farmers.

In the process of building a question-answering system, question classification, also referred to as intent recognition, plays a crucial role. Its core function is to map the natural language questions input by the user into predefined semantic categories, thus laying the foundation for accurately interpreting user needs and generating targeted responses. In the field of agriculture, there are two commonly used text classification methods: machine learning-based methods and deep learning-based methods [7]. In the context of classical machine learning paradigms, algorithms such as support vector machine (SVM) and naive Bayes (NB) have demonstrated significant application results. Notable studies include that by Wei et al. [8], who, after constructing a keyword classification library, performed feature word selection and weight optimization strategies using linear SVM to classify Chinese agricultural texts. Cui et al. [9] used the XGBoost parallelization algorithm within the Spark framework to accurately classify forestry texts. Espejo-Garcia et al. [10], using pesticide usage documents from the Spanish Official Plant Health Products Registration Office as their data source, compared the accuracy of SVM, NB, logistic regression, and random forest models for agricultural regulation classification. Through training, machine learning models can adapt to different text data and classification tasks, demonstrating excellent flexibility and scalability. Although machine learning-based methods can automatically learn from data, they rely heavily on manually defined features, and the quality of these features directly impacts the model's learning efficiency and final performance [11]. Therefore, the careful design and selection of features are crucial for improving the performance of machine learning models.

Deep learning-based methods are currently the most mainstream approach in question classification tasks. Deep learning models such as convolutional neural networks (CNN), gated recurrent units (GRU), and long short-term memory networks (LSTM) can automatically learn complex feature representations from raw text data, thus eliminating the need for tedious manual feature engineering. Jin et al. [12] proposed the BiGRU-MulCNN agricultural question classification model, which achieves good classification performance even on datasets with insufficient data and unclear features. Wang et al. [13] constructed the Attention-DenseCNN model, which establishes dense connections (where each layer receives the outputs of all preceding layers as additional inputs, thereby avoiding the issue of information loss) between upstream and downstream convolutional blocks in CNNs. This effectively enhances feature propagation and gradient flow. Moreover, the attention mechanism assigns higher weights to key features, further improving the model's focus on important information. The model significantly improved the accuracy of rice question classification. Feng et al. [14] further broke through the performance bottleneck of the model through system architecture optimization. Their proposed RIC-Net model deeply integrates a 4-layer residual network with a capsule network. This not only alleviates the gradient vanishing problem through skip connections but also effectively models semantic hierarchical relationships using dynamic routing mechanisms. The model achieved an accuracy of 98.62% on a rice knowledge classification dataset.

Pre-trained models, such as BERT (Bidirectional Encoder Representations from Transformers) [15] and ERNIE (Enhanced Representation through Knowledge Integration) [16], have proven to be effective in various natural language processing tasks and represent a significant breakthrough in the field of deep learning. Agriculture, as a highly specialized domain, involves a vast array of complex terms and proprietary vocabulary. Traditional models, such as Word2Vec and GloVe, are limited by two major drawbacks due to their static nature: first, the generated word vectors are fixed and cannot dynamically capture contextual information; second, they may fail to accurately understand domain-specific terminology in agriculture, thereby impacting task performance. In contrast, pre-trained models, trained on large-scale corpora, can dynamically generate word vectors based on context and possess stronger semantic understanding capabilities, thus effectively overcoming these two major limitations. The BERT-Stacked LSTM model proposed by Li et al. [17] utilizes multiple stacked LSTMs to learn complex semantic information from text, showing significant advantages over six other models in agricultural pest and disease question classification. Li et al. [18] improved the DPCNN model by integrating the word vectors provided by the ERNIE pre-trained model, enabling efficient identification of cotton pest and disease questions. Duan et al. [19] proposed a multimodal agricultural news text classification method, which extracts features from both text and images using ERNIE and Vision Transformer, respectively. An interactive attention mechanism is then applied to compute attention weights between the text and image features, extracting shared features and enhancing cross-modal synergy. Experimental results indicate that this multimodal model outperforms individual text or visual models in classification performance.

In response to the lack of publicly available question-answer datasets in the field of potato pests and diseases as well as the limitations of traditional question classification methods in terms of representational capacity, this paper proposes a dataset construction method that integrates prompt engineering and a question classification model based on gated fusion convolution. Specifically, a structured prompt is used to guide a large language model (LLM) in generating potato pest and disease question templates. These templates are then populated in bulk using entity information extracted from previous named entity recognition (NER) work [20], resulting in the efficient construction of the Potato Pest and Disease Question Classification Dataset (PDPQCD). Moreover, this paper introduces the gated fusion–convolutional neural network (GF-CNN) to optimize the text feature extraction process. The model captures local features through parallel multi-scale convolutions and then employs both max-pooling and average-pooling to extract salient and global statistical features. A gating unit was designed to dynamically fuse these two feature representations, enhancing feature robustness while preserving critical information.

In summary, the main contributions of this paper can be summarized as follows:

(a) The construction of the first question classification dataset for the potato pest and disease domain, which achieves effective integration of LLM with domain knowledge through prompt engineering, outperforming existing datasets in both scale and granularity of categories;
(b) The proposal of the GF-CNN question classification model with dynamic feature fusion capability. This model adapts the feature weights of max-pooling and average-pooling through a gating mechanism, enhancing the model's representational power while maintaining good interpretability;
(c) Experimental results on the PDPQCD, Subj, and THUCNews datasets show that GF-CNN not only accurately classifies potato pest and disease questions into predefined categories but also demonstrates strong generalization ability in cross-domain classification tasks.

2. Materials and Methods

2.1. Prompt Engineering

Prompt engineering aims to improve the quality and relevance of a model's output by carefully designing the textual prompts input into the LLM. When conducting prompt engineering, users must consider how to precisely express their needs, select appropriate keywords, and structure sentences to guide the model in generating more accurate and relevant responses. For example, if the goal is to have the LLM analyze the sentiment of an article, a simple and vague prompt might be "Analyze this article", whereas an optimized prompt might be more specific, such as "Please describe the primary sentiment of this article and provide specific examples to support your viewpoint". Prompt engineering is not only about the skill of asking questions but also involves an understanding of the model's responses. Through precise prompts, one can better leverage the potential of the LLM.

When writing effective prompts, keep the following principles in mind:

(a) Clarity and specificity: Avoid using ambiguous terms. For instance, when dealing with numerical issues, specify exact numbers for the desired output instead of using vague terms like "some" or "a few";

(b) Role-playing: Specify the roles of the LLM and the user. For example, if you are seeking weight loss advice, using the prompt "Assume you are a professional personal fitness trainer, and I am your client. Please use your expertise and consider my physical condition to create a detailed weight loss plan for me" will often yield more professional advice and suggestions than simply saying, "Give me a weight loss plan";
Of course, if the role definition is not clear enough, it may lead to ambiguous or irrelevant outputs from the model. Additionally, when an LLM is assigned a specific role, its responses may become overly constrained by stereotypical assumptions associated with that role. These issues also require the user to make flexible adjustments based on the specific context;

(c) Output formatting: To achieve more consistent output, specify the format in which you would like the LLM to deliver the response. LLMs can handle various common data formats, such as TXT and JSON.

2.2. Classification of Intentions

After the preliminary organization of the data from the team's previous NER work, we classified the questions into 11 categories. This means that if we successfully construct the question-answering system, the system will be able to answer the 11 types of questions shown in Table 1.

Table 1. Category of intent.

Categories	Examples
Disease Symptoms	What are the common symptoms of early blight?
Pest Symptoms	How does the cabbage looper damage potatoes?
Causes of Disease	What are the typical causes of late blight?
Affected Areas	Which parts of the potato are primarily affected by the beet armyworm?
Treatment Agents	What treatment agents can be used to cure potato leaf roll disease?
Alternate Names	What are the alternate names for *Epicauta gorhami*?
Distribution	In which regions is potato cancer disease primarily distributed?
Transmission Vectors	Through which pathways is leaf blight mainly transmitted?
Medication Uses	For which diseases and pests is Bordeaux mixture an effective treatment agent?
Overwintering Sites	Where does *Cercospora solani-tuberosi* overwinter?
Pest Categories	To which biological category does the pea leaf miner belong?

2.3. Dataset Template Generation Based on Prompt Engineering

After defining the question categories, we began generating question classification dataset templates using the LLM. In this study, the ChatGPT 4 model was chosen, and taking the "disease symptoms" category as an example, the prompt and model's response are shown in Figure 1.

○ Role definition

Suppose you are an experienced natural language processing algorithm engineer, and I need to construct a question classification dataset related to potato pests and diseases.

○ Task Description

Currently, there is a lack of training data for the "Disease Symptoms" category, and you are needed to provide additional training templates.

○ Task Requirements

The requirements are as follows: 1) The questions should be diverse in phrasing, as conversational as possible, with a mix of long and short sentences. Avoid repetitive sentence structures. 2) In the training templates, the disease name should be replaced with [Disease] and end with ", disease symptoms." 3) List the questions in bullet points, with at least 30 entries.

○ Output Format

For example:
What are the common symptoms of [Disease]? , disease symptoms
What are the typical symptoms of [Disease]? , disease symptoms

Of course, here are 30 question templates about potato disease symptoms, ensuring diversity and a conversational tone:

1. What are the typical symptoms of [Disease]? , disease symptoms
2. Can you tell me what the symptoms of [Disease] are? , disease symptoms
3. What are the common symptoms of [Disease]? , disease symptoms
4. What are the usual symptoms of [Disease]? , disease symptoms

...

30. What happens to potatoes when they are infected with [Disease]? , disease symptom

These templates cover a variety of lengths and question formats, helping to enhance the natural language processing model's ability to recognize questions related to potato disease symptoms.

Figure 1. Prompt and model replies.

The prompt is divided into four parts: Role Definition, Task Description, Task Requirements, and Output Format. This structure is designed to make the process easier for readers to follow. However, in actual input to the model, these sections should be omitted. In the Role Definition section, the LLM is specified as an experienced NLP engineer. The implicit intention here is to guide the model toward processing NLP-related content and to communicate the real need for constructing a PDPQCD. The Task Description section further refines the requirements, asking the LLM to generate training templates specifically for the "Disease Symptoms" category. In the Task Requirements section, constraints are placed on the content generated by the LLM. The generated questions should simulate real-world scenarios and the way people naturally ask questions, including short, conversational queries and longer, descriptive sentences. Additionally, placeholders like [Disease] are used in place of specific disease names, allowing for the templates to be adapted to any particular disease during practical use. At the end of each question, the label "disease symptoms" is added, directly tagging the question for subsequent classification tasks. Finally, the Output Format section provides two examples to further standardize the model's output, ensuring consistency across all generated content. This step guarantees that the results are structured and ready for use in the intended application, making the training data clean, organized, and effective for building the intent recognition model.

Using prompt engineering to generate dataset templates offers several distinct advantages over manual construction: First, in terms of efficiency and speed, LLMs can generate large volumes of data in a very short time, whereas manual writing may take several hours or even days. Second, in terms of diversity and creativity, manual construction is often limited by an individual's imagination and linguistic habits, whereas LLMs, trained on large-scale text corpora, are capable of producing more diverse and innovative sentence structures, thereby increasing the coverage and complexity of the corpus. Lastly, in terms of customizability, LLMs can flexibly adjust the output based on the input prompts, allowing for the generation of text in specific styles or formats as required.

2.4. Construction of a PDPQCD

After generating various question templates using the LLM, a manual review of each template was conducted to assess alignment with question topics and human linguistic conventions. Templates that were deemed irrelevant were excluded. Finally, we bulk-imported the potato pest and disease entities extracted from the previous NER work to replace the placeholders in the templates. Using the data from the "disease symptoms" category as an example, the algorithmic logic is shown in Figure 2.

```
Input. Disease symptom related entities, LLM-generated question templates
Output: Real data of the "Disease symptoms" category.
1. Diseases = ["Potato early blight", "Potato wart", ...] // Initialize the Diseases list, which contains potato disease-related entities.
2. Templates = [
        "What are the common symptoms of [Disease]? , disease symptoms",
        "What are the typical symptoms of [Disease]? , disease symptoms"
        ...
] // Initialize the Templates list, which contains LLM-generated question templates.
3. Questions = [], id_counter = 1  // Initialize an empty Questions list to store real training data, and set up a counter.
4. for Disease in Diseases:  // Fill in the templates and generate real data.
     for Template in Templates:
        sentence = Replace "[Disease]" in Template with Disease
        Questions.append(f"{id_counter}. {sentence}")
        id_counter += 1
5. for Question in Questions:  // Output the real data.
     Output Question
```

Figure 2. Entity import algorithm logic.

Three lists—Diseases, Templates, and Questions—were initialized. The Diseases and Templates lists store potato disease entities and question templates, respectively, while the Questions list is initially empty and is used to store the generated real data. The algorithm iterates over the Diseases list externally and the Templates list internally. During each iteration, the algorithm identifies placeholders in the templates and replaces them with the corresponding disease entities, generating a real data. Simultaneously, a counter assigns a unique identifier to each question to facilitate the subsequent counting of questions in that category. Finally, by iterating through the Questions list, all real data for the "disease symptoms" category can be obtained.

Through the above method, a dataset containing 20,260 entries was finally constructed, with its distribution shown in Figure 3. As can be seen from Figure 3, the data in the dataset exhibit a certain degree of uneven distribution. This is due to the differing number of entities in each category and the varying number of templates retained for each category, with both factors collectively contributing to this discrepancy.

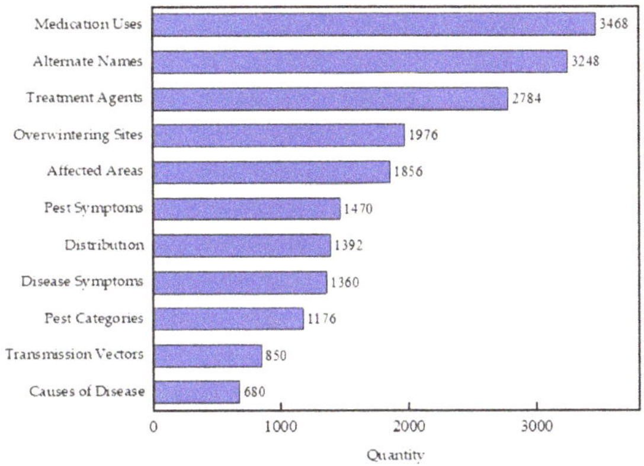

Figure 3. Data distribution.

3. Model Structure

This study utilizes BERT as the pre-trained model, upon which the proposed GF-CNN is employed for feature extraction. Finally, the softmax function is applied for question classification. The overall architecture of the system is illustrated in Figure 4.

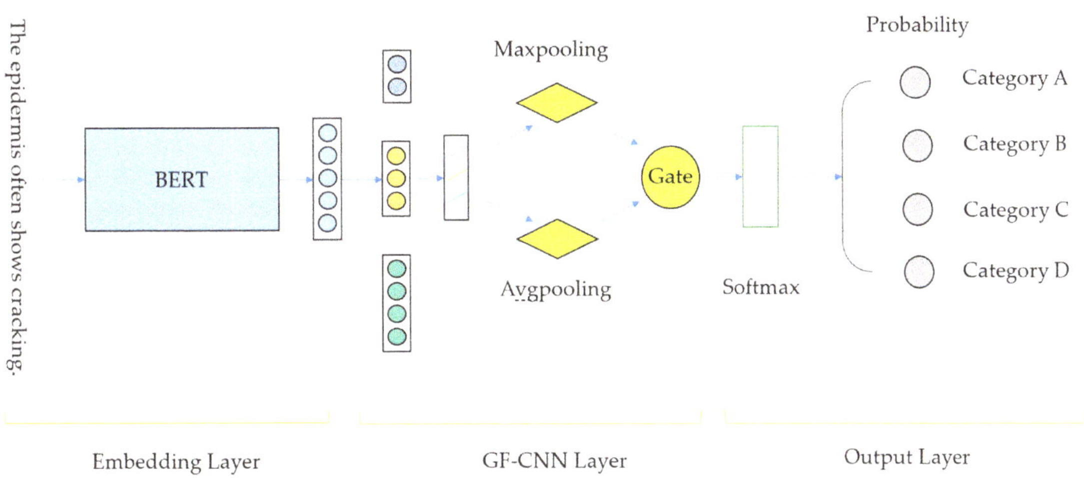

Figure 4. Model structure.

3.1. BERT

BERT, proposed by Devlin et al. [15], is a pre-trained language model. Compared to other language models, such as FastText and GloVe, BERT provides dynamic word embeddings, meaning that the generated word vectors are dynamically adjusted based on the context in which the word appears. This allows the model to distinguish different meanings of the same word in different contexts.

The core innovation of BERT lies in its use of two pre-training tasks: masked language model (MLM) and next-sentence prediction (NSP). In the MLM task, BERT randomly masks some words in the input data and then predicts the original words for these masked positions. This process requires the model to accurately understand and retain contextual information. The NSP task, on the other hand, enhances the model's understanding of sentence relationships by predicting whether the second sentence fragment is a continuation of the first sentence fragment.

3.2. GF-CNN

Since max-pooling efficiently reduces the dimensionality of feature maps without significantly increasing computational cost, most question classification models, such as TextCNN [21] and DPCNN [22], utilize max-pooling for downsampling. Max-pooling is highly effective in scenarios where it is necessary to highlight strong features. However, relying solely on max-pooling has some drawbacks. First, there is the issue of information loss. Max-pooling only retains the maximum value, and any non-maximum values within the feature region, regardless of their importance, are ignored. This can make the model less sensitive to subtle changes in the input data, causing it to lose information that, although not reflected in the local maximum, is important at a global level. Second, max-pooling is particularly sensitive to noise. Since max-pooling focuses on the maximum value, incidental high values in the data might be mistakenly interpreted as important features, which could affect the overall performance and reliability of the model.

To address these issues, this paper proposes the GF-CNN model, which simultaneously incorporates both max-pooling and average-pooling strategies. Through a gating mechanism, it dynamically adjusts the weights of the outputs from the different pooling layers, allowing the model to leverage both local important features and global distributional information, thus achieving a richer data representation. The structure of the GF-CNN model is shown in Figure 5.

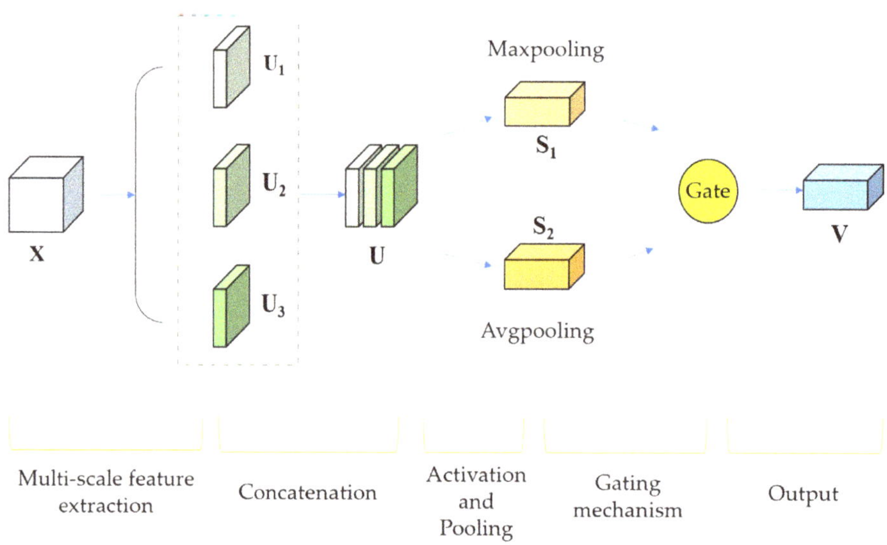

Figure 5. GF-CNN model structure.

For the original input features, $\mathbf{X}$, the model first applies three convolutional kernels of different sizes to extract multi-scale information, resulting in feature maps $\mathbf{U_1}$, $\mathbf{U_2}$, and $\mathbf{U_3}$. Next, $\mathbf{U_1}$, $\mathbf{U_2}$, and $\mathbf{U_3}$ are concatenated together.

$$\mathbf{U} = \mathrm{Concat}(\mathbf{U_1},\mathbf{U_2},\mathbf{U_3}) \tag{1}$$

After activating the fused features, both max-pooling and average-pooling are applied separately.

$$\mathbf{S_1} = \mathcal{P}_{\max}(\mathrm{ReLU}(\mathbf{U})) \tag{2}$$

$$\mathbf{S_2} = \mathcal{P}_{\mathrm{avg}}(\mathrm{ReLU}(\mathbf{U})) \tag{3}$$

The gating fusion unit assigns weights to $\mathbf{S_1}$ and $\mathbf{S_2}$ and performs a weighted fusion.

$$\mathbf{V} = \mathrm{Gated\_fusion}(\mathbf{S_1}, \mathbf{S_2}) \tag{4}$$

The structure of the gating fusion unit is illustrated in Figure 6. Suppose the gating fusion unit receives two inputs, $\mathbf{A_1}$ and $\mathbf{A_2}$. First, two learnable parameters, w_1 and w_2, are initialized for these inputs. Then, the sigmoid function is applied to generate the weight representation $\mathbf{B}$.

$$\mathbf{B} = \mathrm{Sigmoid}(w_1 \cdot \mathbf{A_1} + w_2 \cdot \mathbf{A_2}) \tag{5}$$

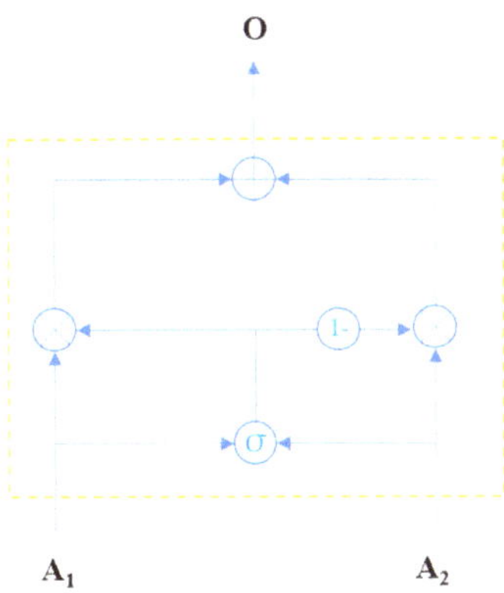

Figure 6. Gating fusion unit.

Subsequently, the two parts are weighted and summed according to the weight representation B, resulting in the final output O.

$$\mathbf{O} = \mathbf{B} \odot \mathbf{A_1} + (1 - \mathbf{B}) \odot \mathbf{A_2} \tag{6}$$

In summary, given inputs **A1** and **A2**, the two are first linearly transformed via learnable parameter matrices w_1 and w_2, respectively. The transformed results are then element-wise added together to generate interaction features. These features are subsequently mapped to the interval (0, 1) using a sigmoid function, forming a weight matrix **B**. This weight reflects the importance of A1 across various feature dimensions, while the weight of **A2** is automatically assigned as **1-B**. The final output **O** is obtained by computing the weighted sum of **A1** and **A2** with weights **B** and **1-B**, respectively. The gated fusion unit models the interaction relationships between inputs in a parametric manner, enabling fine-grained feature adaptive fusion. The normalization constraint further prevents redundancy or conflicts in weight allocation.

GF-CNN extracts and integrates features at multiple levels, enabling the model to respond to various data types more flexibly and effectively, thereby enhancing its generalization ability.

3.3. Softmax Function

The softmax function is an activation function commonly used in multi-class classification tasks. It transforms a real-valued vector into a probability distribution. Softmax is primarily applied to the final layer of a neural network to ensure that the output values represent probabilities, meaning that the sum of all output values equals 1, and each value falls within the range (0, 1).

For an input vector $z = [z_1, z_2, \ldots, z_n]$, the calculation formula of softmax is as follows:

$$\text{Softmax}(z_i) = \frac{e^{z_i}}{\sum_{j=1}^{n} e^{z_j}} \tag{7}$$

In the formula, e^{z_i} represents the exponential operation applied to each input, while the denominator is the sum of the exponentials of all the inputs.

4. Experiments and Analysis

4.1. Datasets

The Potato Pest and Disease Intention Recognition Dataset was divided into three subsets: training set, testing set, and validation set, with a ratio of 7:2:1. The validation set was used for fine-tuning the model's hyperparameters and monitoring its performance during training. Importantly, the three datasets are mutually exclusive, with no overlap or cross-contamination between them. As a result, the experimental results obtained from the testing set served as a reliable benchmark for evaluating the model's performance.

4.2. Experimental Setup

The experiments were conducted on a system running Ubuntu 20.04, with a NVIDIA GeForce RTX 4090D GPU. The programming language used was Python 3.8, and the framework employed for model development was PyTorch 2.0.0. The pre-trained model utilized was bert-base-chinese, which was downloaded from Huggingface. This model consists of 12 layers, with a 768-dimensional hidden state and a 12-head multi-head attention mechanism. Other parameter settings are detailed in Table 2.

Table 2. Experimental parameter settings.

Parameter	Value
learning rate	2×10^{-5}
batch_size	32
epoch	20
optimizer	Adam
text_len	30
filters	256
filter_size	(2, 3, 4) × 768

Precision (P), recall (R), and the $F1$ score were selected as evaluation metrics for the task. The $F1$ score is the harmonic mean of P and R, and it reflects the overall performance of the model. The specific calculation process is shown in Equations (8)–(10):

$$P = \frac{TP}{TP + FP} \tag{8}$$

$$R = \frac{TP}{TP + FN;} \tag{9}$$

$$F1 = \frac{2 \times P \times R}{P + R} \tag{10}$$

where TP denotes the number of samples correctly predicted as the positive class, FP refers to the number of samples incorrectly predicted as the positive class, and FN is the number of samples incorrectly predicted as the negative class.

4.3. Experimental Results and Analysis

In this study, six models—TextCNN [21], DPCNN [22], RCNN [23], LSTM-Attention [24], Stacked-LSTM [17], and Co-LSTM [25]—were compared with the proposed GF-CNN through a series of experiments. All models utilized word embeddings provided by BERT. The experimental results are presented in Table 3.

Table 3. Experimental result.

Model	P (%)	R (%)	$F1$ (%)
TextCNN	99.97	99.95	99.96
DPCNN	99.81	99.63	99.72
RCNN	99.91	99.89	99.90
LSTM-Attention	98.42	100.00	99.20
Stacked-LSTM	99.88	99.84	99.86
Co-LSTM	99.83	99.66	99.74
GF-CNN	100.00	100.00	100.00

As shown in Table 3, all models achieved impressive results, with P, R, and $F1$ all exceeding 98%. This can be attributed to two key factors:

(1) High-quality word embeddings: Since BERT was pre-trained on a large-scale corpus, it captured rich contextual and semantic features, thereby providing high-quality word vector representations for downstream tasks;

(2) High-quality dataset: The PDPQCD dataset underwent thorough scrutiny and filtering after training templates were generated through LLM, ensuring its exceptional quality. This result aligns with our ultimate goal of building an intelligent question-answering system, one that accurately understands user intentions.

Given that the GF-CNN model did not exhibit significant differentiation from other models on the PDPQCD dataset, the next section focused on experiments with several public datasets to further validate the GF-CNN model's effectiveness and generalizability.

5. Experiments on Public Datasets

5.1. Public Dataset

This section selects the Subj and THUCNews datasets for further experimentation. The Subj dataset contains 10,000 sentences, which are used to distinguish between subjective and objective sentences. Of these, 5000 sentences are labeled as subjective, and the other 5000 sentences are labeled as objective. THUCNews, provided by Tsinghua University, consists of 10 topic categories, such as education, entertainment, finance, and so on. Detailed information about both datasets is shown in Table 4.

Table 4. Dataset Details.

Datasets	Categories	Training Set	Validation Set	Test Set	Language
Subj	2	8 k	-	2 k	English
THUCNews	10	180 k	10 k	10 k	Chinese

5.2. Experimental Results

5.2.1. Performance Comparison of Different Models

Table 5 compares the performance of seven text classification models on the Subj and THUCNews datasets. The experimental results show that the GF-CNN model demonstrates significant classification advantages, with $F1$ scores of 96.70% and 93.55% on the two datasets, respectively, leading all models. In contrast, the TextCNN and LSTM-Attention models exhibit strong stability, consistently performing in the second tier across both datasets of different natures. The RCNN and Stacked-LSTM architectures, however, show clear performance limitations, with *F*1 scores of only 95.15% and 94.60% on the Subj dataset and even lower performance than the baseline (92.23% and 91.82%) on the THUCNews dataset. Notably, Co-LSTM and DPCNN exhibit significant data sensitivity; the former performs excellently on Subj but poorly on THUCNews, while the latter shows the opposite trend.

Table 5. Performance comparison of different models.

Models	Subj			THUCNews		
	P (%)	*R* (%)	*F*1 (%)	*P* (%)	*R* (%)	*F*1 (%)
TextCNN	95.79	95.84	95.80	92.52	92.49	92.49
DPCNN	94.46	94.21	94.28	92.74	92.72	92.72
RCNN	95.16	95.21	95.15	92.27	92.23	92.23
LSTM-Attention	95.45	95.50	95.45	92.60	92.56	92.56
Stacked-LSTM	94.59	94.63	94.60	91.89	91.83	91.82
Co-LSTM	95.72	95.60	95.64	92.19	92.15	92.15
GF-CNN	96.70	96.69	96.70	93.55	93.56	93.55

5.2.2. Confusion Matrix

Figures 7a and 7b respectively present the confusion matrices on the Subj and THUCNews datasets. As an important visualization tool for classification tasks, the confusion matrix clearly reflects the model's classification performance across different categories. In these matrices, each row represents the true category, while each column represents the predicted category. The numbers in the cells indicate the frequency with which a specific category was predicted as the corresponding category. Darker colors indicate higher frequencies, while lighter colors represent lower frequencies. In the Subj dataset, the classification results for the "Objective" category are slightly better than those for "Subjective". In the THUCNews dataset, categories such as "Sports" and "Education" are classified with relatively high accuracy, while categories like "Science" and "Stocks" show slightly poorer performance.

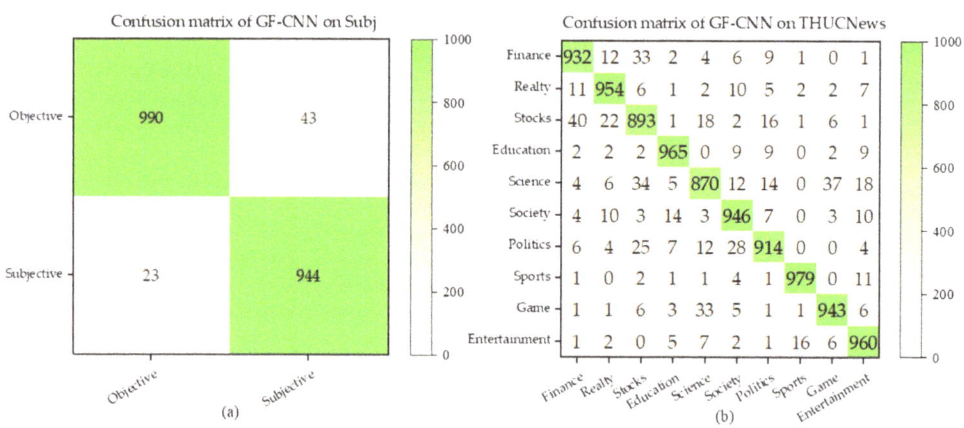

Figure 7. Confusion matrix.

5.2.3. Performance Comparison of Different Feature Fusion Methods

This subsection selects the following three methods to fuse the features processed by maximum pooling and average pooling, and compares them with the gating mechanism.

The experimental results on the Subj and THUCNews datasets are presented in Figures 8a and 8b, respectively.

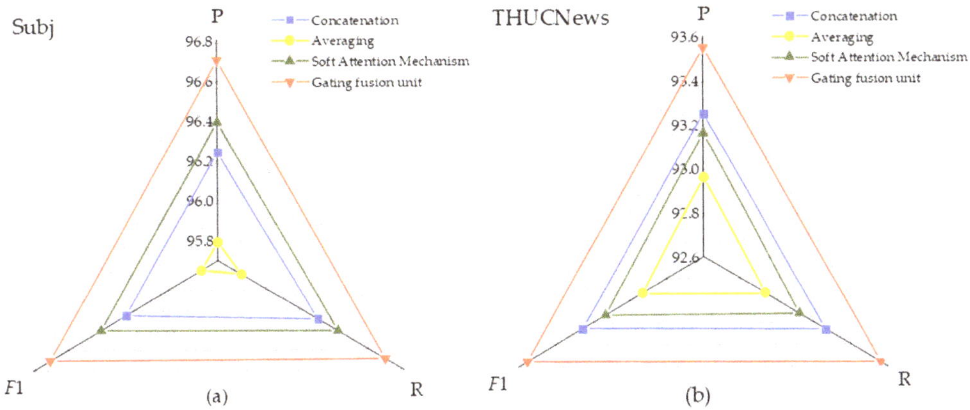

Figure 8. Performance comparison of different feature fusion.

Concatenation: The features processed by max-pooling and average-pooling are 2D tensors of the shape [Batch_size, Embedding_dim]. These tensors are concatenated along the last dimension, resulting in a tensor of the shape [Batch_size, 2×Embedding_dim].

Averaging: The two tensors are summed element-wise, and then, the mean is taken without changing the shape of the tensor.

Soft attention mechanism: First, the two tensors are concatenated along a new dimension, resulting in a three-dimensional tensor with the shape [2, Batch_size, Embedding_dim]. Subsequently, the softmax function is applied along this new dimension to obtain the weight representations for each tensor. These weights are then used to perform element-wise multiplication along the new dimension, followed by summation to obtain the fused tensor. The resulting fused tensor has the shape [Batch_size, Embedding_dim].

From Figure 8, it can be observed that the three alternative feature fusion methods—concatenation, averaging, and soft attention mechanism—all result in lower *F*1 scores compared to the gating fusion unit. Among them, the performance drop is most significant with the averaging method. This is likely because averaging simply takes the mean of the features obtained from max-pooling and average-pooling, disregarding the fact that different features may carry varying levels of importance and information. Despite the decline in performance, these feature fusion methods still outperform other baseline models in most cases. This could be attributed to the complementary nature of max-pooling and average-pooling, which together provide a more comprehensive representation of the data. By effectively combining these two pooling strategies, the extracted features are richer and more robust, even with simpler fusion methods.

5.2.4. Ablation Study

This section presents an ablation experiment to evaluate the impact of different model components on overall performance. The experimental results are shown in Table 6.

From Table 6, we can observe the following: (1) The method of generating word embeddings has a decisive impact on model performance. When BERT was replaced with Word2Vec, the *F*1 scores of the model on the three datasets decreased by 1.43%, 2.5%, and 2.33%, respectively. This significant difference confirms that BERT has a clear advantage over traditional static word embedding methods in capturing contextual semantic information and generating high-quality word vectors. (2) The effectiveness of the multi-scale feature extraction mechanism is thus validated. Experimental results show that using a single convolutional kernel (kernel_size = 3 × 768) led to a decrease in F1 scores compared to the multi-kernel combination (kernel_size = 2 × 768, 3 × 768, 4 × 768). This is attributed to the fact that convolutional kernels at different scales can capture local language features and long-range dependencies in parallel. (3) In the comparison of pooling strategies, max-pooling demonstrated a relative advantage over average-pooling. However, when used individually, both methods result in information loss. Therefore, it is recommended in practice to adopt a dynamic weighted fusion strategy to balance key feature retention and overall information consistency.

Table 6. Ablation study results.

Model Components	**PPDQCD**			**Subj**			**THUCNews**		
	P (%)	*R* (%)	*F*1(%)	*P* (%)	*R* (%)	*F*1(%)	*P* (%)	*R* (%)	*F*1(%)
Word2Vec word embeddings	97.19	100.00	98.57	94.23	94.20	94.20	91.28	91.24	91.26
Using a single convolutional kernel	99.28	100.00	99.64	95.90	95.89	95.89	93.00	93.01	93.00
Using only max-pooling	99.95	99.95	99.95	95.61	95.37	95.44	92.58	92.55	92.55
Using only average-pooling	99.93	99.94	99.94	95.54	95.45	95.44	92.31	92.27	92.27
Our model	100.00	100.00	100.00	96.70	96.69	96.70	93.55	93.56	93.55

6. Discussion

Although this study has made certain progress in classifying questions related to potato pest and disease, there are still some limitations that need to be addressed in future research.

First, at the data construction level, while the dataset generated based on LLM ensures standardization and consistency, it carries the potential risk of homogeneity. These highly standardized training samples may weaken the model's ability to generalize to natural language variations, especially in real-world application scenarios where user inputs often involve non-standard linguistic phenomena such as spelling errors, grammatical mismatches, and semantic ambiguity. However, the models trained in a controlled environment have not yet undergone rigorous multimodal noise testing. In the future, we plan to collect natural language expressions from real-world scenarios and combine data augmentation strategies to construct a hybrid dataset with ecological validity.

Second, in terms of model architecture, although the existing GF-CNN network demonstrates good feature extraction capabilities, there is still a gap between its computational complexity and the real-time response requirements in agricultural production scenarios. The research team plans to implement a dual-path optimization approach: on one hand, we will advance lightweight architecture improvements, reducing the number of parameters through channel pruning and quantization compression; on the other hand, we will explore knowledge distillation techniques to improve computational efficiency while maintaining classification accuracy.

7. Conclusions

This study has addressed the current lack of high-quality question-answer datasets in the field of potato pest and disease and achieved dual breakthroughs at the methodological level. On the one hand, it innovatively constructed the first domain-specific question-answer dataset for this field, and on the other hand, it designed a classification model with feature fusion advantages. Specifically, this research proposes a question classification dataset construction method based on prompt engineering. First, prompts are used to guide the LLM in generating question templates, and then, entities extracted from previous NER work are batch-inserted into the question templates, resulting in the construction of a high-quality PDPQCD. Additionally, this study introduces the GF-CNN question classification model, which employs (a) a parallel dual-pooling structure, using max-pooling to capture key semantic features and average-pooling to retain global statistical information, and (b) a learnable gating fusion module to dynamically adjust the feature weights from the dual channels. Experimental results show that GF-CNN achieves F1 scores of 100.00%, 96.70%, and 93.55% on the PDPQCD, Subj, and THUCNews datasets, respectively, outperforming other models.

Future work includes the completion of a potato pest and disease intelligent question-answering system, with exploration of LLM integration to enhance text-generation capabilities.

Author Contributions: Conceptualization, W.T. and Z.H.; methodology, W.T. and Z.H.; software, W.T.; validation, W.T.; formal analysis, W.T.; investigation, W.T. and Z.H.; resources, W.T.; data curation, W.T. and Z.H.; writing—original draft, W.T.; writing—review and editing, Z.H.; supervision, Z.H.; project administration, Z.H. All authors have read and agreed to the published version of the manuscript.

Funding: Key Discipline Construction of Gannan Normal University (220108), Science and Technology Project of Jiangxi Provincial Department of Education (490164), National Key R&D Program of China (2017YFD0701600), and Gannan Normal University Talent Fund (13SJJ202130).

Institutional Review Board Statement: Not applicable.

Data Availability Statement: The data and code are available at https://github.com/Tracyyytao/Intent-Recognition/tree/master (accessed on 23 February 2025).

Conflicts of Interest: The authors declare no conflicts of interest.

References

1. Liu, Y.; Feng, H.; Yue, J.; Li, Z.; Yang, G.; Song, X.; Yang, X.; Zhao, Y. Remote-sensing estimation of potato above-ground biomass based on spectral and spatial features extracted from high-definition digital camera images. *Comput. Electron. Agric.* **2022**, *198*, 107089. [CrossRef]
2. Meno, L.; Escuredo, O.; Rodríguez-Flores, M.S.; Seiji, M.C. Looking for a sustainable potato crop. Field assessment of early blight management. *Agric. For. Meteorol.* **2021**, *308*, 108617. [CrossRef]
3. Saffer, A.; Tateosian, L.; Saville, A.C.; Yang, Y.; Ristaino, J.B. Reconstructing historic and modern potato late blight outbreaks using text analytics. *Sci. Rep.* **2024**, *14*, 2523. [CrossRef] [PubMed]
4. Zhu, H.; Shi, W.; Guo, X.; Lyu, S.; Yang, R.; Han, Z. Potato disease detection and prevention using multimodal AI and large language model. *Comput. Electron. Agric.* **2025**, *229*, 109824. [CrossRef]
5. Ferro, M.V.; Sørensen, C.G.; Catania, P. Comparison of different computer vision methods for vineyard canopy detection using UAV multispectral images. *Comput. Electron. Agric.* **2024**, *225*, 109277. [CrossRef]
6. Yang, T.; Mei, Y.; Xu, L.; Yu, H.; Chen, Y. Application of question answering systems for intelligent agriculture production and sustainable management: A review. *Resour. Conserv. Recycl.* **2024**, *204*, 107497. [CrossRef]
7. Guo, X.; Wang, J.; Gao, G.; Zhou, J.; Li, Y.; Cheng, Z.; Miao, G. Efficient Agricultural Question Classification with a BERT-Enhanced DPCNN Model. *IEEE Access* **2024**, *12*, 109255–109268. [CrossRef]
8. Wei, F.; Duan, Q.; Xiao, X.; Zhang, L. Classification Technique of Chinese Agricultural Text Information Based on SVM. *Trans. Chin. Soc. Agric. Mach.* **2015**, *46*, 174–179.
9. Cui, X.; Shi, D.; Chen, Z.; Xu, F. Parallel forestry text classification technology based on XGBoost in spark framework. *Trans. Chin. Soc. Agric. Mach.* **2019**, *50*, 280–287.
10. Espejo-Garcia, B.; Martinez-Guanter, J.; Pérez-Ruiz, M.; Lopez-Pellicer, F.J.; Zarazaga-Soria, F.J. Machine learning for automatic rule classification of agricultural regulations: A case study in Spain. *Comput. Electron. Agric.* **2018**, *150*, 343–352. [CrossRef]
11. Tang, W.; Hu, Z. Survey of agricultural knowledge graph. *Comput. Eng. Appl.* **2024**, *60*, 63–76.
12. Jin, N.; Zhao, C.; Wu, H.; Miao, Y.; Li, S.; Yang, B. Classification technology of agricultural questions based on BiGRU_MulCNN. *Trans. Chin. Soc. Agric. Mach.* **2020**, *51*, 199–206.

13. Wang, H.; Wu, H.; Feng, S.; Liu, Z.; Xu, T. Classification technology of rice questions in question answer system based on Attention_DenseCNN. *Trans. Chin. Soc. Agric. Mach.* **2021**, *52*, 237–243.
14. Feng, S.; Xu, T.; Zhou, Y.; Zhao, D.; Jing, N.; Wang, H. Rice Knowledge Text Classification Based on Deep Convolution Neural Network. *Trans. Chin. Soc. Agric. Mach.* **2021**, *52*, 257–264.
15. Devlin, J.; Chang, M.; Lee, K.; Toutanova, K. BERT: Pre-training of Deep Bidirectional Transformers for Language Understanding. *arXiv* **2018**, arXiv:1810.04805.
16. Sun, Y.; Wang, S.; Li, Y.; Feng, S.; Chen, X.; Zhang, H.; Tian, X.; Zhu, D.; Tian, H.; Wu, H. ERNIE: Enhanced Representation through Knowledge Integration. *arXiv* **2019**, arXiv:1904.09223.
17. Li, L.; Diao, L.; Tang, Z.; Bai, Z.; Zhou, H.; Guo, X. Question classification method of agricultural diseases and pests based on BERT_Stacked LSTM. *Trans. Chin. Soc. Agric. Mach.* **2021**, *52*, 172–177.
18. Li, D.; Bai, T.; Xiang, H.; Dai, S.; Wang, Z. Intention recognition of cotton disease and pest questions based on ERNIE and improved DPCNN. *Shandong Agric. Sci.* **2024**, *56*, 143–151.
19. Duan, X.; Li, Z.; Liu, L.; Liu, Y. Multimodal Chinese agricultural news classification method based on interactive attention. *IEEE Access* **2024**, *12*, 161718–161731. [CrossRef]
20. Tang, W.; Wen, X.; Li, M.; Chen, Y.; Hu, Z. ResiAdvNet: A named entity recognition model for potato diseases and pests based on progressive residual structures and adversarial training. *Comput. Electron. Agric.* **2024**, *227*, 109543. [CrossRef]
21. Kim, Y. Convolutional neural networks for sentence classification. *arXiv* **2014**, arXiv:1408.5882.
22. Johnson, R.; Zhang, T. Deep pyramid convolutional neural networks for text categorization. In Proceedings of the 55th Annual Meeting of the Association for Computational Linguistics, Vancouver, BC, Canada, 30 July–4 August 2017; pp. 562–570.
23. Lai, S.; Xu, L.; Liu, K.; Zhao, J. Recurrent convolutional neural networks for text classification. In Proceedings of the AAAI Conference on Artificial Intelligence, Austin, TX, USA, 25–30 January 2015; p. 29.
24. Bai, X. Text classification based on LSTM and attention. In Proceedings of the 2018 Thirteenth International Conference on Digital Information Management, Berlin, Germany, 24–26 September 2018; pp. 29–32.
25. Behera, R.K.; Jena, M.; Rath, S.K.; Misra, S. Co-LSTM: Convolutional LSTM model for sentiment analysis in social big data. *Inf. Process. Manag.* **2021**, *58*, 102435. [CrossRef]

agriculture

MDPI

Article

DCFA-YOLO: A Dual-Channel Cross-Feature-Fusion Attention YOLO Network for Cherry Tomato Bunch Detection

Shanglei Chai [1], Ming Wen [1], Pengyu Li [1], Zhi Zeng [2] and Yibin Tian [1,*]

1 College of Mechatronics and Control Engineering, Shenzhen University, Shenzhen 518060, China; 2450096004@mails.szu.edu.cn (S.C.); wenming@szu.edu.cn (M.W.); 2310294020@email.szu.edu.cn (P.L.)
2 School of Computer and Information Science, Chongqing Normal University, Chongqing 401331, China; zzh406@cqnu.edu.cn
* Correspondence: ybtian@szu.edu.cn

Abstract: To better utilize multimodal information for agriculture applications, this paper proposes a cherry tomato bunch detection network using dual-channel cross-feature fusion. It aims to improve detection performance by employing the complementary information of color and depth images. Using the existing YOLOv8_n as the baseline framework, it incorporates a dual-channel cross-fusion attention mechanism for multimodal feature extraction and fusion. In the backbone network, a ShuffleNetV2 unit is adopted to optimize the efficiency of initial feature extraction. During the feature fusion stage, two modules are introduced by using re-parameterization, dynamic weighting, and efficient concatenation to strengthen the representation of multimodal information. Meanwhile, the CBAM mechanism is integrated at different feature extraction stages, combined with the improved SPPF_CBAM module, to effectively enhance the focus and representation of critical features. Experimental results using a dataset obtained from a commercial greenhouse demonstrate that DCFA-YOLO excels in cherry tomato bunch detection, achieving an mAP50 of 96.5%, a significant improvement over the baseline model, while drastically reducing computational complexity. Furthermore, comparisons with other state-of-the-art YOLO and other object detection models validate its detection performance. This provides an efficient solution for multimodal fusion for real-time fruit detection in the context of robotic harvesting, running at 52fps on a regular computer.

Keywords: cherry tomato bunch detection; robotic harvesting; multimodal image; feature extraction; feature fusion; YOLO network

1. Introduction

With the aging of the global population and the increasing shortage of agricultural labor, modern agriculture is facing unprecedented challenges [1]. Against this backdrop, the rapid development of smart agricultural technology becomes an important driving force for promoting agricultural transformation. As a popular fruit, cherry tomatoes play an important role in global agricultural production due to their appeasing taste and richness in lycopene [2]. However, due to the small size and uneven distribution, their harvesting faces significant challenges. Existing mechanical picking equipment often misses fruits or mistakenly picks leaves because it is difficult to accurately identify the location of the fruit, making it difficult to effectively improve robotic harvesting efficiency and quality. Therefore, developing efficient and intelligent identification and position detection for cherry tomatoes is the key to improving production efficiency and fruit quality.

Fruit detection is one of the core components of fruit-harvesting robots [3]. In recent years, Deep Neural Networks (DNNs) have made significant progress in fruit detection. These algorithms no longer rely on manually designed features and can automatically learn complex patterns in images, effectively overcoming the limitations of traditional fruit detection methods when facing complex lighting and background interference. Object detection methods based on deep learning are mainly divided into single-stage detection and two-stage detection. The single-stage detection directly regresses and predicts the bounding box of the target. Its advantages are a fast detection speed and high efficiency, and it is one of the most widely used methods. Typical single-stage detection methods include the YOLO series and the SSD network [4–6]. For example, Lyu et al. improved the backbone network and loss function of YOLOv5 in YOLOv5-CS for the detection and counting of green citrus in orchard environment [7]. Wang et al. used ShuffleNet v2 as an improved backbone and introduced the CBAM attention mechanism to improve detection accuracy and reduce the model size [8]. Gai et al. improved YOLOv4 by combining it with DenseNet for the maturity detection of small tomatoes [9]. Zhao et al. achieved an accurate detection of grapes and picking points by improving YOLOv4, where the average precision of grape detection reached 93.27% [10]. Another popular single-stage detection method is the SSD model by Yuan et al. [11], who verified its performance on cherry tomatoes using different backbone networks (VGG16, MobileNet, Inception V2). In addition, Fuentes-Peñailillo et al. proposed a seedling counting model that combines traditional image processing with MobileNet-SSD, achieving a maximal precision of 96% [12]. In contrast, two-stage detection processes candidate region extraction and classification in two steps. Although the detection speed is slower, it performs better in precision. Typical two-stage detection methods include the RCNN series [13–15] and SPPNet [16]. For example, Hu et al. used Faster R-CNN combined with color space conversion and fuzzy set method to realize bounding box detection and the segmentation of tomatoes [17]. It performed well in cases where the fruit edges were blurred or overlapped. Song et al. built a Faster R-CNN model based on VGG16 and achieved kiwifruit detection under different lighting conditions [18]. The average detection accuracy reached 87.61%. Gao et al. used Faster R-CNN to detect and classify apples under four different growth conditions, with an average precision of 87.9% [19].

Although single-modal visual data processing based on deep learning has made significant progress in fruit detection, they still face many challenges in complex environments. First, single-modal methods (e.g., RGB only detection) are highly susceptible to illumination changes, occlusion, and background clutter, leading to insufficient detection accuracy and robustness. For instance, in low-light conditions or when fruits are partially occluded by leaves, the performance of these methods degrades significantly. Second, most existing models rely heavily on large-scale annotated datasets, which are labor-intensive and time-consuming to acquire, especially for small and densely distributed fruits like cherry tomatoes. Third, while some studies have attempted to improve detection accuracy by increasing model complexity, this usually comes at the cost of increased computational cost, making them unsuitable for real-time applications, such as robotic harvesting. To address these issues, multimodal fusion has gradually become an important research direction to meet these challenges. Especially when picking, robots are usually equipped with multiple sensors, such as color and depth cameras. Fusing information from different modalities can not only improve the precision of fruit detection but also enhance the system's ability to adapt to environmental changes. In recent years, multimodal methods combining color (RGB) and depth images have been widely used in fruit detection and made remarkable progress. Tu et al. developed a fruit and maturity detection model using RGB-D images [20]. Their ablation experiment showed that the introduction of depth improved the detection accuracy by 3.03%. Cui et al. proposed a single cherry tomato detection method that combines RGB-D inputs. The RGB image is converted into the LAB color space and then fused with the depth map and normal vector map obtained from the point cloud as the inputs of an improved YOLOv7 to detect cherry tomatoes [21,22]. Similarly, Rong et al. optimized YOLOv5 by fusing RGB and depth images to improve the detection performance of cherry tomato clusters/bunches [23]. In addition, Kaukab et al. used multimodal data as the input of YOLOv5 and effectively reduced the impact of depth image noise via a deep fusion method of non-targeted background removal and achieved an precision of 96.4% for apple detection [24].

The fusion methods of multimodal images can be divided into three types, early fusion, mid-term fusion, and late fusion, according to its introduction at different processing stages. In early-fusion methods, images from different modalities are directly integrated at the image level before being fed into the network. For example, Liu et al. significantly improved the average precision of kiwifruit detection to 90.7% by superimposing RGB and near-infrared images to form a four-dimensional tensor as the input of the VGG16 [25]. However, this method is prone to pixel misalignment due to offset between images of different modalities. In contrast, late fusion linearly combines independent predictions at the decision stage. Sa et al. combined the detection results of RGB and NIR images in the last step of the shared feature extraction network Faster R-CNN and improved the detection accuracy from 81.6% to 83.8% [26], especially effective for fruits with similar colors, such as green peppers and melons. However, in late fusion, different modality features cannot learn from each other, resulting in poor interactions between different network branches. As a feature-level fusion method, mid-term fusion combines multimodal feature maps and can effectively balance the depth and efficiency of feature integration. For example, Wei et al. designed a multi-branch backbone network that includes color, infrared, and polarization image inputs [27]. By using operations such as feature connection, dimensionality reduction, and activation, they achieved higher target-recognition precision in complex environments. However, it also significantly reduces the inference speed.

Despite the progress in multimodal fusion, several challenges remain unresolved. First, early-fusion methods often suffer from pixel misalignment due to the spatial offset between RGB and depth images, leading to inaccurate feature extraction. Second, late-fusion methods lack effective interaction between modalities, as they process each modality independently until the final decision stage, which limits their ability to leverage complementary information. Third, mid-term-fusion methods, while achieving better feature integration, often introduce high computational complexity, making them unsuitable for real-time applications such as robotic harvesting. Additionally, most existing multimodal fusion methods do not fully exploit the potential of attention mechanisms to enhance the saliency of critical features and suppress irrelevant information, which is crucial for improving detection accuracy in complex environments. These limitations underscore the need for a more efficient and robust multimodal fusion approach that can address these challenges while maintaining real-time performance.

In order to further optimize the performance of multimodal fusion, in recent years, researchers began to introduce attention mechanisms to highlight critical information and reduce the interference of irrelevant features. Woo et al. introduced self-learning weight parameters in the CBAM module, which effectively improves the weight of the region of interest and suppresses invalid features [28]. Li et al. proposed a solution using DenseBlock combined with the attention mechanism to achieve image denoising when fusing RGB and NIR images for multimodal segmentation tasks [29]. Inspired by their work, this paper proposes a dual-channel cross-fusion attention YOLO (DCFA-YOLO) network to meet the needs of accurate and fast multimodal cherry tomato bunch detection. Specifically, we introduce a parallel attention mechanism in the mid-term-fusion stage. By enhancing the learning ability and saliency of the region of interest, it overcomes the interference introduced by pixel offset in early fusion and the high independence of each modality feature in late fusion. DCFA-YOLO integrates multimodal data of color and depth images and uses dual-channel cross-fusion and attention mechanisms to effectively extract and fuse features of different modalities. At the same time, the model is designed with the goal of being lightweight for real-time applications. Even when multimodal fusion is used, the size of its model parameters remains similar to that of the single-modal method, ensuring its high efficiency and ease of deployment. The design of this model not only enhances the ability to understand and process multimodal data but also takes into account the conservation of computing resources. It provides a reliable and efficient solution for precision agriculture and intelligent fruit harvesting. The contributions of this paper can be summarized as follows:

- A new feature fusion method is developed to combine different modal features effectively. It integrates a dual-channel cross-fusion attention mechanism into YOLO to enhance the fusion of color and depth images in a balanced way, significantly improving fruit detection accuracy and robustness.
- An efficient lightweight design is proposed by replacing the C2f module with the ShuffleNetV2 unit, optimizing the backbone network for faster and more efficient early feature extraction, while reducing computational complexity.
- The attention mechanism is enhanced by introducing a SPPF_CBAM unit to the early feature extraction stage, improving the model's ability to focus on key features through dynamic channel and spatial attention.
- The proposed DCFA-YOLO has been evaluated for cherry tomato bunch detection using a dataset obtained from a commercial greenhouse, achieving an mAP50 of 96.5%, outperforming multiple YOLO models while being relatively lightweight.

The paper is organized as follows: Section 1 provides an overview of related work and existing challenges in fruit detection as well as multimodal object detection. Section 2 describes the proposed DCFA-YOLO method in detail. Section 3 experimentally verifies and analyzes the effectiveness of DCFA-YOLO. Section 4 discusses the results and implications and potential directions for future research, while Section 5 concludes the work.

2. Proposed Methods

The proposed DCFA-YOLO is based on the YOLOv8_n model [30]. Through targeted improvements in the backbone network, feature fusion mechanism, and attention mechanism, the feature extraction and fusion capabilities of multimodal images were significantly improved. At the same time, the computational complexity of the model was effectively reduced. In view of the long-tail characteristics of the target distribution of fruits in complex harvesting environments, the Distribution Focus Loss (DFL) was employed to optimize the model's ability to focus on targets of different sizes. This provided an efficient solution for the multimodal fruit detection task. The overall structure of DCFA-YOLO is shown in Figure 1.

2.1. Cross-Fusion Mechanism

The input to the cross-fusion mechanism consisted of three groups of features initially extracted by the backbone network at different depths (Figure 2): two initial modality features (RGB and depth) and one additional set of features obtained by adding Feat3_rgb and Feat3_depth, followed by upsampling. These three groups served as the basic feature inputs for the cross-fusion process. To enhance the semantic information of these features, the two Feat3 basic feature layers were first added, which were upsampled and cross fused with other two sets of feature layers sequentially. Since the relatively shallow Feat1 and Feat2 contained more detailed information, we directly fused the independent upsampled RGB and depth feature layers using the Concat_BiFPN module.

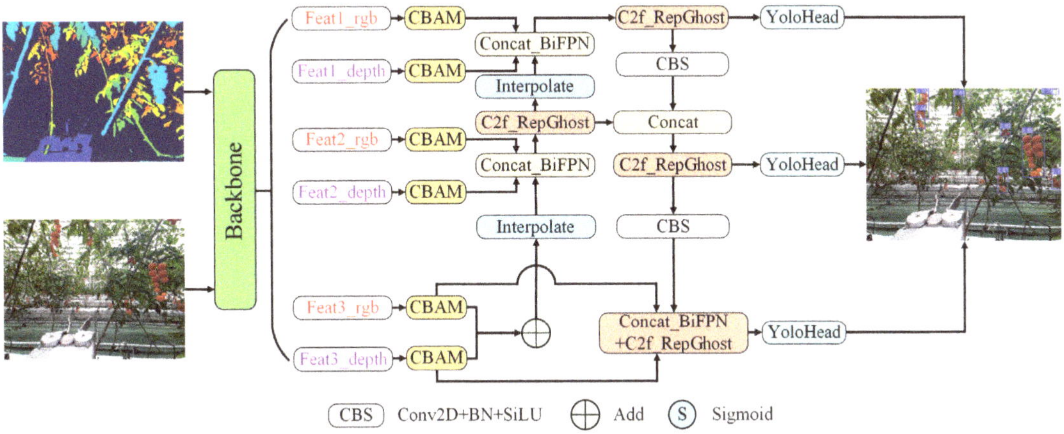

Figure 1. The overall architecture of the proposed DCFA-YOLO.

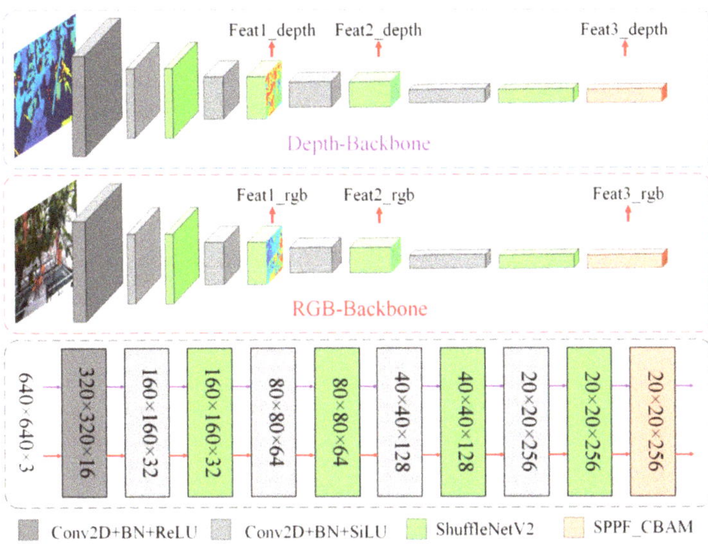

Figure 2. The structure of the backbone network.

The Concat_BiFPN module has many advantages over the traditional BiFPN_Add and simple Concat concatenation [31]. First, unlike BiFPN_Add, Concat_BiFPN not only implements dynamic weighting of features through learnable weight parameters but also retains the independence of each input feature. The BiFPN_Add directly sums the input features after weighting, which may lead to the loss of modal feature details despite the tight fusion. Concat_BiFPN, on the other hand, combines features along the specified dimension by weighting, which can not only integrate multimodal information but also retain the differences of the original features. This provides a richer and more complete feature expression for subsequent processing. Second, compared with the single Concat operation, Concat_BiFPN introduces a dynamic weighting mechanism, which enables the contribution of features from each modality to be adaptively adjusted according to task requirements. This overcomes the feature imbalance problem in simple concatenation. Third, Concat_BiFPN supports a flexible adjustment of the dimension and number of input features, making it more adaptable.

The formula for the Concat_BiFPN module is

$$\hat{x} = \mathrm{concat}(\overline{w_1} \cdot x_1, \overline{w_2} \cdot x_2, \overline{w_3} \cdot x_3, \text{ dim}) \tag{1}$$

where x_1, x_2, and x_3 are input feature maps from different modalities; $\overline{w_1}$, $\overline{w_2}$, and $\overline{w_3}$ are learned weights used to adjust the contribution of different modal features; and dim is the concatenated dimension, which can usually be selected as the channel dimension of the feature map.

$$[\overline{w_1}, \overline{w_2}, \overline{w_3}] = [w_1, w_2, w_3] / (w_1 + w_2 + w_3 + \epsilon) \tag{2}$$

where $w = [w_1, w_2, w_3]$ is the weight vector, ϵ is a small constant used to avoid division by zero, and the value 3 refers to the three input modalities (RGB, depth, and the fused feature) that are being combined in the Concat_BiFPN module. The normalization ensures their contributions are adjusted appropriately during fusion.

It is worth noting that we used bilinear interpolation in the upsampling of feature cross fusion. This allowed for smoother feature map enlargement with more information preservation.

2.2. Model Simplifications

Compared with the networks using a single-modal input, DCFA-YOLO uses multimodal cross fusion and dual-channel feature extraction, leading to a significant increase in the amount of computation. To effectively reduce the computational complexity and parameter size while maintaining high-precision detection performance, we propose targeted optimizations. Specifically, we replaced the C2f module in the backbone with the ShuffleNetV2 unit, the C2f module in cross-feature fusion was replaced by the C2f_RepGhost. These modifications were designed to balance computational efficiency with high-quality feature extraction by considering the distinct roles of the backbone and feature fusion stages.

As shown in Figure 2, the backbone w responsible for preliminary feature extraction. Its goal is to extract global features from the input multimodal data. In order to reduce the computational complexity at this stage, C2f was replaced by the ShuffleNetV2 unit, as shown in Figure 3a. ShuffleNetV2 is a lightweight convolutional neural network that significantly reduces the amount of computation and number of parameters through sophisticated techniques such as Channel Split, grouped convolutions, and multi-scale feature fusion. In addition, Channel Shuffle helped to enhance feature representation by breaking information isolation in group convolutions, resulting in more effective feature extraction with reduced computational overhead. These optimizations allowed for ShuffleNetV2 to significantly reduce GFLOPs and parameter size while maintaining relatively efficient feature extraction.

The feature-cross-fusion stage served as the neck of the network, a crucial component for enhancing feature extraction and improving model performance. As shown in Figure 4, the C2f module was replaced by the C2f_RepGhost module in the feature-cross-fusion stage to achieve efficient feature extraction. It combined depth-wise convolution (DWConv) for spatial feature extraction and pointwise convolution (PWConv) for channel interaction, improving feature expression capabilities while reducing computation. In addition, the module introduced shortcut connections to achieve residual learning, mitigating gradient vanishing during training and promoting feature reuse. Moreover, re-parameterization techniques during the inference phase merged the module's complex structures into efficient inference configurations, further reducing the computational complexity for inference. Through these designs, the C2f_RepGhost module enhanced the deep-level feature expression capability while ensuring computational efficiency.

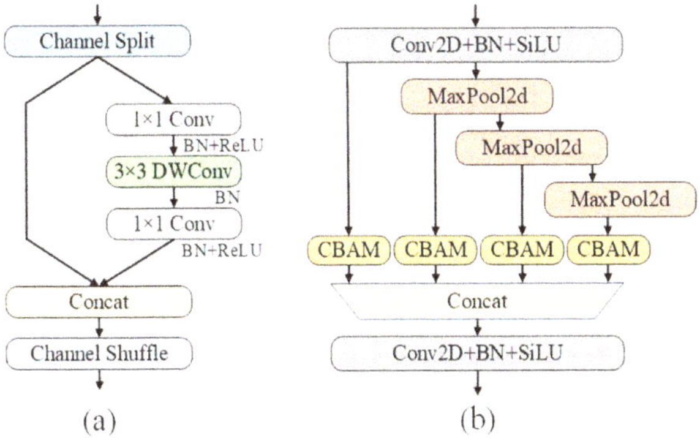

Figure 3. (**a**) The structure of the ShuffleNetV2 unit; (**b**) the structure of the SPPF_CBAM unit.

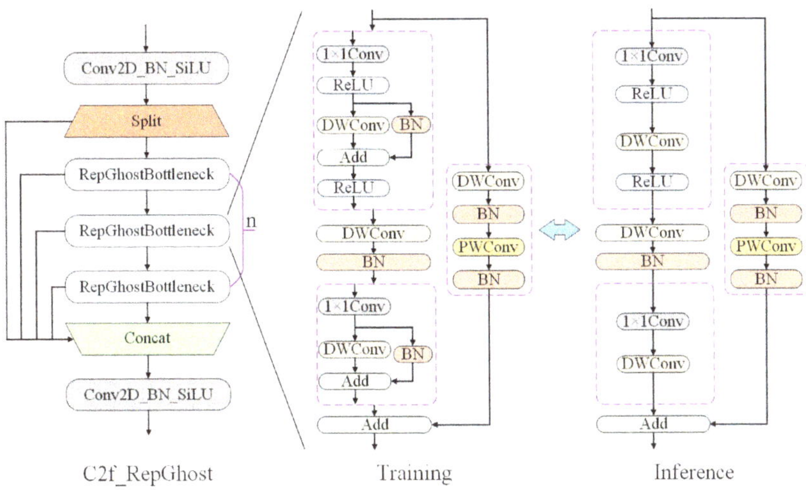

Figure 4. The structure of the C2f_RepGhost module. The module adds skip links during the training phase compared to the inference phase.

These targeted lightweight improvement strategies consider the characteristics of each stage to maximize the respective functions of the backbone and feature fusion. It not only meets the lightweight requirements of preliminary feature extraction but also meets the high performance requirements of enhanced feature fusion.

2.3. Attention Mechanism

To further improve the model's ability to extract and fuse multimodal features, an attention mechanism was introduced into the backbone network. The improvements were reflected in the optimization of the SPPF module and the enhanced attention in the early feature extraction layer. At the last backbone layer, the original SPPF module was replaced by the SPPF_CBAM, as shown in Figure 3b. The details of the CBAM module are shown in Figure 5. Through the parallel channel and spatial attention modules, the weights of different channels and spatial positions in the feature map can be adaptively adjusted. This strengthens key features and suppresses redundant information, improving the feature expression ability of the model.

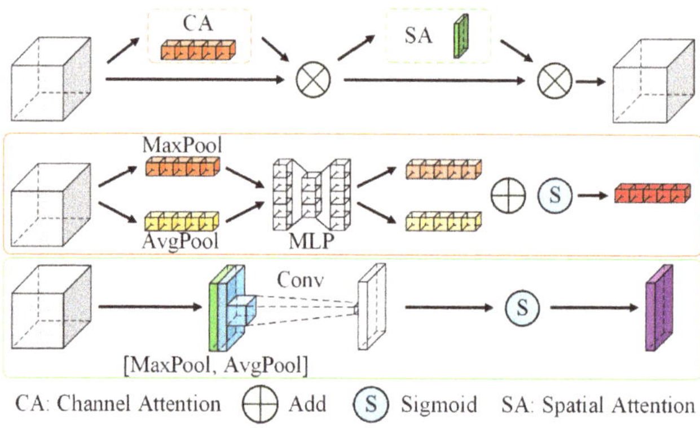

Figure 5. The structure of the CBAM module.

In addition, to fully utilize the advantages of multimodal inputs, the model added CBAM modules to the early-feature layers of RGB and depth maps in the backbone network. This allowed for the features of each modality to be enhanced through the attention mechanism at early stage. The channel attention captured the importance of different feature channels, while the spatial attention focused on the salient regions in the feature map. Ultimately, these enhanced multimodal feature layers were able to fully complement and interact with each other during the fusion stage. This effectively improved the robustness of feature extraction, providing more accurate features for subsequent detection tasks.

2.4. Loss Function

As shown in Figure 6, statistical analysis of the cherry tomato dataset (described below) revealed that the fruit sizes followed a significant long-tail distribution. Among the 22,965 annotated bounding boxes, the average object width and height were 48.2 and 79.9 pixels, while the maximal width and height reached 223 and 387 pixels, respectively, and the minima were close to zero. Looking further at the distribution of widths, 55.47% of the target widths were concentrated in [24.8, 49.6] pixels, while only 0.02% of them exceeded 198.2 pixels. Likewise, 50.44% of the target heights were within [43.9, 86.8] pixels, and only 0.02% of them were above 344.1 pixels. In terms of area distribution, 88.42% of the targets were below 8319 pixels, while only 0.01% of them were above 66,549 pixels. These statistics indicated that a large number of objects had small sizes, while a very small number of objects had significantly larger sizes, resulting in an imbalanced distribution, which poses significant challenges to traditional object detection models, especially in balancing the high detection rate of small objects and the high position precision of large ones.

To address the above problem, we used Distribution Focal Loss (DFL) in the bounding box loss function [32]. Unlike traditional methods that predict bounding boxes as discrete values, DFL represents the bounding box locations as continuous probability distributions. This can more accurately describe the uncertainty of the target location and optimize the model's attention to targets of different sizes. Specifically, DFL converts the predicted bounding boxes into a probability distribution and applies a focal loss on it to address the data class imbalance. It is able to strike a balance between the small objects that dominate the dataset and the large objects that are rare but important. Specifically, the formula of DFL is

$$\mathrm{DFL}(S_i, S_{i+1}) = -((y_{i+1} - y)\log(S_i) + (y - y_i)\log(S_{i+1})) \tag{3}$$

where y_i and y_{i+1} are two adjacent discrete points of the predicted bounding box, y is the true value, and S_i and S_{i+1} are the predicted probabilities corresponding to y_i and y_{i+1}.

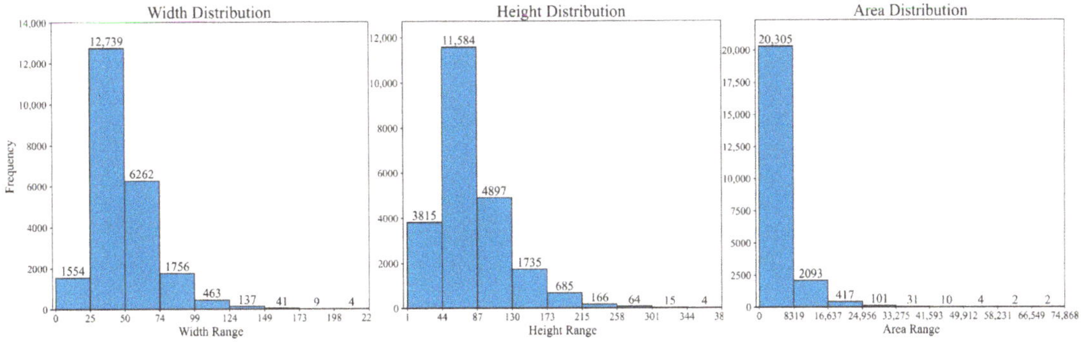

Figure 6. Statistical analyses of the cherry tomato dataset used in the current study (unit: pixel).

3. Experiments and Results

To verify the effectiveness and performance of the proposed DCFA-YOLO, a series of ablation and comparative experiments were designed and implemented. Through ablation experiments, the impact of each component on model performance is analyzed. At the same time, the full DCFA-YOLO is compared with a few state-of-the-art YOLO networks to comprehensively evaluate its performance in terms of detection accuracy, recall, computational complexity, and model parameter quantity.

3.1. Experimental Setup and Dataset

The cherry tomato images were acquired at a greenhouse in a plant science and technology park in Dongguan, China (Figure 7). A Microsoft Azure Kinect DK (Microsoft, Redmond, WA, USA) or Litemaze TOF camera (Litemaze Technology, Shenzhen, China) installed at the end of the robotic arm collected 843 RGB-D image pairs, including RGB and corresponding depth images. The purpose was to simulate the posture and distance of the robotic arm during the picking process. More detailed descriptions of the image acquisition environment and procedures have been reported in a previous study on single cherry tomato detection [22]. The original data were expanded using data augmentation methods such as cropping, flipping, and random brightness enhancement, and finally, 3372 images were obtained. Cherry tomato bunches within 1 m of the camera were labeled using LabelImg (version 1.8.1). Finally, the dataset was divided into the training set (2700 images), the validation set (336 images), and the test set (336 images).

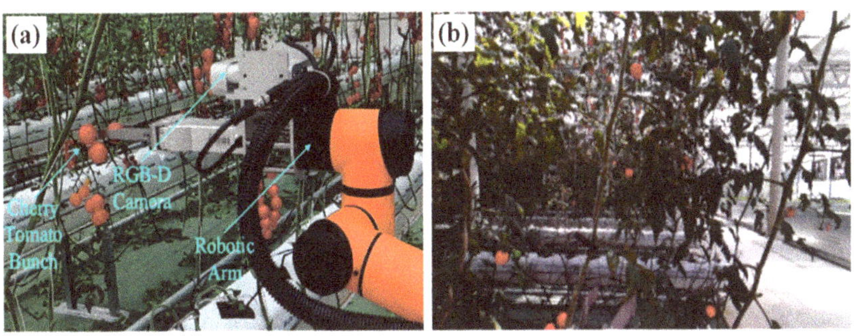

Figure 7. The dataset collection scene. (**a**) Early harvesting season. (**b**) Late harvesting season.

All experimental were carried out on a desktop workstation with Intel(R) Core (TM) i7-14700K CPU (Intel, Santa Clara, CA, USA), 20 cores and 28 threads, 64G memory, and Nvidia GeForce RTX 4070 Ti SUPER GPU (Nvidia, Santa Clara, CA, USA), and the Ubuntu 22.04 operating system. All deep learning algorithms were executed in the same environment of Windows 11, Cuda 12.6, Python 3.8, Pytorch-2.4.1, and Torchvision-0.19.1. All experiments set the image size to 640 × 640, batch_size to 8, and epochs to 200.

The initial learning rate was set to 0.01, with the minimal learning rate reduced to 1% of the initial value. The optimization utilized the Stochastic Gradient Descent (SGD) optimizer, configured with a momentum of 0.937 and a weight decay factor of 0.0005 to regularize the model parameters. The learning rate scheduling followed a cosine decay strategy, dynamically adjusting the learning rate throughout the training process to enhance convergence and performance.

3.2. Evaluation Metrics

We used precision, recall, F1-score, and mAP as the basic detection performance evaluation metrics. The mean average precision (mAP) is an important indicator to measure the global detection performance of the model. AP is defined as the area under the precision–recall curve, which represents the comprehensive evaluation of single-category detection performance [33]. mAP50 represents the average precision value calculated when the Intersection over Union (IoU) threshold is 0.5 and is a commonly used evaluation criterion in object detection tasks. mAP75 is the average precision calculated under a higher IoU threshold (0.75), which requires the model to predict the target position more accurately and can reflect the positioning ability of the model. mAP50-95 represents the average mAP at different IoU thresholds (from 0.5 to 0.95, with an interval of 0.05), which can more comprehensively evaluate the detection performance of the model under different precision requirements. The precision (*Pre*), recall (*Rec*), and F1-score (*FS*) are calculated as

$$\mathrm{Pre} = \frac{\mathrm{TP}}{\mathrm{TP}+\mathrm{FP}},\ \mathrm{Rec} = \frac{\mathrm{TP}}{\mathrm{TP}+\mathrm{FN}},\ \mathrm{FS} = \frac{\mathrm{Pre}\times\mathrm{Rec}}{\mathrm{Pre}+\mathrm{Rec}} \tag{4}$$

where TP, FP, and FN represent the true positives, false positives, and false negatives. The AP and mAP are calculated as follows:

$$\mathrm{AP} = \sum\nolimits_{i=1}^{\mathrm{N}} (\mathrm{R}_i - \mathrm{R}_{i-1})\mathrm{P}_i \tag{5}$$

$$\mathrm{mAP} = \frac{1}{\mathrm{C}}\sum\nolimits_{i=1}^{\mathrm{C}} \mathrm{AP}_i \times 100\% \tag{6}$$

where N is the number of recall levels; R_i and R_{i-1} are consecutive recall levels; P_i is the the precision at recall level; R_i and C is the number of classes.

In addition, we also considered GFLOPs to measure the computational complexity of the model and parameters to reflect the size and complexity of the model. Together, these metrics provided us with a comprehensive perspective to evaluate the precision, robustness, efficiency, and resource consumption of models in object detection tasks.

3.3. Ablation Experiments

The baseline model is the YOLOv8_n implemented by a third party. Compared with the official YOLOv8_n, this version of the model is easier to integrate multimodal feature extraction branches and lightweight improvements. All ablation experiments were conducted based on this model to incrementally verify the effectiveness of each additional module.

The design of the ablation experiment is shown in Table 1. First, the single-modal RGB input was used as the baseline to build the initial detection framework (Single-

RGB). Then, by adding a depth channel, a basic model of dual-modal input (Base_DM) was constructed to verify the performance improvement of multimodal input. On this basis, several improved modules were introduced, including the interpolation module (+Interpolate), the multimodal cross-fusion module (+Concat_BiFPN), the lightweight structure C2f_repghost module (+C2f_repghost), the ShuffleNetV2 lightweight module (+ShuffleNetV2), the channel and spatial attention module CBAM (+CBAM), and the fusion module combining the SPPF and CBAM (+SPPF_CBAM). The last case was the full DCFA-YOLO model.

Table 1. Comparison results of precision, recall, F1-score, mAP50, mAP50–95, mAP75, GFLOPs, and parameters (M) of the ablation experiments. The **bold value** in each column indicate the best performance in the corresponding metric, and the underlined value the second best.

Algorithm	Precision	Recall	F1-Score	mAP50	mAP50–95	mAP75	GFLOPs	Parameters (M)
Single-RGB	0.938	0.886	0.911	0.950	0.650	0.769	8.224	3.011
Base_DM	0.947	0.889	0.917	0.950	0.639	0.745	11.573	4.370
+Interpolate	0.949	0.898	0.923	0.950	0.646	0.765	11.573	4.370
+Concat_BiFPN	**0.959**	0.891	0.924	0.957	0.659	**0.781**	11.573	4.370
+C2f_repghost	0.956	0.887	0.921	0.953	0.655	0.775	10.765	3.968
+ShuffleNetV2	0.949	0.900	0.924	0.958	0.654	0.778	**7.218**	**2.078**
+CBAM	0.951	0.906	0.928	0.958	0.648	0.762	7.225	2.122
+SPPF_CBAM (Full)	0.949	**0.914**	**0.931**	**0.965**	**0.661**	**0.775**	7.589	2.679

The baseline model achieved a basic performance of 0.938 precision, 0.886 recall, and 0.911 F1-score in single mode. With multimodal input, the precision and F1-score increased to 0.947 and 0.917, confirming the positive impact of multimodal input. Adding the interpolation module (+Interpolate) further improved the precision and F1-score to 0.949 and 0.923. The addition of the multimodal cross-fusion module (+Concat_BiFPN) further improved the precision to 0.959, F1-score to 0.924, and mAP50 to 0.957. After further replacing it with the lightweight C2f_repghost module, although GFLOPs dropped to 10.765 and the model complexity was reduced, various metrics only decreased slightly. After adding the ShuffleNetV2 module, GFLOPs further dropped to 7.218, while the precision was 0.949 and F1-score remained at 0.924. The introduction of the CBAM module increased the recall to 0.906 and F1-score to 0.928. Finally, the introduction of the SPPF_CBAM module resulted in the best performance. These results confirm that the improved model achieves a good balance between improving performance and maintaining a lightweight design.

It should be noted that after adding the CBAM module to the backbone's six feature outputs, the performance improves notably with better feature extraction. However, when we further replace the final layer's SPPF module with the SPPF_CBAM module, this small adjustment leads to the best overall performance in terms of precision, recall, F1-score, and mAP50 while still maintaining a lightweight model. This improvement is primarily due to the effective combination of spatial and channel attention from CBAM with the spatial pyramid pooling (SPPF) module, which enhances multi-scale feature extraction, ensuring better performance without a significant increase in computational cost.

Figure 8 shows the heat map comparison for the ablation experiments. The baseline Single-RGB model exhibits focused attention on most targets but struggles with partially occluded cherry tomato clusters. The multimodal Base_DM model improves attention on these targets. Subsequent modules, such as +Interpolate, +Concat_BiFPN, and +C2f_repghost, show increasingly accurate and concentrated attention across all cherry tomato bunches. The +ShuffleNetV2 and +CBAM modules enhance the attention distribution, with +SPPF_CBAM achieving the best focus across all targets, particularly improving the detection of partially occluded clusters.

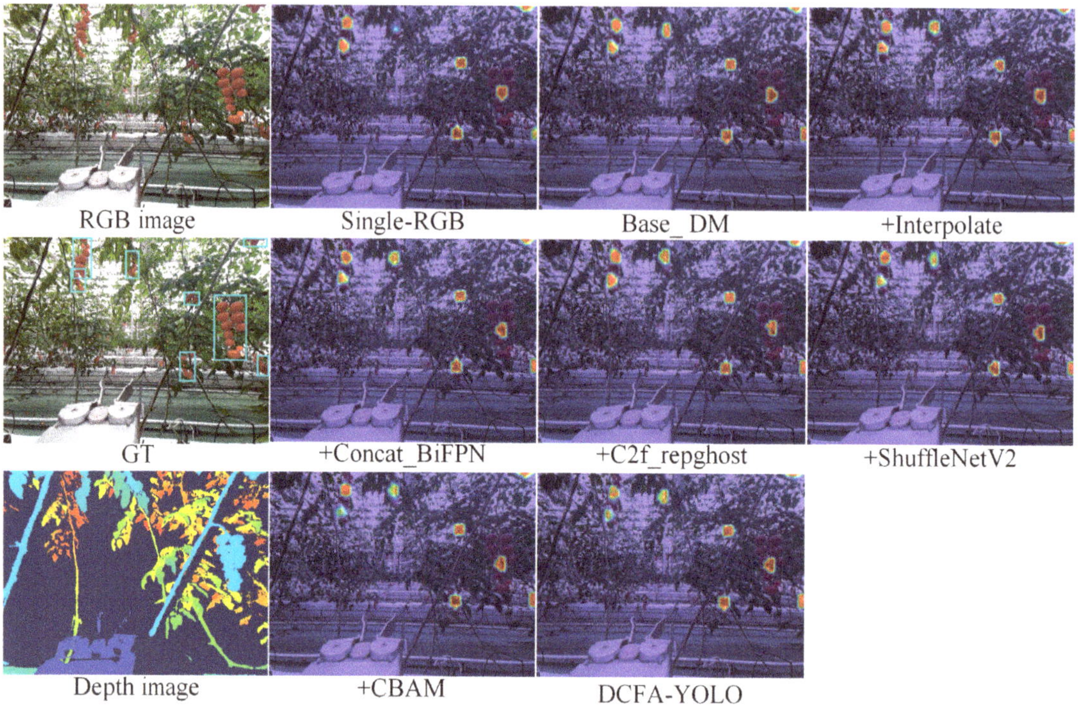

Figure 8. Comparison of heat maps for ablation experiments.

3.4. Comparison of Different Detection Algorithms

To compare the detection performance of DCFA-YOLO to other DNN models, we selected state-of-the-art YOLO models, YOLOv5_n [34], YOLOv8_n [30], YOLOv9_t [35], YOLOv10_n [36], YOLO11_n [37], EfficientDet [31], SSD [6], and CenterNet [38], for quantitative and qualitative comparison in terms of evaluation metrics and visual effects. The quantitative results are summarized in Table 2.

Table 2. Comparison results of precision, recall, F1-score, mAP50, mAP50–95, mAP75, GFLOPs, and parameters (M) of different detection models. The **bold value** in each column indicates the best performance in the metric and the underlined value the second best.

Algorithm	Precision	Recall	F1-Score	mAP50	mAP50–95	mAP75	GFLOPs	Parameters (M)
YOLOv5_n	0.865	0.838	0.851	0.905	0.565	0.626	7.100	2.503
YOLOv8_n	0.871	0.859	0.865	0.923	0.617	0.708	8.100	3.005
YOLOv9_t	0.912	0.879	0.895	0.946	0.656	0.770	7.600	**1.970**
YOLOv10_n	0.888	0.880	0.884	0.930	0.638	0.749	8.200	2.694
YOLO11_n	0.883	0.838	0.861	0.921	0.616	0.705	**6.300**	2.582
EfficientDet	0.948	0.781	0.860	0.900	0.549	0.627	7.401	3.828
SSD	0.870	0.896	0.880	0.929	0.609	0.716	58.363	11.671
CenterNet	0.952	0.797	0.870	0.919	0.572	0.654	109.710	32.665
DCFA-YOLO	**0.949**	**0.914**	**0.931**	**0.965**	**0.661**	**0.775**	7.589	2.679

DCFA-YOLO shows significant improvements in almost all key metrics. Compared with YOLOv5_n, YOLOv8_n, and YOLOv9_t, DCFA-YOLO achieves a notable increase in precision, recall, F1-score, and mAP50. Specifically, its precision improves from 0.871 (YOLOv8_n) to 0.949, and its F1-score rises from 0.865 to 0.931. Even compared with the higher-performance YOLOv9_t, DCFA-YOLO also shows obvious advantages, especially in F1-score and mAP50, with notable improvement in both. Though YOLOv10_n and YOLO11_n have lower GFLOPs and parameters, DCFA-YOLO outperforms them in key metrics like precision and mAP50. Moreover, it surpasses models such as EfficientDet and CenterNet. This further shows that DCFA-YOLO has achieved a good balance between performance and being lightweight, making it an excellent model for both accuracy and computational efficiency.

Figure 9 compares precision–recall curves across various detection models, showing that DCFA-YOLO consistently outperforms other models in terms of both precision and recall. By closely inspecting the PR curve, it is clear that the performance improvements noted in Table 2, especially in precision, recall, and F1-score, are visually reflected, demonstrating the model's strong overall detection capabilities.

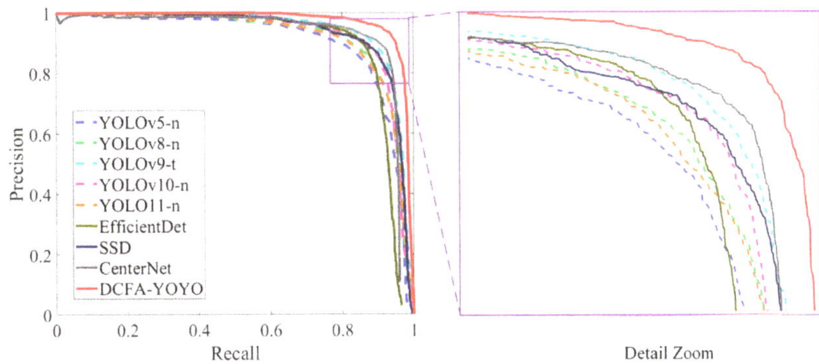

Figure 9. Precision–recall curves of different detection models.

Figures 10–12 present the visualization of detection results under normal lighting, high-light, and low-light scenarios, respectively. Under normal lighting (Figure 10), the compared models exhibit varying degrees of incomplete detection, frequently missing small, or overlapping objects and displaying suboptimal boundary precision. In contrast, DCFA-YOLO can detect cherry tomato bunches more accurately, significantly reducing missed and false detections.

Figures 11 and 12 illustrate detection results under challenging lighting conditions. In the high-light case (Figure 11), excessive brightness causes the image to appear washed out, leading to increased missed detections in most YOLO-based models and over-detection in CenterNet. In the low-light scenario (Figure 12), the darkness of the image exacerbates missed detections for YOLO-based models, while CenterNet again suffers from over-detection. Despite these challenges, DCFA-YOLO demonstrates superior adaptability in both scenarios, maintaining high detection accuracy with minimal false positives or missed targets.

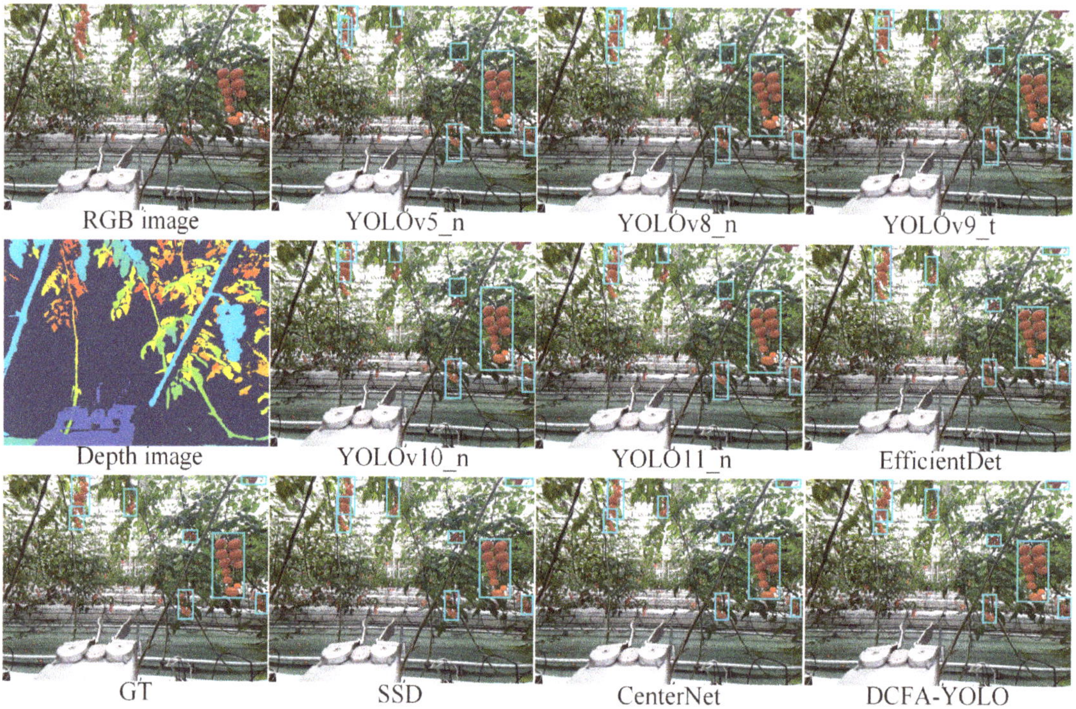

Figure 10. Visualization of detection results of different models under normal lighting conditions.

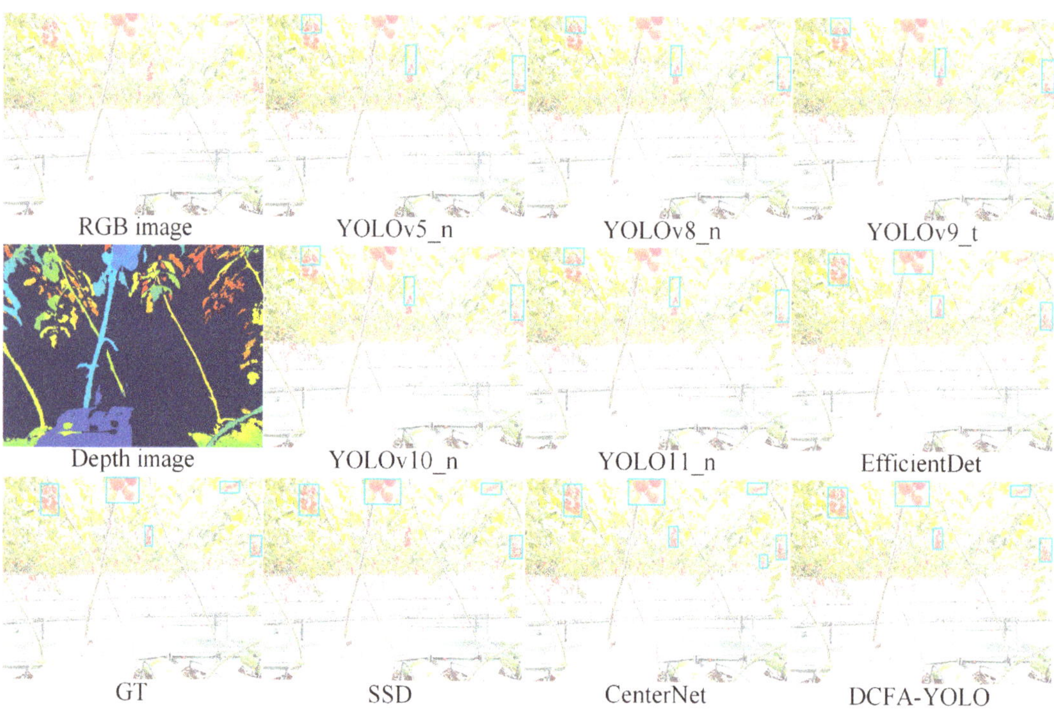

Figure 11. Visualization of detection results of different models under high-light scenarios.

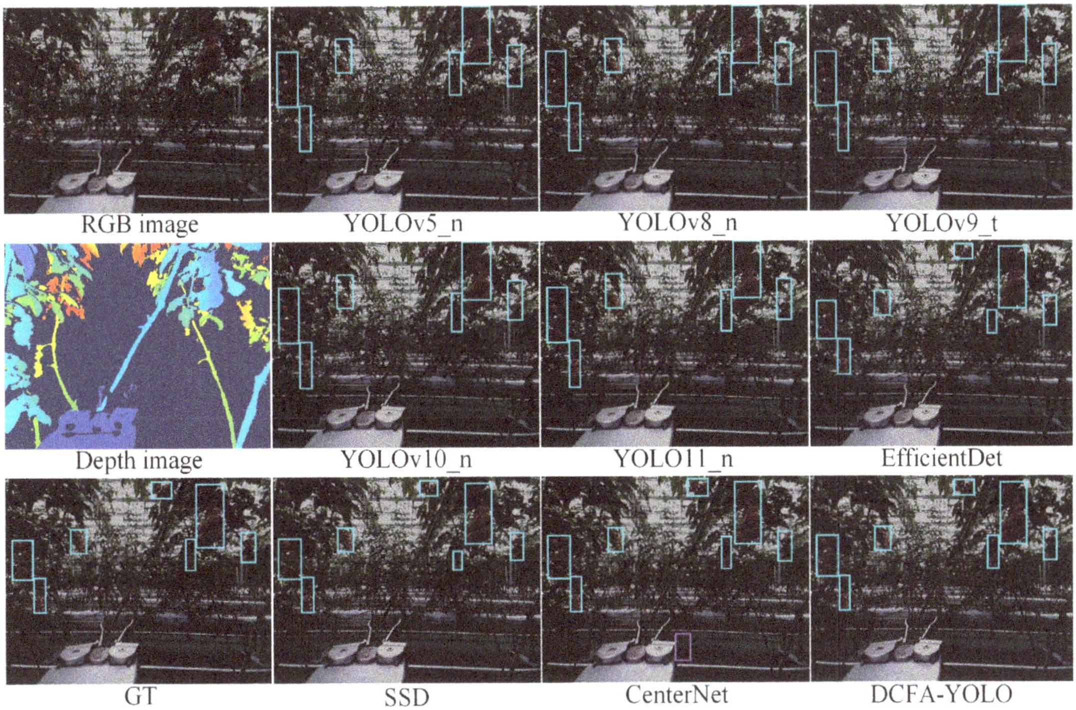

Figure 12. Visualization of detection results of different models under low-light scenarios.

4. Discussion

The proposed DCFA-YOLO method for multimodal cherry tomato bunch detection achieves the best results in six evaluation metrics (precision, recall, F1-score, mAP50, mAP75, and mAP50–95) among nine compared DNNs. It ranks the 5th among the nine models in terms of parameter size, and its GFLOPs rank the 4th lowest. For robotic fruit harvesting, the detection algorithm typically runs on edge computing devices, so the accuracy and computational cost of detection are both important, and we believe DCFA-YOLO strikes an excellent balance between the two. The computing platform in the current study is essentially a regular PC, and DCFA-YOLO runs cherry tomato bunch detection at a frame rate of 52.93 fps, which is sufficient for practical real-time robotic harvesting. By achieving high accuracy with lower computational costs, DCFA-YOLO reduces the burden on hardware, which could lead to lower energy consumption and operational costs in real-world applications. It can be integrated into agricultural systems like robotic harvesters or drones, enhancing their perception for precise and efficient fruit picking and measurement.

In a previous cherry tomato bunch detection study, precision, recall, and an F1-score of 98.9, 92.1, and 95.4 have been reported [39], higher than the corresponding values of 0.949, 0.914, and 0.931 in the current study. However, the two studies were conducted with two very different datasets that were collected by different cameras (Intel RealSense structured light stereo vs. Microsoft Kinect TOF) at different plantations. The code and dataset are not available for apple-to-apple comparisons. To partly address this, we make the code publicly available on Github (see Data Availability Statement). Unfortunately, the dataset cannot be made fully public due to a non-disclosure agreement.

In this study, we only tested cherry tomato bunch detection. For clustered fruits like cherry tomatoes, whether harvesting by single fruits or by bunches depends on the production needs in a commercial setting, and both have been performed in previous studies [22,39,40]. We believe DCFA-YOLO can be applicable to single fruit detection as well, which will be investigated in future work. However, it is important to note that the current model may face challenges in more complex scenarios, such as severe occlusions or highly dense clusters, where the accuracy of detection could be compromised. Future improvements could focus on enhancing the model's robustness to these conditions, potentially through advanced attention mechanisms or multi-scale feature fusion techniques.

We would also like to explore applying DCFA-YOLO to other fruits; for example, it may be partially transferable to detecting other fruits in clusters/bunches, such as grape, longan, or lychee. It should be noted that for these fruits, even at maturity, the color features are less salient compared to cherry tomatoes. As such, we expect that the impact of adding depth information for these fruits is likely to have a more significant impact than for cherry tomatoes.

Moreover, when integrated with harvesting robots, DCFA-YOLO can contribute to agricultural sustainability by improving harvesting efficiency and reducing fruit waste. Its high detection accuracy minimizes missed or damaged fruits, while its real-time capabilities enable faster and more precise harvesting, ultimately supporting more sustainable and resource-efficient agricultural practices.

5. Conclusions

This study introduces DCFA-YOLO, a lightweight fruit detection model with multimodal cross fusion. By integrating the dual-channel cross-fusion mechanism, a dynamically weighted feature-fusion module, and optimized lightweight design alongside attention mechanisms, the model demonstrates notable improvements in efficiency and robustness for multimodal detection tasks. Specifically, the model leverages complementary information from color and depth images through targeted enhancements to the backbone network and feature-fusion stages. These improvements effectively reduce computational complexity and model parameters while maintaining high detection accuracy, addressing the challenges of small fruit detection in complex scenarios. The experimental results on cherry tomato bunch detection further validate the model's superiority, showcasing its potential for practical application in precision agriculture as a core component of robotic cherry tomato harvesting.

Author Contributions: Conceptualization, Z.Z. and Y.T.; methodology, S.C., P.L. and Y.T.; software, S.C.; validation, S.C. and M.W.; formal analysis, S.C. and Y.T.; investigation, S.C. and M.W.; resources, Y.T.; data curation, Z.Z. and P.L.; writing—original draft preparation, S.C.; writing—review and editing, M.W. and Y.T.; visualization, S.C.; supervision, Y.T.; project administration, Y.T.; funding acquisition, M.W. and Y.T. All authors have read and agreed to the published version of the manuscript.

Funding: This study was funded by the Shenzhen Talent Startup Funds (No. 827-000954) and Shenzhen University (Nos. 868/03010315 and 86902/00248).

Institutional Review Board Statement: Not applicable.

Data Availability Statement: The program code is at https://github.com/heitieya/DCFA-YOLO (accessed on 23 January 2025). The dataset cannot be made fully public due to a non-disclosure agreement, but limited samples can be provided at reasonable request.

Conflicts of Interest: The authors declare no conflicts of interest.

References

1. Wang, Z.; Xun, Y.; Wang, Y.; Yang, Q. Review of smart robots for fruit and vegetable picking in agriculture. *Int. J. Agric. Biol. Eng.* **2022**, *15*, 33–54.
2. Agarwal, S.; Rao, A.V. Tomato lycopene and its role in human health and chronic diseases. *CMAJ* **2000**, *163*, 739–744. [PubMed]
3. Hua, X.; Li, H.; Zeng, J.; Han, C.; Chen, T.; Tang, L.; Luo, Y. A review of target recognition technology for fruit picking robots: From digital image processing to deep learning. *Appl. Sci.* **2023**, *13*, 4160. [CrossRef]
4. Jiang, P.; Ergu, D.; Liu, F.; Cai, Y.; Ma, B. A Review of Yolo algorithm developments. *Procedia Comput. Sci.* **2022**, *199*, 1066–1073. [CrossRef]
5. Redmon, J. You only look once: Unified, real-time object detection. In Proceedings of the IEEE Conference on Computer Vision and Pattern Recognition, Las Vegas, NV, USA, 27–30 June 2016.
6. Liu, W.; Anguelov, D.; Erhan, D.; Szegedy, C.; Reed, S.; Fu, C.-Y.; Berg, A.C. SSD: Single shot multibox detector. In Proceedings of the Computer Vision–ECCV 2016: 14th European Conference, Amsterdam, The Netherlands, 11–14 October 2016; pp. 21–37.
7. Lyu, S.; Li, R.; Zhao, Y.; Li, Z.; Fan, R.; Liu, S. Green citrus detection and counting in orchards based on YOLOv5-CS and AI edge system. *Sensors* **2022**, *22*, 576. [CrossRef]
8. Wang, L.; Zhao, Y.; Xiong, Z.; Wang, S.; Li, Y.; Lan, Y. Fast and precise detection of litchi fruits for yield estimation based on the improved YOLOv5 model. *Front. Plant Sci.* **2022**, *13*, 965425. [CrossRef]
9. Gai, R.; Chen, N.; Yuan, H. A detection algorithm for cherry fruits based on the improved YOLO-v4 model. *Neural Comput. Appl.* **2023**, *35*, 13895–13906. [CrossRef]
10. Zhao, R.; Zhu, Y.; Li, Y. An end-to-end lightweight model for grape and picking point simultaneous detection. *Biosyst. Eng.* **2022**, *223*, 174–188. [CrossRef]
11. Yuan, T.; Lv, L.; Zhang, F.; Fu, J.; Gao, J.; Zhang, J.; Li, W.; Zhang, C.; Zhang, W. Robust cherry tomatoes detection algorithm in greenhouse scene based on SSD. *Agriculture* **2020**, *10*, 160. [CrossRef]
12. Fuentes-Peñailillo, F.; Carrasco Silva, G.; Pérez Guzmán, R.; Burgos, I.; Ewertz, F. Automating seedling counts in horticulture using computer vision and AI. *Horticulturae* **2023**, *9*, 1134. [CrossRef]
13. Girshick, R.; Donahue, J.; Darrell, T.; Malik, J. Rich feature hierarchies for accurate object detection and semantic segmentation. In Proceedings of the IEEE Conference on Computer Vision and Pattern Recognition, Columbus, OH, USA, 23–28 June 2014; pp. 580–587.
14. Girshick, R. Fast R-CNN. In Proceedings of the IEEE International Conference on Computer Vision (ICCV), Santiago, Chile, 7–13 December 2015; pp. 1440–1448.
15. Ren, S.; He, K.; Girshick, R.; Sun, J. Faster R-CNN: Towards real-time object detection with region proposal networks. *IEEE Trans. Pattern Anal. Mach. Intell.* **2016**, *39*, 1137–1149. [CrossRef]
16. He, K.; Zhang, X.; Ren, S.; Sun, J. Spatial pyramid pooling in deep convolutional networks for visual recognition. *IEEE Trans. Pattern Anal. Mach. Intell.* **2015**, *37*, 1904–1916. [CrossRef] [PubMed]
17. Hu, C.; Liu, X.; Pan, Z.; Li, P. Automatic detection of single ripe tomato on plant combining faster R-CNN and intuitionistic fuzzy set. *IEEE Access* **2019**, *7*, 154683–154696. [CrossRef]
18. Song, Z.; Fu, L.; Wu, J.; Liu, Z.; Li, R.; Cui, Y. Kiwifruit detection in field images using Faster R-CNN with VGG16. *IFAC-PapersOnline* **2019**, *52*, 76–81. [CrossRef]
19. Gao, F.; Fu, L.; Zhang, X.; Majeed, Y.; Li, R.; Karkee, M.; Zhang, Q. Multi-class fruit-on-plant detection for apple in SNAP system using Faster R-CNN. *Comput. Electron. Agric.* **2020**, *176*, 105634. [CrossRef]
20. Tu, S.; Xue, Y.; Zheng, C.; Qi, Y.; Wan, H.; Mao, L. Detection of passion fruits and maturity classification using Red-Green-Blue Depth images. *Biosyst. Eng.* **2018**, *175*, 156–167. [CrossRef]
21. Cui, B.; Zeng, Z.; Tian, Y. A Yolov7 cherry tomato identification method that integrates depth information. In Proceedings of the Third International Conference on Optics and Image Processing (ICOIP 2023), Hangzhou, China, 14–16 April 2023; pp. 312–320.
22. Cai, Y.; Cui, B.; Deng, H.; Zeng, Z.; Wang, Q.; Lu, D.; Cui, Y.; Tian, Y. Cherry Tomato Detection for Harvesting Using Multimodal Perception and an Improved YOLOv7-Tiny Neural Network. *Agronomy* **2024**, *14*, 2320. [CrossRef]
23. Rong, J.; Zhou, H.; Zhang, F.; Yuan, T.; Wang, P. Tomato cluster detection and counting using improved YOLOv5 based on RGB-D fusion. *Comput. Electron. Agric.* **2023**, *207*, 107741. [CrossRef]
24. Kaukab, S.; Ghodki, B.M.; Ray, H.; Kalnar, Y.B.; Narsaiah, K.; Brar, J.S. Improving real-time apple fruit detection: Multi-modal data and depth fusion with non-targeted background removal. *Ecol. Inform.* **2024**, *82*, 102691. [CrossRef]
25. Liu, Z.; Wu, J.; Fu, L.; Majeed, Y.; Feng, Y.; Li, R.; Cui, Y. Improved kiwifruit detection using pre-trained VGG16 with RGB and NIR information fusion. *IEEE Access* **2019**, *8*, 2327–2336. [CrossRef]
26. Sa, I.; Ge, Z.; Dayoub, F.; Upcroft, B.; Perez, T.; McCool, C. Deepfruits: A fruit detection system using deep neural networks. *Sensors* **2016**, *16*, 1222. [CrossRef] [PubMed]
27. Wei, Z.; Guodong, Y.; Haoji, L.; Keke, G.; Wenhan, H.; Yuan, W.; Hongwei, X. Low-observable Target Detection Method for Autonomous Vehicles Based on Multi-modal Feature Fusion. *China Mech. Eng.* **2021**, *32*, 1114.

28. Woo, S.; Park, J.; Lee, J.-Y.; Kweon, I.S. Cbam: Convolutional block attention module. In Proceedings of the European Conference on Computer Vision (ECCV), Munich, Germany, 8–14 September 2018; pp. 3–19.
29. Li, H.; Wu, X.-J. DenseFuse: A fusion approach to infrared and visible images. *IEEE Trans. Image Process.* **2018**, *28*, 2614–2623. [CrossRef] [PubMed]
30. Sohan, M.; Sai Ram, T.; Reddy, R.; Venkata, C. A review on yolov8 and its advancements. In Proceedings of the International Conference on Data Intelligence and Cognitive Informatics, Tirunelveli, India, 27–28 June 2024; pp. 529–545.
31. Tan, M.; Pang, R.; Le, Q.V. Efficientdet: Scalable and efficient object detection. In Proceedings of the IEEE/CVF Conference on Computer Vision and Pattern Recognition, Seattle, WA, USA, 13–19 June 2020; pp. 10781–10790.
32. Li, X.; Wang, W.; Wu, L.; Chen, S.; Hu, X.; Li, J.; Tang, J.; Yang, J. Generalized focal loss: Learning qualified and distributed bounding boxes for dense object detection. *Adv. Neural Inf. Process. Syst.* **2020**, *33*, 21002–21012.
33. Zheng, L.; Shen, L.; Tian, L.; Wang, S.; Wang, J.; Tian, Q. Scalable person re-identification: A benchmark. In Proceedings of the IEEE International Conference on Computer Vision, Santiago, Chile, 7–13 December 2015; pp. 1116–1124.
34. Jocher, G. YOLOv5 by Ultralytics, Version 7.0; 2020. Available online: https://github.com/ultralytics/yolov5 (accessed on 23 January 2025).
35. Wang, C.-Y.; Yeh, I.-H.; Mark Liao, H.-Y. Yolov9: Learning what you want to learn using programmable gradient information. In Proceedings of the European Conference on Computer Vision, Zurich, Switzerland, 22–23 January 2025; pp. 1–21.
36. Wang, A.; Chen, H.; Liu, L.; Chen, K.; Lin, Z.; Han, J.; Ding, G. Yolov10: Real-Time End-to-End Object Detection. *arXiv* **2024**, arXiv:14458.
37. Khanam, R.; Hussain, M. Yolov11: An Overview of the Key Architectural Enhancements. *arXiv* **2024**, arXiv:17725.
38. Duan, K.; Bai, S.; Xie, L.; Qi, H.; Huang, Q.; Tian, Q. Centernet: Keypoint triplets for object detection. In Proceedings of the IEEE/CVF International Conference on Computer Vision, Seoul, Republic of Korea, 27 October–2 November 2019; pp. 6569–6578.
39. Li, Y.; Feng, Q.; Liu, C.; Xiong, Z.; Sun, Y.; Xie, F.; Li, T.; Zhao, C. MTA-YOLACT: Multitask-aware network on fruit bunch identification for cherry tomato robotic harvesting. *Eur. J. Agron.* **2023**, *146*, 126812. [CrossRef]
40. Chen, W.; Liu, M.; Zhao, C.; Li, X.; Wang, Y. MTD-YOLO: Multi-task deep convolutional neural network for cherry tomato fruit bunch maturity detection. *Comput. Electron. Agric.* **2024**, *216*, 108533. [CrossRef]

agriculture

MDPI

Article

Research and Experiments on Adaptive Root Cutting Using a Garlic Harvester Based on a Convolutional Neural Network

Ke Yang [1], Yunlong Zhou [1], Hengliang Shi [1], Rui Yao [1], Zhaoyang Yu [2], Yanhua Zhang [2], Baoliang Peng [2], Jiali Fan [3] and Zhichao Hu [2,*]

[1] School of Automobile and Rail Transit, Luoyang Polytechnic, Luoyang 471900, China; yk666666yk@126.com (K.Y.); lovechuang@126.com (Y.Z.); shi_hl2013@163.com (H.S.); yaorui@lypt.edu.cn (R.Y.)

[2] Nanjing Institute of Agricultural Mechanization, Ministry of Agriculture and Rural Affairs, Nanjing 210014, China; 99yzy99@163.com (Z.Y.); zhangyanhua@caas.cn (Y.Z.); pengbaoliang@caas.cn (B.P.)

[3] College of Biosystems Engineering and Food Science, Zhejiang University, Hangzhou 310058, China; 12313060@zju.edu.cn

* Correspondence: huzhichao@caas.cn

Abstract: Aimed at the problems of a high leakage rate, a high cutting injury rate, and uneven root cutting in the existing combined garlic harvesting and root-cutting technology, we researched the key technologies used in a garlic harvester for adaptive root cutting based on machine vision. Firstly, research was carried out on the conveyor alignment and assembly of the garlic harvester to realize the adjustment of the garlic plant position and the alignment of the bulb's upper surface before the roots were cut, to establish the parameter equations and to modify the structure of the conveyor to form the adaptive garlic root-cutting system. Then, a root-cutting test using the double-knife disk-type cutting device was carried out to examine the root-cutting ability of the cutting device. Finally, a bulb detector trained with the IRM-YOLO model was deployed on the Jetson Nano device (NVIDIA, Jetson Nano(4GB), Santa Clara, CA, USA) to conduct a harvester field trial study. The pass rate for the root cutting was 82.8%, and the cutting injury rate was 2.7%, which tested the root cutting performance of the adaptive root cutting system and its field environment adaptability, providing a reference for research into combined garlic harvesting technology.

Keywords: garlic; adaptive root cutting; cutting height control; harvester; object detection

1. Introduction

Garlic (*Allium sativum* L.) is a widely grown vegetable seasoning with both edible and medicinal values [1–4]. Studies have shown that garlic consumption can effectively reduce the risk of diseases such as hypertension and diabetes [5,6]. In addition to its disease-preventive properties, garlic has antibacterial, anti-inflammatory and insecticidal properties [7–10]. The organosulfur compounds and polyphenols contained in garlic give it a unique flavor [11]. Garlic is one of the most important flavoring vegetables in the world [12], and China is the number one producer of garlic in the world [13]. At present, garlic harvesting in China is dominated by manual harvesting, and the technological level of mechanized harvesting equipment is low, which hinders the development of the Chinese garlic industry.

Root cutting is an essential process in garlic harvesting. Since the roots of freshly emerged garlic are entrapped within moist soil, failure to remove the roots will result in moldy or even rotting bulbs. Garlic that has been rooted will also look better, which will increase its selling price. Usually, garlic heads vary in size, there are differences in the location of the roots, garlic heads have different postures in the soil, and garlic heads can be watery and have thin skins. Due to their biological characteristics, garlic and garlic roots are easy to dehydrate, which reduces their toughness, meaning that the technical difficulty involved in mechanized root-cutting operations is high. Moreover, cutting the bulb can easily lead to pathogen invasion, causing food safety problems [14–16], and garlic root cutting is mainly carried out manually with a low production efficiency. To improve production efficiency and control food safety, there is an urgent need for garlic harvesters that can realize high-quality root-cutting functions.

At present, garlic harvesters are mainly digging and spreading types or the combined-harvesting type. The combined-harvesting-type garlic harvester can carry out the functions of vibration digging, clamping and conveying, cutting stems, throwing seedlings, and garlic collection and storage at the same time [17–21]. The HZ1 garlic harvester from the YANMAR company in Japan includes a root-cutting function in the process of combined harvesting. It uses a double-disk cutter to cut the garlic root continuously, but the height of the double-disk cutter cannot be adjusted, which has led to the emergence of the problem that the garlic roots cannot be cut cleanly [17,22]. In the mechanized garlic harvesting process, the height of the root knife in the root-cutting device can be adjusted according to differences in the size of the garlic heads; the height of the root-cutting knife in the root-cutting device can be adjusted to improve the rate of cutting the garlic roots cleanly, which is referred to as an adaptive root-cutting technology. The research on adaptive root cutting in garlic harvesters is still in the exploratory and experimental stages [23,24]. Z. Yu et al. [25] developed a floating garlic root-cutting mechanism with a spiral guard fence and a return plate spring, which improved the rate of clean garlic root cutting by increasing the number of root-cutting channels. J. Hou et al. [26] designed a press-type root cutting device to cut the garlic root under the action of a clamping conveyor mechanism and a disk knife, which realized garlic root disk alignment and garlic root cutting through the rotation of a paddle wheel and the deformation of a press. K. Yang et al. [27–29] utilized machine vision and deep learning technology to detect the demarcation line between the garlic bulb and root, move the disk root-cutting knife to the corresponding position according to the detection result, and realize adaptive garlic root cutting based on image recognition technology on a test bench. However, due to the harvester's high functional integration, complexity, efficiency, and compactness [30–36], arranging the adaptive garlic root-cutting system on the harvester and performing the harvesting operation pose significant challenges.

Machine vision and deep learning technologies have been heavily applied to study the information perception aspect of improving the intelligence of harvesters. Q. Zhang et al. [37] acquired image information through the camera installed on a wheat harvester and used the YOLO algorithm and an onboard deep learning system to predict the harvester's feeding volume and realize the closed-loop control of intelligent harvesting operations. Z. Guan et al. [36] used a rice harvester's camera on the seed sampling device to achieve continuous image acquisition, proposed an impurity recognition algorithm based on Mask R-CNN, and evaluated the impurity detection system through bench and field tests. A. Nasirahmadi et al. [38] positioned a camera above the cleaning turbine of the beet harvester to acquire images in real time when the harvester was cleaning the beets and through a deep learning system developed based on CNN to detect beet damage in the harvester automatically. W. Yang et al. [39] installed a camera, edge computing equipment, and related supporting equipment on a dig-and-pull cassava harvester, utilized the improved YOLO algorithm to detect cassava stalks in the field in real time, and adjusted the position of the harvester's clamping and pulling mechanism to align the stalks according to the results of the detection, thus reducing harvesting losses.

Aimed at the problem that existing garlic harvesters do not have an adaptive root-cutting function and the root-cutting quality is low, this study developed a garlic harvester with an adaptive root-cutting function. Combining theoretical analysis, bench tests, field tests, and other methods, we study the adaptive root-cutting process of the garlic harvester conveyor assembly on the role of garlic plants, garlic plant attitude adjustment, root force, and other dynamic characteristics; explore the adaptive root-cutting system matching, quality-influencing factors, and mechanism optimization scheme; provide a reference for the garlic harvester's function and technology improvement.

2. Materials and Methods

2.1. 4DSL-7 Garlic Harvester

Relying on the National Natural Science Foundation of China (NSFC) project, the National Key R&D Program and the Jiangsu Modern Agricultural Machinery Equipment and Technology Demonstration and Promotion Project, Nanjing Agricultural Mechanization Research Institute of the Ministry of Agriculture and Rural Affairs of China (NAMRI) has developed the 4DLB-2 garlic harvester (Nanjing Institute of Agricultural Mechanization, Ministry of Agriculture and Rural Affairs, Nanjing, China), which can harvest two rows at a time, and the 4DSL-7 garlic harvester (Nanjing Institute of Agricultural Mechanization, Ministry of Agriculture and Rural Affairs, Nanjing, China), which can harvest seven rows at a time. The 4DSL-7 garlic harvester (shown in Figure 1) mainly consists of the following components: profiling depth-limiting wheel, splitter, vibrating digging shovel, clamping and conveying device, aligning and cutting seedling mechanism, seedling and fruit collecting and transmitting device, scraper lifting and conveying device, fruit box (which can be unloaded backward), chassis walking mechanism, and so on. Among them, garlic plant sorting alignment, fruit and seedling separation, and seedling and fruit collection and delivery are a vital operation link and key technology. Alignment of seedling cutting mechanism and seedling and fruit collection and delivery device of the structure and operation parameter setting directly determine the operational performance of the machine.

Figure 1. 4DSL-7 garlic harvester. The meaning of the Chinese characters in the picture is 4DSL-7 Garlic Combined Harvester, Jointly developed by Nanjing Institute of Agricultural Mechanization, Ministry of Agriculture and Rural Affairs (Nanjing, China) and Changzhou Han-Sun Machinery Co. (Changzhou, China); National Key R&D Program Project—Research and Development of Intelligent Fine Production Technology and Equipment for Vegetables.

The core component of the 4DSL-7 garlic harvester, the structure of the conveyor alignment assembly, is shown in Figure 2, which mainly consists of a hydraulic motor, clamping conveyor chain, harvesting table welding frame, sorting-alignment device, transmission system, etc. Usually, each conveyor alignment assembly completes one row of garlic harvesting, which can be installed in 3 to 7 groups. Usually, each conveyor alignment assembly completes harvesting one row of garlic, and the prototype of a 4DSL-7 garlic harvester can be installed with 3–7 sets of assemblies to harvest 3–7 rows of garlic at one time. During operation, after the garlic plants are dug out of the soil, they are clamped by the clamping conveyor chain and transported backward (the direction of transport is shown by the blue arrow in Figure 2). The garlic plants are aligned along the upper surface of the bulb (as indicated by the yellow arrow in Figure 2). The garlic plants' position is adjusted by the joint action of the clamping conveyor chain and the toggle finger chain of the sorting-aligning device. Based on bulb alignment, the final separation of the fruit and seedling is realized with the help of a stem-cutting device.

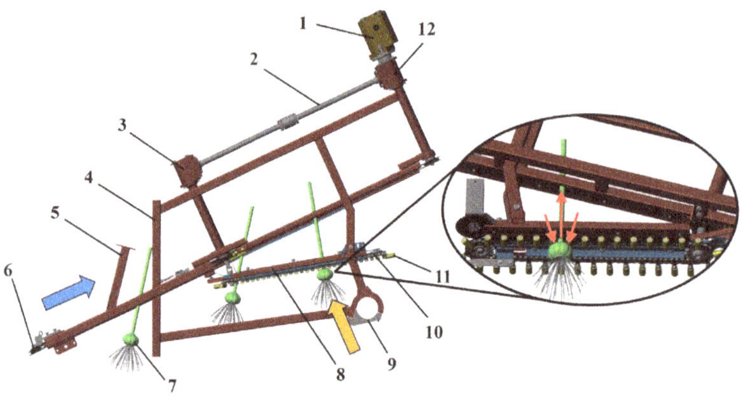

Figure 2. Schematic diagram of the harvesting assembly of 4DSL-7 garlic harvester. 1. Hydraulic motor 2. Drive shaft 3. Transmission box I 4. Harvesting table weld-on 5. Walking wheel mount. 6. Clamping conveyor chain 7. Garlic plant 8. Sorting-aligning device 9. Fixed lugs 10. Finger-toggling chain 11. Finger-toggling 12. Transmission box II. The blue arrows in the figure indicate the direction of transport of the garlic plant; the yellow arrows indicate the position at which the bulb is aligned along the upper surface; and the red arrows indicate the forces applied to the bulb after it is aligned along the upper surface.

The conveyor alignment assembly realizes bulb alignment along the upper surface by forced expansion using an inclined configuration between the gripping conveyor chain and the toggle finger chain of the sorting-aligning device. By changing the speed ratio of the gripping conveyor chain and the toggle finger chain, the position of the garlic plant can be adjusted when the bulbs are aligned along the upper surface so that the adjustment of the position of the garlic plant and the alignment of the bulbs along the upper surface can be accomplished in the front or middle section of the sorting-aligning device. At the same time, when the bulb is aligned along the upper surface, the clamping conveyor chain applies a lifting force in the direction of the garlic bulb, which will cause the upper surface of the bulb to rest on the lower part of the two adjacent fingers of the toggle finger chain, similar to the state of holding the bulb in the hand when cutting the root manually. The force of the bulb is shown by the red arrows in Figure 2. The realization of the above functions allows for adaptive root cutting in garlic.

When a garlic harvester is in operation, the elongation of the traveling wheels determines the initial clamping height of the clamping conveyor chain on the garlic plant, and the initial clamping height of the garlic plant influences the initial position on the sorting-alignment device at the which alignment of the bulbs along the upper surface is accomplished. The initial clamping height of the garlic plant needed to satisfy the bulb alignment along the upper surface was determined by harvesting tests in the pre-tests of the field trials to obtain a suitable amount of traveler wheel elongation, which will not be repeated in this text. Thus, the garlic plant position adjustment and bulb alignment along the upper surface are only related to the speed ratio of the clamping conveyor chain and the finger-toggle finger chain.

2.2. Speed Relationship Between Clamp Conveyor Chain and Toggle Finger Chain

To study the change rule of garlic plant position in the conveyor alignment assembly, to ensure that the garlic plant is perpendicular to the plane of the toggle finger chain, and to realize the alignment of the upper surface of the bulb during the subsequent root-cutting operation, it is necessary to analyze the speed of the clamping conveyor chain V_{JC} and the speed of the toggle finger chain V_{BZ}. Let the ratio of clamping conveyor chain speed V_{JC} and toggle finger chain speed V_{BZ} be K_X.

In Figure 3, α_E is the angle between the garlic plant and the clamping conveyor chain (°); β_A is the angle between the garlic plant and AD (°); γ_{BC} is the angle between the clamping conveyor chain and the toggle finger chain (°); point A is the front rotating center of the toggle finger chain; and point B is the initial position of completing the bulb alignment along the upper surface. The distance between points A and B is l_{AB}, mm. Points D and C are the projection points on the clamping conveyor chain in the vertical direction of points A and B along the sorting-alignment device. The projection point on the clamping conveyor chain, the distance between points D and C is l_{CD}, mm; when the garlic plant enters the sorting-aligning device at an angle of α_E with the clamping conveyor chain, the initial clamping point of the garlic plant on the clamping conveyor chain is point E. The distance between points D and E is l_{DE}, mm; between points A and D is l_{AD}, mm; and between points BC is l_{BC}, mm.

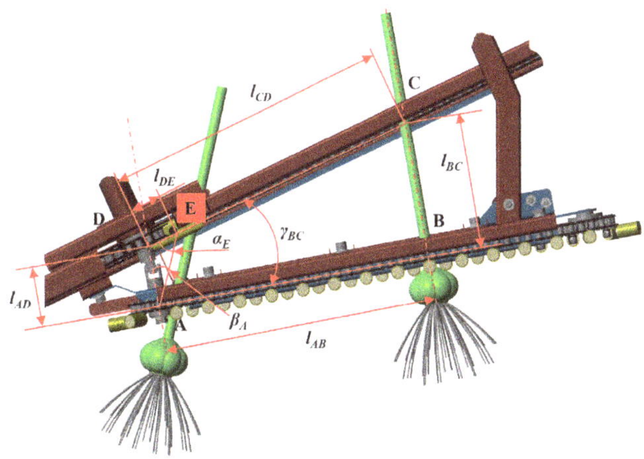

Figure 3. Position parameters of the left-side view. The red arrows in the figure indicate the corresponding lengths and angles.

The distance l_{CD} between the projection points of points A and B along the plane normal to the plane in which the toggle finger chain is located on the clamping conveyor chain is:

$$l_{CD} = \sqrt{l_{AB}^2 + (l_{BC} - l_{AD})^2} \tag{1}$$

The angle γ_{BC} between the clamping conveyor chain and the toggle finger chain is:

$$\gamma_{BC} = \arctan\frac{l_{BC} - l_{AD}}{l_{AB}} \tag{2}$$

In Equations (1) and (2), l_{AB} is the action distance of the sorting-alignment device when the garlic plant is perpendicular to the plane in which the toggle finger chain is located, mm; l_{AD} is the projection distance along the plane of the toggle finger chain where point A is located in the expected direction on the clamping conveyor chain, mm; l_{BC} is the projection distance along the plane of the toggle finger chain where point B is located in the expected direction on the clamping conveyor chain, mm.

When the garlic harvester is in operation, the angle of inclination of the conveyor alignment assembly is different, and the angle α_E between the garlic plant and the clamping conveyor chain after shoveling and pulling out the soil will be different. To realize that the cutting surface of the cut root is perpendicular to the garlic plant, it is required that the garlic plant is perpendicular to the plane where the toggle finger chain is located when the root is cut. Effective matching of the speed ratio between the clamping conveyor chain and the toggle finger chain of the sorting-alignment device ensures that the garlic plant is perpendicular to the plane of the toggle finger chain, so it is necessary to study the ratio of the speed of the clamping conveyor chain, V_{JC}, and the speed of the toggle finger chain, V_{BZ}, to be K_X. To facilitate this study, it is assumed that the digging depth of the excavation shovel is unchanged, the tilt angle of the conveyor alignment assembly is unchanged, and the corresponding angle α_E between the garlic plant and the clamping conveyor chain is also a definite value. Point A is the initial point of action of the toggle finger chain, and the BC line is the target conveying position. Then, the ratio of the speed of the clamping conveyor chain V_{JC} and the speed of the toggle finger chain V_{BZ} is K_X

$$K_X = V_{JC}/V_{BZ} = (l_{CD} - l_{DE})/l_{AB} \tag{3}$$

In the above formula, V_{JC} is the speed of the clamping conveyor chain, m/s; V_{BZ} is the speed of the toggle finger chain, m/s; l_{CD} is the distance between the projection points on the clamping conveyor chain along the plane of the toggle finger chain where points A and B are located in the expected direction, mm; l_{DE} is the amount of adjustment of the sorting-alignment device on the clamping conveyor chain on the clamping point of the garlic plant, mm; l_{AB} is the role of the sorting-alignment device when the garlic plant is perpendicular to the plane of the toggle finger chain, mm. distance, mm.

By the sine theorem

$$l_{DE} = l_{AD}\sin\beta_A/\sin\alpha_E \tag{4}$$

In the above equation, l_{AD} is the projection distance of point A along the normal direction of the plane where the finger-pulling chain is located on the clamping conveyor chain, mm; β_A is the angle between the garlic plant and AD (°); α_E is the angle between the garlic plant and the clamping conveyor chain (°).

From the triangular relationship

$$\beta_A = 90^\circ - (\alpha_E + \gamma_{BC}) \tag{5}$$

In the above formula, γ_{BC} is the clamping conveyor chain and toggle finger chain angle (°).

Equations (3)–(5) can be obtained by the association of

$$K_X = \frac{l_{CD}\sin\alpha_E - l_{AD}\cos(\alpha_E + \gamma_{BC})}{l_{AB}\sin\alpha_E} \tag{6}$$

From the above equation, when K_X satisfies Equation (6), it can be ensured that the garlic plant is perpendicular to the plane where the toggle finger chain is located when cutting the roots [40,41]. Equation (6) also shows that the distance l_{CD} between the projection points, the projection distance l_{AD}, the distance l_{AB}, the angle α_E between the garlic plant and the clamping conveyor chain, and the angle γ_{BC} between the clamping conveyor chain and the toggle finger chain are the common influencing factors. The ratio K_X of the speed V_{JC} of the clamping conveyor chain and the speed V_{BZ} of the finger-pulling chain is determined; i.e., it affects the adjustment of the position of the garlic plant by the sorting-alignment device.

It is the ability of the conveyor alignment assembly to realize the positional adjustment of the garlic plant and the alignment of the upper surface of the bulb that makes it possible to realize adaptive root cutting for a garlic harvester. However, the conveyor alignment assembly is compact. It has no extra installation space, and its structure needs to be modified to arrange a machine vision-based adaptive root-cutting system for garlic harvesters.

2.3. Cutting Device Design and Research

To meet the need for adaptive root cutting, a cutting device containing a camera, a cutting device, and a stepper motor for position adjustment is designed. As shown in Figure 4, the cutting device mainly includes a cutter with a connecting piece, a lifting arm, a micro switch, a stationary seat, a dual-out shaft stepper motor, a clearing servo, a clearing rocker arm, a camera, a damping tension spring, etc. Among them, four lifting arms of equal length, a fixed seat, and a cutter with a connector form a parallel four-bar mechanism. The line between the rotation centers of the upper lifting arm and the lower lifting arm mounted on the fixed seat is perpendicular to the lower bottom surface of the fixed seat. A damping spring increases the lifting force of the parallel four-bar mechanism, and a dual-out shaft stepper motor (KEMING Motor, 57MM44A100, Wenzhou, China) provides the driving force for the two lower lifting arms. A microswitch (Omron, SS-5GL2, Kyoto, Japan) is used to reset the lifting arm; the camera (KINGCENT, KS1A14, Shenzhen, Guangdong, China) has a high-definition 1.0 megapixel and a 60° lens. To address the problem of dust that can cover the camera lens during harvesting, a dirt-clearing rocker arm with a brush was designed to erase the dirt from the transparent lens guard. A servo powers the clearing arm.

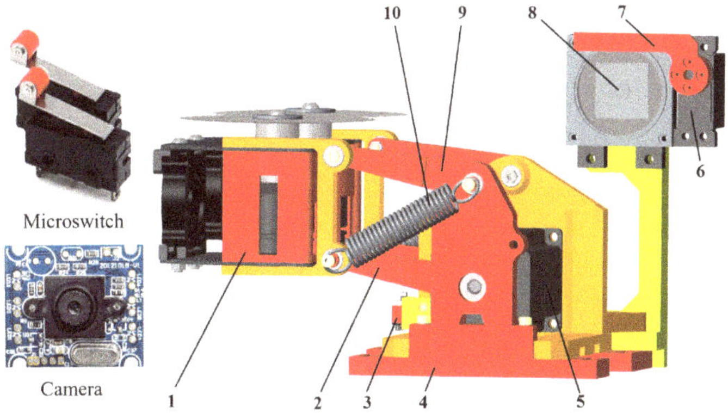

Figure 4. Schematic diagram of the structure of the cutting device. 1. Cutting device with connector 2. Lower lifting arm 3. Microswitch 4. Fixed seat 5. Dual-out shaft stepper motor 6. Clearing servo 7. Clearing Rocker 8. Camera 9. Upper lifting arm 10. Damping tension spring.

When the lower lifting arm triggers the microswitch, the angle of rotation of the dual-out shaft stepper motor is recorded as 0°, at which time all four lifting arms are parallel to the lower bottom surface of the stationary base; when the cutting device completes the reset, and the optical axis of the camera passes through the plane where the circular knife disk of the cutting device is located, the angle of rotation of the dual-out shaft stepper motor is recorded as α_{DJ} (°), as shown in Figure 5. When the dual-out shaft stepper motor makes the first root-cutting rotation to adjust the cutting device's height, the rotation angle is $\Delta\alpha_1$ (°). Then, the angle turned by the dual-out shaft stepper motor is:

$$\theta_1 = \alpha_{DJ} + \Delta\alpha_1 \tag{7}$$

where θ_1 is the angle, in °, that the dual-out shaft stepper motor turns relative to the position of the lower lifting arm, triggering the microswitch.

The distance between the two rotation centers of the upper and lower lifting arms is equal, expressed in l_{JS}, mm.

When $\theta_1 > \alpha_{DJ}$, the height of movement of the cutting device relative to the lower bottom surface of the stationary base is:

$$\Delta h_1' = l_{JS} \cdot (\tan\theta_1 - \tan\alpha_{DJ}) \tag{8}$$

When $\alpha_{DJ} > \theta_1$, the cutting device moves a height relative to the lower bottom surface of the stationary base:

$$\Delta h_1' = l_{JS} \cdot (\tan\alpha_{DJ} - \tan\theta_1) \tag{9}$$

In Equations (8) and (9), $\Delta h'_1$ is the height of the cutting device relative to the lower bottom surface of the stationary base after the first rotation of the dual-out shaft stepper motor for the cutting root, mm; l_{JS} is the distance between the two rotating centers of the lifting arm, mm. Moreover, the cutting device raises $\Delta h'_1$ when $\theta_1 > \alpha_{DJ}$, and the cutting device decreases $\Delta h'_1$ when $\alpha_{DJ} > \theta_1$.

For the n′ + 1st rotation (n′∈N+) of a dual-out shaft stepper motor as a tangent root, there is:

$$\theta_{n'+1} = \theta_{n'} + \Delta\alpha_{n'+1} \tag{10}$$

where $\theta_{n'+1}$ is the angle of rotation of the dual-out shaft stepper motor concerning the bottom surface of the stationary base after the n′ + 1st rotation of the dual-out shaft stepper motor cut the root (°); $\theta_{n'}$ is the angle of rotation of the dual-out shaft stepper motor concerning the bottom surface of the stationary base after the n′th rotation of the dual-out shaft stepper motor to cut the root (°); $\Delta\theta_{n'+1}$ is the n′ + 1st rotation angle of the dual-out shaft stepper motor to cut the root (°).

When $\theta_{n'+1} > \theta_{n'}$, the height of movement of the cutting device concerning the lower bottom surface of the stationary base after the n′ + 1st rotation of the dual-out shaft stepper motor to cut the root is:

$$\Delta h'_{n'+1} = l_{JS} \cdot (\tan\theta_{n'+1} - \tan\theta_{n'}) \tag{11}$$

When $\theta_{n'} > \theta_{n'+1}$, the height of movement of the cutting device concerning the lower bottom surface of the stationary base after the n′ + 1st rotation of the dual-out shaft stepper motor to cut the root is:

$$\Delta h'_{n'+1} = l_{JS} \cdot (\tan\theta_{n'} - \tan\theta_{n'+1}) \tag{12}$$

In Equations (11) and (12), $\Delta h'_{n'+1}$ is the moving height of the cutting device relative to the lower bottom surface of the stationary base after the n′ + 1th rotation of the dual-out shaft stepper motor it cut the root, mm; l_{JS} is the distance between the two rotating centers of the lifting arm, mm. In particular, the cutting device raises $\Delta h'_{n'+1}$; when $\theta_{n'} > \theta_{n'+1}$, the cutting device decreases $\Delta h'_{n'+1}$.

So far, the equations of the motion relationship between the angle of the stepper motor of the double outgoing axis in the cutting device and the change in the height of the cutting device are deduced, which lays the foundation for the design of the adaptive root-cutting control algorithm for garlic based on machine vision.

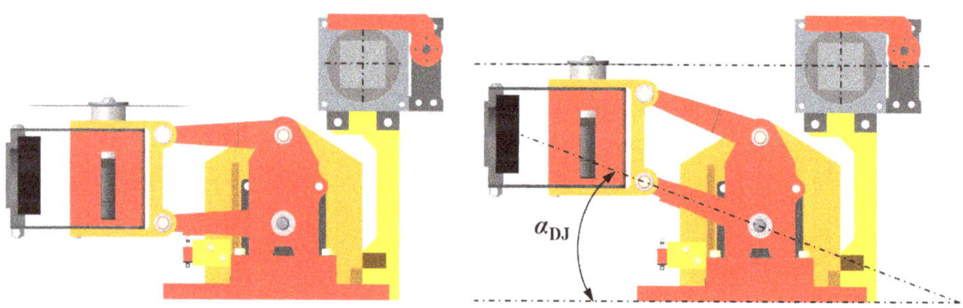

Figure 5. Cutting device reset process. (**a**) Lower lifting arm triggers microswitch; (**b**) the cutting unit reset is completed. The dotted line indicates that the angle between the lower lifting arm and the base plate of the fixed seat when the blade is aligned with the center axis of the camera is α_{DJ}.

2.4. *Adaptive Root-Cutting Conveyor Alignment Assembly Study*

The analysis in Section 2.2 shows that the conveyor alignment assembly can realize the positional adjustment of the garlic plant and the alignment of the upper surface of the bulb, which is a prerequisite for realizing adaptive root cutting on the garlic harvester. Due to the limited space, it is necessary to improve the existing conveyor alignment assembly to arrange the adaptive garlic root-cutting device based on machine vision. The adaptive root-cutting system for the garlic harvester proposed in this study relies on the clamping conveyor chain of the conveyor alignment assembly, the sorting-alignment device, and the harvesting table weld, and it is suggested to remove the garlic roots after completing the adjustment of the position of the garlic plant and the alignment of the upper surface of the bulb.

To carry out adaptive root cutting, the harvesting table weld of the original conveyor alignment assembly was improved, and the space under the lower part of the sorting-alignment device was enlarged without changing the overall structure of the original conveyor alignment assembly, so that the cutting device could be arranged on top of the lower beam of the harvesting table weld and ensure that its disk cutter was parallel to the plane in which the toggle finger chain was located. A laser alignment sensor was mounted on the middle beam of the harvesting table weld (LUOSHIDA, LQD-31NO, Dongguan, China); the adaptive root-cutting conveyor alignment assembly is shown in Figure 6. For adaptive cutting root delivery alignment assembly, in the working state of the laser alignment sensor, with real-time detection of the garlic plant to arrive at the image acquisition position, the garlic plant passes through the laser alignment sensor, triggering the laser alignment sensor to send a signal to the adaptive cutting root control system, allowing the cutting device to work.

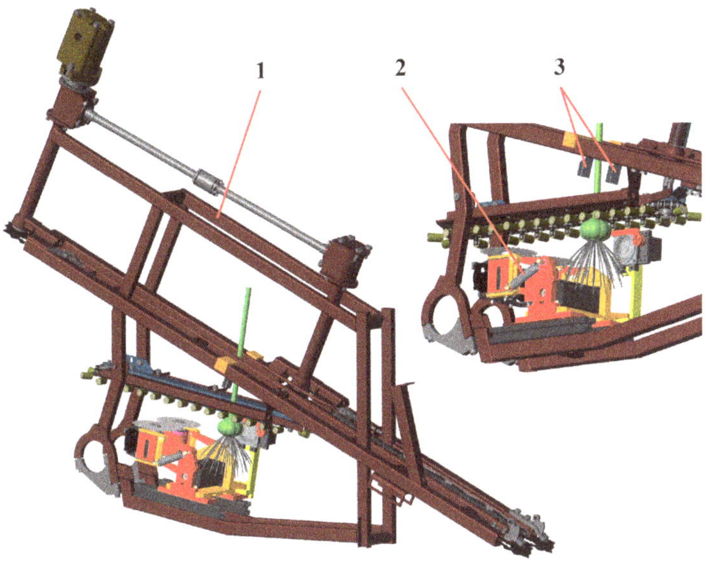

Figure 6. Adaptive root-cutting conveyor alignment assembly. 1. Adaptive root-cutting conveyor alignment assembly 2. Cutting device 3. Laser alignment sensor.

2.5. Study on the Speed Relationship Between Clamped Conveyor Chain and Toggle Finger Chain

Before cutting the roots, the sorting-alignment device of the adaptive root-cutting conveyor alignment assembly adjusts the garlic plant, as shown in Figure 7. The red arrows in Figure 7 indicate the conveying direction of the clamping conveyor chain and the conveying direction of the toggle finger chain, respectively. When the inclined garlic plant starts to enter the sorting-aligning device, point Z_1 is the clamping point of the clamping conveyor chain on the garlic plant, and point B_1 is the point of the toggle finger chain's action on the garlic plant. Next, due to the difference in speeds between the speed of the clamping conveyor chain and the speed of the toggle finger chain, the garlic plant's position is altered during the conveying process. The upper surface of the bulb is aligned.

In terms of surface alignment, the upper surface of the bulb is against the lower surface of the two adjacent toggle fingers by adjusting the clamping conveyor chain speed and the speed of the toggle finger chain speed ratio, to ensure that the upper surface of the bulb is aligned perpendicular to the toggle finger chain located in the plane of the garlic plant, which triggers the laser alignment sensor. At this time, Z_2 is the clamping conveyor chain on the garlic plant of the clamping point, and the B_2 point is the toggle finger chain on the bulb of the equivalent role of the point; with the continuation of the conveying, clamping begins. As the conveying continues, the angle between the clamping conveyor chain and the toggle finger chain makes the garlic plant always subject to the tension along the stalk direction, the upper surface of the bulb is always tightly pressed against the lower surface of the two neighboring toggle fingers, and the anisotropic stalk will undergo bending and deformation, so that the centerline of the bulb can be kept perpendicular to the plane where the toggle finger chain is located under the joint action of the two neighboring toggle fingers. When the root cutting starts, the Z_3 point and the B_3 point are the clamping points of the clamping conveyor chain and the toggle finger chain, respectively, and the equivalent action point of the toggle finger chain on the bulb. Points Z_3 and B_3 are the equivalent action points of the clamping conveyor chain on the garlic plant and the toggle finger chain on the bulb, respectively, when root cutting starts.

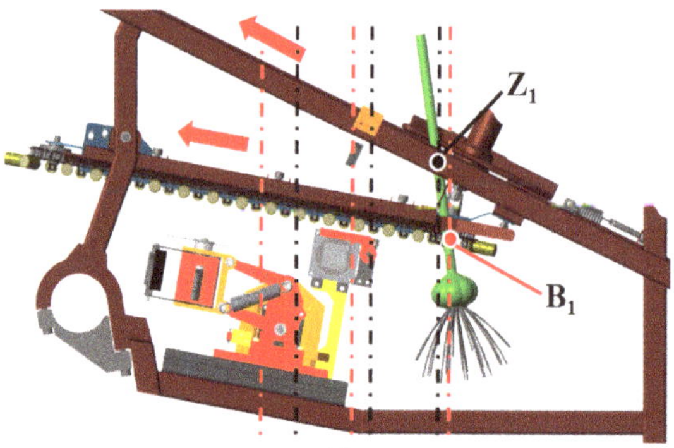

(a)

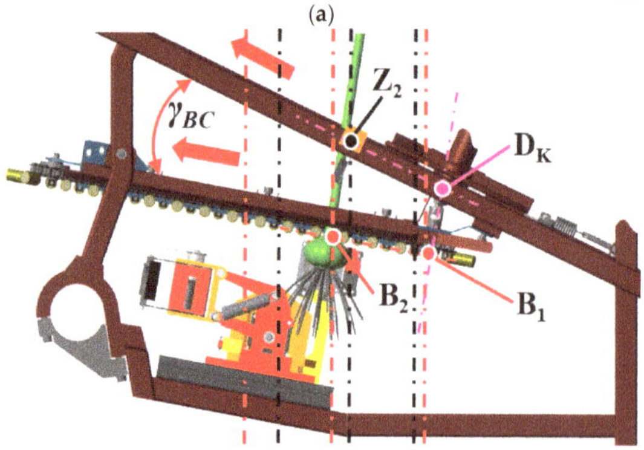

(b)

Figure 7. *Cont.*

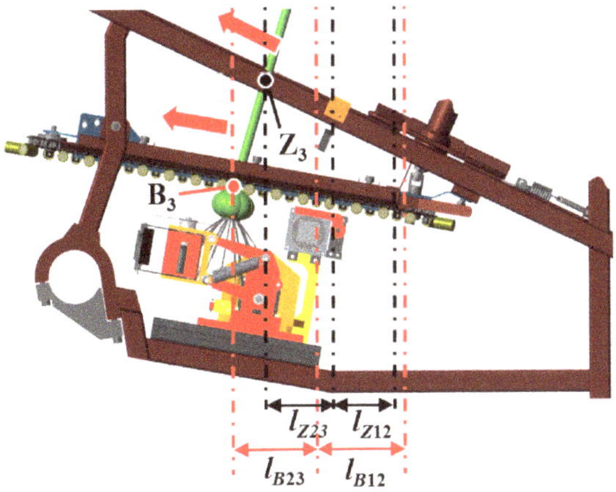

(c)

Figure 7. Sorting-alignment device for garlic plant adjustment. (**a**) Garlic plants begin to enter the sorting-aligning device; (**b**) garlic plant-triggered laser-aligned sensors; (**c**) root cutting is started after completion of the adjustment of the cutting device. The red arrows in (**a**–**c**), indicate the conveying direction of the conveyor chain and the toggle finger chain, respectively.

According to the clamping conveyor chain and the toggle finger chain uniform speed, movement can be seen.

$$\frac{l_{Z12}}{l_{Z23}} = \frac{l_{B12}}{l_{B23}} \tag{13}$$

In the above equation, l_{Z12} is the projection in the horizontal direction of the movement distance between the garlic plant at the clamping points Z_1 and Z_2, mm; l_{Z23} is the projection in the horizontal direction of the movement distance between the garlic plant at the clamping points Z_2 and Z_3, mm; l_{B12} is the projection in the horizontal direction of the distance between the point of action of the finger-toggle finger chain on the garlic plant, B_1, and the point of action of the finger-toggle finger chain on the bulb equivalent action point, B_2, mm; and l_{B23} is the projection in the horizontal direction of the distance between the finger-toggle finger chain is the projection in the horizontal direction of the distance between the equivalent action point B_2 of the toggle finger chain on the bulb and the equivalent action point B_3 of the toggle finger chain on the bulb, mm.

After the sorting-alignment device process to adjust the position of the garlic plant, it is necessary to study the speed ratio between the speed of the clamping conveyor chain and the speed of the toggle finger chain to realize the completion of the alignment of the upper surface of the bulb perpendicular to the plane of the toggle finger chain of the garlic plant, triggering the laser alignment sensor.

From Equation (6) and Figure 7b, the ratio K_G between the speed of the gripping conveyor chain and the speed of the toggle finger chain required to achieve the triggering of the laser alignment sensor by the garlic plant perpendicular to the plane in which the toggle finger chain is located to accomplish the alignment of the upper surface of the bulb is:

$$K_G = \frac{l_{D_K Z_2} \sin \alpha_E - l_{B_1 D_K} \cos(\alpha_E + \gamma_{BC})}{l_{B_1 B_2} \sin \alpha_E} \tag{14}$$

where α_E is the angle between the garlic plant and the clamping conveyor chain when it enters the sequencing-alignment device (°); γ_{BC} is the angle between the clamping conveyor chain and the finger-dialing chain (°); l_{DKZ2} is the distance moved from the clamping point of the clamping conveyor chain of the garlic plant when the garlic plant is kept perpendicular to the plane where the finger-dialing chain is located to the point of action of the clamping conveyor chain of the triggering laser alignment sensor, mm; l_{B1DK}

is the distance between the clamping point of the distance between the gripping point of the clamping conveyor chain and the point of action of the toggle finger chain when the garlic plant enters the sorting-aligning device when the garlic plant remains perpendicular to the plane in which the toggle finger chain is located, mm; l_{B1B2} is the distance between the point of action of the toggle finger chain on the garlic plant and the equivalent point of action of the toggle finger chain on the bulb from the time when the garlic plant enters the sorting-aligning device to the triggering of the laser alignment sensor, mm.

Substituting the value of each distance, the angle between the clamping conveyor chain and the toggle finger chain into Equation (14), we have

$$K_G = \frac{150\sin\alpha_E - 84\cos(\alpha_E + 15^\circ)}{140\sin\alpha_E} \tag{15}$$

where α_E is the angle between the garlic plant and the clamping conveyor chain when entering the sorting-alignment device (°).

When the harvester operates after the digging depth is fixed by adjusting the travel wheels, it can be considered that the angle α_E between the plucked garlic plant and the clamping conveyor chain is fixed. The angle α_E can be deduced from the inclination angle of the conveying alignment assembly, thus determining the ratio of the speed of the clamping conveyor chain and the speed of the toggle finger chain, K_G [40,41]. During harvesting operation, the inclination angle of the conveyor alignment assembly is usually 11°~34° [13,19,42]. The garlic plants in the farmland are approximately perpendicular to the ground, so the deduction shows that the angle αE is 56°~79°, and the value of KG is 0.836~1.114 by substituting into Equation (15). Then, K_G takes the intermediate value of 0.975 during the combined garlic harvesting.

2.6. Garlic Adaptive Root-Cutting System Working Principle

According to the need to establish the relationship between the pixel coordinate system and the image coordinate system, the adaptive root-cutting system of the garlic harvester resets the cutting device first when it starts to work. The adaptive root-cutting conveying and aligning assembly pulls the garlic plant dug by the digging shovel out of the soil using the clamping conveyor chain. The garlic plant enters the sorting-aligning device along the clamping conveyor chain. It continues to be conveyed under the joint action of the clamping conveyor and finger-pulling chains. The clamping point moves along the direction of the garlic plant's stem due to the structure of the clamping chain, the long bristle brush, and the garlic plant's position being adjusted. The bulb's upper surface is aligned under the finger-pulling chain action, and the stem is aligned with the laser alignment sensor, triggering the laser alignment sensor, and the cutting device camera takes a bulb image. The Matlab-based IRM-YOLO bulb detector [27,28] was converted to NVIDIA CUDA code by GPU Coder (NVIDIA, Santa Clara, CA, USA) and deployed in the edge computing device Jetson Nano. The Jetson Nano (NVIDIA, Jetson Nano(4GB), Santa Clara, CA, USA) deployed with the bulb detector detects the bulb by convolutional neural network to obtain the classification and pixel position information of the target, i.e., to obtain the pixel position information of the cutting line. Based on the cutting device's control algorithm and the cutting line's pixel position information, we obtain the digital command code to control the rotation angle of the dual-out shaft stepper motor. The Jetson Nano processor sends the digital command code to the STM32F103 development board (ALENTEK, Guangzhou, China) through serial communication, and the dual-out shaft stepper motor will be rotated to the corresponding angle to complete the adjustment of the cutting height under the control of the STM32F103 development board. After that, the cutting height will be adjusted, and the garlic root is removed by rotating the circular knife disk.

The IRM-YOLO bulb detection convolutional neural network model, shown in Figure 8 [27,28], accurately, quickly, and reliably detects bulb targets in images with variable light and soil by learning bulb features.

Five hundred randomly selected pairs of captured bulb images constitute the test set [27], and the confidence score of the bulb detector trained using the IRM-YOLO algorithm model is 0.98228, with an AP of 99.2%; the detection time is 0.0356 s. The detection results are shown in Figure 9. The space the detector occupies is 24.2 MB. The detection results of IRM-YOLO are shown in Figure 10. The results show that the proposed IRM-YOLO has excellent detection performance. Firstly, IRM-YOLO can work generally under different environmental brightness conditions without supplemental light treatment, which reduces the detection cost. Secondly, the soil adhering to the bottom of the bulb does not affect the detection performance of IRM-YOLO, which can still make correct and fast detection. Finally, IRM-YOLO is an intelligent means of detection, with the ability to predict the cutting position of bulbs of different shapes.

The detection results of the bulb target provide a basis for the cutting device to adjust the cutting height. Adaptive root cutting based on machine vision can avoid cutting bulbs and improve the clean-cutting rate of garlic roots.

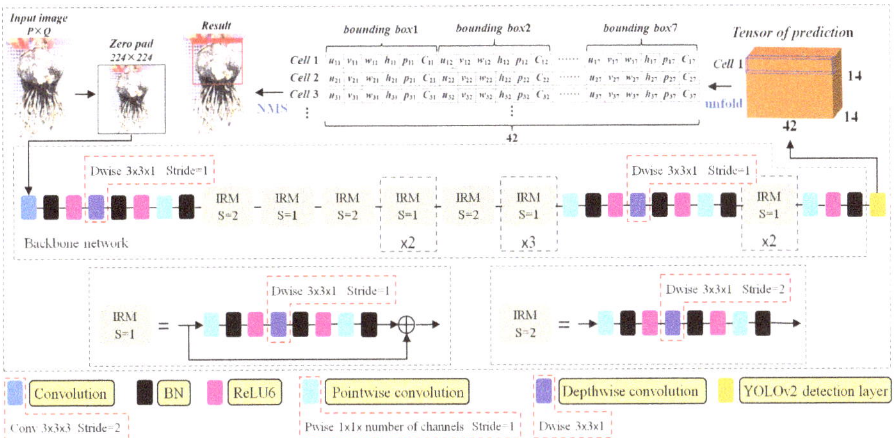

Figure 8. IRM-YOLO model structure.

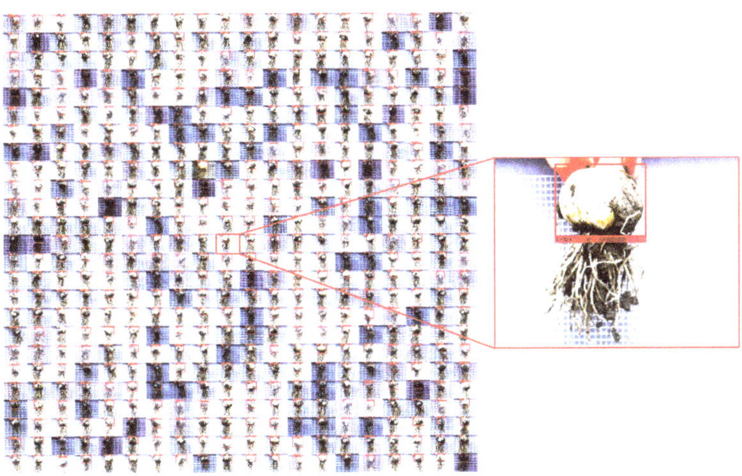

Figure 9. Results of the IRM-YOLO detector on the test set.

Figure 10. Detection results of IRM-YOLO detector.

2.7. System Design and Software Synergy

The adaptive root-cutting control system's hardware consists of the edge computing device Jetson Nano, power module, camera, STM32F103 development board, laser alignment sensors, dual-out shaft stepper motor, soil clearing servo, and stepper motor driver, as shown in Figure 11.

Edge computing device Jetson Nano is a high-performance processor developed by NVIDIA with powerful performance, compact size, and suitable for mobile equipment, with specific specifications and parameters shown in Table 1, capable of running a variety of algorithms and AI frameworks, such as TensorFlow, Keras, PyTorch, Caffe, and other applications in the fields of image categorization, target detection, face recognition, speech processing, object recognition tracking, etc. Jetson Nano includes a rich set of peripheral interfaces, including SPI, UART, USB, etc., which are suitable for controlling many different types of peripherals, as well as a comprehensive NVIDIA JetPack SDK containing acceleration libraries for computer vision, deep learning, graphics, multimedia, and more.

Using GPU Coder, it is possible to transcode Matlab code to CUDA code, automatically cross-compile and deploy the resulting code to the Jetson Nano, and accelerate deep learning networks and other computationally intensive parts of the algorithm.

Table 1. Jetson Nano device specifications list.

Configuration	Parameters
AI algorithm	472GFLOPS
GPU	NVIDIA Maxwell build, 128 NVIDIA CUDA cores
CPU	Quad-core ARM Cortex-A57 MP Core processor
Display memory	4 GB 64-bit LPDDR4
Power input	5 V-4 A DC/5 V-3 A micro USB
Size	100 × 80 × 29 mm

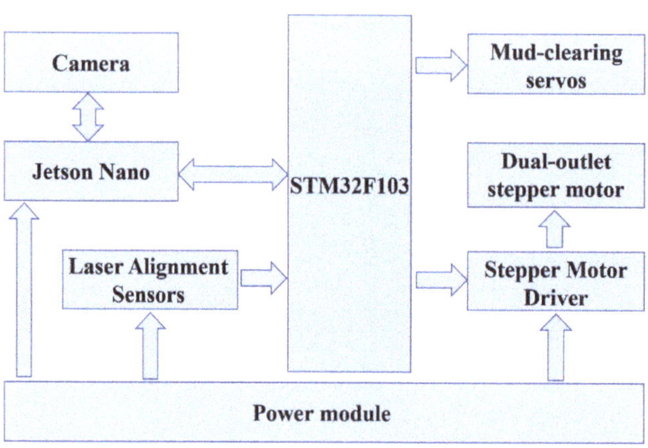

Figure 11. Adaptive root-cutting control system hardware structure diagram.

UART communication is used between the Jetson Nano and the STM32F103 development board. The baud rate is set to 115,200 b/s, and the control commands are transmitted using decimal numeric command codes, with values ranging from 0 to 199. The meanings of the numeric command codes are shown in Table 2.

Table 2. Turning angle digital command code.

Items	Value	Meaning
a_1	0	Dual-out shaft stepper motor angle reduction
	1	Dual-out shaft stepper motor corner increase
a	0~99	Decrease in angle of rotation, angle of rotation
	100~199	Increased angle of rotation, angle of rotation

Note: a_1 is the hundredth value of a.

The STM32F103 development board receives the digital command code, as shown in Table 2, to control the rotation of the dual-out shaft stepper motor. Adjusting the subdivision and setting the stepper motor drive to 5000 pulse/rev give the pulse-width-modulated signal a single pulse turn angle of 0.072°. Using STM32CubeMX and Keil5 development software, set the clock of the STM32F103ZET6 chip as 72 MHz; obtain the pulse width modulation signal period $T_{\min}$ as 0.08 ms by configuring timer, counter, and interrupt controller; and stipulate that five consecutive single pulses corresponding to 1 minimum rotational angle $\alpha_{\min}$ so that the rotation accuracy of dual-out shaft stepper motor is 0.36°, and the time consumed for one minimum rotation angle $\alpha_{\min}$ is 0.4 ms. The maximum angle range of each rotation of the dual-out shaft stepper motor is [−35.64, 35.64] (unit: °), but the actual cutting of the root is to avoid the collision of the cutting device with other parts, which restricts the range of the rotation angle of the dual-out shaft stepper motor.

2.8. Adaptive Root-Cutting Control Algorithm Design

The adaptive root-cutting control algorithm is more complex than the trial bench root-cutting control algorithm because the rotation of the dual-out shaft stepper motor of the cutting device regulates the cutting height. However, the logic of the adaptive root-cutting control algorithm is the same as that of the test bench root-cutting control algorithm. After the relationship between the pixel coordinate system and the image coordinate system is established, the rotation angle of the dual-out shaft stepper motor is calculated based on the pixel position information of the cutting line. Then, the cutting device is moved to the appropriate cutting height to cut the root.

After completing the previous root cutting, the upper computer system converts the pixel coordinates of the cutting line according to the pixel coordinates of the cutting line predicted to obtain the image coordinates of the cutting line. It feeds back to the motor control system to adjust the movement of the cutting device along the height direction. Before and after the two cuts, the movement of the cutting device along the height direction is:

$$\Delta h_{n+1} = (k_{n+1} - k_n)\rho_h'' \qquad (\mathrm{n} \in \mathrm{N}+) \tag{16}$$

where Δh_{n+1} is the coordinate difference between the n + 1st cut root and the projection point of the nth cut root cutting line along the height direction in the image coordinate system, and its value is positive when the cutting device moves upward along the height direction and harmful when the cutting device moves downward along the height direction, mm:

k_n is the nth predicted cut line pixel vertical coordinate;

k_{n+1} is the n + 1th predicted pixel vertical coordinate of the cut line;

ρ_h'' is the height of each pixel point in the detected image, mm/pixel.

In addition, a relationship exists when the roots are cut for the first time:

$$\Delta h_1 = (k_1 - k_0)\rho_h'' \tag{17}$$

where k_0 is the pixel vertical coordinate of the cutting surface of the cutting device after completing the reset.

Based on Equations (7)–(12), (16), and (17), and pixel coordinate information of the predicted cut line position, the following equations can be obtained.

When adjusting the cutting height, the angle $\Delta\alpha_1$ of the first rotation of the dual-out shaft stepper motor is:

When $k_0 > k_1$, the cutting height of the cutting device should rise after the rotation of the dual-out shaft stepper motor

$$\Delta\alpha_1 = \arctan\left[\frac{l_{\mathrm{JS}}\tan\alpha_{\mathrm{DJ}} + (k_0 - k_1)\rho_h''}{l_{\mathrm{JS}}}\right] - \alpha_{\mathrm{DJ}} \tag{18}$$

When $k_1 > k_0$, the cutting height of the cutting device should decrease after the rotation of the dual-out shaft stepper motor

$$\Delta\alpha_1 = \arctan\left[\frac{l_{\mathrm{JS}}\tan\alpha_{\mathrm{DJ}} - (k_1 - k_0)\rho_h''}{l_{\mathrm{JS}}}\right] - \alpha_{DJ} \tag{19}$$

When adjusting the cutting height, the angle $\Delta\alpha_{\mathrm{n}'+1}$ of the n$'$ + 1st rotation of the dual-out shaft stepper motor is (n$'\in$N+):

When $k_{\mathrm{n}'} > k_{\mathrm{n}'+1}$, the cutting height of the cutting device should rise after the dual-out shaft stepper motor is rotated.

$$\Delta\alpha_{n'+1} = \arctan\left[\frac{l_{\mathrm{JS}}\tan\theta_{n'} + (k_{n'} - k_{n'+1})\rho_h''}{l_{\mathrm{JS}}}\right] - \theta_{n'} \tag{20}$$

When $\mathrm{k}_{\mathrm{n}'+1} > \mathrm{k}_{\mathrm{n}'}$, the cutting height of the cutting device should decrease after the rotation of the dual-out shaft stepper motor

$$\Delta\alpha_{n'+1} = \arctan\left[\frac{l_{\mathrm{JS}}\tan\theta_{n'} - (k_{n'+1} - k_{n'})\rho_h''}{l_{\mathrm{JS}}}\right] - \theta_{n'} \tag{21}$$

In Equations (18)–(21)

k_0 is the pixel longitudinal coordinate of the cutting surface of the cutting device after completing the reset, 300 pixels in this paper (the camera resolution is 800 × 600 pixels);

$k_{\mathrm{n}'}$ is the n$'$th predicted pixel vertical coordinate of the cutting line, pixels;

$k_{\mathrm{n}'+1}$ is the n$'$th + 1th predicted pixel vertical coordinate of the cutting line, pixels;

α_{DJ} is the angle of rotation of the dual-out shaft stepper motor for the completion of the rest of the cutting device (°);

$\Delta\alpha_1$ is the angle of the first rotation of the stepper motor for cutting root (°);

$\Delta\alpha_{\mathrm{n}'+1}$ is the angle of the n$'$ + 1th rotation of the dual-out shaft stepper motor for root cutting (°);

$\theta_{\mathrm{n}'}$ is the angle of rotation of the dual-out shaft stepper motor relative to the lower bottom surface of the stationary base after the n$'$th rotation of the dual-out shaft stepper motor to cut the root (°);

l_{JS} is the distance between the two rotation centers of the lifting arm, mm in mm.

So far, the parameter equation system between the change in the longitudinal coordinates of the pixels of the cutting line predicted by the target detection and the rotation angle of the dual-out shaft stepper motor has been established, and the adaptive root-cutting control algorithm was designed according to this parameter equation system.

It is to be noted that the number of pulse width modulation signal pulses $N_{\mathrm{n}'}$ is related to the rotation angle $\Delta\alpha_{\mathrm{n}'}$ of the dual-out shaft stepper motor when controlling the dual-out shaft stepper motor for the n$'$th root-cutting rotation as follows

$$\Delta\alpha_{n'} = \frac{1}{5}N_{n'}\alpha_{min} + \mu_{n'} \tag{22}$$

The time taken for the n′th root-cutting rotation of the dual-out shaft stepper motor is $t_{n'}$

$$t_{n'} = N_{n'} T_{min} \tag{23}$$

In Equations (22) and (23), $\Delta\alpha_{n'}$ is the rotation angle of the n′th root-cutting rotation of the dual-out shaft stepper motor (°); $N_{n'}$ is the number of pulse width modulation signal pulses $N_{n'}$controlling the dual-out shaft stepper motor to perform the n′th root-cutting rotation; α_{min} is the minimum rotation angle of the dual-out shaft stepper motor (°); $\mu_{n'}$ is the error in the n′th root-cutting (°); $t_{n'}$ is the time consumed for the n′th tangential rotation of the dual-out shaft stepper motor, s; T_{min} is the pulse width modulation signal period, s. The pulse width modulation signal period is the time required for the n′th tangential rotation of the dual-out shaft stepper motor.

The error $\mu_{n'}$ will be small enough to eliminate the cumulative error by resetting the cutting device again since the rotation accuracy of the dual-out shaft stepper motor is 0.36°. The calibration coefficient of the adaptive root-cutting system $\rho_h'' = 30$ mm/113 pixels = 0.2655 mm/pixel was measured.

2.9. Validation of the Execution Speed of an Adaptive Root-Cutting System

The machine vision-based root-cutting function is the core of the adaptive root-cutting system. The machine vision-based adaptive root-cutting process includes the adaptive root-cutting conveyor alignment assembly for garlic plant position adjustment and bulb upper-surface alignment, the cutting line position predicted by the target detection model, the cutting height adjustment by the cutting device, and the cutting device for removing the garlic root, which requires precise coordination of multiple systems. Among them, whether the cutting height can be quickly adjusted according to the predicted cutting line position is the key to the success of adaptive root cutting. Hence, the execution speed of the adaptive root-cutting system needs to be verified.

The harvester clamping conveyor chain speed VJC is 0.75 m/s; according to Section 2.5, it can be seen that the ratio of clamping conveyor chain speed and toggle finger chain speed K_G is 0.975, and the toggle finger chain speed V_{BZ} is 0.77 m/s. Generally, garlic plant spacing l_{ZJ} is more than 110 mm [19,26,42–44]. After measurement, when the adaptive root-cutting system works, the distance l_{LJ} from image capture to the beginning of cutting the root bulb moving along the toggle finger chain is more significant than 100 mm. Taking the minimum value of the plant distance l_{ZJ} and the distance l_{LJ} of the bulb moving along the toggle finger chain $l_{LJ} = 100$ mm, the target detection model predicts that the allowable time for the position of the cutting line and the cutting device to adjust the height of the cut, t_{YX}, is 0.129 s. Section 2.7 shows that the maximum value of time consumed by each dual-out shaft stepper motor rotation, t_{MDJ}, is 0.016 s. After testing the IRM-YOLO detection model deployed on the Jetson Nano, the detection time, t_{MJC}, is within 0.1 s, and by taking $t_{MJC} = 0.1$ s, the following relationship is observed:

$$t_{MDJ} + t_{MJC} < t_{YX} \tag{24}$$

In the formula, t_{MDJ} is the maximum value of time consumed for each rotation of the dual-out shaft stepper motor during the operation of the adaptive root-cutting system, s; t_{MJC} is the maximum detection time of the IRM-YOLO detection model deployed on the Jetson Nano, s; t_{YX} is the permissible time for the target detection model to predict the position of the cutting line and for the cutting device to adjust the height of the cut, s.

Through the above analysis, it can be seen that the cutting device adopts a parallel four-bar mechanism to effectively shorten the time for executing the cutting height adjustment so that the time consumed by the target detection model for predicting the position of the cutting line and the cutting device for adjusting the cutting height is less than the allowable time. The adaptive root-cutting system can complete the root-cutting process based on the target detection of the convolutional neural network.

3. Results

3.1. Cutting Device Root-Cutting Test

A bench test was designed to verify the root-cutting ability of the cutting device. The cutting device adopts four equal-length lifting arms, a fixed base, and a cutting device with connectors to form a parallel four-bar mechanism. It uses a damping spring to increase the lifting force of the parallel four-bar mechanism, which also enhances the ability of the parallel four-bar mechanism to resist impact forces. As a result of the lifting arm, the cutting height adjustment amount corresponding to the rotation angle of the dual-out shaft stepper motor is amplified, thereby effectively shortening the time required for the cutting device to adjust the cutting height. All garlic plants were obtained from the planting base in Jinxiang County, Nanjing Institute of Agricultural Mechanization, Ministry of Agriculture and Rural Affairs, in this experiment. To retain the moisture of the garlic roots, after the garlic was excavated from the planting base, the roots were wrapped entirely with soil. The treated test samples were delivered to the laboratory within 3 days for root-cutting tests. Random selection was used in the sample pool to determine the test subjects for the root-cutting test [45].

Before the test, the cutting device was fixed on the garlic root-cutting test bench so that the bottom surface of the fixed seat of the cutting device was parallel to the surface of the garlic root-cutting test bench, and the cutting device was reset before the root-cutting test was carried out, as shown in Figure 12. The conveying speed of the X-axis slide was set to 0.8 m/s, and the rotational speed of the cutting device was 1200 r/min. Three groups of root-cutting tests were conducted, with 100 samples in each group. Examining the cut root qualified rate index, we can calculate the cut root qualified rate calculation formula

$$\alpha_{HG} = \frac{N_H}{N_Z} \times 100\% \tag{25}$$

where α_{HG} is the qualified rate of the cut root, %;

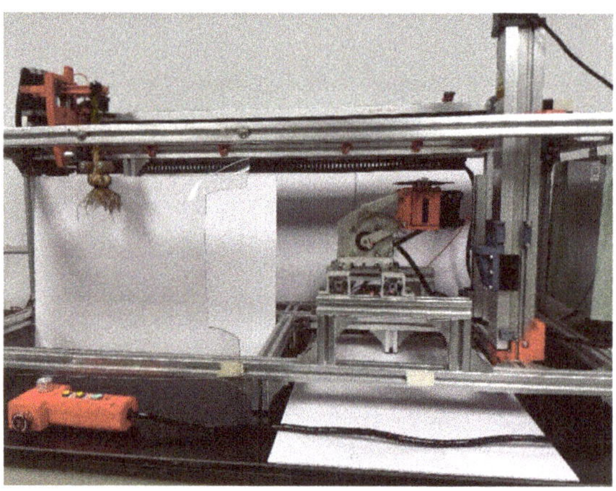

Figure 12. Improved cutting device root-cutting test.

N_{H} is the number of garlic plants with qualified cut roots;

N_{Z} is the total number of garlic plants.

A comparison between before and after root cutting is shown in Figure 13. It can be seen that the cutting device can effectively remove the garlic root and keep the cutting height unchanged during the cutting process. The root-cutting pass rates of the three groups of tests were 99%, 96%, and 97%, and the average value of the root-cutting pass rate was 97.33%, with a total of eight samples in which the residual garlic root was too long.

(**a**) (**b**) (**c**)

Figure 13. Comparison before and after root cutting. (**a**) Before root cutting; (**b**) during root cutting; (**c**) after root cutting.

The root-cutting ability of the cutting device was examined through the root-cutting test. The test results show that the cutting device has good consistency in terms of the root-cutting effect, and damping springs effectively improve the lifting force and reliability of the parallel four-bar mechanism for lifting the cutting device.

3.2. Field Trial

A field trial of combined garlic harvesting was conducted in May 2022 at the planting test field in Jinxiang County, Shandong Province, to test the adaptive root-cutting system. The adaptive root-cutting harvesting system was installed on the harvester. The adaptive root-cutting conveyor alignment assembly was arranged at the front of the harvester to complete the digging and pulling conveyance of the garlic plant, the positional adjustment of the garlic plant and the alignment of the upper surface of the bulb, the prediction of the cutting line position by the bulb detector, and the adaptive root cutting to realize the precise synergy of the multi-systems; the harvester finally conveyed the completed root-cutting bulbs to the fruit box. The garlic plants in the experimental field had good uprightness, with garlic root lengths ranging from 76 to 163 mm and bulb heights ranging from 39.26 to 51.12 mm.

The adaptive root-cutting system for the field trial was mainly composed of an electronic control box, a welded harvesting table, a background board, a cutting device, and a blade speed control board, as shown in Figure 14. The upper computer is a Jetson Nano, the lower computer is an STM32F103 development board, and the stepper motor driver (Ruite Technology, DM542, Shenzhen, China) and other components are installed in the electronic control box.

A garlic harvester installed with the adaptive root-cutting system, as shown in Figure 14, was used for continuous harvesting and root-cutting tests to verify the root-cutting capability of the machine vision-based adaptive root-cutting system for garlic harvesters. In the preparation stage of the test, the adaptive root-cutting conveyor alignment assembly, electronic control box, background board, and cutter speed control board were installed. The cutting device was fixed to the harvesting table weldment by bolts, with the lower-bottom surface of the fixing base of the cutting device parallel to the plane of the toggle finger chain. The camera's position corresponded to the position for completing the adjustment of the position of the garlic plant and the alignment of the upper surface of the bulb, as shown in Figure 7b. The laser alignment sensors were fixed to the harvesting table weldment by bolts. We installed the laser alignment sensor on the harvesting table weld joint so that the camera could capture the bulb image when the garlic plant triggered the laser alignment sensor; the camera was connected to the Jetson Nano through the extended soft row cable; the UART communication between the Jetson Nano and the STM32F103 development board was established; the clearing servo, the laser alignment sensor, and the stepper motor driver were connected to the STM32F103 development board; and the

DC brushless motor of the cutting device was connected to the STM32F103 development board. The DC brushless motor of the cutting device was connected to the cutter speed control board. The wires of the required specifications were used to connect each device to the power supply. The adaptive root-cutting system was tested before the subsequent tests were carried out, and the preparation is shown in Figure 15.

Figure 14. Composition of the adaptive root-cutting system for a field test. 1. Electric control box 2. Harvesting table welding joint 3. Background plate 4. Laser counter sensor 5. Cutting device 6. Cutter speed control board.

Figure 15. Test preparation.

During the adaptive root-cutting system test, the cutting device is first reset to establish the relationship between the pixel coordinate system and the image coordinate system. Then, the clamping conveyor chain and toggle finger chain of the adaptive root cutting and conveying and aligning assembly are activated, with the speed of the clamping conveyor chain at 0.75 m/s and the speed of the toggle finger chain at 0.77 m/s. The harvester advances with a uniform speed of 0.36 m/s. The garlic plant is dug out of the ground under the action of the excavating shovel and the shovel pulling of the clamping conveyor chain. Under the action of the digging shovel and the gripping conveyor chain, the garlic plant is excavated out of the soil, and the garlic plant is transported along the direction of the gripping conveyor chain; after the sorting-aligning device starts to act on the garlic plant, the position adjustment of the garlic plant and the alignment of the bulb along the upper surface are accomplished under the joint action of the gripping conveyor chain and the finger-pulling chain, so that the upper surface of the bulb is close to the lower surfaces of the two neighboring finger-pulling fingers and the centerline of the bulb is perpendicular to the plane where the finger-pulling chain is located; the garlic plant triggers the laser-aligning sensor, and the camera captures the bulb. The garlic plant triggers the laser-projected sensor, the camera captures the bulb image, and the IRM-YOLO-based bulb detector deployed in Jetson Nano predicts the position of the cutting line; the cutting height of the cutting device is dynamically adjusted based on the adaptive root-cutting control algorithm, and the disk knife rotates at high speed to excise the garlic root; the harvesting machine transports the bulb to the fruit box. After the harvester stops, the power supply of the adaptive root-cutting system is cut off. The performance of the adaptive root-cutting system is statistically analyzed based on the bulbs collected in the fruit box, and the test site is shown in Figure 16.

The cut rate α_{QS} is calculated as

$$\alpha_{QS} = \frac{N_{QS}}{N_{LZ}} \times 100\% \tag{26}$$

where α_{QS} is the rate of cut wounding, %;

N_{Qs} is the number of bulbs cut and wounded;

N_{LZ} is the total number of bulbs.

Figure 16. Garlic harvesting field test. The meaning of the Chinese character in the figure is garlic.

Three groups of combined harvesting adaptive root-cutting field trials were conducted; 186 bulbs were harvested under standard transportation of garlic plants, of which 154 bulbs were qualified for root cutting, 5 were cut, 27 residual garlic roots were too long, and 0 were missed. The qualified rate of root cutting was 82.8%, and the rate of cutting injury was 2.7%. The test results show that the adaptive root-cutting system effectively reduces the cut rate of the bulb by adjusting the cutting height through the cutting device. The qualified rate of root cutting is above 80%, which verifies the feasibility of the adaptive root-cutting system of the garlic harvester based on machine vision and the multi-system precise cooperative operation through the field test. In the field test, the qualified rate of root cutting was lower than that in Section 3.1 because the uneven ground in the test field caused the angle α_E between the garlic plant and the clamping conveyor chain to change.

There was a situation in which the bulb did not realize the alignment of the upper surface when the camera captured the image.

4. Discussion

(1) In this paper, we studied garlic bulbs grown in large fields, with garlic root lengths ranging from 75 to 168 mm and bulb heights ranging from 38 to 54 mm. The garlic bulb images, including different sunlight conditions, were collected from 8:00 to 19:00, and 2500 images were collected for bulb detector training and testing. The detection results of the detector described in this study may vary for bulb target detection in different planting locations and under different sunlight conditions.

(2) The operational performance of the adaptive root-cutting system on a garlic harvester in different regions and under different soil moisture and hardness conditions must be further investigated. In addition, the interaction between various components of the garlic harvester needs to be studied.

(3) The joint application of the adaptive root-cutting system and garlic harvester must also consider different clamping conveyor chain speeds, forward speeds, harvester types, etc., which must be studied.

(4) In the literature [25], the cutting rate of the bench test was 2.78%, and the root-cutting pass rate was 93.17%; in comparison, the cutting device in this study had a root-cutting pass rate of 97.33% in the bench test, and the harvesting machine field test had a cutting rate of 2.7% and a root-cutting pass rate of 82.8%. Compared to the bench test, there was a decrease in the passing rate of root cutting in the field test of the harvester. The possible reason for the lower pass rate was that the field ground was uneven, and there were cases where the bulbs did not realize the alignment of the upper surface when the camera obtained the image.

(5) In the literature [46], an improved apple detection model used in orchards was deployed using Jetson Nano. The authors of [47] used a computer vision-based detection model in strawberry harvesting machinery deployed via Jetson Nano. It can be seen that the Jetson family of low-power edge computing devices is often used to deploy AI detectors on agricultural mobile platforms.

(6) With the further development of information sensing technology and its application in agriculture [48], the root-cutting operation during garlic harvesting also requires new technological breakthroughs to improve the intelligence of garlic harvesting machines.

5. Conclusions

(1) The existing garlic mechanized root-cutting technology has problems, with a high leakage rate, high cutting injury rate, and uneven root cutting. We propose a non-contact adaptive root-cutting method for bulbs based on the YOLO algorithm based on a 4DSL-7 garlic harvester.

(2) In this study, for the structural characteristics and working principle of the conveyor alignment assembly of the 4DSL-7 garlic harvester, a cutting device integrating a camera, a circular cutter disk, a dual-out shaft stepper motor, a microswitch, and a lifting arm is designed, and an adaptive root-cutting conveyor alignment assembly is created. The cutting device can realize image acquisition functions, drive the circular knife disk rotation, reset the cutting device, adjust the cutting height, etc. The equations of the relationship between the angle of the dual-out shaft stepper motor and the cutting height are deduced. In the adaptive root-cutting and conveying alignment assembly, the action mechanism of the adaptive root-cutting system, the garlic plant bulb, is studied to realize the bulb alignment along the upper surface and adjust the garlic plant's position. Through dynamic analysis, the ratio of the clamping conveyor chain's speed and the toggle finger chain's speed was deduced to be 0.975.

(3) The confidence score of the bulb detector trained based on the IRM-YOLO algorithm model is 0.98228, with an AP of 99.2%; the detection time is 0.0356 s, and the storage space occupied by the detector is 24.2 MB. By setting the digital command code, the

STM32F103 development board controls the dual-out shaft stepper motor, and the angle of a single rotation is in the range of [−35.64, 35.64] (unit: °). A set of parametric equations between the predicted change in the longitudinal coordinates of the pixels of the cutting line and the rotation angle of the dual-out shaft stepper motor was established, and an adaptive cut-root control algorithm was designed. The allowable time for adjusting the cutting height was calculated to be 0.129 s, and the execution speed of the adaptive root-cutting system was verified.

(4) The cutting device bench test was designed to verify the cutting device's root-cutting capability. When the *X*-axis slide conveyor speed of the test bench was 0.8 m/s and the cutter speed of the cutting device was 1200 r/min, the average value of the qualified root-cutting rate was 97.33%.

(5) The adaptive root-cutting system was installed on a garlic harvester, the bulb detector was deployed on a Jetson Nano, and the STM32F103 development board was used as the lower computer for a garlic harvesting field test. The speed of the clamping conveyor chain was 0.75 m/s, the speed of the finger-pulling chain was 0.77 m/s, and the harvester moved at a uniform speed of 0.36 m/s. The passing rate of root cutting was 82.8%, and the cutting injury rate was 2.7%. The feasibility of the adaptive root-cutting system and the multi-system precision cooperative operation were tested.

Author Contributions: K.Y.: Conceptualization, methodology, software, visualization, writing—original draft. Y.Z. (Yunlong Zhou): methodology, resources, supervision. H.S.: supervision, methodology, software. R.Y.: software, methodology. Z.Y.: investigation, methodology, resources. Y.Z. (Yanhua Zhang): data curation, project administration. B.P.: methodology, supervision. J.F.: software, investigation. Z.H.: conceptualization, funding acquisition, resources, writing—review and editing. All authors have read and agreed to the published version of the manuscript.

Funding: This work was supported by the National Key R&D Program of China (2017YFD0701305-02), the Projects funded by the Jiangsu Modern Agricultural Machinery Equipment and Technology Demonstration and Extension (NJ2020-24), and the Luoyang Polytechnic School Fund Program (2024056).

Institutional Review Board Statement: Not applicable.

Data Availability Statement: Data are reported within the article.

Conflicts of Interest: The authors declare that they have no known competing financial interests or personal relationships that could have appeared to influence the work reported in this paper.

References

1. Tao, Y.; Zhang, J.; Jiang, S.; Xu, Y.; Show, P.; Han, Y.; Ye, X.; Ye, M. Contacting Ultrasound Enhanced Hot-air Convective Drying of Garlic Slices: Mass Transfer Modeling and Quality Evaluation. *J. Food Eng.* **2018**, *235*, 79–88. [CrossRef]
2. Furdak, P.; Pieńkowska, N.; Bartosz, G.; Sadowska-Bartosz, I. Extracts of Common Vegetables Inhibit the Growth of Ovary Cancer Cells. *Foods* **2022**, *11*, 2518. [CrossRef] [PubMed]
3. Recinella, L.; Gorica, E.; Chiavaroli, A.; Fraschetti, C.; Filippi, A.; Cesa, S.; Cairone, F.; Martelli, A.; Calderone, V.; Veschi, S.; et al. Anti-Inflammatory and Antioxidant Effects Induced by *Allium sativum* L. Extracts on an Ex Vivo Experimental Model of Ulcerative Colitis. *Foods* **2022**, *11*, 3559. [CrossRef] [PubMed]
4. Thomas, A.; Boobyer, C.; Borgonha, Z.; van den Heuvel, E.; Appleton, K.M. Adding Flavours: Use of and Attitudes towards Sauces and Seasonings in a Sample of Community-Dwelling UK Older Adults. *Foods* **2021**, *10*, 2828. [CrossRef] [PubMed]
5. Han, C.; Liu, J.; Chen, K.; Lin, Y.; Chen, C.; Fan, C.; Lee, H.; Liu, D.; Hou, W. Antihypertensive Activities of Processed Garlic on Spontaneously Hypertensive Rats and Hypertensive Humans. *Bot. Stud.* **2011**, *52*, 277–283.
6. Wang, J.; Zhang, X.; Lan, H.; Wang, W. Effect of Garlic Supplement in the Management of Type 2 Diabetes Mellitus (T2DM): A Meta-analysis of Randomized Controlled Trials. *Nutr. Res.* **2017**, *61*, 1377571. [CrossRef]
7. Phan, A.D.T.; Netzel, G.; Chhim, P.; Netzel, M.E.; Sultanbawa, Y. Phytochemical Characteristics and Antimicrobial Activity of Australian Grown Garlic (*Allium sativum* L.) Cultivars. *Foods* **2019**, *8*, 358. [CrossRef]
8. Shrestha, D.; Sapkota, H.; Baidya, P.; Basnet, S. Antioxidant and antibacterial activities of allium sativum and allium cepa. *Bull. Pharm. Res.* **2016**, *6*, 50–55. [CrossRef]
9. Ghani, M.I.; Ali, A.; Atif, M.J.; Ali, M.; Amin, B.; Anees, M.; Cheng, Z. Soil Amendment with Raw Garlic Stalk: A Novel Strategy to Stimulate Growth and the Antioxidative Defense System in Monocropped Eggplant in the North of China. *Agronomy* **2019**, *9*, 89. [CrossRef]

10. Nascimento, V.F.; Auad, A.M.; de Resende, T.T.; Visconde, A.J.M.; Dias, M.L. Insecticidal Activity of Aqueous Extracts of Plant Origin on *Mahanarva spectabilis* (Distant, 1909) (Hemiptera: Cercopidae). *Agronomy* **2022**, *12*, 947. [CrossRef]
11. Netzel, M.E. Garlic: Much More Than a Common Spice. *Foods* **2020**, *9*, 1544. [CrossRef] [PubMed]
12. Pocketbook, F. *World Food and Agriculture*; Food and Agriculture Organization: Rome, Italy, 2016.
13. Chen, Y. Development of Vegetable Mechanization in China from 2022 to 2023. *China Veg.* **2023**, *1*, 5–10. [CrossRef]
14. Katsigiannis, A.S.; Bayliss, D.L.; Walsh, J.L. Cold Plasma for the Disinfection of Industrial Food-contact Surfaces: An Overview of Current Status and Opportunities. *Compr. Rev. Food Sci. Food Saf.* **2022**, *21*, 1086–1124. [CrossRef] [PubMed]
15. Erichson, M.C.; Liao, J.; Cannon, J.L.; Ortega, Y.R. Contamination of Knives and Graters by Bacterial Foodborne Pathogens During Slicing and Grating of Produce. *Food Microbiol.* **2015**, *52*, 138–145. [CrossRef]
16. Kwok, T.Y.; Ma, Y.; Chua, S.L. Biofilm Dispersal Induced by Mechanical Cutting Leads to Heightened Foodborne Pathogen Dis-semination. *Food Microbiol.* **2022**, *102*, 103914. [CrossRef]
17. Wang, H.; Li, T.; Wu, Y.; Geng, A.; Li, Y.; Hou, J. Research Status and Prospects of Garlic Harvesting Machinery. *J. Chin. Agric. Mech.* **2018**, *39*, 102–107.
18. Yang, H.; Hu, Z.; Peng, B.; Wang, B.; Zhang, Y.; Yu, Z. Design and Test of Collecting and Conveying device of Garlic Multi-row Combine Harvester. *J. Chin. Agric. Mech.* **2021**, *42*, 8–12.
19. Peng, B.; Lü, X.; Wang, H.; Hu, Z. Half-feed and Self-propelled Garlic Combine Harvester. *Trans. Chin. Soc. Agric. Mach.* **2011**, *42*, 138–141, 146.
20. Zhao, D.; Cai, D.; Qin, L.; Gao, X.; Huang, W.; Liu, C. Design and Experiment of Modularized Garlic Combine Harvester. *Trans. Chin. Soc. Agric. Mach.* **2020**, *51*, 95–102. [CrossRef]
21. Li, S.; Li, H.; Qi, X.; Li, B.; Ding, Y. Research Status and Prospect of Garlic Harvester. *Acta Agric. Jiangxi* **2020**, *32*, 99–104. [CrossRef]
22. Hou, J.; Li, C.; Zhang, Z.; Li, T.; Li, Y.; Wu, Y. Design and Test of Double-row Walking Garlic Combine Harvester. *Trans. Chin. Soc. Agric. Eng. (Trans. CSAE)* **2021**, *37*, 1–11. [CrossRef]
23. Yao, D.; Lu, Y.; Chen, L.; Li, H.; Lu, Z. Research and Design of a New Type Garlic Harvester. *J. Agric. Mech. Res.* **2024**, *46*, 126–130. [CrossRef]
24. Xu, G.; Li, H.; Qi, X.; Fu, J.; Wang, Y.; Sun, X. Design and Experiment of Non-aligned Garlic Combined Harvester. *J. Chin. Agric. Mech.* **2024**, *45*, 20–25. [CrossRef]
25. Yu, Z.; Hu, Z.; Yang, K.; Peng, B.; Zhang, Y.; Yang, M. Operation Mechanism Analysis and Parameter Optimization of Garlic Root Floating Cutting Device. *Trans. Chin. Soc. Agric. Mach.* **2021**, *52*, 111–119. [CrossRef]
26. Hou, J.; Li, C.; Lou, W.; Li, T.; Li, Y.; Zhou, K. Operation Mechanism Analysis and Test of Press Root Cutting Device for Garlic Combine Harvester. *Trans. Chin. Soc. Agric. Mach.* **2022**, *53*, 167–174. [CrossRef]
27. Yang, K.; Yu, Z.; Gu, F.; Zhang, Y.; Wang, S.; Peng, B.; Hu, Z. Experimental Study of Garlic Root Cutting Based on Deep Learning Application in Food Primary Processing. *Foods* **2022**, *11*, 3268. [CrossRef]
28. Yang, K.; Peng, B.; Gu, F.; Zhang, Y.; Wang, S.; Yu, Z.; Hu, Z. Convolutional Neural Network for Object Detection in Garlic Root Cutting Equipment. *Foods* **2022**, *11*, 2197. [CrossRef]
29. Zhai, Y.; Hu, D.; Yuan, C.; Wang, J.; Zhang, X.; Liu, Y. Design and Experiments of Phalaenopsis Seedling Cutting System Using Visual Servo. *Trans. Chin. Soc. Agric. Eng. (Trans. CSAE)* **2022**, *38*, 148–156. [CrossRef]
30. Ye, S.; Wang, X.; Zhang, C.; Zhang, J.; Wang, J.; Zheng, D. Design and Experiments of a Roll-Knife Pickup for a Buckwheat Pickup Harvester. *Agronomy* **2024**, *14*, 1944. [CrossRef]
31. Wang, H.; Zhao, W.; Sun, W.; Liu, X.; Shi, R.; Zhang, H.; Chen, P.; Gao, K. The Design and Experimentation of a Wheeled-Chassis Potato Combine Harvester with Integrated Bagging and Ton Bag-Lifting Systems. *Agriculture* **2024**, *14*, 1461. [CrossRef]
32. Ling, J.; Shen, H.; Gu, M.; Hu, Z.; Zhao, S.; Wu, F.; Xu, H.; Gu, F.; Zhang, P. The Design and Optimization of a Peanut-Picking System for a Fresh-Peanut-Picking Crawler Combine Harvester. *Agriculture* **2024**, *14*, 1332. [CrossRef]
33. Sun, J.; Zhou, J.; He, Y.; Jia, H.; Rottok, L.T. Detection of Rice Panicle Density for Unmanned Harvesters via RP-YOLO. *Comput. Electron. Agric.* **2024**, *226*, 109371. [CrossRef]
34. Liang, Z.; Qin, Y.; Su, Z. Establishment of a Feeding Rate Prediction Model for Combine Harvesters. *Agriculture* **2024**, *14*, 589. [CrossRef]
35. Schalch, W.; Colton, J. Analysis of Manufacturing Methods and Die Design for Rice Transplanter Claws and Combine Harvester Blades in Bangladesh. *Machines* **2023**, *11*, 219. [CrossRef]
36. Guan, Z.; Li, H.; Chen, X.; Mu, S.; Jiang, T.; Zhang, M.; Wu, C. Development of Impurity-Detection System for Tracked Rice Combine Harvester Based on DEM and Mask R-CNN. *Sensors* **2022**, *22*, 9550. [CrossRef]
37. Zhang, Q.; Chen, Q.; Xu, W.; Xu, L.; Lu, E. Prediction of Feed Quantity for Wheat Combine Harvester Based on Improved YOLOv5s and Weight of Single Wheat Plant without Stubble. *Agriculture* **2024**, *14*, 1251. [CrossRef]
38. Nasirahmadi, A.; Wilczek, U.; Hensel, O. Sugar Beet Damage Detection during Harvesting Using Different Convolutional Neural Network Models. *Agriculture* **2021**, *11*, 1111. [CrossRef]
39. Yang, W.; Xi, J.; Wang, Z.; Lu, Z.; Zheng, X.; Zhang, D.; Huang, Y. Embedded Field Stalk Detection Algorithm for Digging–Pulling Cassava Harvester Intelligent Clamping and Pulling Device. *Agriculture* **2023**, *13*, 2144. [CrossRef]
40. Zhang, X.; Qiu, L.; Sun, Q.; Jing, Y.; Zhao, Y.; Yao, P. Design and Test of a Single-Row Harvesting and Cutting Integrated Handheld Garlic Harvester. *Appl. Sci.* **2023**, *13*, 7077. [CrossRef]

41. Zhu, Z.; Li, W.; Wen, F.; Chen, L.; Xu, Y. Towards Optimizing Garlic Combine Harvester Design with Logistic Regression. *Appl. Sci.* **2022**, *12*, 6015. [CrossRef]
42. Lü, X.; Peng, B.; Yu, X.; Gao, X.; Hu, Z. Working Performance Test on Garlic Bulb Combine Harvester in Field. *J. Yangzhou Univ. (Agric. Life Sci. Ed.)* **2012**, *33*, 77–80. [CrossRef]
43. Hou, J.; Li, C.; Lou, W.; Zhou, K.; Li, Y.; Li, T. Design and Test of Floating Clamping Device for Garlic Combine Harvester. *Trans. Chin. Soc. Agric. Mach.* **2023**, *54*, 137–145. [CrossRef]
44. Li, Y.; Liu, Q.; Li, T.; Wu, Y.; Niu, Z.; Hou, J. Design and Experiments of Garlic Bulbil Orientation Adjustment Device Using Jetson Nano processor. *Trans. Chin. Soc. Agric. Eng. (Trans. CSAE)* **2021**, *37*, 35–42. [CrossRef]
45. Zhou, X.; Condori-Apfata, J.A.; Liu, X.; Condori-Pacsi, S.J.; Valencia, M.V.; Zhang, C. Transcriptomic Changes Induced by Drought Stress in Hardneck Garlic during the Bolting/Bulbing Stage. *Agronomy* **2021**, *11*, 246. [CrossRef]
46. Fu, H.; Guo, Z.; Feng, Q.; Xie, F.; Zuo, Y.; Li, T. MSOAR-YOLOv10: Multi-Scale Occluded Apple Detection for Enhanced Harvest Robotics. *Horticulturae* **2024**, *10*, 1246. [CrossRef]
47. Zhang, Y.; Yu, J.; Chen, Y.; Yang, W.; Zhang, W.; He, Y. Real-time Strawberry Detection Using Deep Neural Networks on Embedded System (rtsd-net): An edge AI application. *Comput. Electron. Agric.* **2023**, *215*, 108445. [CrossRef]
48. Nie, P.; Qian, C.; Qin, R.; Deng, S.; Sun, C.; He, Y. Development status and trends of space-air-ground integrated information sensing and fusion technology. *J. Intell. Agric. Mech.* **2023**, *4*, 1–11. [CrossRef]

agriculture

MDPI

Article

Detection and Classification of *Agave angustifolia* Haw Using Deep Learning Models

Idarh Matadamas [1,*], Erik Zamora [2] and Teodulfo Aquino-Bolaños [1]

[1] Centro Interdisciplinario de Investigación para el Desarrollo Integral Regional, Unidad Oaxaca, Instituto Politécnico Nacional, Hornos No. 1003, Colonia Noche Buena, Municipio de Santa Cruz Xococotlán 71230, Oaxaca, Mexico; taquino@ipn.mx

[2] Centro de Investigación en Computación, Instituto Politécnico Nacional, Avenida Juan de Dios Batiz esquina Miguel Othón de Mendizábal, Colonia Nueva Industrial Vallejo, Gustavo A. Madero, Ciudad de México 07738, Mexico; ezamorag@ipn.mx

* Correspondence: idarhmatadamas@gmail.com

Abstract: In Oaxaca, Mexico, there are more than 30 species of the *Agave* genus, and its cultivation is of great economic and social importance. The incidence of pests, diseases, and environmental stress cause significant losses to the crop. The identification of damage through non-invasive tools based on visual information is important for reducing economic losses. The objective of this study was to evaluate and compare five deep learning models: YOLO versions 7, 7-tiny, and 8, and two from the Detectron2 library, Faster-RCNN and RetinaNet, for the detection and classification of *Agave angustifolia* plants in digital images. In the town of Santiago Matatlán, Oaxaca, 333 images were taken in an open-air plantation, and 1317 plants were labeled into five classes: sick, yellow, healthy, small, and spotted. Models were trained with a 70% random partition, validated with 10%, and tested with the remaining 20%. The results obtained from the models indicate that YOLOv7 is the best-performing model, in terms of the test set, with a *mAP* of 0.616, outperforming YOLOv7-tiny and YOLOv8, both with a *mAP* of 0.606 on the same set; demonstrating that artificial intelligence for the detection and classification of *Agave angustifolia* plants under planting conditions is feasible using digital images.

Keywords: artificial intelligence; computer vision; pattern recognition; convolutional networks; agave plants

1. Introduction

According to data from the National Commission for the Knowledge and Use of Biodiversity (Comision Nacional para el Conocimiento y Uso de la Biodiversidad, CONABIO), the number of species of the Agave genus distributed in the Mexican territory is 159, which represents 75% of the total species in the world [1]. Agave angustifolia, also known as "agave espadin", is the species with the greatest distribution in North America and can be found from Costa Rica to the northern part of Mexico, with Oaxaca being one of the entities where it is cultivated thanks to the existence of countless microclimates that facilitate its production and yield [2]. According to data from the Mexican Regulatory Council for the Quality of Mezcal (Consejo Mexicano Regulador de la Calidad del Mezcal, COMERCAM), mezcal production in Mexico in 2022 was 14,165,505 L, with the agave espadin being the most widely used variety for this production with 81.08% [3].

Recently, damage has been reported to plantations of different agave species caused by pests resulting from the presence of the agave weevil (*Scyphophorus acupunctatus* Gyllenhal) [4] or the rhinoceros beetle (*Strategus aloeus*) [5], which are described by visual characteristics such as wilting, yellowing of the plant's leaves, and reduced growth [6]; or diseases like 'sooty mold', which produces black spots on the leaves due to the presence of fungi from the Bipolaris genus [7], compromising yield and, consequently, economic benefits.

Artificial intelligence in agriculture offers the possibility of improving and automating activities such as irrigation planning, orchard monitoring, and greenhouse automation; all thanks to the implementation of techniques such as robotics and the use of unmanned aerial vehicles [8,9]. In addition to this, the field of machine learning offers an option to perform tasks such as the classification of fruit varieties using the visual information contained in the images of their leaves [10], the detection of pests [11,12] and diseases [13,14], as well as counting plants [15,16]. This is done by locating the objects of interest within the image, enclosing them in a space defined by a bounding box, and subsequently estimating the class to which they belong [17]. Traditionally, disease identification in plantations is done through visual inspection by people considered experts [18,19], however, ref. [20] reported cases where deep learning achieved results of up to 95% in the execution of this task.

Different studies report the use of techniques such as digital image processing, pattern recognition, and machine learning for the identification of plants and fruits, mostly focused on counting and classification. Ref. [21] implemented a modified version of an Inception-ResNet architecture for fruit counting, trained using a set of synthetic images to later test it on real images; the results showed 91% accuracy in the detection of fruits in real images and 93% in synthetic images. In this same sense, ref. [22] tested the capacity of deep learning models to develop leaf counting tasks; the tested models were trained using synthetic images, demonstrating that a data set of this type is sufficient to achieve counting in real images. The detection of corn plants using versions 4 and 5 of the model YOLO was reported by [23] and they obtained the best results with YOLOv5 with an average of 73.1% accuracy.

In applications related to agave plants, ref. [24] trained a convolutional neural network for the detection and counting of agave plants, showing that the performance of the network is higher with an $F1$ of 0.96, compared to the 0.57 obtained with the Haar algorithm. Ref. [25] implemented an algorithm based on mathematical morphology and parallel computing for the detection of agave plants in images obtained by an unmanned aerial vehicle, reporting results between 83% and 89% accuracy. In this same sense, ref. [16] implemented a deep learning algorithm based on YOLOv5 to automate the counting of agave plants in images taken from an unmanned aerial vehicle. More recently, ref. [26] presented a method for the segmentation and classification of *Agave tequilana* Weber blue crop maturity in very high-resolution satellite images, achieving an accuracy of 95% in the classification of test images. However, no results have been reported from studies focused on the detection and classification of agave plants in photographs taken directly in the field.

The main objective of this work is the implementation and evaluation of five deep learning models for the detection and classification of *Agave angustifolia* H. plants in photographs taken in open-air planting plots, with the aim of providing an alternative in the design of automatic evaluation and diagnostic systems directly in the field, for the identification of possible deficiencies and early diagnosis of diseases.

Five object detection models were used: Faster-RCNN, RetinaNet, Yolov8n (hereinafter YOLOv8), and two versions of the YOLOv7 model [27] were used: YOLOv7 and YOLOv7-tiny. During 2023 and 2024, image labeling, training, validation, and results analysis activities were developed. The main contributions of the work can be listed as: (i) the generation of a database of *Agave angustifolia* images under planting conditions; (ii) the demonstration of the use of deep learning models for the detection and classification of agave plants and (iii) the comparison of the performance of five deep learning models for the detection of agave plants: Faster-RCNN, RetinaNet, Yolov8, YOLOv7, and YOLOv7-tiny.

2. Materials and Methods

2.1. Image Colletion and Capture Site

To create the digital image database, a plot was visited in the Central Valleys within the so-called "mezcal region" in the town of Santiago Matatlan, in the district of Tlacolula in the state of Oaxaca (16°51′13.7″ N 96°23′25.1″ W), with an average altitude of 1725 m above sea level and an annual rainfall of 166 mm. The plot with dimensions of 35 m × 215 m corresponds to a plantation of *Agave angustifolia* H., with an age of four years. The digital images of the agave plants were obtained with a Canon® Rebel T3i digital camera (Canon Inc., Tokyo, Japan), between 12:00 h and 13:30 h on 25 August 2022, with a partially cloudy sky. The capture angle was a "normal angle" (camera lens parallel to the ground at the same level as the photographed object). A total of 333 primary images of agave plants planted in the open air and under monoculture conditions were obtained, and the dimensions of the images were 2592 × 1728 px in RGB color space.

2.2. Image Annotation and Classification

The VGG image annotator tool [28] was used to define the regions of interest in each of the images. The image annotation process was carried out manually, through rectangles that delimit the existence of an agave plant within the image, thus obtaining a total of 1317 sub-images that represent an individual plant, Figure 1:

Figure 1. Selection criteria for agave samples in the images obtained in the field. **A**: Complete agaves in the foreground of the image, **B**: partially recognizable agaves, **C**: agaves in the background whose shape can be fully recognized, and **D**: agaves in the third background that are partially or totally recognizable.

Each of the detections was assigned to one of five classes identified with a visual characteristic that distinguishes it from the others, as seen in Figure 2:

- Sick agave: plants with most of their leaves withered with a brown color, thin, and small in size;
- Yellow agave: plants that have a yellowish color on most of their leaves;
- Healthy agave: plants of normal size compared to the size of the plants present in the plot, with a homogeneous green color on all the leaves;
- Small agave: plants of small size compared to the rest of the plants found in the plot;
- Spotted agave: plants of normal size compared to the size of the plants present in the plot, which have dark-colored spots on some or all of their leaves.

Figure 2. Examples of images (detections) of the different classes in the labeling process. (**A**) Sick agave; (**B**) yellow agave; (**C**) healthy agave; (**D**) small agave; (**E**) spotted agave.

2.3. Training of the Models

You Only Look Once (YOLO) is an advanced method used for object detection using a convolutional neural network that divides the input image into a grid and calculates for each of the cells a bounding box and the probability that the object of interest is located in that space [29,30]. YOLO, in its version 7 (YOLOv7), offers three basic models designed for Edge GPUs, regular GPUs, and Cloud GPUs: YOLOv7-tiny, YOLOv7, and YOLOv7-W6, respectively. Here, the YOLOv7-tiny and YOLOv7 models were selected because they are the smallest architectures, as the number of trainable parameters were 6.2 M and 36.9 M, respectively, in addition to sharing the same set of training hyperparameters [31].

Detectron2 is a library that provides the capability to perform object detection and segmentation tasks. It includes various available models such as Mask R-CNN, Faster R-CNN, Fast R-CNN, RetinaNet, TridentNet, DensePose, Cascade R-CNN, and TensorMask [32–34]. In this work, two models were selected: RetinaNet and Faster R-CNN, as they specialize in object detection tasks.

Three random partitions were obtained from the image database with a distribution of 70% for the training set and 20% for the test set of the detection model. The validation set (Val) consisted of the remaining 10% (Table 1). The number of iterations used for training was 1000.

Table 1. Number of agave sub-images per class in each of the training sets of the YOLOv7 model.

Set	Sick	Yellow	Healthy	Small	Spotted	Total
Training	109	213	345	173	100	940
Test	34	60	77	55	35	261
Validation	7	24	43	27	15	116
Total	150	297	465	255	150	**1317**

All the models were trained under the same conditions in a programming environment based on the Python language accelerated by CUDA. The configuration of the training environment in terms of hardware and software is shown in Table 2. The size of the input images to the models was 640 × 640 px.

Table 2. Environment settings and initial values of the training hyperparameters of the models.

Item/Element	Model/Version
CPU	AMD Ryzen 5 3600 G
GPU	NVIDIA GeForce RTX 3050
RAM	Adata Xpg Spectrix D50 8 GB 3200 Mhz (2) [1]
OS	Ubuntu Linux 22.04.2 LTS
Programming Language	Python 3.10.9
CUDA Toolkit	12.1
NVIDIA Driver	530.30.02
Epochs	1000

[1] Number of units used.

2.4. Models Performance Metrics

To evaluate the detection capacity of the models, the metrics Precision (*Pr*), which represents the number of true positives out of a total of elements identified as positive, and Recall (*Rc*), which represents the sensitivity and provides information on the number of true positives that the model classified based on the total number of positive values, were used. The Precision-Recall (*P-R*) curve presents the *Pr* along the vertical axis and the *Rc* on the horizontal axis. The area under this curve is used to calculate a scalar value that represents the average precision (*AP*); a high value indicates that the classifier's performance is high [35]. For each of the classes present in the data set, it is possible to calculate the *P-R* curve and its corresponding *AP*. *mAP* represents the average of the *APs* of all the classes:

$$Pr = \frac{TP}{TP + FP} \tag{1}$$

$$Rc = \frac{TP}{TP + FN} \tag{2}$$

$$AP = \int_0^1 Pr(Rc)d(Rc) \tag{3}$$

$$mAP = \frac{1}{n}\sum_{k=1}^{k=n} AP_k \tag{4}$$

where:

TP (True Positives): is the number of items correctly detected as positives;
FP (False Positives): is the number of items incorrectly detected as positives;
FN (False Negatives): is the number of items incorrectly detected as negatives;
AP_k: *AP* corresponding to class *k*;
n: number of classes

3. Results

The performance curves of the models were obtained in the last epoch for the loss or error in classification in training and validation (Figure 3). The YOLOv7 model obtained a loss of 0.00018 in training and one of 0.032 in validation after running 1000 epochs. On the other hand, the YOLOv7-tiny model obtained a loss of 0.001 and 0.056 in training and validation, respectively, after the same number of epochs. YOLOv8 achieved a loss of 0.248 during training and 1.7545 during validation; for Faster-RCNN, the loss was 0.328 during training, compared to a validation value of 0.316; similar values were observed with the RetinaNet model, with 0.2154 for training and 0.3017 for validation.

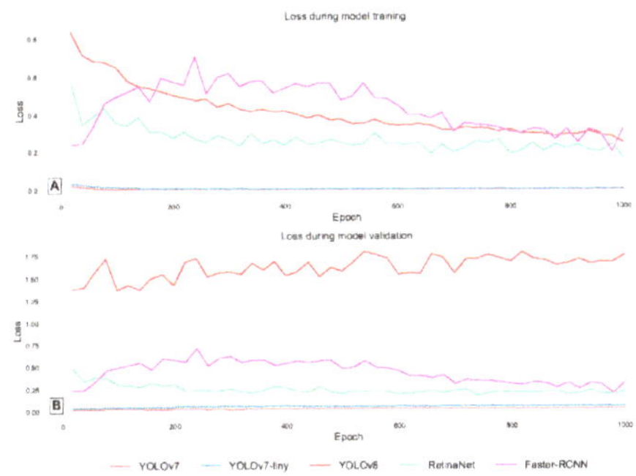

Figure 3. Comparison of the loss behavior of the YOLOv7, YOLOv7-tiny, YOLOv8, Faster-RCNN, and RetinaNet models during the training and validation processes. (**A**) Training process; (**B**) validation process.

The confusion matrix (Figure 4) calculated for each of the models with the validation set shows that the class with the highest classification accuracy was "sick" with the highest value reached by the YOLOv8 model at 0.86, while Faster R-CNN had the lowest value at 0.43. The class with the lowest classification value in the YOLOv7, YOLOv7-tiny, and YOLOv8 models was "small", with values ranging from 0.033 to 0.5, while for RetinaNet, the class with the lowest value was "yellow" at 0.29. Faster R-CNN was the model with the worst accuracy in four classes: "yellow", "healthy", "small", and "spotted", with values of 0 for all four classes.

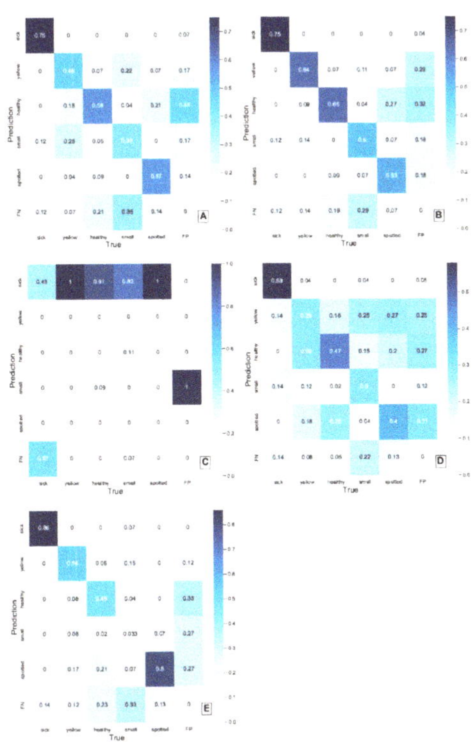

Figure 4. Confusion matrix of the performance in predicting images of agave espadin. (**A**) YOLOv7 model; (**B**) YOLOv7-tiny model; (**C**) Faster R-CNN model; (**D**) RetinaNet model; (**E**) YOLOv8 model.

The performance metrics during model validation were recorded in order to later calculate the *AP* per class (Table 3) and determine the overall *mAP*.

Table 3. Performance in *AP* metric for the prediction in images of agave espadin of the models with the validation set.

	AP				
Class/Model	**YOLOv7**	**YOLOv7-Tiny**	**YOLOv8**	**Faster-RCNN**	**RetinaNet**
Sick	**0.714**	**0.678**	**0.791**	**0.287**	0.246
Yellow	0.554	0.56	0.671	0	**0.404**
Healthy	0.552	0.573	0.61	0.063	0.327
Small	0.397	0.577	0.577	0.043	0.216
Spotted	0.546	0.542	0.567	0	0.316

The YOLOv7 model stands out primarily in the "sick" class with an *AP* of 0.714, which indicates good detection of plants belonging to this class. Additionally, the results show that the classes sharing similar characteristics have very similar performance metrics, with *AP* values of 0.554, 0.552, and 0.546 for the "yellow", "healthy" and "spotted" classes, respectively. The class with the lowest AP value is "small" with 0.397, making it the class with the worst performance for the model.

The YOLOv7-tiny model, overall, performs similarly to YOLOv7, with the "sick" class still being the best detected, achieving an *AP* of 0.678. In the "healthy" class, it reaches an *AP* of 0.573, slightly higher than YOLOv7, while in the "small" class, it achieves 0.577, showing improvement. However, its performance in the "sick" class, with an *AP* of 0.678, is lower than that of YOLOv7, highlighting a loss of precision compared to the previous model. In this same metric, the lowest value was obtained by the "spotted" class, with an *AP* of 0.542.

The YOLOv8 model shows the best performance, achieving high *AP* values in most classes. It performs the best in the "sick" class, with an *AP* of 0.791, making it the most effective model for detecting plants in this class. Additionally, it achieved similar values for the "yellow" and "healthy" classes, with 0.671 and 0.61, respectively. However, although the values obtained for the "spotted" class with 0.567 and the "small" class with 0.577 are similar to those of the previous models, the performance in these classes is lower than that of the YOLOv7-tiny model.

The Faster-RCNN model performs worse compared to the YOLO family models across all classes. The *AP* scores are consistently low, reaching a minimum value in the "yellow" and "spotted" classes, both with *AP* = 0, suggesting an inability to detect plants in these categories. In the "sick" class, it achieves an *AP* of 0.287, which is low compared to the YOLO models. The "healthy" class also shows a very low *AP* of 0.063, indicating very poor performance in detecting healthy plants.

Finally, the RetinaNet model, in the *AP* metric, shows a performance similar to that of Faster-RCNN, with poor results in most of the evaluated classes. Like the previous model, it is unable to efficiently detect most of the classes, with "sick" and "small" having the lowest performance at 0.246 and 0.216, respectively. Unlike the other models, the best-performing class for this model is "small", although overall, the value is low compared to the other evaluated models.

The models were evaluated using the set of weights that showed the best performance during the training periods, employing a test set of 261 images that were not used in training or validation; images unknown to the models; see Table 4.

Table 4. Performance in *AP* metric for prediction in images of agave espadin of the models with the test set.

Class/Model	YOLOv7	YOLOv7-Tiny	*AP* YOLOv8	Faster-RCNN	RetinaNet
Sick	**0.799**	**0.892**	**0.833**	**0.342**	**0.421**
Yellow	0.630	0.583	0.568	0	0.301
Healthy	0.574	0.585	0.558	0.013	0.259
Small	0.474	0.404	0.474	0.026	0.170
Spotted	0.602	0.567	0.596	0	0.294

The YOLOv7 model showed good performance in most classes. In the "sick" class, it achieved an average precision of 0.799, indicating a good ability to detect sick plants. It also performed well in the "spotted" class with an *AP* of 0.602. Although its performance in the "yellow" class with 0.630 and the "healthy" class with 0.574 was not as high as in other classes, it remains comparable to other models evaluated in this study. The "small" class, with an *AP* of 0.474, demonstrates that YOLOv7 is able to identify small plants, though with low accuracy.

The YOLOv7-tiny model stands out in the "sick" class, with a precision of 0.892, suggesting that it is effective in detecting sick plants. However, in other classes, the YOLOv7-tiny showed a decrease in the *AP* metric compared to the full YOLOv7 model. In "yellow", it achieved an *AP* of 0.583, in "healthy", 0.585, and in "small", 0.404, reflecting its average performance in those categories.

YOLOv8 showed results similar to YOLOv7 in several classes, with good performance in the "sick" class, where it achieved a precision of 0.833, the highest for the model. This result is slightly below that obtained by YOLOv7 and YOLOv7-tiny, but it is still good. In the "yellow" class, it achieved a precision of 0.568, slightly lower than YOLOv7, and in the "healthy" class, it reached 0.558, indicating moderate performance in detecting healthy plants. However, the model had its lowest value in the "small" class, where it obtained an *AP* of 0.474, similar to YOLOv7. The "spotted" class was also detected with a precision of 0.596, indicating that YOLOv8 is a model with comparable performance, although it did not outperform the previous models.

Faster-RCNN performed worse than the YOLO family models in almost all classes. In the "yellow" class, it achieved a result of 0.0, indicating that it was unable to adequately detect plants with yellow leaves in the test set. In the "healthy" and "small" classes, the results were also very low, barely reaching 0.013 and 0.026, suggesting that Faster-RCNN had significant difficulties in identifying healthy and small plants. In the "spotted" class, it also achieved a precision of 0.0, reflecting the model's inability to detect plants with spots. In summary, Faster-RCNN showed poor performance across all classes.

RetinaNet, like Faster-RCNN, did not show good performance in detecting agave plants with this dataset. Although its results were better than those of Faster-RCNN, they were still inferior to those of the YOLO models. In the "sick" class, it achieved a precision of 0.421, which was notably low compared to YOLOv7 and YOLOv7-tiny. In the "yellow" class, its performance was also low, with an *AP* of 0.301, and in "healthy", it reached an *AP* of 0.259. In the "small" class, it achieved an *AP* of 0.170, indicating difficulties in detecting small plants. The "spotted" class was detected with a precision of 0.294, better than Faster-RCNN's performance. Overall, RetinaNet did not reach the precision level of the YOLO models.

The *mAP* metric was calculated for each of the evaluated models. These results reflect the performance of the detection models in the task of identifying agave plants (Figure 5).

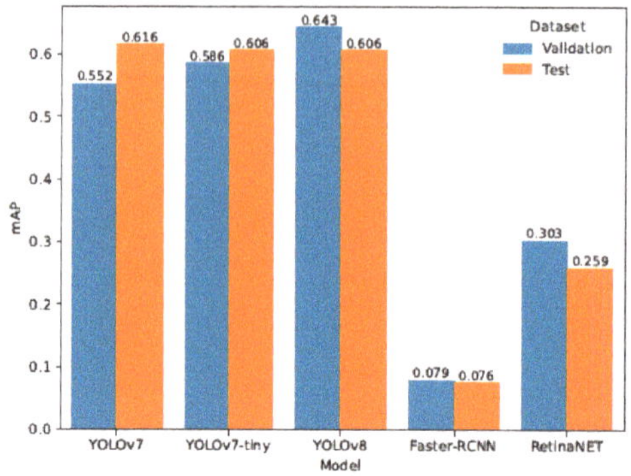

Figure 5. *mAP* metric calculated for the five evaluated models using the validation and test sets.

The YOLOv7 model shows similar performance across both datasets. In the validation set, it achieved a *mAP* of 0.552, indicating a reasonable ability to generalize over the validation data. In the test set, it improved its performance, reaching a *mAP* of 0.616, demonstrating that the model has good predictive ability on previously unseen data.

YOLOv7-tiny showed slightly lower performance compared to the YOLOv7 model. In the validation set, it achieved a *mAP* of 0.586, which is a good result considering the reduced size and complexity of the model. In the test set, its *mAP* was 0.606, similar to the validation performance, demonstrating that YOLOv7-tiny is capable of maintaining good accuracy despite being a more efficient model in terms of speed and resource consumption.

YOLOv8 showed the best performance in the validation set, with a *mAP* of 0.643, indicating its capability to detect agave plants in the validation data. However, in the test set, its *mAP* decreased to 0.606, which is slightly lower than that of YOLOv7.

The Faster-RCNN model showed very low performance in both sets, with a *mAP* of 0.079 in validation and 0.076 in the test. These results indicate that Faster-RCNN has serious difficulties in detecting agave plants in this dataset. This poor performance could be related to several factors, such as the lack of data or the model's inability to adapt to the specific characteristics of the plants.

Similarly, RetinaNet showed low results in both datasets, achieving a *mAP* of 0.303 in validation, and its *mAP* decreased to 0.259 in the test. Although RetinaNet's results were better than Faster-RCNN's, they are still inferior to those of the YOLO models.

Since the models from the YOLO family showed the best results in terms of detection accuracy, their performance in FLOPs (floating point operations per second) was calculated to determine which one exhibits the highest efficiency in terms of computational cost. YOLOv7 reaches a value of 72.26 GFLOPs, which is higher than the 36.13 GFLOPs of the YOLOv7-tiny model. Finally, the YOLOv8 model achieved 2.87 GFLOPs, making it the model with the lowest performance among the three during the inference process on the same image.

4. Discussion

The Table 1 data show that the class with the largest number of images is "healthy" and the ones with the lowest representation in the sets are "sick" and "spotted", resulting in an unbalanced database; for this reason, it is common for the models to over-qualify the classes that dominate the database, wrongly classifying those that present a minority [36]. In many cases, this problem is treated by increasing images by scaling, adding noise, translation, rotation, or changing contrast [37]; however, in our case, the class that obtained the highest percentage of correct classification for the validation and test sets was "sick" in both trained models, since, in visual appearance, this class is visually more distant from the others, making it easier to classify for the tested models.

Although the use of computer vision techniques in agave cultivation has grown [16,24–26], their specific application for the detection and classification of symptoms that may indicate the presence of diseases or pests in a crop with high economic relevance in Mexico, particularly in the state of Oaxaca, is relatively novel.

The YOLOv7 and YOLOv7-tiny models showed the best performance in terms of accuracy, with a *mAP* of 0.616 and 0.606 on the test set, respectively. These models exhibit a favorable balance between speed and accuracy, making them suitable for implementation in real-time automatic assessment systems in open-field plots.

Although the YOLOv8 model showed a slight improvement in accuracy during validation with a *mAP* of 0.643, during testing it achieved a *mAP* of 0.606, similar to that of YOLOv7-tiny on the test set. This suggests that, in this context, it does not offer a significant advantage.

On the other hand, Faster-RCNN and RetinaNet showed notably lower performance, especially on the test set, with *mAP* scores of 0.076 and 0.259, respectively, indicating that these models are not suitable for the specific task of detecting and classifying *Agave angustifolia*.

One of the critical factors limiting the effectiveness of the detection models tested in this work is the relatively small size and the imbalance of the image dataset. While the results obtained are promising, a larger and more diverse dataset could significantly improve the performance of the models in terms of generalization and precision.

The YOLOv7 model has a higher number of FLOPs, indicating greater complexity compared to YOLOv7-tiny and YOLOv8. This difference is attributed to the higher number of parameters [31,38]. The results highlight the typical trade-off between precision and efficiency: YOLOv7 is the most powerful in terms of FLOPs and also delivers the best performance in detection. Therefore, it emerges as the most suitable option for the task of detecting Agave angustifolia plants, using the development environment employed in this study.

5. Conclusions

This article presents the implementation of five different deep learning models to detect *Agave angustifolia* H. plants under open-field planting conditions. The plants can be classified into one of five classes based on their visible characteristics. These features may suggest the presence of a disease or pest that affects the optimal development of the plants.

The creation of the image dataset, although currently limited, represents a significant advancement in research applied to the monitoring, evaluation, and diagnosis of Agave plants. This dataset will enable researchers to develop more accurate computer vision models to detect issues such as diseases, pests, and environmental stress, contributing to more efficient and sustainable agricultural practices.

This study compares advanced versions of YOLO, one of the most popular models for real-time object detection, along with models from the Detectron2 library, specifically Faster-RCNN and RetinaNet. This comparative approach is an important contribution, as it allows for the evaluation of which of these models is more efficient and accurate for detecting agaves in images, considering the real-world conditions of a crop field.

Although no specific modifications were made to the parameters during this study, its development focused on demonstrating that, by using default configurations, the selected models, especially those belonging to the YOLO family, are effective for monitoring and classifying agave plants, which is an important contribution to the field of precision agriculture.

6. Future Works

In future work, we will expand the dataset with more representative images of the classes presented here, aiming to have a larger and more balanced database. This will allow us to train more robust models, as well as help reduce overfitting and improve the models' ability to detect maguey plants in real-world scenarios like the one presented in this study.

Although the evaluation in terms of computational efficiency was not the primary objective of this work, the results obtained provide a broader perspective on the performance of the tested models for the task at hand. In future research, we will conduct a detailed comparison of these models and their implementation on specific devices, such as Raspberry Pi, NVIDIA Jetson, mobile devices, or drones, to assess their performance in real-world conditions. This approach will allow us to gather crucial information about inference time and FLOPS (floating point operations per second), which will be essential for understanding how the models behave in practical scenarios, where resources such as memory, processor, and battery may be limited. In this way, we will gain a more comprehensive view of computational efficiency and detection accuracy, contributing to a better understanding of the challenges associated with the practical implementation of these models in specific agricultural tasks.

Author Contributions: Conceptualization, I.M., E.Z. and T.A.-B.; methodology, I.M. and E.Z.; validation, I.M. and E.Z.; investigation, I.M., E.Z. and T.A.-B.; resources, T.A.-B. and E.Z.; writing—original draft preparation, I.M.; writing—review and editing, E.Z. and T.A.-B.; visualization, I.M.; supervision, E.Z. and T.A.-B. All authors have read and agreed to the published version of the manuscript.

Funding: This work was supported by the Instituto Politécnico Nacional through the Secretaría de Investigación y Posgrado (SIP20220002, 20230232, 20240108), and the Mexican National Council of Humanities, Science, and Technology CONAHCyT under a doctorate grant.

Institutional Review Board Statement: Not applicable.

Data Availability Statement: The data associated with the paper will be made available to readers upon request post-publication, through contact with any of the authors.

Acknowledgments: The authors thank Rafael Hernández Martínez for providing the facilities to take the photographs on his plot in Santiago Matatlán, Oaxaca. I. Matadamas thanks CONAHCYT for the scholarship granted to continue his postgraduate studies.

Conflicts of Interest: The authors declare no conflicts of interest.

References

1. Palma, F.; Pérez, P.; Vinicio, M. Diagnóstico de la Cadena de Valor Mezcal en las Regiones de Oaxaca. 2016. Available online: https://www.oaxaca.gob.mx/coplade/wp-content/uploads/sites/29/2017/04/Perfiles/AnexosPerfiles/6.%20CV%20MEZCAL.pdf (accessed on 28 November 2024).
2. Mariles-Flores, V.; Ortiz-Solorio, C.A. Las clases de tierras productoras de maguey mezcalero en la Soledad Salinas, Oaxaca* Classes maguey mezcal producing land in La Soledad Salinas, Oaxaca. *Rev. Mex. De Cienc. Agrícolas* **2016**, *7*, 1199–1210. [CrossRef]
3. COMERCAM. Informe Estadístico 2023. 2023. Available online: https://comercam-dom.org.mx/wp-content/uploads/2023/05/INFORME-2023_PUBLICO.pdf (accessed on 28 November 2024).
4. Aquino-Bolaños, T.; Parraguirre-Cruz, M.A.; Ruiz-Vega, J. Scyphophorus acupunctatus (=interstitialis) Gyllenhal (Coleoptera: Curculionidae). Pest of agave mezcalero: Losses and damage in Oaxaca, Mexico. *Rev. Científica UDO Agrícola* **2007**, *7*, 175–180.
5. Aquino-Bolaños, T.; Aquino-Lopez, T.; Ruiz-Vega, J.; Bautista-Cruz, A. Strategus aloeus (Coleoptera: Scarabaeidae) damage in two agave species and its management based on entomopathogenic fungi in oil suspensions. *Rev. Colomb. Entomol.* **2024**, *50*, e12865. [CrossRef]
6. CESAVEG. Manual de Plagas y Enfermedades. Available online: http://cesaveg.org.mx/divulgacion/agave/manual_agave.pdf (accessed on 11 November 2024).
7. Romero-Cortes, T.; Pérez España, V.H.; Pescador-Rojas, J.A.; Rangel-Cortés, E.; Armendaríz-Ontiveros, M.M.; Cuervo-Parra, J.A. First Report of Leaf Spot Disease ("Negrilla") on *Agave salmiana* Otto Ex Salm-Dyck (ssp. *salmiana) Plants Caused by Bipolaris zeae Zivan in Mexico. Agronomy* **2024**, *14*, 623. [CrossRef]
8. De Oliveira, R.C.; e Silva, R.D.d.S. Artificial Intelligence in Agriculture: Benefits, Challenges, and Trends. *Appl. Sci.* **2023**, *13*, 7405. [CrossRef]
9. Wakchaure, M.; Patle, B.K.; Mahindrakar, A.K. Application of AI techniques and robotics in agriculture: A review. *Artif. Intell. Life Sci.* **2023**, *3*, 100057. [CrossRef]
10. Ayala-Niño, D.; González-Camacho, J.M. Evaluation of machine learning models to identify peach varieties based on leaf color. *Agrociencia* **2022**, *4*, 21–179. [CrossRef]
11. Agnihotri, V. Machine Learning Based Pest Identification in Paddy Plants. In Proceedings of the 2019 3rd International conference on Electronics, Communication and Aerospace Technology (ICECA), Coimbatore, India, 12–14 June 2019; pp. 246–250. [CrossRef]
12. Park, Y.-H.; Choi, S.H.; Kwon, Y.-J.; Kwon, S.-W.; Kang, Y.J.; Jun, T.-H. Detection of Soybean Insect Pest and a Forecasting Platform Using Deep Learning with Unmanned Ground. *Veh. Agron.* **2023**, *13*, 477. [CrossRef]

13. Ambrosio, J.P.A.; Camacho, J.M.G.; Aguilar, A.R.; Paniagua, D.H.d.V. Identification of disease in tomato leaves using machine learning classifiers and digital images. *Agrociencia* **2023**, *8*, 2462. [CrossRef]
14. Joseph, D.S.; Pawar, P.M.; Pramanik, R. Intelligent plant disease diagnosis using convolutional neural network: A review. *Multimed. Tools Appl.* **2022**, *82*, 21415–21481. [CrossRef]
15. Ribera, J.; Chen, Y.; Boomsma, C.; Delp, E.J. Counting plants using deep learning. In Proceedings of the 2017 IEEE Global Conference on Signal and Information Processing (GlobalSIP), Montreal, QC, Canada, 14–16 November 2017; pp. 1344–1348. [CrossRef]
16. Omar, H.C.; Florián, F.; Sánchez, M.G.; Guadalupe, S.M.; Ávila-George, H.; Departamento de Ciencias Computacionales e Ingenierías. Conteo de plantas de agave usando redes neuronales convolucionales e imágenes adquiridas desde un vehículo aéreo no tripulado. *RISTI—Rev. Ibérica De Sist. E Tecnol. De Informação* **2022**, *1*, 64–76. [CrossRef]
17. Gündüz, M.; Işik, G. A new YOLO-based method for real-time crowd detection from video and performance analysis of YOLO models. *J. Real-Time Image Process.* **2023**, *20*, 5. [CrossRef] [PubMed]
18. Shahi, T.B.; Xu, C.Y.; Neupane, A.; Guo, W. Recent Advances in Crop Disease Detection Using UAV and Deep Learning Techniques. *Remote Sens.* **2023**, *15*, 2450. [CrossRef]
19. Toda, Y.; Okura, F. How Convolutional Neural Networks Diagnose Plant Disease. *Plant Phenomics* **2019**, *2019*, 9237136. [CrossRef] [PubMed]
20. Saleem, M.H.; Potgieter, J.; Arif, K.M. Plant Disease Detection and Classification by Deep Learning. *Plants* **2019**, *8*, 468. [CrossRef]
21. Rahnemoonfar, M.; Sheppard, C. Deep Count: Fruit Counting Based on Deep Simulated Learning. *Sensors* **2017**, *17*, 905. [CrossRef]
22. Ubbens, J.; Cieslak, M.; Prusinkiewicz, P.; Stavness, I. The use of plant models in deep learning: An application to leaf counting in rosette plants. *Plant Methods* **2018**, *14*, 6. [CrossRef]
23. Mota-Delfin, C.; López-Canteñs, G.d.J.; López-Cruz, I.L.; Romantchik-Kriuchkova, E.; Olguín-Rojas, J.C. Detection and Counting of Corn Plants in the Presence of Weeds with Convolutional Neural Networks. *Remote Sens.* **2022**, *14*, 4892. [CrossRef]
24. Flores, D.; González-Hernández, I.; Lozano, R.; Vazquez-Nicolas, J.M.; Hernandez Toral, J.L. Automated Agave Detection and Counting Using a Convolutional Neural Network and Unmanned Aerial Systems. *Drones* **2021**, *5*, 4. [CrossRef]
25. Calvario, G.; Alarcón, T.E.; Dalmau, O.; Sierra, B.; Hernandez, C. An Agave Counting Methodology Based on Mathematical Morphology and Images Acquired through Unmanned Aerial Vehicles. *Sensors* **2020**, *20*, 6247. [CrossRef]
26. Sánchez, A.; Nanclares, R.; Quevedo, A.; Pelagio, U.; Aguilar, A.; Calvario, G.; Moya-Sánchez, E.U. Agave crop segmentation and maturity classification with deep learning data-centric strategies using very high-resolution satellite imagery. *arXiv* **2023**, arXiv:2303.11564. [CrossRef]
27. Wang, M.; Yang, B.; Wang, X.; Yang, C.; Jie, X.; Mu, B.; Xiong, K.; Li, Y. YOLO-T: Multitarget Intelligent Recognition Method for X-ray Images Based on the YOLO and Transformer Models. *Appl. Sci.* **2022**, *12*, 11848. [CrossRef]
28. Dutta, A.; Zisserman, A. The VIA Annotation Software for Images, Audio and Video. In Proceedings of the 27th ACM International Conference on Multimedia, Nice, France, 21–25 October 2019; pp. 2276–2279. [CrossRef]
29. Redmon, J.; Divvala, S.; Girshick, R.; Farhadi, A. You Only Look Once: Unified, Real-Time Object Detection. In Proceedings of the 2016 IEEE Conference on Computer Vision and Pattern Recognition (CVPR), Las Vegas, NV, USA, 27–30 June 2016; pp. 779–788. [CrossRef]
30. Shinde, S.; Kothari, A.; Gupta, V. YOLO based Human Action Recognition and Localization. *Procedia Comput. Sci.* **2018**, *133*, 831–838. [CrossRef]
31. Wang, C.Y.; Bochkovskiy, A.; Liao, H.Y.M. YOLOv7: Trainable bag-of-freebies sets new state-of-the-art for real-time object detectors. *arXiv* **2022**, arXiv:2207.02696.
32. Rani, A.; Ortiz-Arroyo, D.; Durdevic, P. Defect Detection in Synthetic Fibre Ropes Using Detectron2 Framework. *Appl. Ocean Res.* **2024**, *150*, 104109. [CrossRef]
33. de Almeida, G.P.S.; dos Santos, L.N.S.; da Silva Souza, L.R.; da Costa Gontijo, P.; de Oliveira, R.; Teixeira, M.C.; De Oliveira, M.; Teixeira, M.B.; do Carmo França, H.F. Performance Analysis of YOLO and Detectron2 Models for Detecting Corn and Soybean Pests Employing Customized Dataset. *Agronomy* **2024**, *14*, 2194. [CrossRef]
34. Butt, M.; Glas, N.; Monsuur, J.; Stoop, R.; de Keijzer, A. Application of YOLOv8 and Detectron2 for Bullet Hole Detection and Score Calculation from Shooting Cards. *AI* **2024**, *5*, 72–90. [CrossRef]
35. Liu, K.; Sun, Q.; Sun, D.; Yang, M.; Wang, N.; Peng, L. Underwater target detection based on improved YOLOv7. *J. Mar. Sci. Eng.* **2023**, *11*, 77. [CrossRef]
36. Ferreira, U.E.C.; Camacho, J.M.G. Clasificador de red neuronal convolucional para identificar enfermedades del fruto de aguacate (*Persea americana* Mill.) a partir de imágenes digitales. *Agrociencia* **2021**, *55*, 695–709. [CrossRef]
37. Kumar, N.; Nagarathna; Flammini, F. YOLO-Based Light-Weight Deep Learning Models for Insect Detection System with Field Adaption. *Agriculture* **2023**, *13*, 741. [CrossRef]
38. Sohan, M.; Sai Ram, T.; Rami Reddy, C.V. A Review on YOLOv8 and Its Advancements. In *Data Intelligence and Cognitive Informatics*; ICDICI 2023. Algorithms for Intelligent Systems; Springer: Singapore, 2024. [CrossRef]

agriculture

MDPI

Article

AI-Based Monitoring for Enhanced Poultry Flock Management

Edmanuel Cruz [1,2], Miguel Hidalgo-Rodriguez [1], Adiz Mariel Acosta-Reyes [1], José Carlos Rangel [2,3,*] and Keyla Boniche [4]

1 Centro Regional Veraguas, Universidad Tecnológica de Panamá, Atalaya 0901, Panama; edmanuel.cruz@utp.ac.pa (E.C.); miguel.hidalgo@utp.ac.pa (M.H.-R.); adiz.acosta@utp.ac.pa (A.M.A.-R.)
2 Sistema Nacional de Investigación (SNI), SENACYT, Panama City 0816-02852, Panama
3 Facultad de Ingeniería de Sistemas Computacionales, Universidad Tecnológica de Panamá, Panama City 0819-07289, Panama
4 Facultad de Ingeniería Mecánica, Universidad Tecnológica de Panamá, Panama City 0819-07289, Panama; keyla.boniche@utp.ac.pa
* Correspondence: jose.rangel@utp.ac.pa; Tel.: +507-560-3928

Abstract: The exponential growth of global poultry production highlights the critical need for efficient flock management, particularly in accurately counting chickens to optimize operations and minimize economic losses. This study advances the application of artificial intelligence (AI) in agriculture by developing and validating an AI-driven automated poultry flock management system using the YOLOv8 object detection model. The scientific objective was to address challenges such as occlusions, lighting variability, and high-density flock conditions, thereby contributing to the broader understanding of computer vision applications in agricultural environments. The practical objective was to create a scalable and reliable system for automated monitoring and decision-making, optimizing resource utilization and improving poultry management efficiency. The prototype achieved high precision (93.1%) and recall (93.0%), demonstrating its reliability across diverse conditions. Comparative analysis with prior models, including YOLOv5, highlights YOLOv8's superior accuracy and robustness, underscoring its potential for real-world applications. This research successfully achieves its objectives by delivering a system that enhances poultry management practices and lays a strong foundation for future innovations in agricultural automation.

Keywords: poultry monitoring; computer vision; artificial intelligence; flock management; agriculture automation

1. Introduction

Chicken meat, one of the world's most consumed proteins, contributes billions of euros annually to the global economy [1,2]. In 2024, global poultry production is projected to reach approximately 138.75 million metric tons, marking a significant increase from 115 million metric tons in 2016. This growth aligns with a broader trend in agriculture, where poultry has surpassed other meat types in production volume [3], as illustrated in Figure 1.

As demand increases, poultry farms face mounting pressure to optimize operations, particularly in flock management, to prevent economic losses caused by miscounts and resource mismanagement. Accurate chicken counts are crucial, as errors in these data can have significant economic and operational repercussions.

First, inaccurate counts can lead to financial losses. Misrecording the number of chickens may result in incorrect inventory decisions, such as over-purchasing or under-purchasing essential supplies like feed or medication. These errors not only increase operational costs but also disrupt financial planning and undermine long-term profitability.

Second, data inaccuracies directly affect productivity. Precise counts are essential for adjusting feed quantities to match the actual flock size, thereby preventing both overfeeding and underfeeding. Such discrepancies can negatively impact flock growth and operational efficiency, hindering the optimization of production processes.

Lastly, imprecise data compromises decision-making quality. Poor data can lead to suboptimal choices in genetic selection or management practices, reducing competitiveness and productivity. This cascading effect diminishes operational efficiency and the quality of the final product.

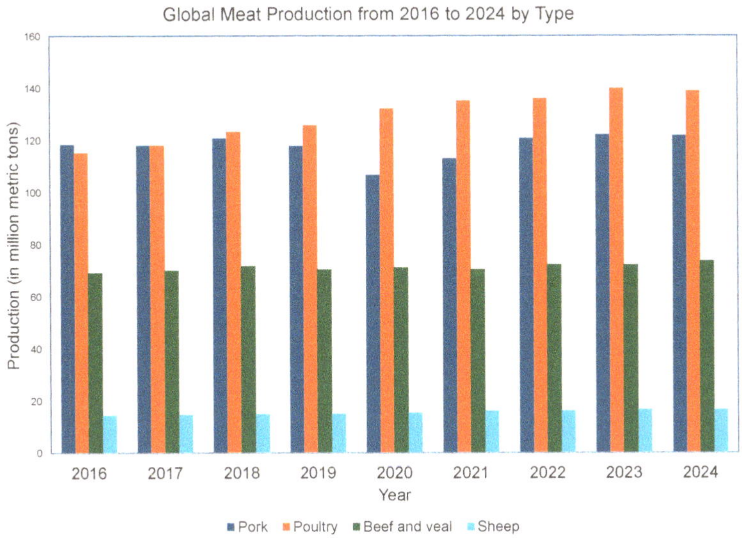

Figure 1. Production of meat worldwide from 2016 to 2024 by type (in million metric tons). The data for 2024 are estimated as of the writing of this article. Source: FAO; OECD; 2016 to 2023.

Such inaccuracies can lead to suboptimal decisions in flock management, affecting various aspects of poultry farming, from the health and welfare of chickens to the overall productivity of the farm [4]. Inaccurate counts may result in overstocking or understocking resources, mismanagement of feed and water, and inefficiencies in disease control, all of which directly impact the profitability and sustainability of poultry operations.

To address these challenges, the poultry industry is increasingly embracing technological innovations to enhance operational efficiency. Among these, artificial intelligence (AI) has emerged as a transformative tool with the potential to revolutionize poultry farming. Specifically, advanced object detection algorithms within AI offer a promising solution for automating the chicken counting process. These algorithms can process real-time data from cameras installed in poultry houses, accurately identifying and counting individual chickens with remarkable precision [5,6].

As AI technology continues to evolve, its role in poultry farming is expected to expand, aligning with broader agricultural trends toward automation and data-driven decision-making [7]. This shift not only boosts productivity and improves chicken welfare but also enables the industry to meet the growing global demand for chicken meat while maintaining high standards of efficiency and sustainability.

Considering the above, the research problem is identified as the lack of an effective method for accurately counting poultry, which hinders efficient management during their stay in breeding sheds. This issue stems from the limitations of existing manual and semi-automatic methods, which are incapable of effectively processing the large volumes of data generated in such environments or adapting to the specific conditions of chicken coops.

To address this gap, this research aims to develop a computer vision-based counting system capable of providing reliable and automated estimations of the poultry population. This approach not only addresses the challenges of accurate counting but also facilitates the management and processing of large data volumes within the unique environmental conditions of breeding sheds.

Accordingly, the research question guiding this study is the following: How can computer vision be leveraged to enable reliable and efficient chicken counting while managing the data volume and environmental challenges of breeding sheds?

This study contributes to advancements in poultry management by developing an AI-driven system for precise chicken counting using the YOLOv8 model. The model is trained on custom datasets of poultry images captured in real-world environments. It effectively addresses challenges like occlusions (instances where chickens overlap or obstruct one another) and dense flock conditions (highly crowded environments).

The primary aim of this research is to develop and validate an AI-driven automated poultry flock management system utilizing the YOLOv8 model. The scientific objective is to advance the application of artificial intelligence in poultry management by overcoming critical challenges like occlusions, lighting variability, and dense flock conditions, thereby enriching the broader understanding of computer vision techniques in complex agricultural settings. Concurrently, the practical objective is to design a reliable and scalable system that enhances flock management by enabling automated monitoring and data-driven decision-making in real-world farming environments. This system aims to reduce economic losses, improve resource efficiency, and promote sustainable agricultural practices. By achieving these objectives, it offers a scalable and impactful solution to enhance the accuracy and efficiency of poultry management while addressing broader challenges in agriculture.

The remainder of this article is organized as follows: Section 2 provides a review of relevant background information. Section 3 outlines the methodology employed in this study. Section 4 presents and analyzes the experimental results. Section 5 discusses the practical implications and limitations of the research. Finally, Section 6 concludes with a summary of the key findings.

2. Background

Object detection in poultry farming, particularly in chicken detection, has seen significant advancements with the emergence of deep learning models. These models have demonstrated exceptional capabilities in identifying and classifying objects within images and video streams, paving the way for more accurate and scalable systems for real-time monitoring in commercial poultry farms.

2.1. AI's Impact on Agriculture

The application of artificial intelligence (AI) in agriculture has revolutionized traditional farming practices by increasing efficiency, promoting sustainability through optimized resource use, and enhancing responsiveness to global food demands. AI technologies—including machine learning, computer vision, and robotics—have introduced precision agriculture, a data-driven approach that improves resource allocation and decision-making. This transformation aligns with the broader trend of "Agriculture 4.0" [8,9], which integrates digital tools, data analytics, and automation to advance agricultural productivity and sustainability.

Recent advancements in spatial–spectral image classification, which combines spatial and spectral data for improved image analysis, further illustrate AI's potential in agricultural monitoring. For example, reference [10] employed a spatial–spectral classification technique with edge-preserving chromaticity mapping, significantly enhancing the accuracy of remotely sensed imagery. This method integrates colorimetric edge preservation with principal component analysis to incorporate spatial context, demonstrating that maintaining spatial details improves classification reliability. In agriculture, such techniques enable precision monitoring by enhancing image clarity and data accuracy, directly aligning with the objectives of AI-based poultry monitoring models. By optimizing image quality and data precision, these methods support AI-driven agricultural solutions, particularly in high-density settings where accurate detection is critical.

AI-driven systems have also transformed crop management through the integration of data from IoT sensors (for soil and climate monitoring), drones (for aerial crop health assessments), and satellite imagery (for large-scale field analysis). These technologies provide real-time, actionable insights into crop health, soil conditions, and climate variables, enabling precision interventions in irrigation, pest control, and fertilization. Such interventions significantly reduce chemical use and minimize environmental impacts [11,12]. Furthermore, automating tasks like crop health monitoring and disease detection with advanced machine learning models has improved both efficiency and accuracy compared to traditional methods, such as manual inspections and visual assessments [13,14].

In livestock management, AI applications—particularly deep learning models for behavior and health monitoring—have significantly enhanced the ability to track animal welfare. Convolutional neural networks (CNNs), known for their accuracy in image-based analysis, facilitate the early detection of health issues by monitoring animal behavior, enabling proactive management practices [12,13]. Additionally, AI's predictive capabilities in yield forecasting and climate adaptability optimize resource use, reduce waste, and improve yield predictions, helping agriculture adapt more effectively to climate variability and resource constraints [9,14].

Beyond productivity gains, advancements in AI address broader social and ethical considerations, including data privacy, labor impacts, and the promotion of fair and unbiased AI applications. By fostering resource-efficient and environmentally friendly farming practices, AI technologies help farmers transition to sustainable methods, a critical step toward meeting the growing global food demand driven by population growth [9,14]. However, challenges remain in areas such as data quality—ensuring datasets are diverse and representative—model interpretability for transparent decision-making, and ethical considerations. Addressing these challenges is essential for the continued development and responsible deployment of AI in agriculture [11,12].

2.2. *Classical Machine Learning Approaches*

Early efforts in chicken detection primarily relied on traditional computer vision techniques, which utilized handcrafted features and classical machine learning classifiers. Techniques such as HOG (Histogram of Oriented Gradients) [15] were widely employed to detect objects by identifying edges and shapes, while SIFT (Scale-Invariant Feature Transform) and SURF (Speeded-Up Robust Features) [16,17] were used for feature extraction to identify distinct characteristics of chickens in images.

To classify objects within images, these feature descriptors were often combined with machine learning algorithms such as Random Forests and SVM (Support Vector Machines) [18]. While these approaches achieved some success in controlled environments, they proved less effective in real-world poultry farming conditions. Factors such as environmental variability, inconsistent lighting, and occlusions significantly hindered the accuracy and robustness of these methods, limiting their applicability for real-time or large-scale monitoring in dynamic settings like poultry farms.

2.3. *Deep Learning Algorithms for Chicken Detection*

Deep learning has revolutionized object detection, offering more accurate and scalable solutions for applications like chicken detection. Convolutional Neural Networks (CNNs) [19] form the backbone of many object detection tasks by efficiently processing image data. In the context of chicken detection, CNN-based architectures such as VGGNet [20] and ResNet [21] have been employed to extract hierarchical features from images, enabling the effective distinction of chickens from their surroundings. The deep feature extraction layers in these models allow for greater detail and precision in detection.

Building on the capabilities of CNNs, Region-Based Convolutional Neural Networks (R-CNNs) [22] introduced a significant advancement by dividing images into regions and applying CNNs to each region for object identification. Faster R-CNN [23], a widely adopted variant, strikes a balance between speed and accuracy, making it particularly effective for detecting chickens. Its architecture incorporates a Region Proposal Network (RPN) that generates candidate regions for object detection, enabling the identification of multiple chickens within a single image. The integration of region-based methods with deep learning has demonstrated high efficacy in poultry monitoring, addressing the complexities of dense and dynamic farm environments.

2.4. YOLO (You Only Look Once) Models

The YOLO (You Only Look Once) family of object detectors is among the most widely adopted models for chicken detection, valued for its real-time performance and high accuracy. Since the introduction of YOLO [24], successive versions, from YOLOv3 to the latest YOLOv8, have been successfully applied in poultry monitoring. These models offer significant speed advantages, making them particularly well-suited for real-time detection in dynamic environments such as commercial chicken farms.

YOLOv8, the latest iteration, introduces architectural refinements that enhance both detection accuracy and processing efficiency. One of YOLO's key strengths lies in its ability to detect multiple chickens within a single image by processing the entire image in one pass. This capability is critical for monitoring large flocks efficiently, enabling the identification of abnormal behaviors, health issues, or other anomalies in real time. By combining speed, accuracy, and scalability, YOLO models provide an effective solution for poultry monitoring in modern agricultural settings.

2.5. Single Shot Detectors (SSDs) and Transformer-Based Models

Single Shot Detectors (SSDs) [25] represent another deep learning framework employed in chicken detection. Known for their speed and lightweight architecture, SSD models are particularly advantageous in environments with limited computational resources. The combination of SSD with MobileNet is especially well-suited for deployment on mobile and embedded devices, providing a cost-effective solution for implementing detection systems in resource-constrained farm settings.

Recent advancements in transformer-based models have also begun to influence object detection tasks, including chicken detection. Detection Transformers (DETR) [26], for instance, eliminate the need for traditional region proposals and anchors. Instead, they leverage transformers to model relationships across different parts of an image, offering a more efficient and robust approach to object detection. DETR has demonstrated promise in addressing the challenges of chicken detection, particularly in complex farm environments characterized by occlusions and visual clutter. These capabilities make transformer-based models a valuable addition to the toolkit for poultry monitoring in modern agricultural systems.

2.6. Recent Advances in AI-Driven Poultry Monitoring

The application of artificial intelligence in poultry management has progressed rapidly, with innovative approaches enhancing precision, efficiency, and sustainability. Recent studies highlight AI's transformative potential across various aspects of poultry farming, including flock monitoring, environmental control, and disease detection.

Reference [13] investigated the integration of YOLOv8 with DeepSORT for tracking behaviors in cage-free hens, achieving a high multi-object tracking accuracy (MOTA) of 94%. This system effectively detects critical behavioral patterns, such as smothering and piling, which are vital indicators of poultry welfare in high-density environments. Such real-time behavioral monitoring enables early intervention, ultimately improving animal welfare and operational efficiency in commercial poultry farms.

Another notable advancement is the Poultry-Edge-AI-IoT system developed in [27], which combines edge computing and IoT technologies to monitor environmental factors within poultry houses. This system uses a gated recurrent unit (E-GRU) deep learning model to predict hazardous levels of gases like ammonia (NH_3) and carbon dioxide (CO_2), as well as temperature fluctuations. By providing real-time predictions of harmful conditions, the system safeguards poultry health and enhances productivity through precise environmental control.

AI has also proven valuable in disease detection. Reference [28] developed a diagnostic system leveraging the EfficientNetB7 model to identify poultry diseases via fecal image analysis. Their system achieves a classification accuracy of 97.07% in distinguishing healthy and diseased conditions. This non-invasive diagnostic tool facilitates early detection of diseases, mitigating the risk of outbreaks, reducing economic losses, and ensuring better flock health management.

These advancements underscore AI's critical role in automating poultry monitoring processes. By providing accurate, real-time data on flock behavior, environmental conditions, and health indicators, AI-driven systems enhance decision-making, support early interventions, and promote sustainable poultry farming practices.

2.7. Challenges in Chicken Detection

Chicken detection presents several challenges, with one of the most significant being occlusion, where chickens overlap or cluster together, making it difficult to detect each individual accurately. Variations in chicken size, pose, and orientation further complicate detection tasks, as these factors affect the accuracy and reliability of detection systems.

Understanding chicken poses is particularly critical for behavioral analysis. The Multi-Chicken Pose (MCP) estimation system introduced in [29] provides a novel solution by leveraging transfer learning to estimate multiple chicken poses simultaneously. This approach addresses occlusion-related challenges while offering valuable insights into poultry welfare monitoring, thus enhancing the effectiveness of behavior analysis in dynamic farm environments.

To further mitigate detection challenges, techniques such as Non-Maximum Suppression (NMS) [30] have been employed to reduce multiple detections of the same object, ensuring more accurate identification. Additionally, Focal Loss [31] has been introduced to address class imbalance by focusing on harder-to-detect chickens, improving the overall robustness of detection systems.

Multi-scale feature detection, implemented in networks like Faster R-CNN and YOLOv5, enhances the ability to handle objects of varying sizes, a common issue in dynamic farm environments. These advancements collectively contribute to overcoming the inherent challenges of chicken detection, enabling more precise monitoring and analysis in poultry management systems.

2.8. Datasets for Chicken Detection

Publicly available datasets specifically tailored for chicken detection remain scarce. However, general-purpose datasets such as COCO [32] and ImageNet [33] are frequently used to pre-train models, which can then be fine-tuned on chicken-specific data to improve performance.

In most cases, researchers and commercial entities working in poultry detection rely on custom datasets, often generated from real-world farm environments. These datasets typically consist of images or real-time video captured from cameras installed in poultry houses. Farm-generated data provide a valuable resource for training detection models as they reflect the variability and challenges of actual farm conditions. To prepare these datasets, chickens are annotated with bounding boxes or other labels, creating the ground truth data required for supervised learning. Such custom datasets are critical for addressing the unique challenges of chicken detection, including occlusion, variations in size and pose, and diverse environmental conditions.

2.9. Practical Applications and Systems

Real-time poultry monitoring systems are increasingly leveraging YOLO-based models for practical applications in commercial farming. These systems are utilized to count chickens, track their movement, and monitor behavior, enabling the detection of abnormal patterns that may indicate health issues or welfare concerns.

The integration of Internet of Things (IoT) devices [34] has further enhanced the deployment of these detection models on edge devices, such as Raspberry Pi or NVIDIA Jetson. This advancement facilitates real-time detection and analysis directly at the farm level, reducing latency and dependence on centralized processing. By combining AI-based detection with IoT-enabled edge computing, these systems offer scalable, cost-effective solutions that improve operational efficiency and promote proactive poultry management.

2.10. Background Conclusions

Advancements in machine learning, particularly through deep learning models such as YOLO and Faster R-CNN, have revolutionized real-time monitoring in poultry farming. These models have proven effective in managing large flocks, tracking movement, and identifying health issues, thereby addressing key challenges in modern poultry management.

The next frontier in chicken detection involves the integration of object detection with behavioral analysis and IoT technologies. By combining these innovations, future systems can enable more intelligent, automated monitoring solutions that not only improve operational efficiency but also enhance animal welfare and sustainability in commercial poultry farming.

3. Materials and Methods

3.1. Dataset Creation

The dataset creation process was critical to ensuring the accuracy and robustness of the chicken counting model. This process involved several key stages designed to capture the entire chicken production cycle, from the chick stage to slaughter age.

The chickens used in this study were commercial broiler chickens, selected for their rapid growth and high feed conversion efficiency, characteristics that make them ideal for research focused on optimizing poultry management systems. Observations spanned the complete production cycle, from 1-week-old (beginning of the experiment) to 6-week-old (end of the experiment), providing comprehensive coverage of growth stages. At the start of the study, the birds' average weight was approximately 150 g, increasing to 2.54 kg by the final week.

The stage of growth significantly influenced detection accuracy. During the early stages (1–3 weeks), smaller bird size and higher activity levels posed challenges, such as increased movement and overlapping of birds. However, these issues were effectively managed by the YOLOv8 model, demonstrating its robustness in complex environments. In contrast, during later stages (4–6 weeks), larger bird size and reduced activity improved detection accuracy. These variations were incorporated into the model training process to ensure consistent performance across all growth stages. By accounting for these developmental differences, the model was able to achieve reliable results throughout the production cycle.

3.1.1. Chicken Coop Setup

A chicken coop was constructed with dimensions of 6 m in length and 4 m in width, providing a controlled environment suitable for housing up to 100 chickens. This setup replicated real poultry production conditions, enabling continuous observation of the chickens' growth and development. Figure 2 shows the exterior view of the chicken coop used during the experiment.

The pen provided a total usable area of 24 m^2. By the final week of the study, the average weight of the 100 chickens housed in the coop was 2.54 kg per bird, resulting in a stocking density of 10.58 kg/m^2. This density is significantly below the regulatory maximum of 33 kg/m^2, as outlined in animal welfare standards [35], ensuring compliance and optimal conditions for poultry welfare throughout the experiment.

Figure 2. Exterior view of the chicken coop.

The coop featured rice husk bedding, selected for its excellent absorbent properties. To maintain cleanliness and prevent moisture buildup, the bedding was replaced weekly. Environmental conditions inside the coop were carefully controlled, with temperatures maintained between 22 and 32 °C and adequate ventilation to ensure a comfortable atmosphere. Lighting was provided for 12 h through natural daylight, supplemented with 8 h of artificial lighting daily, in accordance with poultry welfare standards.

3.1.2. Data Collection and Processing

Surveillance cameras were installed on the ceiling of the chicken coop to capture images and continuously record the production process. The cameras used were IP ColorVu Lite models, selected for their ability to capture high-resolution images (1920 × 1080 pixels) even in low-light conditions. Figure 3 illustrates the camera arrangement, with four cameras positioned to provide full coverage of the coop. Each camera was oriented to point directly at the ground, as marked within the red box in Figure 3.

During real-time video processing, occasional errors were observed, where frames appeared incomplete or corrupted, displaying as black or gray. These issues were traced to data transfer errors or packet loss during transmission. To address packet loss and data errors, the system included functionality to automatically detect faulty frames. When a faulty frame was identified, the system executed a frame reading restart to resume sequence processing without significantly disrupting analysis continuity. This mechanism minimized the loss of critical data and ensured high-quality input for the deep learning algorithm during the detection and counting of chickens.

Data collection was conducted weekly over six weeks, capturing images at different growth stages to comprehensively observe the chickens' development. A total of 1300 images were extracted from the recorded videos and manually annotated. Boundaries around each chicken in the images were marked to create ground truth labels for training the detection model. Figure 4 shows an example of an annotated image from the dataset.

Figure 3. Layout of the four cameras used for data collection during the experiment. Cameras used for capturing the data are marked with a red box. The white device in the center of the ceiling was not used for this project.

Figure 4. Image from the dataset captured using the surveillance camera.

The maximum number of animals the algorithm can process in a single image depends on factors such as image resolution and bird density. High-resolution images provide greater pixel detail for more accurate individual identification but require increased computational power. Conversely, when bird density is high, contour overlap becomes a challenge, complicating accurate detection. As shown in Figure 4, overlapping contours can make individual segmentation more difficult, particularly when chickens are in close proximity.

To balance detection accuracy and computational efficiency, images were processed at a resolution of 640 × 480 pixels, as recommended by the Ultralytics YOLOv8 documentation. At this resolution, the algorithm can detect up to 300 birds in a single image, defined by the `max_det` parameter in the YOLOv8 framework. For this study, `max_det=300` was selected to ensure efficient processing under experimental conditions, as the coop housed only 100 birds. This setup provided sufficient accuracy while optimizing computational requirements.

3.1.3. Dataset Division and Validation

To train the YOLOv8 model, the labeled dataset was divided into 70% for training and 30% for validation. This split was chosen to ensure that the model could be thoroughly evaluated during training while maintaining a sufficient amount of data for training the model. The validation set played a critical role in fine-tuning the model's hyperparameters and providing an objective measure of its performance, helping to prevent overfitting.

A separate test set was not employed during this phase, as the format recommended by Ultralytics emphasizes the use of training and validation datasets for iterative evaluation. The validation set consisted of data not seen by the model during training, ensuring unbiased evaluation and enabling necessary adjustments throughout the training process.

The dataset organization adhered to the directory structure recommended in the Ultralytics YOLO guidelines. A main directory was created with subdirectories `train` and `val`, each containing two folders: `images` and `labels`. For example, training images were stored in `train/images/`, while their corresponding label files, containing bounding box coordinates and class annotations, were stored in `train/labels/`.

A YAML configuration file was used to define the dataset structure, including paths to the `train` and `val` directories and the number of classes in the dataset. This configuration file was essential to ensuring that YOLOv8 could correctly interpret and process the training and validation data, allowing for efficient and accurate training of the model.

3.2. Model Training

The training of the YOLOv8 model was an iterative process designed to optimize its parameters for accurate chicken detection in the dataset. The following steps outline the key components of the training process.

3.2.1. Dataset Annotation

To prepare the dataset for model training, a manual annotation process was conducted using the labelImg annotation tool [36] widely used in the computer vision community. Each chicken in the images was marked with a bounding box. Special attention was given to ensure accuracy and consistency in annotations, especially in cases with occlusions or lighting variations. To maintain label quality, the following were implemented:

- Cross-validation: Annotations were reviewed by at least two annotators to minimize errors.
- Clear guidelines: Specific criteria were established for annotation, including marking only visible chickens and accurately outlining their body boundaries.

3.2.2. YOLOv8 Model Architecture

The YOLOv8 architecture was selected for its superior speed and accuracy. This state-of-the-art model was trained on the annotated dataset to optimize its parameters for precise chicken detection and localization. Its design integrates anchor-free detection and enhanced feature extraction, making it particularly suitable for complex environments like poultry houses.

3.2.3. Why YOLOv8? A Comparison with SSD and Faster R-CNN

The choice of YOLOv8 over SSD and Faster R-CNN was based on several factors.

- Real-time detection and speed: YOLOv8 operates as a one-stage detector, where classification and bounding box regression are performed simultaneously. This results in faster inference compared to two-stage models like Faster R-CNN [37,38], making it ideal for real-time applications in poultry management.
- High accuracy in complex environments: YOLOv8 achieves superior precision and recall (often exceeding 90%) in crowded settings, such as poultry houses [37,38]. In contrast, SSD sacrifices some accuracy for speed [39], and Faster R-CNN, while accurate, is slower and less suited for real-time tasks [39].

- Anchor-free detection: YOLOv8's anchor-free approach simplifies detection and improves accuracy in environments with overlapping objects such as clustered chickens [37]. This feature is particularly advantageous over SSD, which can struggle in such scenarios [39].
- Efficiency in training and data processing: YOLOv8 reduces the complexity of object detection pipelines by eliminating the need for region proposal networks seen in Faster R-CNN. This makes it more efficient in handling large datasets [40].

3.2.4. Hyperparameter Selection

The training process involved careful selection of hyperparameters to balance computational efficiency and model accuracy. Table 1 provides a summary of the key hyperparameters used during training.

- Model and Data Configuration: The model `yolov8m.pt` was used, alongside a custom dataset configuration file (`dataset.yaml`) specifying paths to data and the number of classes.
- Epochs and Early Stopping: Training ran for 150 epochs, with an early stopping patience of 100 to prevent overfitting.
- Batch Size and Image Size: A batch size of 8 was selected to accommodate GPU memory limitations, while images were resized to 640 × 480, balancing detail and computational load.
- Learning Rate and Momentum: An initial learning rate (`lr0`) of 0.01 and a momentum of 0.937 were chosen to ensure smooth and efficient convergence.

These hyperparameters were refined based on guidance from Ultralytics [41] and experimental results to achieve optimal performance.

Table 1. Key hyperparameters for YOLOv8 training.

Hyperparameter	Value	Description
`mode`	train	Training mode selected.
`model`	yolov8m.pt	Pretrained model file specifying structure and weights.
`data`	dataset.yaml	Path to the dataset configuration file.
`epochs`	150	Total number of training epochs.
`patience`	100	Early stopping patience for validation metrics.
`batch`	8	Batch size per training iteration.
`imgsz`	640	Target image resolution for training.
`device`	0	Computational device (e.g., GPU or CPU).
`lr0`	0.01	Initial learning rate.
`lrf`	0.01	Final learning rate fraction for adjustment.
`momentum`	0.937	Momentum factor for optimization.
`weight_decay`	0.0005	L2 regularization to prevent overfitting.

3.2.5. Training Process

During the training process, several optimization techniques were employed to enhance the model's accuracy in detecting and counting chickens. The YOLOv8 model was trained iteratively, with performance metrics such as precision, recall, and overall accuracy used to evaluate its performance at each stage. These metrics guided adjustments to the model's parameters, ensuring continuous improvement and robustness.

As illustrated in Figure 5, the training scheme involves feeding the YOLOv8 model a large dataset of annotated images containing chickens and their precise locations. This process teaches the model to detect and localize chickens in new, unseen images effectively.

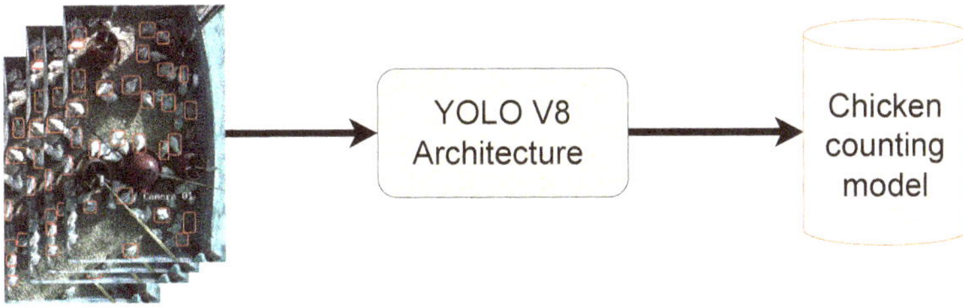

Figure 5. Training scheme using YOLOv8. The model is trained on a large dataset of images containing chickens, annotated with their exact positions, to enable accurate detection in new images.

4. Results

4.1. Experiment Setup

The experiment was conducted with a flock of broiler chickens ranging in age from 1 week to 6 weeks, with an average flock size of 100 chickens. Environmental conditions were carefully controlled, maintaining temperatures between 22 °C and 32 °C. Chickens were exposed to natural daylight during the day and artificial lighting for 8 h at night.

For the annotation process, images were manually labeled using the labelImg tool (version 1.8.6) [36], a widely adopted tool in computer vision applications. Each chicken was enclosed within a bounding box, with particular attention to challenging scenarios such as occlusions and varying lighting conditions. The annotations were saved in YOLO format, adhering to consistent and accurate guidelines. To ensure label quality, a cross-validation approach was implemented, with annotations reviewed by at least two independent annotators. Clear guidelines emphasized marking only visible chickens and accurately delineating their body boundaries.

Training was performed on a workstation equipped with an NVIDIA RTX 3080 Ti GPU (12 GB VRAM), Intel i7-12700K CPU, and 32 GB RAM, using PyTorch 1.13 and the Ultralytics YOLOv8 framework. The training process took approximately 5 h to complete 100 epochs.

Throughout training, key metrics such as box loss, confidence loss, and classification loss were monitored to evaluate the model's progress. The model's performance was assessed using precision, recall, mAP@0.5, and mAP@0.5:0.95. The YOLOv8m (medium) version was selected for its balance of accuracy and speed. The model utilized the default YOLOv8 loss function, which combines localization, confidence, and classification losses. Training employed the Adam optimizer, with an initial learning rate of 0.001 that decreased progressively over the course of training.

A batch size of 16 images per iteration was used, with input images resized to 640×640 pixels. Images were extracted from videos recorded at 25 frames per second (fps). Data augmentation techniques, including random rotation, scaling, cropping, and adjustments to brightness and contrast, were applied to improve model generalization.

The breeding and care of the poultry were overseen by experienced poultry farming professionals hired specifically for this project. These experts ensured compliance with animal welfare standards, including daily health monitoring and behavioral observations. No significant health issues or mortalities were reported during the study, ensuring consistent and reliable conditions for data collection.

4.2. Training Results

The training of the YOLOv8 model demonstrated significant performance improvements in chicken detection accuracy. Key metrics, including precision, recall, mAP@0.5, and mAP@0.5:0.95, highlight the model's effectiveness in detecting chickens across varied farm environments.

To provide a comprehensive understanding of the model's behavior during training, three performance curves are presented, illustrating key aspects of its detection capabilities.

- Recall–Confidence Curve (Figure 6a): The Recall–Confidence curve illustrates the trade-off between recall and confidence thresholds. As confidence thresholds increase, recall gradually decreases, reflecting the model's precision-focused behavior at higher confidence levels.
- Precision–Confidence Curve (Figure 6b): Precision improves with increasing confidence thresholds, stabilizing at 1.0 around a threshold of 0.935. This trend demonstrates the model's strong predictive power when high-confidence predictions are prioritized.
- Precision–Recall Curve (Figure 6c): The Precision–Recall curve, with a mean Average Precision (mAP) of 0.931 at an IoU threshold of 0.5, illustrates the balance between precision and recall, achieving high precision while maintaining adequate recall.

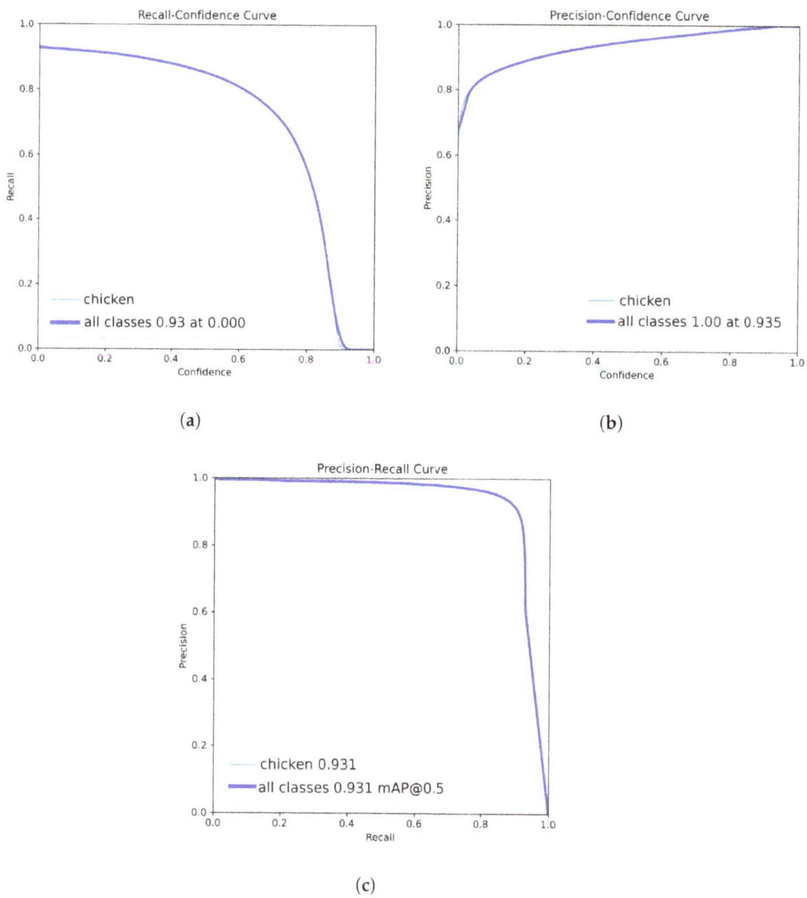

Figure 6. Performance curves of the trained YOLOv8 model. (**a**) Recall–Confidence Curve. (**b**) Precision–Confidence Curve. (**c**) Precision–Recall Curve.

These curves complement the quantitative metrics and emphasize the model's robustness, particularly in high-confidence scenarios where accurate detection is crucial.

Additionally, if available, loss curves and epoch-wise performance improvements are presented to highlight the model's convergence and training stability. These visualizations illustrate the reduction in training and validation losses over epochs, providing insight into the model's optimization process and confirming its steady performance gains throughout training.

The combination of these metrics and visualizations underscores the YOLOv8 model's suitability for real-time poultry monitoring, even under diverse and challenging conditions. Its high precision and recall demonstrate its potential for accurate and efficient chicken detection in commercial farming applications.

Performance Metrics

Table 2 presents the performance metrics of the YOLOv8 model after additional training. The model achieved high precision (93.1%) and recall (93.0%), demonstrating a strong balance between detection accuracy and thoroughness. Furthermore, the model's mAP@0.5 improved to 93.1%, and its mAP@0.5:0.95 increased to 62.5%, significantly outperforming earlier versions and highlighting its capability in handling varying Intersection over Union (IoU) thresholds.

Table 2. Performance metrics of the YOLOv8 model on the training and test sets.

Metric	Training (%)	Test (%)
Precision	93.5	93.1
Recall	85.0	93.0
F1-score	88.9	91.0
mAP@0.5	90.0	93.1
mAP@0.5:0.95	58.2	62.5

The high F1-score on the test set (91.0%) reflects the model's ability to maintain a balance between precision and recall, ensuring reliable detection even in challenging conditions. The increase in mAP@0.5:0.95 to 62.5% further underscores the model's improved performance across a range of IoU thresholds, indicating enhanced generalization capability for detecting objects of varying sizes and overlaps.

These results affirm the YOLOv8 model's suitability for real-time poultry monitoring, providing a robust and scalable solution for detecting and counting chickens under diverse environmental conditions.

4.3. *Performance Analysis*

The YOLOv8 model's performance was evaluated under various conditions to ensure its robustness and applicability in realistic farm environments.

4.3.1. Inference Process

Test images captured by surveillance cameras were processed through the trained YOLOv8 model to predict bounding boxes around each chicken. This inference process facilitated the accurate counting of chickens within each image. The evaluation utilized the inference scheme illustrated in Figure 7.

The inference process demonstrated the model's ability to reliably detect and count chickens across various scenarios, including images with occlusions and variable lighting conditions. These results affirm the model's potential for practical deployment in real-time poultry monitoring systems.

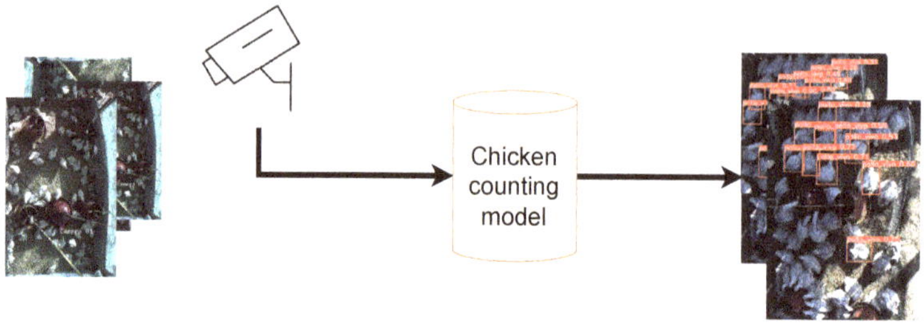

Figure 7. Inference process using the YOLOv8 trained model. The model processes input images captured from surveillance cameras and outputs detected chickens with bounding boxes.

4.3.2. Performance Under Different Conditions

The YOLOv8 model was evaluated under varying lighting and occlusion conditions to assess its robustness. Results demonstrated that YOLOv8 maintained consistent performance even in challenging scenarios, such as when chickens were partially occluded or in low-light environments. This level of reliability is crucial for practical deployment in poultry farms, where environmental conditions can vary significantly.

Despite its strong performance, some instances of missed detections were observed, which can be attributed to several factors.

- Occlusion: Occlusion remains one of the primary challenges, as chickens often cluster together, partially or fully covering one another. This issue is particularly prevalent in crowded areas of the coop where movement is unpredictable, increasing the likelihood of missed detections. Multi-view camera setups or depth-sensing technologies could help address this limitation.
- Lighting Variations: Shadows, reflections, and poor lighting within the coop can affect the model's ability to detect chickens. While YOLOv8 performs well under controlled lighting conditions, the dynamic lighting often found in real-world environments such as poultry farms can pose challenges. Bright reflections can confuse the model, while poorly lit areas may obscure chickens from view. Incorporating adaptive lighting techniques or infrared imaging could enhance performance in such scenarios.
- Image Resolution: The resolution of input images significantly impacts detection accuracy. Although YOLOv8 is optimized for 640 × 640 pixels, higher resolutions (e.g., 1920 × 1080) provide more detail, improving accuracy, particularly in complex scenes. Conversely, lower-resolution images (e.g., 320 × 320) may lack the detail needed to correctly identify chickens, especially in crowded or cluttered environments. Employing higher-resolution images in critical areas could mitigate this limitation.

YOLOv8 demonstrated robust detection capabilities across different conditions, highlighting its suitability for real-world poultry monitoring. However, addressing these limitations through strategies such as higher-resolution imaging, depth-sensing technologies, or multi-camera setups could further enhance detection accuracy and reliability, particularly in complex environments.

4.3.3. Visual Evaluation

In addition to quantitative metrics, a visual evaluation of the model's detections was conducted. Output images were analyzed to verify the accuracy of chicken detection and counting. The results showed that the model accurately identified most chickens present in the images. Figure 8 presents examples of the system's detection results during its operation.

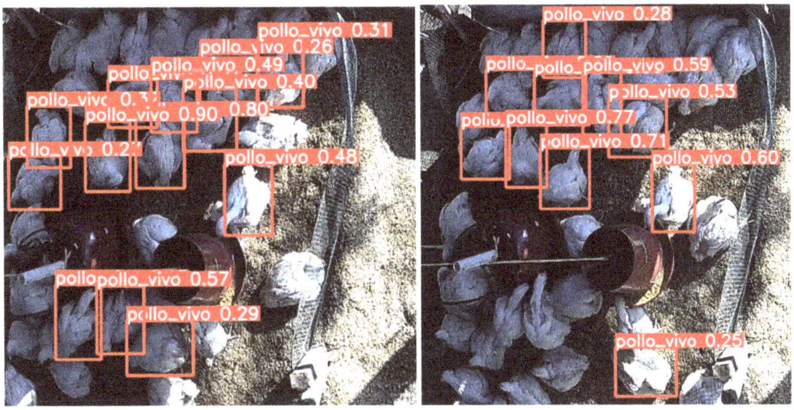

Figure 8. Labeling results achieved by the trained model. Red bounding boxes indicate detected chickens along with the confidence values generated by the model.

4.4. Comparison with Existing Methods

To contextualize the performance of the YOLOv8 model, its results were compared with previous methods employed in chicken counting tasks.

4.4.1. Comparison with YOLOv5 Model

The performance of the YOLOv8 model was benchmarked against the YOLOv5 model, which has been previously utilized for chicken counting tasks. YOLOv8 outperformed YOLOv5 in both precision and inference speed, demonstrating significant advancements in handling occlusions and lighting variations. These enhancements mark a notable improvement in the robustness and accuracy of detection models within the poultry farming context. As outlined in Table 3, YOLOv8 exhibited higher accuracy and better adaptability to variable illumination conditions compared to YOLOv5.

Table 3. Comparison of YOLOv5 and YOLOv8 on chicken counting tasks.

Criteria	YOLOv5	YOLOv8
Accuracy	High	Higher
Inference Speed	Moderate	Fast
Real-time Detection	Yes	Yes
Handling Occlusions	Moderate	High
Handling Illumination Variations	Moderate	High

Table 4 provides a detailed comparison of key performance metrics between YOLOv5 and YOLOv8 on the test set. YOLOv8 demonstrated significant improvements across all evaluated metrics, particularly in recall and mean Average Precision (mAP). These metrics underscore YOLOv8's superior ability to accurately detect chickens under challenging conditions, such as occlusion and variable lighting.

Table 4. Comparison of performance metrics between YOLOv5 and YOLOv8 on the test set.

Metric	YOLOv5 (%)	YOLOv8 (%)	Improvement (%)
Precision	90.2	93.1	+2.9
Recall	55.0	93.0	+38.0
F1-score	68.7	91.0	+22.3
mAP@0.5	60.0	93.1	+33.1
mAP@0.5:0.95	35.0	62.5	+27.5

The results demonstrate that YOLOv8 significantly surpasses YOLOv5 in precision, recall, and overall detection accuracy. Its improvements in recall (+38.0%) and mAP@0.5 (+33.1%) are particularly noteworthy, indicating that YOLOv8 is highly effective in detecting chickens even in complex scenarios. These enhancements solidify YOLOv8 as a reliable and efficient model for real-time poultry monitoring applications.

4.4.2. Innovations Introduced

The application of YOLOv8 introduced several key innovations in the task of chicken counting, significantly enhancing its practicality for real-world use. These include:

- Real-time Detection: YOLOv8's architecture enables efficient real-time detection of chickens, a critical feature for monitoring dynamic environments in poultry farms.
- Robustness to Occlusions: The model demonstrates improved handling of occlusions, effectively detecting chickens even in crowded conditions where individuals may be partially obscured.
- Adaptability to Lighting Variations: YOLOv8 exhibits strong performance under variable lighting conditions, including low-light scenarios and environments with shadows or reflections, surpassing YOLOv5 in these aspects.

4.5. Practical Implications

The experimental results validate the YOLOv8 model's effectiveness for practical applications in poultry management, demonstrating its potential to enhance operational efficiency and animal welfare.

4.5.1. Applications in Poultry Management

The model's capability to accurately count chickens in real time enables farmers to make informed decisions about flock management. This improves resource allocation and overall operational efficiency while promoting better animal welfare. Agricultural producers can adopt the proposed method by deploying a complete system that integrates both hardware and software components. The hardware setup involves installing surveillance cameras, such as IP ColorVu Lite models, in poultry houses to continuously capture images of the animals. These cameras are connected to a computing system running the YOLOv8-based object detection algorithm, which processes the video feeds in real time to count the number of animals present. The software is designed to be user-friendly, featuring automated error detection, data logging, and a streamlined interface for consistent and accurate flock monitoring. By integrating this system into their operations, producers can optimize critical resources, such as feed and medication, while improving flock management efficiency. This approach not only reduces waste and operational costs but also supports sustainable farming practices by ensuring precise and timely interventions based on real-time data.

4.5.2. Resource Optimization

Automating the chicken counting process reduces labor intensity and minimizes human errors, leading to more efficient resource utilization in poultry farms. By accurately tracking flock size in real time, farmers can better allocate critical resources such as feed, water, and medication, ensuring that supplies are neither underused nor wasted. This precision not only lowers operational costs but also supports sustainable farming practices by optimizing resource distribution and reducing environmental impact.

5. Discussion

5.1. Model Accuracy and Robustness

The results validate the high accuracy and robustness of the YOLOv8 model, highlighting its potential for applications in the poultry industry. The model achieved a precision of 93%, demonstrating its ability to accurately count chickens even under challenging conditions such as occlusions and lighting variations. These findings are particularly notable compared to traditional manual methods, which are prone to errors and require considerable labor. Moreover, the model's consistent performance across varying environmental conditions reinforces its suitability for deployment in real farm settings, where conditions frequently fluctuate.

5.2. Advantages over Traditional Methods

In flock management, traditional manual counting methods, which rely on initial and final verifications during the production cycle, have significant limitations in terms of both accuracy and efficiency. By contrast, the AI-based approach developed in this study provides substantial improvements. Leveraging computer vision algorithms, the system enables continuous and precise counting throughout the flock's life cycle. This overcomes the challenges faced by manual methods, such as high chicken density, overlaps, variable lighting conditions, and human fatigue.

Flock handlers typically only have exact chicken counts at two points: the start of the flock (as chicks) and at the end (before slaughter), limiting monitoring capabilities during intermediate stages. While quantitative data directly comparing the accuracy of manual and AI-based methods are currently unavailable, prior research indicates that AI systems can outperform manual methods by 10–20% in accuracy, particularly in high-density environments. For instance, Lempitsky and Zisserman (2010) demonstrated the effectiveness of computer vision systems in counting tasks under complex conditions [42]. Similarly, Kumar et al. (2023) validated the high precision of computer vision systems in industrial settings, emphasizing their adaptability to challenging environments [43]. Additionally, Zhang et al. (2016) addressed density variability using a multi-column convolutional neural network, further supporting the adaptability of AI-based systems in dense and overlapping scenarios [44].

Future research will focus on detailed quantitative comparisons between manual and AI-based methods in real operational environments. Developing a robust dataset for such comparisons will further validate the advantages of AI systems over traditional approaches and provide a clearer understanding of their impact on flock management efficiency.

5.3. Practical Applications and Benefits

The development of an AI-based poultry monitoring system was complemented by the creation of a user-friendly prototype web application. This application serves as the primary interface for managing and monitoring multiple chicken coops, enabling farm managers to make data-driven decisions efficiently. Built using Flask [45], the application integrates directly with the YOLOv8 model and processes video feeds from surveillance cameras installed within the chicken coops.

5.3.1. System Overview

Figure 9 illustrates the camera view interface, where detected chickens are annotated with bounding boxes and IDs, enabling immediate visual confirmation of counting results. This real-time monitoring capability forms the core functionality of the system.

Access to the application is secured through a login interface (Figure 10), ensuring only authorized users can manage the system. Once logged in, users are greeted with a dashboard (Figure 11), where they can navigate through various sections, including coop management, system configuration, and report generation.

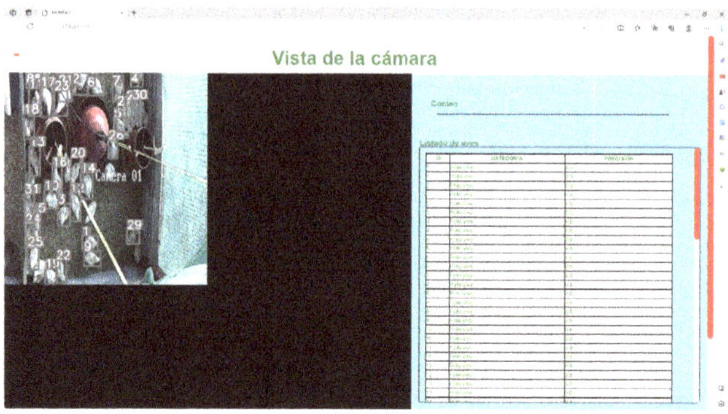

Figure 9. Camera view interface integrated into the system. Detected chickens are displayed with bounding boxes and confidence values.

Figure 10. Login window for the application. Secure access ensures data integrity during poultry management.

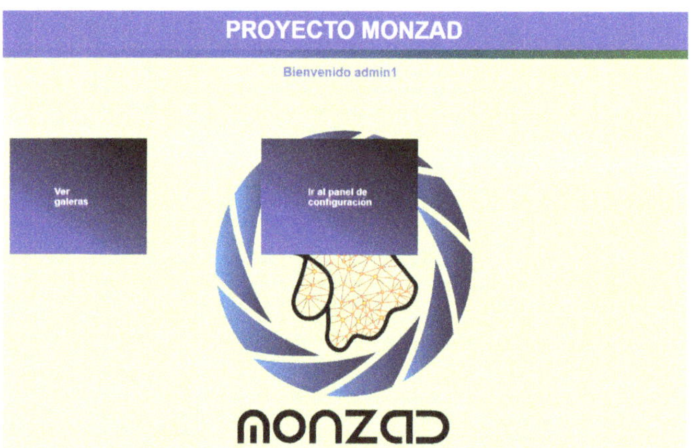

Figure 11. Dashboard interface. Users can access main features, including coop management and system configuration.

5.3.2. Multi-Coop Management and Reporting

A critical feature of the application is its ability to manage multiple coops simultaneously. Figure 12 displays the coop management interface, allowing users to select and monitor different coops. This functionality is essential for large-scale operations requiring continuous observation of multiple flocks.

Figure 12. Coop management interface. Users can view and manage connected coops.

Detailed reporting tools are also included, as shown in Figure 13. Users can generate comprehensive reports summarizing data, such as the number of live and possibly dead chickens over specific periods. These reports are instrumental in monitoring trends, optimizing resource allocation, and improving flock management.

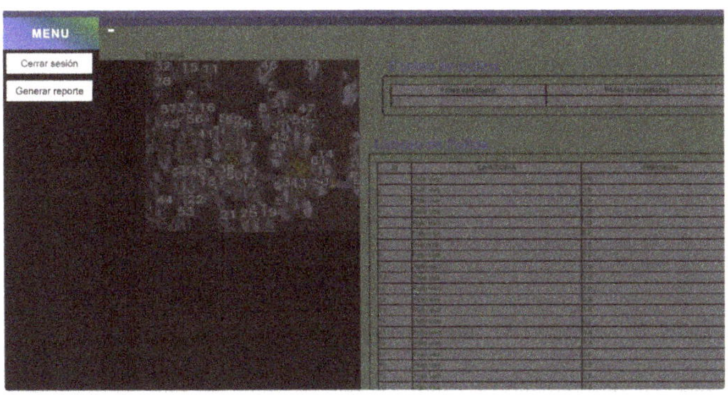

Figure 13. Report generation interface. Provides detailed summaries of poultry data, including live and possibly dead chickens.

5.3.3. User Management and System Configuration

The application includes a robust user management system, enhancing data security by ensuring only authorized personnel can access sensitive information. Figures 14–16 showcase the menu interface, user registration panel, and account management panel, respectively.

The system configuration panel, illustrated in Figure 17, allows users to link the trained YOLOv8 model and configure the database connection. This flexibility ensures the system can be updated with new models or datasets as AI technology advances.

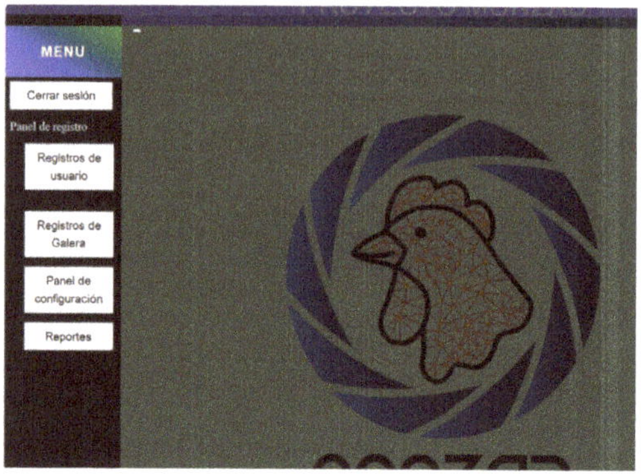

Figure 14. Menu interface. Allows access to various sections, including report generation.

Figure 15. User registration panel. Ensures secure access by limiting usage to authorized personnel.

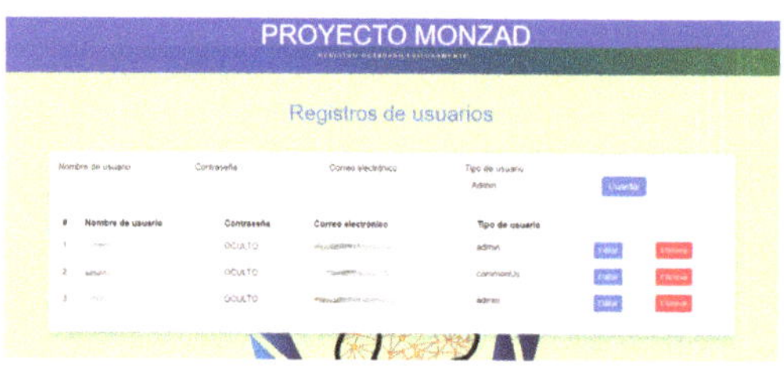

Figure 16. Account management panel. Enables administrators to manage registered users.

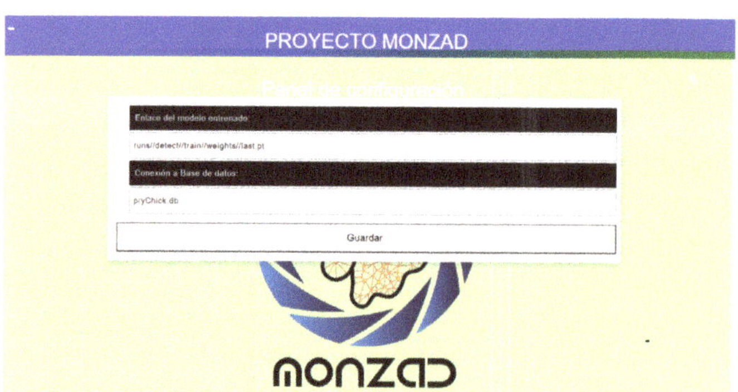

Figure 17. System configuration panel. Allows linking of trained YOLOv8 models and database setup.

5.3.4. Historical Data and Records

The poultry study records interface, shown in Figure 18, provides a comprehensive overview of each flock's historical data. Users can navigate through records and generate reports for specific dates, supporting long-term analysis and decision-making.

Figure 18. Poultry study records interface. Enables navigation through historical data and report generation.

5.3.5. Practical Benefits

The prototype application enhances the practicality and usability of the AI-based poultry monitoring system. By offering real-time monitoring, multi-coop management, and comprehensive reporting capabilities, the system empowers farm managers to optimize operations, reduce human error, and improve animal welfare. The inclusion of secure user management and customizable settings further underscores the application's value in modern poultry farming.

5.4. Limitations and Challenges

Despite the promising results, several limitations must be acknowledged to provide a balanced understanding of the potential and constraints associated with deploying this model in real-world poultry farm environments.

One primary limitation is scalability. Expanding this AI-based system to monitor larger flocks or multiple locations introduces significant computational and resource demands. For example, higher-resolution input data or larger datasets would necessitate more powerful GPU resources, potentially impacting accessibility and increasing costs. Memory and storage requirements would also rise due to the need for processing and managing greater volumes of data.

Furthermore, achieving real-time detection across large-scale farms could result in latency issues, which might compromise the practical usability of the system. To address these challenges, techniques such as model pruning, quantization, or deploying lightweight variants (e.g., YOLO-Tiny) could be explored. These strategies aim to maintain performance efficiency while reducing computational overhead. By addressing these scalability considerations, future implementations could achieve a balance between accuracy, resource efficiency, and real-time applicability, enhancing the system's adaptability for broader applications in poultry farm management.

5.5. Future Opportunities and Expansion

Beyond species-specific applications, the system presents opportunities for adaptation across diverse environmental contexts. For instance, deploying the model in outdoor open-range farms would allow testing under more complex conditions, including variable lighting, terrain, and weather patterns. Conversely, controlled indoor environments, such as high-density automated farms with artificial lighting, could provide additional insights into optimizing system accuracy in specialized settings.

Integrating advanced sensors and multi-view cameras could further enhance detection capabilities in these environments, offering a comprehensive tool for agricultural monitoring. The system also has potential for applications beyond poultry farming. For example, similar methodologies could be applied to other livestock or species, enabling cross-species adaptability.

In addition, scaling the system to a commercial poultry setting with the same chicken species represents a significant opportunity. Testing the algorithm in commercial environments with higher-density flocks and larger enclosures would enable evaluation of its robustness and accuracy under more demanding conditions. Addressing challenges such as variability in lighting, flock density, and movement patterns in these real-world scenarios would provide valuable insights into the system's scalability and adaptability. These advancements would be instrumental in driving the broader adoption of AI-driven solutions in commercial poultry management.

5.6. Integration with Poultry Management Systems

Integrating this system with existing poultry management platforms offers the potential to create a comprehensive solution for flock management. Such integration would enable farmers to fully leverage the benefits of automation, facilitating a data-driven approach to resource allocation, health monitoring, and operational efficiency.

Moreover, incorporating this technology into broader farm management systems could drive innovation in the poultry industry. The development of interconnected tools and solutions could further enhance productivity and animal welfare while encouraging sustainable farming practices. By enabling seamless integration with current platforms, the system can serve as a foundation for future advancements in agricultural automation, solidifying its role as a transformative technology in modern poultry farming.

5.7. Adaptability of the Proposed System

The proposed system comprises two main components: the administration module, which serves as the primary user interface, and the AI-driven model, responsible for detecting and counting broiler chickens within the coop. While this study focuses on broiler chickens, the underlying concept and infrastructure presented are adaptable to other poultry breeds and even livestock species in environments with similar conditions. Adapting the system to new settings requires modifications to the AI model, specifically retraining it with a dataset tailored to the target breed or species.

AI models, particularly those utilizing computer vision, rely heavily on training data that accurately reflect the characteristics and environmental conditions they encounter. For instance, transitioning from broiler chickens to layer hens or other poultry breeds necessitates capturing and annotating new data from the target group. This process follows the steps outlined in this study for data collection, model training, and validation.

Although the current model is optimized for broiler chicken farming, the flexible design of the administration module allows seamless integration with new AI models. This adaptability significantly broadens the system's potential applications, making it suitable for deployment in various environments or for managing other farm animals.

The modular design ensures compatibility with diverse datasets and models, facilitating its use in open-range poultry farming, automated livestock management, or monitoring other species. By leveraging this adaptability, the system can contribute to more sustainable and efficient farming practices across different agricultural domains. Future research should focus on extending the system's applicability to various species and environmental contexts, thereby enhancing the versatility and impact of AI-driven solutions in agriculture.

6. Conclusions

This study successfully developed and validated an AI-based automated chicken counting system using the YOLOv8 model. The system demonstrates high accuracy, achieving a 93% precision rate under real farm conditions, significantly addressing the limitations of traditional manual methods. By automating the counting process, the system reduces human error, enhances operational efficiency, and enables more effective management of poultry flocks.

The results underscore the transformative potential of adopting advanced AI technologies in agriculture, particularly in poultry management. The system's ability to perform real-time counts and adapt to varying environmental conditions ensures its applicability across a broad range of productive scenarios. Moreover, the innovations introduced in this study pave the way for applying similar technologies in other areas of agriculture and livestock management.

However, scalability remains a key challenge, particularly for implementation in larger farms or operations with significantly higher flock sizes. To address this, future work could explore strategies such as integrating distributed processing techniques, where multiple processing units (e.g., edge servers or cloud nodes) collaborate to manage large data volumes in real time. Additionally, deploying lighter AI models optimized for low-power devices could reduce computational demands and increase infrastructure flexibility. Data compression algorithms and federated learning techniques could also facilitate large-scale data management while maintaining model accuracy across diverse locations and conditions.

Despite these challenges, the system represents a significant advancement in automated poultry management. It provides a practical and efficient tool for improving productivity and animal welfare. With continued development and broader application, this approach has the potential to significantly transform the poultry industry and drive innovation across agricultural sectors.

Author Contributions: E.C.: Conceptualization, Funding acquisition, Project administration, Resources, Investigation, Writing—original draft, Writing—review and editing. J.C.R.: Investigation, Methodology, Visualization, Writing—original draft. M.H.-R.: Software, Data curation, Formal analysis, Methodology, Writing—review and editing. A.M.A.-R.: Conceptualization, Resources, Validation, Writing—review and editing. K.B.: Data curation, Formal analysis, Writing—original draft, Writing—review and editing. All authors have read and agreed to the published version of the manuscript.

Funding: This research was funded by Project IDDS22-28 of the National Secretariat for Science, Technology, and Innovation (SENACYT), Panama. José Carlos Rangel and Edmanuel Cruz were supported by SENACYT's National Research System (SNI). Keyla Boniche received a scholarship from SENACYT's Program for the Strengthening of National Graduate Programs for a Master's Degree in Mechanical Engineering Sciences.

Institutional Review Board Statement: Not applicable.

Data Availability Statement: Data available on request due to restrictions

Acknowledgments: The authors thank the IDDS Program and the National Research System (SNI) of SENACYT, as well as CEMCIT, the Veraguas Regional Center of the Universidad Tecnológica de Panamá, and Agropecuaria El Buen Pastor for their invaluable support.

Conflicts of Interest: The authors declare that the research was conducted in the absence of any commercial or financial relationships that could be construed as a potential conflicts of interest.

References

1. Connolly, G.; Clark, C.M.; Campbell, R.E.; Byers, A.W.; Reed, J.B.; Campbell, W.W. Poultry consumption and human health: How much is really known? A systematically searched scoping review and research perspective. *Adv. Nutr.* **2022**, *13*, 2115–2124. [CrossRef] [PubMed]
2. Day, J. *Global Chicken Market Report 2023: Rising Consumption of Poultry Worldwide to Boost Growth*; Poultry Producer: Stirling-Rawdon, ON, Canada, 2023.
3. Shahbandeh, M. *Global Production of Meat 2016–2024*; Statista: Hamburg, Germany, 2024.
4. Taleb, H.M.; Mahrose, K.; Abdel-Halim, A.A.; Kasem, H.; Ramadan, G.S.; Fouad, A.M.; Khafaga, A.F.; Khalifa, N.E.; Kamal, M.; Salem, H.M.; et al. Using artificial intelligence to improve poultry productivity—A review. *Ann. Anim. Sci.* **2024**. [CrossRef]
5. Khanal, R.; Choi, Y.; Lee, J. Transforming Poultry Farming: A Pyramid Vision Transformer Approach for Accurate Chicken Counting in Smart Farm Environments. *Sensors* **2024**, *24*, 2977. [CrossRef] [PubMed]

6. Neethirajan, S. Automated tracking systems for the assessment of farmed poultry. *Animals* **2022**, *12*, 232. [CrossRef] [PubMed]
7. Shukla, P.P.; Bhattachayya, A.; Sharma, A. *Artificial Intelligence in Poultry Production*; SR Publications: Vijayawada, India, 2024.
8. Eissa, M.E. Precision Agriculture using Artificial Intelligence and Robotics. *J. Res. Agric. Food Sci.* **2024**, *1*, 35–52. [CrossRef]
9. Araújo, S.O.; Peres, R.S.; Ramalho, J.C.; Lidon, F.; Barata, J. Machine Learning Applications in Agriculture: Current Trends, Challenges, and Future Perspectives. *Agronomy* **2023**, *13*, 2976. [CrossRef]
10. Merugu, S.; Tiwari, A.; Sharma, S.K. Spatial—Spectral image classification with edge preserving method. *J. Indian Soc. Remote Sens.* **2021**, *49*, 703–711. [CrossRef]
11. Sharma, R. Artificial Intelligence in Agriculture: A Review. In Proceedings of the International Conference on Intelligent Computing and Control Systems, Madurai, India, 6–8 May 2021; pp. 937–945. [CrossRef]
12. Lamping, C.; Kootstra, G.; Derks, M. Uncertainty Estimation for Deep Neural Networks to Improve the Assessment of Plumage Conditions of Chickens. *Smart Agric. Technol.* **2023**, *5*, 100308. [CrossRef]
13. Yang, X.; Bist, R.; Paneru, B.; Chai, L. Monitoring Activity Index and Behaviors of Cage-Free Hens with Advanced Deep Learning Technologies. *Poult. Sci.* **2024**, *103*, 104193. [CrossRef]
14. Ryan, M. The Social and Ethical Impacts of Artificial Intelligence in Agriculture: Mapping the Agricultural AI Literature. *AI Soc.* **2023**, *38*, 2473–2485. [CrossRef]
15. Dalal, N.; Triggs, B. Histograms of Oriented Gradients for Human Detection. In Proceedings of the IEEE Conference on Computer Vision and Pattern Recognition (CVPR), San Diego, CA, USA, 21–23 September 2005; pp. 886–893.
16. Lowe, D.G. Distinctive Image Features from Scale-Invariant Keypoints. *Int. J. Comput. Vis.* **2004**, *60*, 91–110. [CrossRef]
17. Bay, H.; Tuytelaars, T.; Gool, L.V. Speeded-Up Robust Features (SURF). *Comput. Vis. Image Underst. (CVIU)* **2008**, *110*, 346–359. [CrossRef]
18. Cortes, C.; Vapnik, V. Support-Vector Networks. *Mach. Learn.* **1995**, *20*, 273–297. [CrossRef]
19. LeCun, Y.; Bottou, L.; Bengio, Y.; Haffner, P. Gradient-Based Learning Applied to Document Recognition. *Proc. IEEE* **1998**, *86*, 2278–2324. [CrossRef]
20. Simonyan, K.; Zisserman, A. Very Deep Convolutional Networks for Large-Scale Image Recognition. *arXiv* **2014**, arXiv:1409.1556.
21. He, K.; Zhang, X.; Ren, S.; Sun, J. Deep Residual Learning for Image Recognition. In Proceedings of the IEEE Conference on Computer Vision and Pattern Recognition (CVPR), Honolulu, Hawaii, 21–26 July 2016; pp. 770–778.
22. Girshick, R.; Donahue, J.; Darrell, T.; Malik, J. Rich feature hierarchies for accurate object detection and semantic segmentation. In Proceedings of the IEEE Conference on Computer Vision and Pattern Recognition (CVPR), Columbus, OH, USA, 24–27 June 2014; pp. 580–587.
23. Ren, S.; He, K.; Girshick, R.B.; Sun, J. Faster R-CNN: Towards Real-Time Object Detection with Region Proposal Networks. In *Advances in Neural Information Processing Systems (NIPS). Proceedings of the Annual Conference on Neural Information Processing Systems 2015, Montreal, QC, Canada, 7–12 December 2015*; Curran Associates, Inc.: Red Hook, NY, USA, 2015; pp. 91–99.
24. Redmon, J.; Divvala, S.; Girshick, R.B.; Farhadi, A. You Only Look Once: Unified, Real-Time Object Detection. In Proceedings of the IEEE Conference on Computer Vision and Pattern Recognition (CVPR), Las Vegas, NV, USA, 27–30 June 2016; pp. 779–788.
25. Liu, W.; Anguelov, D.; Erhan, D.; Szegedy, C.; Reed, S.; Fu, C.Y.; Berg, A.C. SSD: Single Shot MultiBox Detector. In *Computer Vision—ECCV 2016, Proceedings of the 14th European Conference, Amsterdam, The Netherlands, 11–14 October 2016*; Springer: Cham, Switzerland, 2016; pp. 21–37.
26. Carion, N.; Massa, F.; Synnaeve, G.; Usunier, N.; Kirillov, A.; Zagoruyko, S. End-to-End Object Detection with Transformers. In *Computer Vision—ECCV 2020, Proceedings of the 16th European Conference, Glasgow, UK, 23–28 August 2020*; Springer: Cham, Switzerland, 2020; pp. 213–229.
27. Jebari, H.; Hayani Mechkouri, M.; Rekiek, S.; Reklaoui, K. Poultry-Edge-AI-IoT System for Real-Time Monitoring and Predicting by Using Artificial Intelligence. *Int. J. Interact. Mob. Technol.* **2023**, *17*, 149–152. [CrossRef]
28. Vandana.; Yogi, K.K.; Yadav, S.P. Chicken Diseases Detection and Classification Based on Fecal Images Using EfficientNetB7 Model. *Evergr. Jt. J. Nov. Carbon Resour. Sci. Green Asia Strategy* **2024**, *11*, 314–330. [CrossRef]
29. Fang, C.; Wu, Z.; Zheng, H.; Yang, J.; Ma, C.; Zhang, T. MCP: Multi-Chicken Pose Estimation Based on Transfer Learning. *Animals* **2024**, *14*, 1774. [CrossRef]
30. Neubeck, A.; Gool, L.V. Efficient Non-Maximum Suppression. In Proceedings of the International Conference on Pattern Recognition (ICPR), Hong Kong, China, 20–24 August 2006; pp. 850–855.
31. Lin, T.Y.; Goyal, P.; Girshick, R.; He, K.; Dollár, P. Focal Loss for Dense Object Detection. In Proceedings of the IEEE International Conference on Computer Vision (ICCV), Venice, Italy, 22–29 October 2017; pp. 2980–2988.
32. Lin, T.Y.; Maire, M.; Belongie, S.; Hays, J.; Perona, P.; Ramanan, D.; Dollár, P.; Zitnick, C.L. Microsoft COCO: Common Objects in Context. In *Computer Vision—ECCV 2014, Proceedings of the 13th European Conference, Zurich, Switzerland, 6–12 September 2014*; pp. 740–755.
33. Deng, J.; Dong, W.; Socher, R.; Li, L.J.; Li, K.; Fei-Fei, L. ImageNet: A large-scale hierarchical image database. In Proceedings of the IEEE Conference on Computer Vision and Pattern Recognition (CVPR), Miami, FL, USA, 20–25 June 2009; pp. 248–255.
34. Atzori, L.; Iera, A.; Morabito, G. The Internet of Things: A survey. *Comput. Netw.* **2010**, *54*, 2787–2805. [CrossRef]

35. Ministerio de la Presidencia. Real Decreto 692/2010, de 20 de Mayo, Por el que se Establecen las Normas Mínimas Para la Protección de los Pollos Destinados a la Producción de Carne y se Modifica el Real Decreto 1047/1994, de 20 de mayo, Relativo a las Normas Mínimas para la Protección de Terneros. In *BOE*; Ministerio de la Presidencia, Gobierno de España: Madrid, Spain, 2010; Núm. 135, Sección I; pp. 47986–47995.
36. Tzutalin. LabelImg. Git Code, 2015. Available online: https://github.com/tzutalin/labelImg (accessed on 16 April 2023).
37. Kaliappan, V.K.; V, M.S.; Shanmugasundaram, K.; Ravikumar, L.; Hiremath, G.B. Performance Analysis of YOLOv8, RCNN, and SSD Object Detection Models for Precision Poultry Farming Management. In Proceedings of the 2023 IEEE 3rd International Conference on Applied Electromagnetics, Signal Processing, & Communication (AESPC), Bhubaneswar, India, 24–26 November 2023; pp. 1–9. [CrossRef]
38. Li, M.; Zhang, Z.; Lei, L.; Wang, X.; Guo, X. Agricultural Greenhouses Detection in High-Resolution Satellite Images Based on Convolutional Neural Networks: Comparison of Faster R-CNN, YOLO v3 and SSD. *Sensors* **2020**, *20*, 4938. [CrossRef]
39. ah Kim, J.; Sung, J.Y.; ho Park, S. Comparison of Faster-RCNN, YOLO, and SSD for Real-Time Vehicle Type Recognition. In Proceedings of the 2024 IEEE International Conference on Deep Learning for Autonomous Vehicles (DLAV), Vienna, Austria, 5–7 June 2024; pp. 50–60.
40. Sapkota, R.; Ahmed, D.; Karkee, M. Comparing YOLOv8 and Mask R-CNN for instance segmentation in complex orchard environments. *Artif. Intell. Agric.* **2024**, *13*, 84–99. [CrossRef]
41. Ultralytics. *Train Settings—Ultralytics Documentation*; Ultralytics: Ballenger Creek, MD, USA, 2024.
42. Lempitsky, V.; Zisserman, A. Learning To Count Objects in Images. In Proceedings of the European Conference on Computer Vision (ECCV), Crete, Greece, 5–11 September 2010.
43. Kumar, K.; Kumar, P.; Kshirsagar, V.; Bhalerao, R.H.; Shah, K.; Vaidhya, P.K.; Panda, S.K. A Real-Time Object Counting and Collecting Device for Industrial Automation Process Using Machine Vision. *IEEE Sens. J.* **2023**, *23*, 13052–13065. [CrossRef]
44. Zhang, Y.; Zhou, D.; Chen, S.; Gao, S.; Ma, Y. Single-Image Crowd Counting via Multi-Column Convolutional Neural Network. In Proceedings of the IEEE Conference on Computer Vision and Pattern Recognition (CVPR), Las Vegas, NV, USA, 27–30 June 2016; pp. 589–597.
45. Grinberg, M. *Flask Web Development: Developing Web Applications with Python*; O'Reilly Media, Inc.: Sebastopol, CA, USA, 2018.

agriculture

MDPI

Article

FRESH: Fusion-Based 3D Apple Recognition via Estimating Stem Direction Heading

Geonhwa Son [1], Seunghyeon Lee [2] and Yukyung Choi [2,*]

[1] Department of Artificial Intelligence and Robotics, Sejong University, Seoul 05006, Republic of Korea; ghson@rcv.sejong.ac.kr

[2] Department of Convergence Engineering for Intelligent Drone, Sejong University, Seoul 05006, Republic of Korea; shlee@rcv.sejong.ac.kr

* Correspondence: ykchoi@rcv.sejong.ac.kr

Abstract: In 3D apple detection, the challenge of direction for apple stem harvesting for agricultural robotics has not yet been resolved. Addressing the issue of determining the stem direction of apples is essential for the harvesting processes employed by automated robots. This research proposes a 3D apple detection framework to identify stem direction. First, we constructed a dataset for 3D apple detection that considers the 3-axis rotation of apples based on stem direction. Secondly, we designed a 3D detection algorithm that not only recognizes the dimensions and location of apples, as existing methods do, but also predicts their 3-axis rotation. Furthermore, we effectively fused 3D point clouds with 2D images to leverage the geometric data from point clouds and the semantic information from images, enhancing the apple detection performance. Experimental results indicated that our method achieved AP@0.25 89.56% for 3D detection by considering apple rotation, surpassing the existing methods. Moreover, we experimentally validated that the proposed loss function most effectively estimated the rotation among the various approaches we explored. This study shows the effectiveness of 3D apple detection with consideration of rotation, emphasizing its potential for practical application in autonomous robotic systems.

Keywords: smart agriculture; deep learning; 3D detection; stem direction estimation

1. Introduction

Apples are one of the most consumed fruits in the world [1], and ensuring their quality is essential to meet consumer requirements. Producing high-quality apples requires continuous management throughout the growing process and harvesting methods that preserve the quality of the fruit. The various stages of apple cultivation are traditionally managed by humans and rely heavily on human labor. Additionally, fruits like apples need different management depending on the growth stage [2] and demand significant labor, especially during the harvest season. To address these issues, research is being conducted on the automation of fruit cultivation [3].

In order to automate apple cultivation, extensive research has been conducted on the entire process, from pruning and harvesting to evaluating the quality of the harvested crop. Automation research has grown beyond simple algorithms to involve tasks that require human judgment, including the assessment of crop quality, the analysis of growth conditions, and the monitoring of fruit development [4–7]. In particular, advancements in image-based computer vision technology [8–10] have allowed the automation of tasks such as monitoring pests and diseases and measuring crop maturity [11,12]. All of these tasks require fruit location accuracy. The precision of these tasks has been significantly enhanced by employing approaches such as detection and segmentation for estimating the locations of fruits in an image [13–17].

However, the harvesting process in apple cultivation necessitates additional considerations. The standard method of harvesting apples is to wrap them and twist them in a direction centered on the stem, to separate them from the branch [18]. Without considering the stem direction, the apple could crash into the others, reducing its quality, or fail to detach from the branch during the harvesting process. Thus, when detecting apples, it is crucial not just to recognize their location but also to verify the direction of their stems. Therefore, automating apple cultivation requires recognition technology that accounts for the actual harvesting methods of fruits. These topics represent extensive research areas in agriculture [19], covering apples [3] as well as other fruits like strawberries [20,21], melons [22], and tomatoes [23–25].

The existing approach [26] estimates the direction of the apple stem by considering simple rotation on a 2D plane. However, this has limitations, in that it generates arbitrary 3D data based on results predicted from the image, which fails to reflect the actual volume. Furthermore, it is only possible to compute the rotation of each axis separately rather than account for the actual 3-axis rotation toward the stem. To address these limitations, this paper aimed to perform 3D apple detection using the stem direction, to accurately recognize apples in a real orchard environment.

Although 3D detection of apples has been studied recently (e.g., [27]), it remains limited, since it overlooks the stem direction, which is important during harvesting. Similarly, apple datasets [28–30] do not contain information regarding the direction of apples in an orchard environment. Existing 3D apple detection datasets offer 3D bounding box data at the scene level within orchard environments. However, effective 3D detection requires either paired RGB-point cloud data or separate point cloud data, which divide the entire scene into smaller portions. Therefore, we propose a novel 3D apple detection framework and dataset that consider stem direction during the harvesting process. Figure 1 shows a flowchart representing the work conducted in this study.

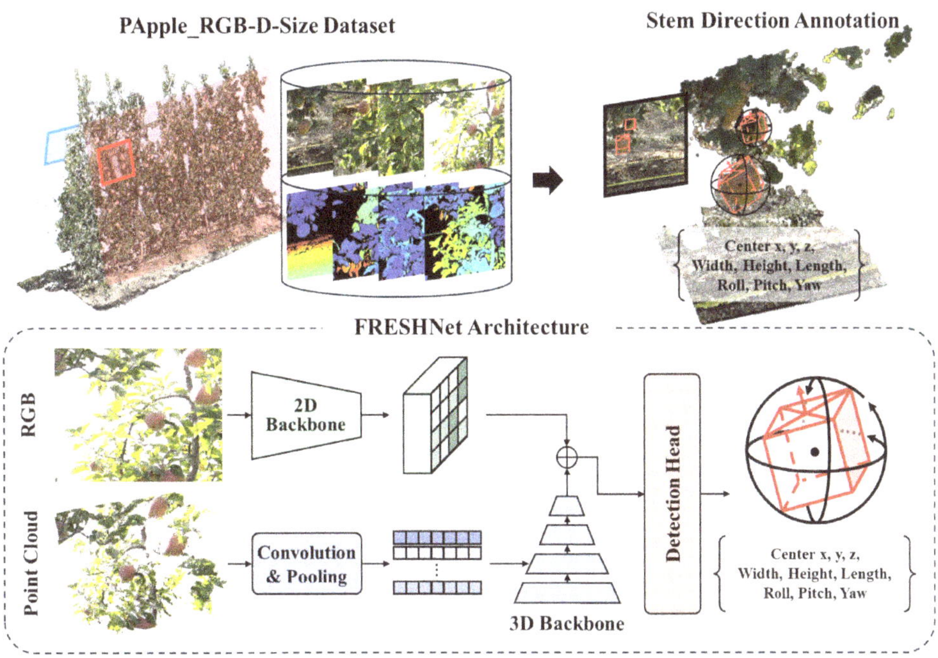

Figure 1. Overall flowchart of the 3D apple detection task considering the stem direction. Reprocessing the PApple_RGB-D-Size [30] dataset to account for stem direction and designing a 3D detection method that takes into account 3-axis rotation.

First, we reprocessed existing apple datasets, Fuji-SfM [28,29] and PApple_RGB-D-Size [30]. The 3D apple detection datasets were reconstructed by annotating the stem direction, considering the potential application. We also designed a 3D detection framework named FRESHNet, introducing a loss function and modified fusion strategy.

Through extensive and comprehensive experiments, we demonstrated the potential applications of our restructured dataset and the effectiveness of our methodology. The main contributions of our proposed framework can be summarized as follows:

- We established a dataset that includes 3-axis rotation information of the stem direction, designed to account for current harvesting techniques. The dataset provides 3-axis rotation information to facilitate 3D apple detection, considering current harvesting techniques.
- To the best of our knowledge, the proposed approach, named FRESHNet, is the first to introduce a loss function that considers apple 3-axis rotation.
- We demonstrated the utility of the established dataset and experimentally validated the effectiveness of FRESHNet.

2. Materials and Methods

2.1. Dataset Reprocessing

The rolling method [18] is a standard harvesting technique that involves rotating apples around their stems to remove them from the branch. In addition, branch pruning and other activities need to know the stem direction; thus, considering apple stems in recognition systems is necessary for automated cultivation. Nevertheless, prior research has not adequately accounted for the process of harvesting apples. Considering the standard apple picking method, we performed additional processing on an existing dataset for 3D perception of stem direction.

This study employed the PApple_RGB-D-Size dataset [30] collected from real orchards in October 2018 and July 2020, to account for orchard environments. The PApple_RGB-D-Size dataset is composed of images and corresponding depth maps. It provides ground-truth diameter measurements from real apples, along with both modal segmentation masks for the visible regions and amodal segmentation masks that account for occlusions such as leaves. In this study, we reprocessed the PApple_RGB-D-Size dataset to annotate the stem direction, utilizing semantic information from the images and point clouds derived from both images and depth maps.

2.1.1. 3D Reconstruction

In this section, we aimed to generate dense point clouds aligned with the images. To achieve this, intrinsic parameters were required, to reconstruct point clouds from the PApple_RGB-D-Size dataset. Accordingly, we obtained intrinsic parameters from images by applying SfM and utilized these parameters to reconstruct 3D point clouds from the images and depth maps. We also applied filtering techniques to eliminate unnecessary noise during the apple detection process. To ensure the accuracy of the intrinsic parameters, we validated them by verifying the diameters of apple instances in the reconstructed 3D point clouds. We then automatically generated an initial 3D bounding box using the segmentation mask and diameter information of the existing dataset, followed by manual bounding box adjustment and stem direction annotation. A description of this process is presented in Section 2.1.2.

2.1.2. Stem Direction Annotation

In order to obtain information about the stem direction of the apples, we manually annotated 3D bounding boxes and stem directions. Due to self-occlusion, where parts of the apple were partially obscured by other parts of the apple, the reprocessed point cloud was unable to acquire information about the back of the apple. Therefore, we determined the volume information based on the actual diameter, automatically generated an initial 3D bounding box from the segmentation mask position, and refined it to accurately annotate the location of the 3D bounding box within the point cloud, as well as the stem direction. The step-by-step procedure for annotating the stem direction is shown in Figure 2.

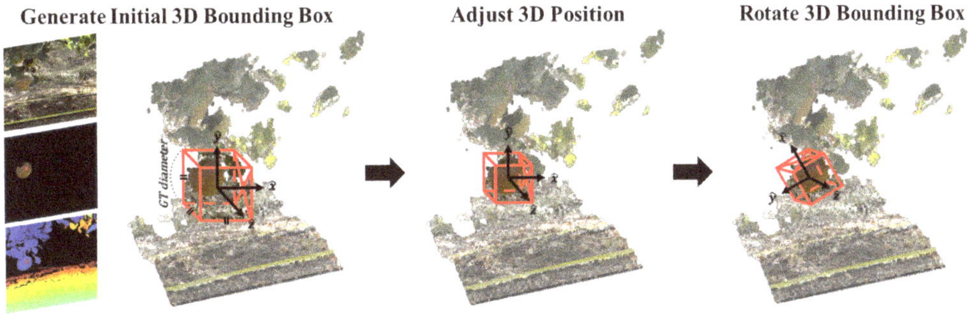

Figure 2. The annotation process for 3D bounding boxes and stem directions.

To annotate the 3D bounding box and stem direction, we followed these procedures: (1) Generate the initial 3D bounding box. To generate the initial 3D bounding box, we found the smallest value of the depth map corresponding to the modal segmentation mask area; that is, the nearest surface. Next, to set the center of the apple, we placed a point on the z-axis at a distance corresponding to the radius of the apple, and used the diameter value to set the bounding box size. (2) Adjust the 3D position. We modified the three-dimensional location of the apple to ensure the point cloud aligned with the front of the bounding box. (3) Rotate the 3D bounding box. To represent the stem direction, we rotated the bounding box around the three axes, so that the x-axis of its body frame aligned with the stem. Throughout these processes, the labelCloud [31] tool (version 1.1.1) was employed to manually adjust the 3D bounding box and provide information about the stem direction. And we used the corresponding image as a reference for the stem direction, excluding apples whose leaves and branches made it difficult to determine the stem direction.

Next, we projected the 3D bounding box to generate a 2D bounding box. In order to prevent cases in which the 2D bounding box fit loosely around the apple, 3D rotation was not applied during this step. This is demonstrated in Figure 3, which illustrates that the 2D bounding box that was obtained by applying 3D rotation (Figure 3b) was not as tightly bound as without applying 3D rotation (Figure 3a). In summary, the dataset consists of 3D point clouds and image pairs of apples in an orchard, and the label information includes parameters for both the 3D bounding box (center coordinates, dimensions, and rotations) and the 2D bounding box (minimum and maximum pixel coordinates).

2.1.3. Dataset Analysis

The reprocessed dataset included 3381 images and point cloud pairs, divided into training, testing, and validation sets based on the split of the original dataset. The resulting dataset included 1995 for training, 685 for testing, and 701 for validation. With this split, the training set consisted of the data from the west side orchard of the PApple_RGB-D-Size dataset, while the east side data belonged to the test and validation sets, maintaining a distinct separation between the training and evaluation data. Additionally, the total number of annotated apples in the entire dataset was 9044, and additional statistics related to the organization of the data can be found in Table 1.

Figure 3. Comparison of 2D bounding boxes with and without applying rotation. (**a**) Illustrates examples of 2D bounding boxes projected without 3D rotation, while (**b**) presents examples of 2D bounding boxes after applying 3D rotation. The colored side of the box corresponds to the direction of the stem.

Table 1. Summary of the dataset. This table provides the number of data pairs, the number of apple objects, and the average diameter of apples for each split.

	Num. of Images	Num. of Apples	Diameter [mm] Average (Variance)
Train	1995	4976	67.30 (0.215)
Test	685	2016	71.41 (0.181)
Val	701	2052	69.95 (0.194)
Total	3381	9044	68.82 (0.206)

Figure 4 is a visualization of the KDE distribution of the stem direction of the annotated apples, and it shows that the upright direction had the densest distribution, but there was a wide range of other stem directions. Apple harvesting necessitates identifying, cutting, or rolling the stem according to its direction. However, as shown in Figure 4, the direction of the stem is inconsistent, making it unfeasible to conduct batch work based on the stem of apples during crop cultivation. Therefore, apple detection that considers the stem direction of the fruit tree is necessary to automate the entire process of cultivation.

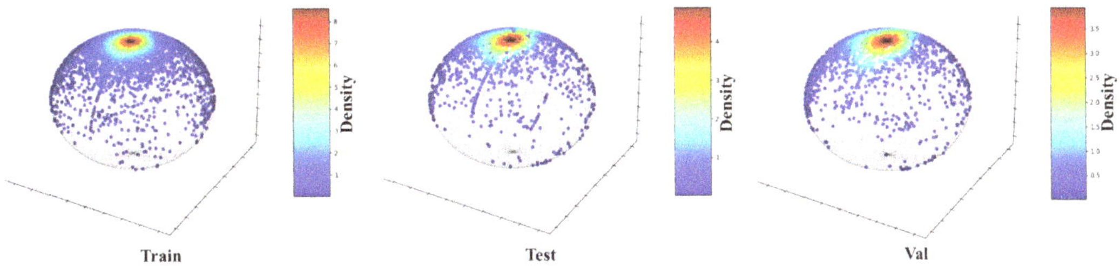

Figure 4. KDE distribution of stem directions on a sphere for reprocessed dataset. Color intensity shows the direction density: red for high density and blue for low density, with a spherical grid for spatial reference.

2.2. Preliminary

Research on 3D object detection has been extensively explored in both indoor and outdoor autonomous driving scenarios. Recognizing obstacles on the road is essential in outdoor autonomous driving, making research on 3D perception critical. The common assumption in this domain is that objects are located on a flat plane. Since distance information is crucial, researchers often project a 3D point cloud onto a plane from a birds-eye-view perspective, thereby suppressing the height dimension [32,33]. While this assumption reduces the computational load, it poses a challenge for accurately estimating the 3D position of objects when their heights are not uniform, due to the loss of height information.

In contrast to outdoor autonomous driving scenarios that suppress the height dimension, indoor autonomous robots are required to recognize the locations of various objects with varied height dimensions [34,35]. In agricultural environments, where estimating the position of fruits is important, considering the height information of each fruit is crucial. Therefore, we adopted the indoor approach for 3D object detection.

Based on whether datasets provide rotation angles, researchers have divided research on 3D perception in indoor environments into two bounding box representation methods: the Axis-aligned bounding box (AABB) method, which does not consider rotation, and the oriented bounding box (OBB) method, which only estimates the heading angle around one axis. The absence of datasets considering 3-axis rotation in 3D perception research has impeded the advancement of approaches for predicting such rotations. This research required both a dataset that provided 3-axis information and an algorithm capable of predicting 3-axis rotation, as it was important to consider the 3-axis rotation for the apple stem direction. Therefore, utilizing the reprocessed dataset described in Section 2.1, we designed a framework that, in contrast to prior research, incorporates 3-axis rotation to determine the direction of apple stems.

2.3. FRESHNet

FRESHNet, a framework for axial 3D apple detection, was constructed by adopting TR3D [36], the SOTA method for 3D detection algorithms. We designed the network by fusing high-level features extracted from point clouds and images, respectively, for apple detection. The architecture of FRESHNet is illustrated in Figure 5.

The input data consist of a point cloud and an image. A 3D backbone processes the points, while a 2D backbone and FPN process the image to extract features. The 3D features given by the 3D backbone are fused with image features in the 3D neck, resulting in a feature that integrates information from both the image and points. Details of the fusion method are provided in Section 2.3.1. The detection head uses this fusion feature as input to predict the class, bounding boxes, and 3-axis rotation. The output is a set of 3D object bounding boxes $(x, y, z, w, h, l, roll, pitch, yaw)$, where (x, y, z) represent the center coordinates of the bounding box; (w, h, l) indicate its width, height, and length; and $(roll, pitch, yaw)$ denote the rotation angles around the x, y, and z axes, respectively.

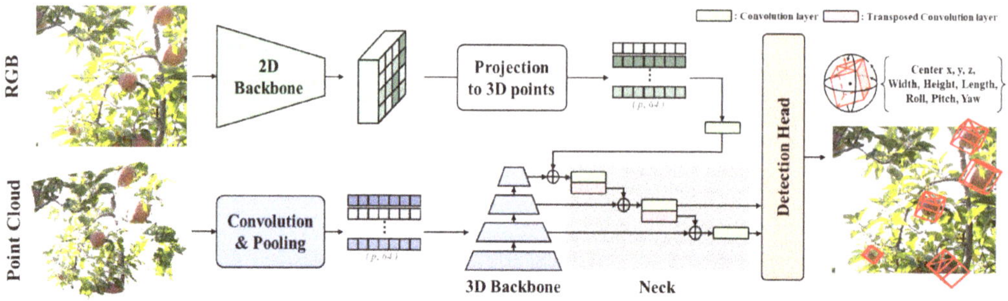

Figure 5. FRESHNet architecture. A 3D apple detection network architecture for predicting the direction of stems and effectively integrating image and point cloud features.

2D and 3D Backbone. The backbone extracts high-level features from the input point cloud and image. We use the ResNet50+FPN network for the image input, which is a network pretrained on a 2D object detection task. For the point cloud input, we use MinkResNet [37], which replaces all convolutions in ResNet with sparse convolutions. A sparse convolution is a structure that processes point clouds effectively by operating only on the voxels containing points, without utilizing the empty space generated during the voxelization of raw point clouds. This approach enables the processing of point clouds, without increasing the sparsity of the input data.

Neck. The precise center point of an object cannot be known only from the input points, as point clouds are designed to provide information about the surfaces of objects. To achieve this, the decoder applies transposed convolution to iteratively upsample the input points, thereby expanding the points to create denser regions. We then connect the resulting multi-level features to the next detection head.

Detection Head. The detection head operates with two parallel sparse convolution layers. Its output consists of classification probabilities, box predictions, and rotation angle predictions.

2.3.1. Multimodal Feature Fusion

Three-dimensional point clouds have a sparse nature compared to images, due to their inherent data properties. Point clouds provide geometric data on objects but are less successful at conveying contextual information compared to images. While images provide semantic and contextual information, they struggle to capture the precise geometric details of an object. Therefore, to enhance 3D apple recognition, we designed a network that effectively integrates the complementary elements of both data types. Previous research has employed the summation of image features provided via an image backbone as input to a 3D backbone. To optimally leverage the individual features of the two modalities, we fuse the image features into the first 3D feature map transmitted from the backbone to the neck.

The output features of the 2D backbone of an RGB image are used for feature fusion. However, since the output feature map is at the image level, it cannot be fused with 3D data features directly. The feature map of the images is transformed into a 3D feature map for fusion. First, the 3D point cloud data are projected onto the image dimension using the intrinsic camera parameters provided by the dataset. Next, the image features corresponding to each projected point are sampled and combined.

An image sparse tensor T_{I3D} is generated in 3D data format by combining image features that match points and converting them to a sparse tensor. As described in Equation (1), we add a convolution block to T_{I3D} in order to obtain 3D image features. To fuse the 3D data with the image features in an effective manner, we fuse the first block feature F_{B3D}

that is transmitted to the neck between the four levels of backbone blocks extracted from the 3D backbone network.

$$F_{I3D} = ReLU(BN(Conv(T_{I3D}))) \tag{1}$$

The MinkowskiReLU, MinkowskiBN, and MinkowskiConv structures in the Minkowski are consistent with the 3D backbone network architecture. According to Equation (1), the 3D image feature F_{I3D} of T_{I3D} is extracted. As shown in Equation (2), the fused feature F_{fusion} is generated through element-wise operations with multi-modality features.

$$F_{fusion} = F_{B3D} + F_{I3D} \tag{2}$$

We show that the structure of F_{fusion} efficiently transfers semantic information from the images to the 3D data, resulting in an improved performance compared to previous feature fusion algorithms [36], as shown in the experimental Section 2.

2.3.2. Training Loss

BBox Loss. BBox loss estimates the center coordinates (x, y, z) and dimension (w, h, l). The BBox loss is defined as follows:

$$\mathcal{L}_{bbox} = 1 - IoU + \frac{\rho^2(b, \hat{b})}{c^2} \tag{3}$$

As shown in Equation (3), $\mathcal{L}_{bbox}$ uses the DIoU loss, where b and $\hat{b}$ denote the ground truth (GT) box and the predicted box, respectively. IoU calculates the overlapping region between the two boxes. We adopt an Axis-aligned IoU approach that considers only the center point and size of the boxes. We calculate the volumes of both b and $\hat{b}$. Next, we determine the maximum and minimum coordinates of the overlapping region to calculate the intersection area. The IoU is calculated by dividing the intersection area by the union area of the two boxes. $\rho^2(b, \hat{b})$ denotes the Euclidean distance between the centers of b and $\hat{b}$. c^2 represents the diagonal length of the smallest box enclosing both boxes.

Stem direction Loss. The model predicts the rotation angles for the 3 axes, which are roll, pitch, and yaw. The aim is to find the direction of the stem for apple harvesting, leveraging these angles. Therefore, we designed the stem direction loss to guide the predicted 3-axis rotation angles to effectively learn the stem direction. First, we calculate the stem direction vector. Given that the stem direction in the dataset corresponds with the x-axis of the bounding box, we transform the GT 3-axis rotation angle $A = (roll, pitch, yaw)$ and the predicted 3-axis rotation angle $\hat{A} = (\hat{roll}, \hat{pitch}, \hat{yaw})$ into a rotation matrix R and $\hat{R}$, respectively. We calculate this using the x-axis unit vector u_x, as shown in Equation (4).

$$v = u_{stem}@R, \hat{v} = u_{stem}@\hat{R} \tag{4}$$

The stem direction vectors v and $\hat{v}$ are obtained by multiplying the x-axis unit vector u_x with each of R and $\hat{R}$. The proximity of v to $\hat{v}$ enhances the accuracy of the model in predicting the stem direction. To achieve this, we formulate the angle loss to maximize the dot product of the two vectors, targeting a value near 1. The stem direction loss is given by the following Equation (5).

$$\mathcal{L}_{stem} = 1 - v \cdot \hat{v}, \mathcal{L}_{stem} \in [0, 2] \tag{5}$$

The loss function $\mathcal{L}_{stem}$ is optimized to minimize the loss by guiding the dot product of the two vectors to approach 1. This enables the model to precisely predict the 3-axis rotation angles that represent the direction of the stem. The total loss, including $\mathcal{L}_{stem}$, is defined as shown in Equation (6).

$$\mathcal{L} = \frac{1}{n_{pos}}(\mathcal{L}_{cls}(p, \hat{p}) + \mathcal{L}_{bbox}(b, \hat{b}) + \mathcal{L}_{stem}(A, \hat{A})) \tag{6}$$

The classification loss $\mathcal{L}_{cls}(p, \hat{p})$ is focal loss, where p and $\hat{p}$ denote the GT class and the predicted class, and n_{pos} denotes the number of positive samples of the prediction.

2.4. Training Details

We trained and tested on a single NVIDIA A100 GPU with 40 GB of memory. The operating system was Ubuntu 18.04. Our network trained end-to-end from scratch based on the experimental settings of TR3D. The optimizer was the Adam model, with a learning rate of 10^{-4}. We implemented the 3D object detection framework based on MMDetection3D.

2.5. Evaluation Metrics

Evaluation for 3D detection. To evaluate the performance of the 3D apple detection network, we utilized the average precision (AP) and average recall (AR), which are commonly used metrics in 3D object detection for indoor environments. We calculated the AP using precision (P), which is the ratio of correctly detected objects, and recall (R), which is the detection rate among all objects. The evaluation of detection performance typically uses AP due to the trade-off between P and R. It is defined as the area under the curve in a P–R graph. AR is the average of R, and in this paper, it corresponded to the R performance for the single apple class. AR measures the capability of detecting all objects and aims to evaluate undetected instances. Improving the AP performance to ensure the quality of apples through high-precision detection is important for agricultural automation. Enhancing the AR performance is also crucial, to minimize non-detection cases and maximize production. AP and AR were evaluated by calculating the intersection over union (IoU) between the predicted 3D bounding box and the GT 3D bounding box. A sample with an IoU over the threshold was defined as a true positive (TP), but a sample with an IoU below the threshold was regarded as a false positive (FP), and failure to identify an existing object was categorized as a false negative (FN). In the 3D detection task, previous research has typically employed thresholds of @0.25 and @0.5. Detailed mathematical formulas are given in Equations (7)–(10).

$$P = \frac{TP}{TP + FP} \tag{7}$$

$$R = \frac{TP}{TP + FN} \tag{8}$$

$$AP = \int_0^1 P \cdot R\, dR \tag{9}$$

$$AR = \frac{1}{N} \sum_{i=1}^{N} R_i \tag{10}$$

In Equation (10), N is the total number of detected target categories. In this paper, we detected a singular *APPLE* class, so N equaled 1.

IoU with stem direction. In this context, evaluating the IoU between two boxes was similar to traditional 3D detection; however, assessing the predicted bounding box B_{pred} and the GT bounding box B_{gt} necessitates calculating the IoU considering 3-axis rotation. To calculate the IoU for these volumetric boxes, we first find the eight corner points for both B_{pred} and B_{gt}. The corner points are subsequently adjusted to reflect the rotation angles $(yoll, pitch, yaw)$ across the 3 axes, indicating their locations when the box is rotated. The obtained corner coordinates are connected to create a volumetric bounding box, and the IoU is determined by assessing the overlap between the two volumetric boxes.

Evaluation for rotation. This work aimed to predict the stem direction of apples, enabling adaptive recognition of apples with various directions in agricultural task automation. We analyzed true positive cases of previously recognized items to evaluate the predicted 3-axis rotation. We computed the average of the error for the predicted values for each axis and the quaternion distance for all parameters. The quaternion distance was used for a comprehensive evaluation of the 3-axis rotation by expressing the GT and predicted rotations in quaternions, which is shown in Equation (11). The average angular error (in degrees) between the predicted and GT values was calculated for each axis, enabling the assessment of the prediction accuracy on all axes.

$$QuaternionDistance = ||q - \hat{q}|| \quad (11)$$

$\hat{q}$ and q represent the GT and predicted rotation values, respectively, expressed as quaternions, while $||\cdot||$ denotes the Euclidean norm.

3. Results

In this study, we verified our 3D apple detection network, FRESHNet, using stem direction. In Section 3.1, we validate the accuracy of the apple detector considering 3-axis rotation and show that it is possible to make accurate predictions with a rotation-conscious detection network. In Section 3.2, we evaluate the 3D apple detection performance and rotation error by varying the loss function for rotation learning.

3.1. Results of 3D Object Detection

This experiment demonstrated the 3D detection performance of our proposed framework, which estimates stem direction by predicting the 3-axis rotation of apples. Additionally, the FPS (the number of scenes processed per second) was measured to assess the real-time detection capability for 3D detection. The experimental results are shown in Table 2.

Table 2. Results of 3D detection for reprocessed dataset. Comparison between existing methods (axis-aligned) and the proposed method considering 3-axis rotation (3-axis).

Setting		Method	AP [%]		AR [%]		FPS
Axis	Input		@0.25	@0.5	@0.25	@0.5	
Axis aligned	P	VoteNet [35]	71.99	33.25	83.28	49.85	6.58
		FCAF3D [34]	84.60	48.62	94.94	68.11	9.79
		TR3D [36]	85.76	57.92	95.09	71.83	17.13
		Point-GCC+TR3D [38]	82.30	37.16	92.01	53.76	17.56
	I+P	ImVoteNet [39]	77.21	41.87	85.91	57.19	9.02
		TR3D+FF [36]	85.88	57.03	95.78	71.48	12.92
3-axis	P	rot-Point-GCC+TR3D	86.34	75.13	97.62	86.31	17.56
	I+P	rot-TR3D+FF	89.18	77.22	98.91	88.59	12.92
		FRESHNet (Ours)	**89.56**	**78.29**	**99.16**	**89.19**	13.17

Axis settings are divided into Axis-aligned and 3-axis, based on the consideration of 3-axis rotation. Axis-aligned methods predict 3D bounding boxes in the form (x, y, z, w, h, l). The 3-axis predictions proposed in this work predicted the following parameters: (x, y, z, w, h, l, roll, pitch, yaw). In our evaluation of the latest 3D object detection method, we selected Point-GCC as a comparison model. This model performed pre-training on a custom dataset through self-supervised learning, followed by fine-tuning. To compare the performance of considering 3-axis rotation with existing methods, we added rot-Point-GCC+TR3D, rot-TR3D+FF as an additional comparative model, which included our proposed stem direction loss in the Point-GCC+TR3D [38], TR3D+FF [36] framework.

Firstly, the FRESHNet achieved outstanding performance for all metrics. Notably, rot-TR3D+FF achieved a significant performance improvement of 20.19% for AP@0.5 compared to TR3D+FF. Similarly, rot-Point-GCC+TR3D showed a 37.97% enhancement in AP@0.5 relative to Point-GCC+TR3D, with notable improvements in AR also measured. This suggests that identifying the stem direction of apples provides a more precise comprehension of their location, particularly in the context of harvesting. In addition, FRESHNet showed an improvement of 0.38% for AP@0.25 and 1.07% for AP@0.5 compared to rot-TR3D+FF. This demonstrated the effectiveness of the fusion of semantic information from images with the point cloud. Furthermore, our method achieved an FPS of 13.17, the fastest among the image fusion methods, demonstrating improvements in real-time detection performance with a simple yet effective image fusion approach. Additionally, we visualized a PR-curve for each method, as shown in Figure 6.

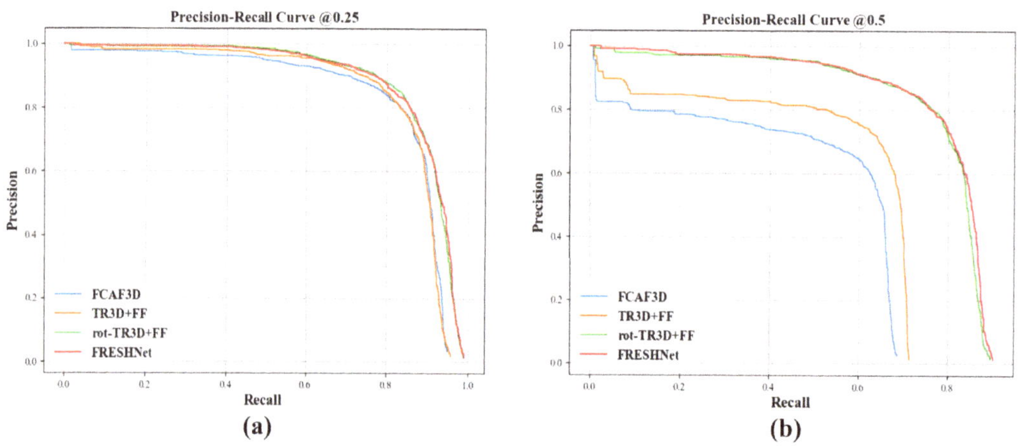

Figure 6. PR-curve for 3D apple detection results of Axis-aligned and 3-axis methods. (**a**) is PR-curve of FCAF3D, TR3D+FF, rot-TR3D+FF, and FRESHNet at AP@0.25, (**b**) is PR-curve of FCAF3D, TR3D+FF, rot-TR3D+FF, and FRESHNet at AP@0.5.

This curve illustrates the precision in relation to a varying recall. At AP@0.25, the Axis-aligned methods showed a precision trend similar to the 3-axis methods as the recall increased. In contrast, at AP@0.5, the Axis-aligned methods not only reached a recall of 71.83% but also exhibited a significant decrease in precision as the recall increased. On the other hand, the 3-axis methods maintained a relatively stable and accurate apple detection, even as the recall increased.

3.2. Ablation Study

We performed an ablation study to validate the impact of the stem direction loss proposed in this paper on the 3D object detection results and to analyze the rotation prediction performance. Table 3 shows the result of the ablation study, demonstrating the effect of the various rotation loss types on the 3D object detection precision. Table 4 presents a comparative analysis of the rotation prediction performance for each type of rotation loss.

Table 3. Experimental results of the ablation study. We used the inner product of the stem direction and compared it to the mean square error (MSE) loss for Euler angles and quaternion-based rotation representations.

Rotation Loss	AP [%]		AR [%]	
	@0.25	@0.5	@0.25	@0.5
Euler+MSE	88.59	77.79	98.76	88.24
Quaternion+MSE	88.80	76.39	99.06	87.65
Stem direction (Ours)	**89.56**	**78.29**	**99.16**	**89.19**

Table 4. Rotation error in the ablation study. We measured the rotation prediction error for our methodology utilizing the interior of the stem direction versus our methodology utilizing the MSE.

Rotation Loss	Rotation Error ↓							
	Quaternion Distance		Roll [*Degree*]		Pitch [*Degree*]		Yaw [*Degree*]	
	@0.25	@0.5	@0.25	@0.5	@0.25	@0.5	@0.25	@0.5
Euler+MSE	0.01301	0.01295	58.93	59.23	21.86	22.49	**33.01**	**32.79**
Quaternion+MSE	0.01277	0.01285	61.43	61.18	**21.34**	**21.86**	33.48	32.80
Stem direction	**0.00717**	**0.00709**	**1.21**	**1.15**	23.65	23.88	35.12	34.43

The Euler+MSE loss computed the error between the rotations $(roll, pitch, yaw)$ and their corresponding GT rotation. Quaternion+MSE means converting the rotations into quaternion format instead of using $(roll, pitch, yaw)$ directly, then calculating the MSE between the GT and predicted quaternions. Lastly, the stem direction denotes the stem direction loss proposed in this paper.

The experimental results are presented in Table 3, which demonstrate that the stem direction achieved the best performance across all metrics for 3D object detection. This result suggests that the stem direction loss, which effectively predicted stem direction using 3-axis rotation, performed better for apple detection than Euler+MSE loss and Quaternion+MSE loss, which simply aimed to reduce the error with the GT.

As shown in Table 4, stem direction loss achieved the lowest error rate for both quaternion distance and roll prediction. In contrast, the pitch and yaw error rates were lowest with the Quaternion+Mse and Euler+MSE methods, respectively. However, when comparing with stem direction loss, the differences in pitch and yaw were within a small range of approximately 2°. Significant performance improvements were achieved for quaternion distance and roll error, whereas the differences for pitch and yaw remained minimal. Notably, stem direction loss had the lowest error for quaternion distance, which compares rotation performance as a single vector considering all 3 axes. This suggests that using the loss function for predicting rotations based on stem direction is an effective approach.

Figure 7 provides a qualitative assessment of the efficacy of the stem direction loss design. As seen in Figure 7b–d, the results trained with our proposed stem direction loss predicted the 3-axis rotation of the apples more effectively than those trained with Euler+MSE and Quaternion+MSE loss, thereby accurately estimating the stem direction. The results indicate that our stem direction loss function effectively estimated the direction of the apple stems. Consequently, we finally chose the stem direction loss function.

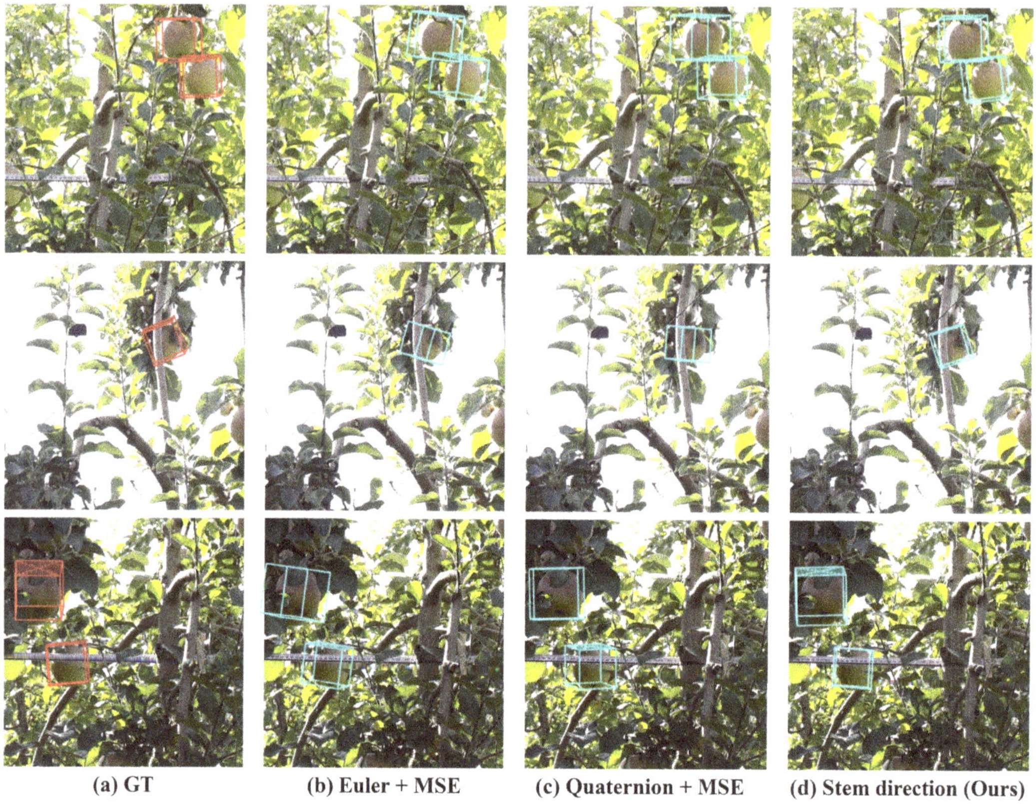

Figure 7. Qualitative results from an ablation study trained with various rotation loss functions. The colored side of the box corresponds to the direction of stem.

4. Discussion

In this section, we discuss the strengths and limitations of the proposed framework. First, our approach takes into account the 3-axis rotation of the stem, enabling the model to more precisely recognize the location of apples by considering various directions of the stem. Figure 8 shows that although FCAF3D [34] and TR3D+FF [36] detected the apples, they did not accurately estimate their dimension and location in relation to the GT. In contrast, both rot-TR3D+FF and FRESHNet could detect the apples more precisely by estimating the stem direction. FRESHNet predicts rotational bounding boxes by considering the stem direction, leading to tighter and more precise predictions of the apple dimensions and location compared to methods that overlook 3-axis rotation. Our approach can enhance the precision of apple 3D vision in automated robot harvesting in orchard environments.

Secondly, our approach considers both detection precision and inference speed. Achieving superior performance in 3D detection while balancing detection accuracy and speed is a challenge. Nonetheless, considering the importance of real-time performance in practical agricultural applications, we designed our algorithm to enhance the detection performance, while maintaining a real-time capability. Table 2 illustrates that, whereas the comparative methods showed a superior detection speed compared to our method, they performed worse in terms of detection precision. We achieved the fastest real-time performance among the image fusion methods. Ultimately, by optimally balancing the trade-off between detection accuracy and speed, we confirmed the potential applicability of our approach for real agricultural automation applications.

Figure 8. Qualitative results of 3D apple detection for axis-aligned method and 3-axis rotation method. The colored side of the box corresponds to the direction of stem.

Finally, as shown in Figure 9, our analysis found that our system encountered limitations under severe occlusion conditions. The dataset included instances where leaves and branches significantly concealed apples. Due to the absence of a specialized framework to address severe occlusions, our method had difficulty in instances of severe occlusion. To address this limitation, future research will aim to improve the model to be robust even under severe occlusion conditions, leveraging the visibility of our dataset for evaluation. This improvement aims to guarantee robust performance with the complex scenes of real-world scenarios.

Figure 9. Qualitative comparison with GT in severe occlusion environments. The colored side of the box corresponds to the direction of the stem.

5. Conclusions

In this paper, we aimed to achieve 3D apple detection by considering the stem direction for harvesting automation. We constructed a dataset that offered 3-axis rotation data aligned with the apple stem. In addition, we designed a 3D apple detection framework named FRESHNet that predicted 3-axis rotation to determine the stem direction, successfully leveraging the constructed dataset.

Our dataset and FRESHNet each have two advantages. Firstly, the dataset complements the existing 3D apple dataset that only provides dimension and location information by accurately reflecting the actual position of apples, considering their growth direction from the stem. Secondly, the dataset also includes processing data and annotations, which can be used in future 3D apple detection research. Next, the FRESHNet model addresses the absence of methods that fully leverage the 3-axis information offered by the dataset. We optimized a loss function to predict 3-axis rotations, taking the stem directions into account. Additionally, our proposed FRESHNet performs simple point cloud and image fusion to effectively convey semantic information about apples in an image. Comprehensive experiments demonstrated that our proposed FRESHNet method was significantly more accurate in apple 3D detection than approaches that did not consider rotation information. Furthermore, we showed the effectiveness of the FRESHNet method in predicting apple rotation by evaluating various rotation loss functions for rotation precision.

In summary, we performed 3D apple perception by considering autonomous harvesting methods in orchard environments, which was achieved through the construction of a dataset and algorithm. We believe that our proposed dataset and method can enhance automated harvesting robots by improving the accuracy and practicality of perception systems, effectively reflecting real apple harvesting techniques. We also expect that our deep learning based approach will simultaneously enable 3D apple detection and stem direction estimation for harvesting, making an importnant scientific contribution to automated systems. This study contributes to advancements in 3D object detection tasks and anticipates the effective implementation of computer vision tasks in agriculture, considering real-world working scenarios.

Author Contributions: Conceptualization, Y.C.; methodology, G.S. and S.L.; software, G.S. and S.L.; validation, S.L.; formal analysis, G.S.; investigation, G.S.; resources, Y.C.; data curation, S.L. and G.S.; writing—original draft preparation, G.S. and S.L.; writing—review and editing, S.L., G.S. and Y.C.; visualization, S.L. and G.S.; supervision, Y.C.; project administration, Y.C.; funding acquisition, Y.C. All authors have read and agreed to the published version of the manuscript.

Funding: This work was partly supported by an Institute of Information & communications Technology Planning & Evaluation (IITP) grant funded by the Korea government(MSIT) (RS-2023-00262891, Development of AI-based HD map building and crop image analysis for smart farm agricultural automation robots, 50%) and Unmanned Vehicles Core Technology Research and Development Program through the National Research Foundation of Korea (NRF) and Unmanned Vehicle Advanced Research Center (UVARC) funded by the Ministry of Science and ICT, the Republic of Korea (NRF-2023M3C1C1A01098408, 30%) and the Technology Innovation Program(RS-2024-00421828, Development of artificial intelligence software for unseen object manipulation that integrates prompt and situation-specific unseen object recognition and arbitrary gripper shape analysis through gripper self-observation, 20%) funded by the Ministry of Trade Industry & Energy (MOTIE, Republic of Korea).

Institutional Review Board Statement: Not applicable.

Data Availability Statement: Data underlying the results presented in this paper are not publicly available at this time but may be obtained from the authors upon reasonable request.

Conflicts of Interest: The authors declare no conflicts of interest.

References

1. Vasylieva, N.; Harvey, J. Production and trade patterns in the world apple market. *Innov. Mark.* **2021**, *17*, 16. [CrossRef]
2. Kuden, A.; Imrak, B.; Comlekcioglu, S.; Küden, A. Introductory Chapter: Apple Cultivation–Recent Advances. *Apple-Cultiv.-Recent Adv.* **2023**. [CrossRef]
3. Mhamed, M.; Zhang, Z.; Yu, J.; Li, Y.; Zhang, M. Advances in apple's automated orchard equipment: A comprehensive research. *Comput. Electron. Agric.* **2024**, *221*, 108926. [CrossRef]
4. Sheel, K.; Sharma, A. Intelligent Orchard monitoring: IoT integrated Fuzzy Logic based real-time apple disease prediction encompassing environmental factors. *J. Integr. Sci. Technol.* **2024**, *12*, 795–795. [CrossRef]
5. Sitharthan, R.; Rajesh, M.; Vimal, S.; Kumar, S.; Yuvaraj, S.; Kumar, A.; Raglend, J.; Vengatesan, K. A novel autonomous irrigation system for smart agriculture using AI and 6G enabled IoT network. *Microprocess. Microsyst.* **2023**, *101*, 104905.
6. Tian, Y.; Yang, G.; Wang, Z.; Wang, H.; Li, E.; Liang, Z. Apple detection during different growth stages in orchards using the improved YOLO-V3 model. *Comput. Electron. Agric.* **2019**, *157*, 417–426. [CrossRef]
7. Sekharamantry, P.K.; Melgani, F.; Malacarne, J. Deep learning-based apple detection with attention module and improved loss function in YOLO. *Remote Sens.* **2023**, *15*, 1516.
[CrossRef]
8. Dosovitskiy, A. An image is worth 16 × 16 words: Transformers for image recognition at scale. *arXiv* **2020**, arXiv:2010.11929.
9. Chen, L.C.; Zhu, Y.; Papandreou, G.; Schroff, F.; Adam, H. Encoder-decoder with atrous separable convolution for semantic image segmentation. In Proceedings of the European Conference on Computer Vision (ECCV 2018), Munich, Germany, 8–14 September 2018; pp. 801–818.
10. Zhang, W.; Zhao, W.; Li, J.; Zhuang, R.; Sun, H.; Xu, Y.; Li, C. CVANet: Cascaded visual attention network for single image super-resolution. *Neural Netw.* **2024**, *179*, 622–634. [CrossRef]
11. Singh, S.; Gupta, I.; Gupta, S.; Koundal, D.; Aljahdali, S.; Mahajan, S.; Pandit, A.K. Deep learning based automated detection of diseases from Apple leaf images. *Comput. Mater. Contin.* **2022**, *71*, 1.
12. Abd El-aziz, A.A.; Darwish, A.; Oliva, D.; Hassanien, A.E. Machine learning for apple fruit diseases classification system. In Proceedings of the International Conference on Artificial Intelligence and Computer Vision (AICV 2020), Cairo, Egypt, 8–10 April 2020; pp. 16–25.
13. Yang, R.; He, Y.; Hu, Z.; Gao, R.; Yang, H. CA-YOLOv5: A YOLO model for apple detection in the natural environment. *Syst. Sci. Control Eng.* **2024**, *12*, 2278905. [CrossRef]
14. Liu, Z.; Abeyrathna, R.R.D.; Sampurno, R.M.; Nakaguchi, V.M.; Ahamed, T. Faster-YOLO-AP: A lightweight apple detection algorithm based on improved YOLOv8 with a new efficient PDWConv in orchard. *Comput. Electron. Agric.* **2024**, *223*, 109118. [CrossRef]
15. Sapkota, R.; Ahmed, D.; Karkee, M. Comparing YOLOv8 and Mask R-CNN for instance segmentation in complex orchard environments. *Artif. Intell. Agric.* **2024**, *13*, 84–99. [CrossRef]
16. Jurewicz, M.; Swiderski, B.; Kurek, J. Application of Mask R-CNN Algorithm for Apple Detection and Semantic Segmentation. *Prz. Elektrotechniczny* **2024**, *2024*, 5. [CrossRef]
17. Xiao, F.; Wang, H.; Xu, Y.; Zhang, R. Fruit detection and recognition based on deep learning for automatic harvesting: An overview and review. *Agronomy* **2023**, *13*, 1625. [CrossRef]
18. Dobrzański, B.; Rabcewicz, J.; Rybczyński, R. *Handling of Apple*; Institute of Agrophysics Polish Academy of Sciences (IA PAS) Press: Lublin, Poland, 2006; pp. 1–13.
19. Chen, Z.; Lei, X.; Yuan, Q.; Qi, Y.; Ma, Z.; Qian, S.; Lyu, X. Key Technologies for Autonomous Fruit-and Vegetable-Picking Robots: A Review. *Agronomy* **2024**, *14*, 2233. [CrossRef]
20. Fujinaga, T. Cutting point detection for strawberry fruit harvesting and truss pruning by agricultural robot. In Proceedings of the 2023 IEEE 19th International Conference on Automation Science and Engineering (CASE 2023), Auckland, New Zealand, 26–30 August 2023; pp. 1–6
21. Fujinaga, T. Strawberries recognition and cutting point detection for fruit harvesting and truss pruning. *Precis. Agric.* **2024**, *25*, 1262–1283. [CrossRef]
22. Edan, Y.; Rogozin, D.; Flash, T.; Miles, G. Robotic melon harvesting. *IEEE Trans. Robot. Autom.* **2000**, *16*, 831–835. [CrossRef]
23. Yoshida, T.; Fukao, T.; Hasegawa, T. Cutting point detection using a robot with point clouds for tomato harvesting. *J. Robot. Mechatronics* **2020**, *32*, 437–444. [CrossRef]
24. Ci, J.; Wang, X.; Rapado-Rincón, D.; Burusa, A.K.; Kootstra, G. 3D pose estimation of tomato peduncle nodes using deep keypoint detection and point cloud. *Biosyst. Eng.* **2024**, *243*, 57–69. [CrossRef]
25. Zheng, X.; Rong, J.; Zhang, Z.; Yang, Y.; Li, W.; Yuan, T. Fruit growing direction recognition and nesting grasping strategies for tomato harvesting robots. *J. Field Robot.* **2024**, *41*, 300–313. [CrossRef]
26. Gao, R.; Zhou, Q.; Cao, S.; Jiang, Q. An algorithm for calculating apple picking direction based on 3D vision. *Agriculture* **2022**, *12*, 1170. [CrossRef]
27. Liu, T.; Wang, X.; Hu, K.; Zhou, H.; Kang, H.; Chen, C. FF3D: A Rapid and Accurate 3D Fruit Detector for Robotic Harvesting. *Sensors* **2024**, *24*, 3858. [CrossRef] [PubMed]

28. Gené-Mola, J.; Sanz-Cortiella, R.; Rosell-Polo, J.R.; Morros, J.R.; Ruiz-Hidalgo, J.; Vilaplana, V.; Gregorio, E. Fuji-SfM dataset: A collection of annotated images and point clouds for Fuji apple detection and location using structure-from-motion photogrammetry. *Data Brief* **2020**, *30*, 105591. [CrossRef] [PubMed]
29. Gené-Mola, J.; Sanz-Cortiella, R.; Rosell-Polo, J.R.; Morros, J.R.; Ruiz-Hidalgo, J.; Vilaplana, V.; Gregorio, E. Fruit detection and 3D location using instance segmentation neural networks and structure-from-motion photogrammetry. *Comput. Electron. Agric.* **2020**, *169*, 105165. [CrossRef]
30. Ferrer-Ferrer, M.; Ruiz-Hidalgo, J.; Gregorio, E.; Vilaplana, V.; Morros, J.R.; Gené-Mola, J. Simultaneous fruit detection and size estimation using multitask deep neural networks. *Biosyst. Eng.* **2023**, *233*, 63–75. [CrossRef]
31. Sager, C.; Zschech, P.; Kühl, N. labelcloud: A lightweight domain-independent labeling tool for 3d object detection in point clouds. *arXiv* **2021**, arXiv:2103.04970.
32. Zhou, Z.; Zhao, X.; Wang, Y.; Wang, P.; Foroosh, H. Centerformer: Center-based transformer for 3d object detection. In Proceedings of the European Conference on Computer Vision (ECCV 2022), Tel Aviv, Israel, 23–27 October 2022; pp. 496–513.
33. Shi, S.; Guo, C.; Jiang, L.; Wang, Z.; Shi, J.; Wang, X.; Li, H. Pv-rcnn: Point-voxel feature set abstraction for 3d object detection. In Proceedings of the IEEE/CVF Conference on Computer Vision and Pattern Recognition (CVPR 2020), Seattle, WA, USA, 14–19 June 2020; pp. 10529–10538.
34. Rukhovich, D.; Vorontsova, A.; Konushin, A. Fcaf3d: Fully convolutional anchor-free 3d object detection. In Proceedings of the European Conference on Computer Vision (ECCV 2022), Tel Aviv, Israel, 23–27 October 2022; pp. 477–493.
35. Ding, Z.; Han, X.; Niethammer, M. Votenet: A deep learning label fusion method for multi-atlas segmentation. In Proceedings of the Medical Image Computing and Computer Assisted Intervention (MICCAI 2019), Shenzhen, China, 13–17 October 2019; pp. 202–210.
36. Rukhovich, D.; Vorontsova, A.; Konushin, A. Tr3d: Towards real-time indoor 3d object detection. In Proceedings of the 2023 IEEE International Conference on Image Processing (ICIP 2023), Kuala Lumpur, Malaysia, 8–11 October 2023; pp. 281–285.
37. Choy, C.; Gwak, J.; Savarese, S. 4d spatio-temporal convnets: Minkowski convolutional neural networks. In Proceedings of the IEEE/CVF Conference on Computer Vision and Pattern Recognition (CVPR 2019), Long Beach, CA, USA, 15–20 June 2019; pp. 3075–3084.
38. Fan, G.; Qi, Z.; Shi, W.; Ma, K. Point-gcc: Universal self-supervised 3D scene pre-training via geometry-color contrast. In Proceedings of the 32nd ACM International Conference on Multimedia (ACMMM 2024), Melbourne, Australia, 28 Octobor–1 November 2024; pp. 4709–4718.
39. Qi, C.R.; Chen, X.; Litany, O.; Guibas, L.J. Imvotenet: Boosting 3d object detection in point clouds with image votes. In Proceedings of the IEEE/CVF conference on computer vision and pattern recognition (CVPR 2020), Seattle, WA, USA, 14–19 June 2020; pp. 4404–4413.

agriculture

MDPI

Article

Intelligent Fault Diagnosis of Inter-Turn Short Circuit Faults in PMSMs for Agricultural Machinery Based on Data Fusion and Bayesian Optimization

Mingsheng Wang [1,*], Wuxuan Lai [2], Hong Zhang [1], Yang Liu [1] and Qiang Song [2,*]

[1] College of Mechanical Electrification Engineering, Tarim University, Alar 843300, China; zhghog@163.com (H.Z.); 120150012@taru.edu.cn (Y.L.)
[2] National Engineering Laboratory for Electric Vehicles, Beijing Institute of Technology (BIT), Beijing 100081, China; 3120195251@bit.edu.cn
* Correspondence: wangmingsheng_08@163.com (M.W.); songqiang@bit.edu.cn (Q.S.)

Abstract: The permanent magnet synchronous motor (PMSM) plays an important role in the power system of agricultural machinery. Inter-turn short circuit (ITSC) faults are among the most common failures in PMSMs, and early diagnosis of these faults is crucial for enhancing the safety and reliability of motor operation. In this article, a multi-source data-fusion algorithm based on convolutional neural networks (CNNs) has been proposed for the early fault diagnosis of ITSCs. The contributions of this paper can be summarized in three main aspects. Firstly, synchronizing data from different signals extracted by different devices presents a significant challenge. To address this, a signal synchronization method based on maximum cross-correlation is proposed to construct a synchronized dataset of current and vibration signals. Secondly, applying a traditional CNN to the data fusion of different signals is challenging. To solve this problem, a multi-stream high-level feature fusion algorithm based on a channel attention mechanism is proposed. Thirdly, to tackle the issue of hyperparameter tuning in deep learning models, a hyperparameter optimization method based on Bayesian optimization is proposed. Experiments are conducted based on the derived early-stage ITSC fault-severity indicator, validating the effectiveness of the proposed fault-diagnosis algorithm.

Keywords: agricultural mechanization; fault diagnosis; permanent magnet synchronous motors (PMSM); inter-turn short circuit (ITSC) fault; data fusion

1. Introduction

As an agricultural powerhouse, China feeds nearly 20% of the world's population with only 9% of the world's arable land [1]. Against this backdrop, the level of agricultural mechanization in China has continuously improved alongside the rapid development of the national industrial level [2]. Agricultural mechanization plays a crucial role in promoting agricultural modernization and sustainable development, making intelligent fault-diagnosis research in agricultural mechanization especially important [3]. Agricultural machinery is widely used in various stages of modern agricultural production, including sowing, fertilization, tillage, and harvesting [4]. Permanent magnet synchronous motors (PMSMs) are key power components in agricultural machinery. Due to their advantages of high power density, high efficiency, and excellent control performance, they are widely applied in sowing machines, harvesters, seeders, electric tractors, spraying equipment, and tillage [5]. However, the harsh working environments faced by agricultural machinery—such as extreme temperatures, humidity, corrosion, and dust—along with complex and variable operating conditions—including multimodal vibrations and shocks, fluctuating loads, high loads, frequent starts and stops, and overload operations—pose significant challenges to the durability and safe operation of PMSMs, leading to faults [6]. Common faults in PMSMs include short circuit faults, mechanical faults, and permanent magnet failures. Among these, an ITSC fault is one of the most common short circuit faults in PMSMs [7]. These faults not only pose significant hazards but are also difficult to detect. The occurrence of ITSC faults creates a new closed loop at the short circuit point, generating a large fault current. This not only disrupts the magnetic field distribution in the air gap, increasing motor vibrations but also produces a considerable amount of heat, further threatening the insulation of nearby windings. If not detected and addressed promptly, this can lead to further deterioration of the fault severity, potentially causing the motor and agricultural machinery to lose control, resulting in serious losses [8]. Early diagnosis of ITSC faults not only helps protect equipment and improve production efficiency but also reduces costs and ensures safety, making it highly significant.

Existing fault-diagnosis approaches mainly focus on model-based methods and data-driven approaches [9]. Model-based methods are developed based on the analysis of signal features in different domains, primarily including time domain features, frequency domain features, and time-frequency domain features. These methods demonstrate high accuracy when agricultural machinery operates under stable conditions; however, their effectiveness is often limited in dynamic operating conditions. Data-driven methods are typically used for more complex agricultural machinery fault-diagnosis problems. They rely on machine learning (ML) algorithms to achieve fault recognition and classification. Commonly used algorithms include artificial neural networks (ANN), random forests (RF), extreme learning machines (SVM), and support vector machines (SVM). Nevertheless, these methods share a common drawback: the effectiveness of the algorithms largely depends on the quality of the extracted fault features, which are typically manually extracted. This process requires not only a strong background in expertise but also involves a degree of subjectivity, and it can be time-consuming.

Deep learning models possess the self-learning capability of distributed features, enabling them to automatically abstract and extract the relationships and hierarchical structures among vast amounts of data [10]. In the application of fault diagnosis, deep learning can achieve end-to-end feature extraction and fault classification, effectively overcoming the drawbacks of the aforementioned methods [11]. Xu et al. addressed identifying and diagnosing faults in a tractor's transmission system [12]. They proposed a fault-diagnosis model that combines transformer networks and time generative adversarial networks (Time GANs), achieving high accuracy in fault diagnosis. Xie et al. applied deep learning to the fault diagnosis of rolling bearings in agricultural machinery [1]. Their method employs energy spectrum and singular value decomposition for noise reduction of the vibration signals and then combines ResNet and Vision Transformer to achieve fault diagnosis of the bearings. Lee et al. proposed a method for diagnosing ITSC faults using current signals and rotational speed signals. This approach employs a recurrent neural network (RNN) for fault-feature extraction and incorporates an attention mechanism to assess the severity of ITSC faults [13]. Zhu et al. proposed an intelligent fault-diagnosis method based on principal component analysis (PCA) and deep belief networks (DBNs), which, according to experimental results, is more effective and easier to implement compared to other approaches [14]. Zhu et al. applied a novel capsule network for bearing fault diagnosis, integrating deep learning and short-time Fourier transform (STFT) to convert one-dimensional signals into time-frequency maps, with validation results demonstrating that this model outperforms traditional methods in terms of generalization ability [15]. Husari et al. proposed a hybrid model architecture for diagnosing early ITSC faults, which uses current as input, employs a CNN for feature extraction, and utilizes long short-term memory (LSTM) networks and gated recurrent units (GRUs) for fault diagnosis and severity identification, achieving an accuracy exceeding 97%, thereby outperforming single deep learning models [16,17].

From the above analysis, it is evident that although deep learning-based methods for ITSC fault diagnosis have shown promising results, such as fault recognition accuracy exceeding 97%, an examination of the confusion matrix on the validation set reveals that the model still faces issues related to false alarms and missed detections [18]. This indicates that there is still significant room for improvement in the model's feature-extraction capability. In motor fault diagnosis, a single type of signal often fails to capture all the characteristic information about the motor's operating condition. Sufficiently rich features are a necessary condition for achieving high recognition accuracy in fault diagnosis [19]. Recent studies have shown that diagnostic systems using multi-sensor resources and sensor fusion technologies can provide superior and more robust diagnostic results. Based on this analysis, to further improve the accuracy of ITSC fault diagnosis and reduce the model's false alarm and missed detection rates, this paper proposes a CNN model based on multi-source data fusion for identifying the severity of ITSC faults. This method employs both current and vibration signals for feature learning to enhance the richness of the fault-feature space and uses the fused features for subsequent fault diagnosis to improve the accuracy of the model's fault classification. The main contributions of this paper are summarized as follows:

(1) An indicator suitable for early-stage ITSC fault-severity analysis is derived from the equivalent circuit. This indicator cannot be directly used for fault diagnosis of ITSC, but it can serve as a guide for setting the fault severity during the experimental process.
(2) A feature-level multi-source data-fusion algorithm based on CNNs is proposed. This algorithm employs Bayesian optimization for model hyperparameter tuning, fusing current and vibration signal features to enhance the richness of the fault-feature space, thereby improving the accuracy of ITSC fault-severity identification.
(3) A signal synchronization method is proposed to construct a synchronized signal dataset for current and vibration signals. By calculating the maximum cross-correlation of synchronized signals collected by two devices, synchronization of the current and vibration signals acquired by both devices is achieved.
(4) The effectiveness of the proposed multi-source data-fusion algorithm is validated through experiments. The experimental results indicate that compared to three other methods, the proposed algorithm not only demonstrates higher training efficiency but also superior model performance, highlighting its advantages.

The remainder of this paper is arranged as follows. Section 2 analyzes the mechanism of ITSC faults. Section 3 introduces the details of the proposed algorithm. Section 4 describes the experimental equipment used and the settings required for simulating fault tests. In Section 5, the experiment results are presented to demonstrate the superiority of the proposed algorithm. Finally, Section 6 summarizes this article.

2. ITSC Fault in PMSMs

Estimating ITSC faults is crucial to ensure the safe operation of PMSMs, primarily because it enhances the safety and reliability of the motor, reduces potential safety hazards such as fires, and effectively protects the equipment from damage, thus avoiding costly repairs [20]. Previous research has lacked effective indicators specifically designed for the early diagnosis of ITSC faults. This paper presents an equivalent circuit model for ITSC faults based on the winding coil structure, and on this basis, derives an indicator to guide the setting of early ITSC fault severity in experiments.

The winding of a PMSM is typically composed of multiple coils arranged in series or parallel. This paper primarily studies the winding structure of multiple coils in series. To ensure the uniformity of the air gap magnetic field, reduce harmonic content, and improve the efficiency and performance of the motor, the winding of the PMSM generally adopts a distributed winding structure. The coils are wound into appropriate shapes and installed in two stator slots with a certain spacing between them. When a coil in a certain slot experiences an ITSC fault, the related turns in the corresponding slot are also affected by the ITSC fault, as shown in Figure 1. Figure 1 shows a cross-sectional view illustrating the winding structure of an 8-pole, 36-slot PMSM with an ITSC fault. Each turn of wire within the slot is uniquely marked in the form of Pc-t. For example, A1–4 indicates that the fourth turn of the first coil in the A winding has experienced an ITSC fault. The red section in

Figure 1 indicates the location of the ITSC fault, while the corresponding enlarged view shows the position and number of the shorted wire turns within the slot.

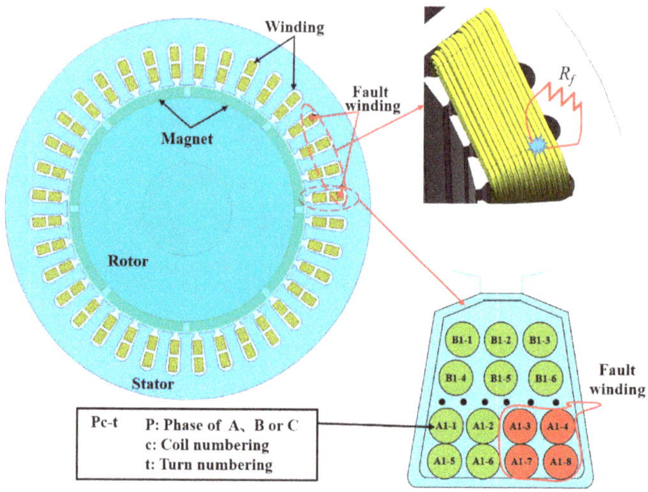

Figure 1. Schematic diagram of a PMSM with ITSC fault.

The ITSC fault in the winding can lead to changes in the structure of the faulty phase winding in a PMSM. At the shorted point, the faulty phase winding is divided into a faulty part and a healthy part. Parameters related to the faulty phase winding, such as resistance, inductance, and magnetic flux, will also change accordingly. Additionally, the shorted point will create a new closed loop, if the resistance at that point is low enough, the fault current can increase significantly and generate a large amount of heat. If heat dissipation is not timely, it may cause further damage to the adjacent insulation, exacerbating the severity of the fault. Changes in the winding structure and the presence of fault current can lead to an imbalance in the air-gap magnetic field and introduce higher harmonic components, thereby exacerbating the unbalanced magnetic pull between the stator and rotor, which in turn increases motor vibrations [21–23]. Additionally, the disruption of the current balance among the windings results in larger torque fluctuations, further intensifying the motor's vibration [24].

Therefore, ITSC faults not only cause changes in the three-phase current but also exacerbate motor vibrations and alter their vibration characteristics [25]. This means that the fault features are reflected in both the current and vibration signals. The characteristic information contained in different signals varies, and their sensitivity to the motor's operating conditions is also different [26,27]. To enrich the extracted feature space and improve the accuracy of ITSC fault identification, this study employs both current and vibration signals for feature learning, thereby increasing the diversity of the feature space. The fused features are then used for subsequent fault diagnosis to enhance the accuracy of the model's fault classification.

Assuming that the first coil of winding A experiences an ITSC fault, the schematic of the equivalent circuit model is shown in Figure 2. From the figure, it can be observed that after the fault occurs, the faulty phase winding is divided into two parts: the yellow part represents the shorted section, while the green part represents the remaining healthy section. The red part indicates the newly formed closed loop at the short circuit point and its fault current. In the fault current loop, the fault phase current i_a is divided into two components: one is the current i_f flowing through the fault resistance R_f, and the other is the current $i_a - i_f$ flowing through the shorted winding.

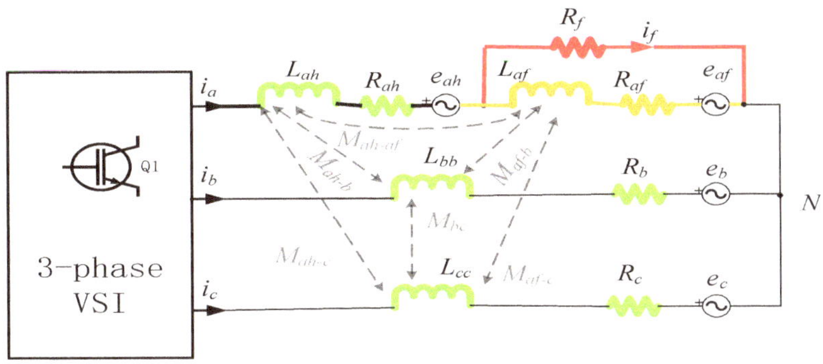

Figure 2. Equivalent circuit diagram of the PMSM with an ITSC fault.

Let N_s be the number of turns that are shorted in the A-phase winding, N_t be the number of turns in each coil, and N_c be the number of coils in each phase winding. The proportion of the shorted turns relative to a single coil and the total number of coils can be expressed as follows:

$$\mu = \frac{N_s}{N_c N_t} \tag{1}$$

where η represents the proportion of the shorted turns in the coil relative to the total number of turns in that coil, and μ represents the proportion of the shorted turns in the faulty phase relative to the total number of turns in that phase winding. Based on the above analysis, the equivalent circuit model is derived as follows:

$$\boldsymbol{V}_{abcn} = \boldsymbol{R}_{abcf}\boldsymbol{I}_{abcf} + \frac{d}{dt}(\boldsymbol{L}_{abcf}\boldsymbol{I}_{abcf}) + \boldsymbol{e}_{abcf} \tag{2}$$

where

$$\boldsymbol{V}_{abcn} = \begin{bmatrix} v_{an} & v_{bn} & v_{cn} & 0 \end{bmatrix}^T$$

$$\boldsymbol{R}_{abcf} = \begin{bmatrix} R_{ah}+R_{af} & & & R_{af} \\ & R_b & & \\ & & R_c & \\ R_{af} & & & R_{af}+R_f \end{bmatrix}$$

$$\boldsymbol{I}_{abcf} = \begin{bmatrix} i_a & i_b & i_c & -i_f \end{bmatrix}^T$$

$$\boldsymbol{L}_{abcf} = \begin{bmatrix} L_{ah}+L_{af}+2M_{ahf} & M_{ahb}+M_{afb} & M_{ahc}+M_{afc} & L_{af}+M_{ahf} \\ M_{ahb}+M_{afb} & L_{bb} & M_{bc} & M_{afb} \\ M_{ahc}+M_{afc} & M_{bc} & L_{cc} & M_{afc} \\ L_{af}+M_{ahf} & M_{afb} & M_{afc} & L_{af} \end{bmatrix}$$

$$\boldsymbol{e}_{abcf} = \begin{bmatrix} e_{ah}+e_{af} & e_b & e_c & e_{af} \end{bmatrix}^T = \tfrac{d}{dt}\begin{bmatrix} \psi_{fah}+\psi_{faf} & \psi_{fb} & \psi_{fc} & \psi_{faf} \end{bmatrix}^T$$

In Equation (2), R_{af}, R_{ah}, R_f, R_b, and R_c indicate the resistance of the shorted part of the faulty phase winding A, the resistance of the remaining healthy part, and the fault resistance at the shorted point, the resistance of the phase winding B, and the resistance of the phase winding C, respectively. The i_a, i_b, i_c, and i_f represent the currents flowing through the A, B, and C phase windings, as well as the fault current at the shorted point of the fault resistance, respectively. L_{ah}, L_{af}, M_{ahf}, M_{ahb}, M_{afb}, M_{ahc}, and M_{afc} denote the self-inductance of the healthy portion and the shorted portion of faulted phase winding A, the mutual inductance between these two portions, the mutual inductance between these two portions with phase B, and the mutual inductance between these two portions with phase C, respectively. Ψ_{fah}, Ψ_{faf}, Ψ_{fb}, and Ψ_{fc} represent the permanent magnet flux linkage of the healthy portion of the faulted phase winding A, the permanent magnet flux linkage of the faulted portion, the permanent magnet flux linkage of phase B winding, and the permanent magnet flux linkage of phase C winding, respectively.

The resistance and permanent magnet flux linkage are directly proportional to the number of turns in the winding; therefore, the various parts of the faulted phase winding can be expressed as:

$$\begin{aligned} R_{af} &= \mu R_a \\ R_{ah} &= (1-\mu)R_a \\ \psi_{af} &= \mu\psi_{fa} \\ \psi_{ah} &= (1-\mu)\psi_{fa} \end{aligned} \tag{3}$$

where R_a and Ψ_f stand for the phase resistance and permanent magnet flux linkage of phase A winding under the condition of a healthy state.

Since the focus of the study is on the early fault diagnosis of ITSCs, only the case of faults occurring within a single coil is considered. To streamline the analysis, it is also assumed that the ITSC fault affects the other phase windings symmetrically. The mutual inductance between coils within the same phase winding is ignored [28]. The relationship between the inductance of different parts of the winding and the degree of the shorted ratio can be expressed as [29]:

$$\begin{aligned} L_{af} &= \mu^2 L_{aa} \\ L_{ah} &= (1-\mu)^2 L_{aa} \\ M_{afh} &= \mu(1-\mu)L_{aa} \\ L_{af} + 2M_{afh} + L_{ah} &= L_{aa} \\ M_{ahb} &= (1-\mu)M_{ab} \\ M_{afb} &= \mu M_{ab} \\ M_{ahc} &= (1-\mu)M_{ac} \\ M_{afc} &= \mu M_{ac} \end{aligned} \tag{4}$$

According to Kirchhoff's current law, the sum of currents flowing into the same node is zero; thus, we can conclude the following:

$$i_a + i_b + i_c = 0 \tag{5}$$

By substituting Equations (3) and (4) into (2) and combining with Equation (5), the expression for the fault current can be derived as follows:

$$i_f = \frac{\mu(v_a - v_n) + (\mu(L_{af} + M_{ahf}) - L_{af})\frac{di_f}{dt}}{\mu R_a + Rf - \mu^2 R_a} \tag{6}$$

From Equation (4), it can be calculated that $\mu(L_{af} + M_{ahf}) - L_{af} = 0$. In the early stages of an ITSC fault, the amplitude of v_n can be considered negligible compared to that of v_a, thus $v_a \approx v_a - v_n$. Let $v_a = V_a\, sin(\omega t)$, Thus, an approximate expression for the fault current amplitude can be derived as follows:

$$I_f \approx \frac{\mu V_a}{\mu R_a + R_f - \mu^2 R_a} \tag{7}$$

It is widely recognized that the voltage amplitude of the stator winding in a PMSM has a positive correlation with the motor speed [30]. Then, Equation (7) can be represented as follows:

$$I_f \approx K\frac{\mu\omega_r}{\mu R_a + R_f - \mu^2 R_a} \tag{8}$$

where K represents a constant coefficient that can be considered a known quantity, while ω_r denotes the mechanical speed of the rotor.

From Equation (8), it can be seen that K can be considered a known quantity, while R_a represents the inherent parameter of the motor, which can also be regarded as known. The remaining parameters will directly affect the fault current i_f. The parameters μ and R_f are both related to the severity of the ITSC fault, whereas ω_r is independent of the fault severity. By removing ω_r from the right side of Equation (8), only the parameters related to the severity of the ITSC fault remain, expressed as follows:

$$\frac{I_f}{\omega_r} \approx K\frac{\mu}{\mu R_a + R_f - \mu^2 R_a} := FI \tag{9}$$

The right side of the equation contains only known quantities and parameters related to the severity of the ITSC fault. We define the right side expression as a representation of the fault severity, denoted by the symbol *FI*. When the motor is in a healthy state, this fault-severity indicator is 0, conversely, when the faulted phase winding of the motor is completely shorted and the fault resistance at the shorted point is 0, this indicator becomes infinite. The left side of the equation represents the ratio of the magnitude of the fault current to the mechanical speed of the rotor. From the derivation process, it can be seen that this indicator is only applicable for analyzing the fault severity in the early stages of ITSC faults, during which the indicator is generally unaffected by speed. It increases as the fault resistance R_f decreases or the shorted ratio μ increases, and vice versa. In practice, it is difficult to directly measure R_f and μ during the motor's operation. Therefore, this indicator is not suitable for the direct estimation of ITSC fault severity. However, it can be used as a fault-severity indicator in experiments, guiding the setting of fault levels for ITSC faults.

3. Proposed Algorithm

3.1. Fault-Diagnosis Methods Based on Multi-Source Data Fusion

Multi-source data fusion, also known as information fusion, is a technology that enables the automated processing of integrated information. This technology originated in the military domain and has gradually been widely applied in the civilian sector after years of development. Today, it has achieved rapid advancements in various fields, including robot control, autonomous driving, and fault diagnosis [31]. The application of data fusion in fault diagnosis relies on the research object's ability to collect information from multiple types of sensors. Additionally, to achieve a comprehensive analysis and accurate assessment of the fault status, it is necessary to utilize various signal processing techniques to obtain a rich feature space. By combining different intelligent algorithms, multi-level data fusion can be achieved, leading to the final assessment results.

According to the different levels of data fusion, data-fusion methods can be classified into three categories: data-level fusion, feature-level fusion, and decision-level fusion. When selecting different fusion levels, it is necessary to consider the balance between fusion performance and implementation cost [11].

Data-level fusion, also known as pixel-level fusion, refers to the direct integration of raw data, which maximally preserves the original information and exhibits superior fusion performance. Xia et al. proposed a CNN-based fault-diagnosis method for rotating machinery that combines sensor fusion with spatiotemporal information to achieve automatic feature extraction [32]. Chen et al. proposed a fault-diagnosis method for gearboxes based on DCNN, which integrates the raw data of vertical and horizontal vibration signals to achieve automatic feature extraction [33]. However, data-level fusion models do not possess the ability to correct errors, and their performance is poor when the sensor types are different or when there are significant differences in magnitude.

Feature-level fusion can achieve a more refined integration of information through the dimensionality reduction of the data. Compared to data-level fusion, feature-level fusion offers higher robustness, less information redundancy, greater flexibility, and better real-time performance, allowing for a more comprehensive representation of the data's characteristics. Azamfar et al. proposed a novel two-dimensional CNN architecture that integrates features extracted from multiple current sensors to monitor gearbox faults under different operating conditions and speeds [34]. Parai et al. proposed a feature-level fusion method for circuit fault diagnosis, which employs wavelet analysis for fault-feature extraction from multiple signals and uses PCA for feature fusion. Ultimately, circuit fault-type diagnosis is achieved based on a support vector machine [9]. In recent years, the combination of feature-level fusion and deep learning models has gradually become a common approach to achieve better diagnostic results.

Decision-level fusion typically occurs after each model or sensor completes its processing independently, followed by a decision merger. This approach has a high degree of information integration and lower computational complexity, making it suitable for various signal sources or data types. Common decision-level fusion methods include the Dempster–Shafer (D-S) theory, decision tree fusion, and weighted voting. However, the complexity of decision-level fusion methods is relatively high, with significant loss of original data and various challenges in choosing fusion strategies, which is why they are less frequently used in conjunction with deep learning models.

Based on the advantages and disadvantages of different levels of multi-source data fusion described earlier, and in conjunction with the content of this study, a multi-source data-fusion algorithm based on a CNN model has been proposed. This algorithm is applied to the identification of the severity of ITSC faults, aiming to enhance feature learning from various types of sensor information. By enriching the feature space and integrating multi-faceted fault characteristics, the algorithm seeks to improve the diagnostic accuracy of ITSC fault identification under various complex operating conditions.

3.2. *Convolutional Neural Networks*

A CNN is a significant type of deep neural network that can achieve end-to-end fault classification by automatically extracting local features [35]. Compared to traditional artificial neural networks, a CNN has convolutional layers that feature "local connectivity" and "weight sharing", which substantially decreases the number of parameters in the model and lowers the training difficulty [36]. CNN models typically consist of multiple hidden layers that can automatically extract various features from the input signals. The lower hidden layers focus on learning the basic characteristics of the input signals, while the higher layers abstract and re-extract these basic features to form more complex high-dimensional features, resulting in more accurate classification [37].

The convolutional layer typically needs to work in conjunction with functional layers such as pooling layers, normalization layers, activation layers, and dropout layers to enhance the feature-extraction capability of the convolutional module. A typical structure of a convolutional module is shown in Figure 3a. In this module, the convolutional layer filters out redundant information from the input signals through convolution operations, reinforcing important features related to fault classification. The pooling layer usually follows the convolutional layer to perform feature dimensionality reduction while maintaining the translation invariance of the features. The activation function accelerates the convergence of the model and helps mitigate the vanishing gradient problem to some extent. The dropout layer is primarily used to prevent overfitting, thereby improving the model's generalization ability.

In this paper, to maintain the high resolution of the extracted fault features, reduce downsampling operations, and enhance the model's ability to capture multi-level complex information, a dilated CNN is employed. Its structural diagram is illustrated in Figure 3b, and the expression for dilated convolution is shown in Equation (10) [38].

$$F(x) = (S *_d f)(x) = \sum_{i=0}^{k-1} f(i) \cdot S_{x-d \cdot i} \tag{10}$$

where $F(x)$ stands for the dilated convolution operation. The input signal $S \in R^n$ is convolved using the operator *, with x indicating the specific element of the input signal involved in the convolution. The dilation factor is denoted by d, while f: $\{0, 1, \ldots, k-1\} \rightarrow R$ represents the set of weight values applied during the convolution. The parameter k defines the size of the weight matrix, and $x - d \cdot i$ indicates the i-th element of the input signal undergoing the convolution operation.

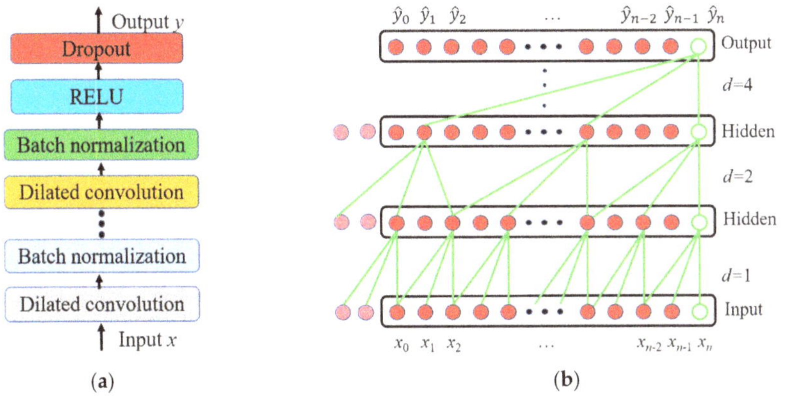

Figure 3. (**a**) Schematic diagram of a conventional convolutional module. (**b**) Schematic diagram of a dilated convolutional module.

3.3. The Attention Mechanism

Due to its significant impact on deep learning models, attention mechanisms have garnered widespread attention in recent years. The core of this mechanism lies in adjusting weights to guide the model in filtering out redundant information that is irrelevant to the task, thereby focusing attention on the features that are more critical for achieving the task objectives [35]. In CNN models, common attention mechanisms include channel attention, spatial attention, and hybrid attention. This paper employs a channel attention mechanism to adjust the weights of input features extracted from different signals under varying fault levels and operating conditions, reallocating the model's attention and enhancing the contribution of each channel feature to improve the performance of the ITSC fault-diagnosis model.

The typical channel attention mechanism is known as SENet (Squeeze and Excitation Network), and its core structure mainly includes three steps: squeeze, excitation, and scaling, as illustrated in Figure 4 [39]. During the training process of the fault-diagnosis model, the feature weights extracted by the model are adjusted and redistributed after undergoing the squeeze and excitation operations of SENet. The weights of features that are relevant and sensitive to fault level classification are enhanced, while irrelevant features are suppressed or weakened. The core three steps of SENet can be expressed as follows:

$$
\begin{aligned}
z_c &= F_{sq}(u_c) = \frac{1}{W \times H} \sum_{i=1}^{W} \sum_{j=1}^{H} u_c(i,j) \\
s &= F_{ex}(z_c, W_i) = \delta(W_2 \sigma(W_1 z_c)) \\
\widetilde{x}_c &= F_{scale}(u_c, s_c) = s_c \cdot u_c
\end{aligned}
\tag{11}
$$

In SENet, the first step involves applying a squeeze operation to the information from each input channel, which sets the stage for adjusting the weights of the different channels. Assuming the input feature map has dimensions $W \times H \times C$, where $\times$ denotes scalar multiplication, this operation converts the input features into global features of size $1 \times 1 \times C$ through global average pooling. In Equation (11), F_{sq} denotes the squeeze operation, u_c represents the entire input feature map, and Z_c refers to the global features obtained after the squeeze operation.

The second step is the excitation operation, which aims to capture the relationships between the features input from different channels. The key step of the excitation operation is to input the global features of size $1 \times 1 \times C$ obtained from the squeeze operation into a fully connected layer of dimension $C \div r \times C$, where r represents the scaling factor, primarily used to reduce the number of channels, which in turn lowers the computational complexity of the model. The output from the previous step is passed through a ReLU activation layer to a second fully connected layer, where the number of feature channels is restored. Subsequently, the output feature information is normalized to the range of (0, 1) through a Sigmoid activation layer, completing the readjustment of the weights for the relevant fault features across different channels. In Equation (11), S represents the adjusted weights of the fault features for each channel, F_{ex} denotes the excitation operation, while W_1 and W_2 represent the operations of the two fully connected layers. σ and δ correspond to the ReLU activation layer and the Sigmoid activation layer, respectively.

The third step is the scaling operation, which primarily involves multiplying the fault features in each channel by the redistributed weights to recalibrate the fault features and achieve the overall adjustment of the attention mechanism. As shown in Equation (11), $\widetilde{x}_c$ denotes the features after channel attention adjustment, and F_{scale} denotes the scaling operation.

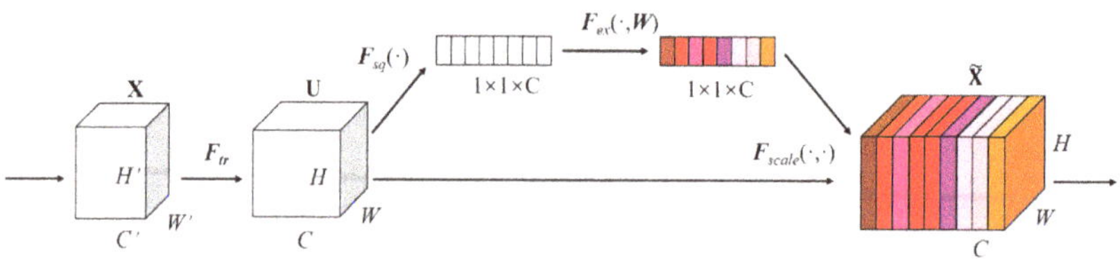

Figure 4. Principle of channel attention mechanism.

3.4. Bayesian Optimization Algorithm

The CNN-based ITSC fault-diagnosis model features a flexible and variable structure, requiring numerous hyperparameters. Different combinations of hyperparameters can significantly impact the model's training efficiency and final validation accuracy. To achieve a robust ITSC fault-diagnosis model, it is essential to optimize various hyperparameter combinations to identify the best set. Common optimization algorithms include grid search, random search, and Bayesian optimization. Random search and grid search are methods of random enumeration and exhaustive search, respectively, which can lead to a degree of blindness in the optimization process [40]. This may result in a significant waste of computational resources on unsuitable hyperparameter combinations. Consequently, under limited computational resources, these two methods often struggle to yield satisfactory results without extensive prior experience. In contrast to these two algorithms, Bayesian optimization is a global optimization technique that employs a sequential search process [41]. It effectively utilizes the prior information of known data points to autonomously adjust its optimization strategy. Therefore, this paper adopts Bayesian optimization to optimize the hyperparameters of the ITSC fault-diagnosis model.

Bayesian optimization is a global optimization strategy named after the famous Bayes' theorem (as shown in Equation (12)) used in its framework. This algorithm constructs a probabilistic model of the objective function to effectively select the most promising evaluation points in each iteration, making it particularly suitable for scenarios where the objective function is expensive or difficult to evaluate.

$$p(f \mid D_{1:t}) = \frac{p(D_{1:t} \mid f)p(f)}{p(D_{1:t})} \tag{12}$$

where f represents the objective function, which is typically difficult to express directly in functional form, in the fault-diagnosis model, it reflects the overall performance of the model. $D_{1:t}$ denotes the sample points of the hyperparameter combinations to be optimized, with the number of sample points being t. The process of hyperparameter tuning for the ITSC fault-diagnosis model based on Bayesian optimization mainly includes the following key steps:

(1) Initialization

Select several initial values as the starting points for Bayesian optimization based on the given hyperparameter combinations and the value ranges for each hyperparameter.

(2) Constructing the probabilistic model

Establish a probabilistic model for the objective function using a Gaussian process model based on the evaluated sample points. This model can predict the objective function values and their corresponding uncertainties for untested points.

(3) Selecting evaluation points

Consider both the predicted values and uncertainties of the model and select the next set of promising evaluation points from the probabilistic model using the expected improvement acquisition function.

(4) Objective function evaluation

Evaluate the objective function at the newly obtained sample points and compute the function values.

(5) Updating the probabilistic model

Combine the results from the new evaluation points with previous data to update the probabilistic model, thereby enhancing its accuracy. Repeat (3) to (5) until the specified termination criteria are met. Finally, select the hyperparameter combination corresponding to the model with the best performance as the output result.

The hyperparameter tuning process of the ITSC fault-diagnosis model is illustrated in Figure 5. As shown in the figure, the hyperparameter adjustment for the entire fault-diagnosis model is primarily divided into two parts. One part is the constructing and training process of the fault-diagnosis model, as indicated in the black box. When the model meets the termination criteria for training, the final testing accuracy will be passed to the Bayesian optimization process. The other part involves hyperparameter optimization based on the Bayesian optimization algorithm, as depicted in the green box. In this entire process, Bayesian optimization is responsible for finding the optimal hyperparameter combinations, which mainly includes the initialization and optimization of the hyperparameter combinations. Throughout the process, the model training and hyperparameter optimization cycle back and forth until the termination criteria for optimization are met. Ultimately, the optimal hyperparameter combination and its validation accuracy are selected as the output results.

3.5. Multi-Source Data-Fusion Algorithm Based on Bayesian Optimization and a CNN

The above analysis indicates that the occurrence of ITSC faults not only affects the three-phase currents of the windings but also alters the distribution of the air gap magnetic field. This leads to the generation of unbalanced radial magnetic forces within the motor, which exacerbates the production of vibration signals and alters the vibration characteristics of the motor. Therefore, this paper proposes a fault-diagnosis method for ITSCs based on the fusion of three-phase current signals and vibration signals. This method utilizes both three-phase current signals and vibration signals as sources for fault diagnosis. Using a CNN model, it extracts features from the current and vibration signals separately and then performs a feature-level fusion of the two types of signals. The channel attention mechanism is employed to adjust the weights of the fault features from different signals and channels. Finally, to improve the training efficiency of the model and enhance its performance, Bayesian optimization is used to fine-tune the training hyperparameters of the model.

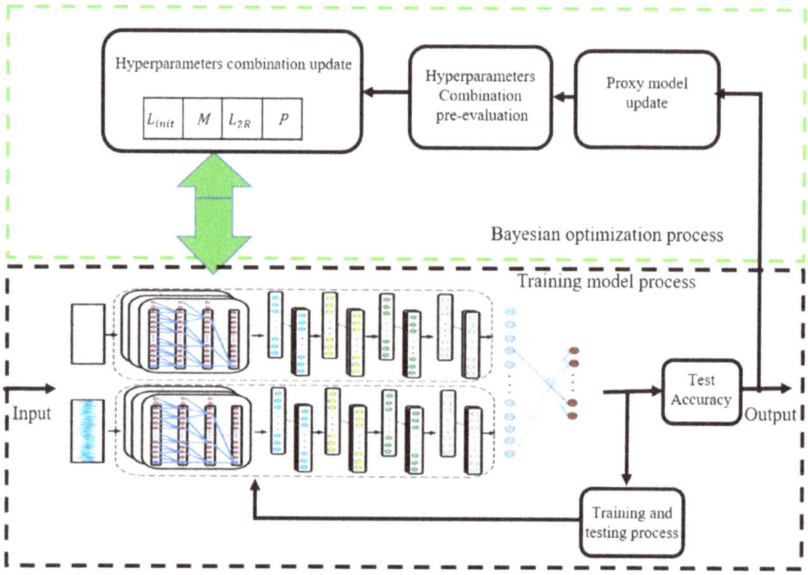

Figure 5. Schematic of hyperparameter optimization for ITSC fault-diagnosis model.

The flowchart of the entire process is shown in Figure 6, which can be divided into five specific steps.

(1) Data collection. Experiments are conducted with varying degrees of ITSC faults, collecting three-phase current signals and vibration signals simultaneously under different operating conditions.
(2) Dataset preparation. The collected current and vibration signals undergo data synchronization, followed by a series of preprocessing steps including filtering, normalization, downsampling, data slicing, and grouping. Ultimately, the processed current and vibration signals are organized into a dataset suitable for a multi-source data-fusion model.
(3) Model construction and initialization. This study employs a multi-stream high-level feature fusion model based on CNNs. The two types of signals are fed into different CNN branches for feature extraction, and then the extracted features are fused at the high level of the network. Fault-severity recognition is achieved through fully connected layers and classification layers. The model's training hyperparameters are initialized using Bayesian optimization.
(4) Model training and optimization. The dataset constructed in Step 2 and the multi-source data-fusion model built in Step 3 are used for training. The performance of the trained model is evaluated using a test set, and the Bayesian optimization algorithm updates the hyperparameter combinations based on the results, repeating the process until the optimization iterations or model performance reaches the termination criteria.
(5) Output results. When the model reaches the optimization termination condition, the best ITSC model is selected for output, which includes the optimal hyperparameter combinations and the corresponding accuracy results for ITSC fault diagnosis.

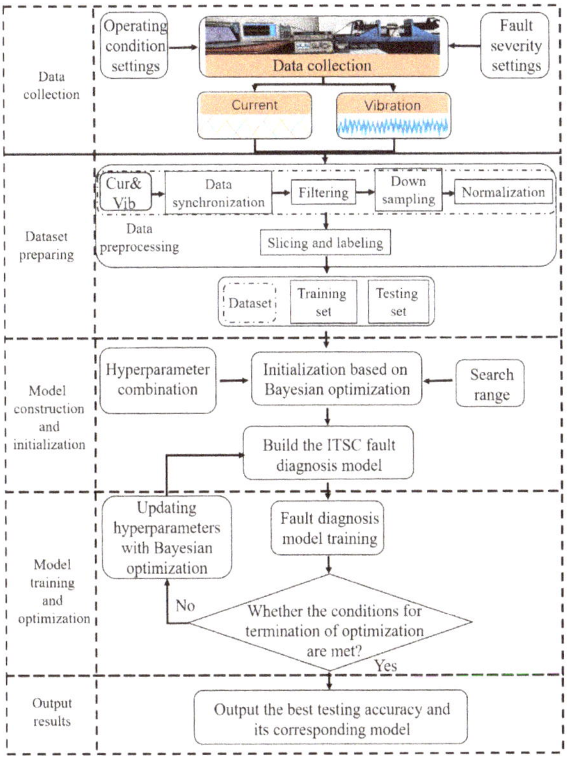

Figure 6. Flowchart of ITSC fault-diagnosis algorithm based on multi-source data fusion.

4. Experimental Setup and Data Description

To validate the effectiveness of the ITSC fault-diagnosis algorithm based on the fusion of current and vibration signal fault characteristics, as well as the Bayesian-optimized ITSC fault-diagnosis model, a simulation test for ITSC faults in PMSM was conducted. The experiment was carried out on the test bench for the dual-drive motor, as shown in Figure 7. The main test equipment included the motor under test and its controller, a dynamometer, torque and speed sensors, vibration sensors, current sensors, and a data acquisition system. The motor under test was an 8-pole, 36-slot PMSM with a star-connected winding structure. The specific parameters are listed in Table 1.

The torque sensor used in the experiment was the HCNJ-101, with a measurement accuracy of $\pm 0.1\%$; the current sensor was the ETA-5301B, with a measurement accuracy of 3% RD; the vibration sensor was the KS78B100 from MMF Germany, with an IEPE interface, a sensitivity of 100 mV/g and an accuracy of 2% RD. To prevent low-frequency interference during data acquisition, the sampling frequency for the current signal was set to 1 MHz. Since the maximum sampling frequency for the vibration signal acquisition using the NI cRIO9068 was also 1 MHz, different devices were used for data collection. The data acquisition device for the current signal was an oscilloscope with a sampling frequency of 1 MHz, while the vibration signal was acquired using the NI9401 card within the NI cRIO9068, at a sampling frequency of 10.24 kHz.

To ensure synchronization between the data acquisition devices, a signal generator was used to create a sweep frequency signal with a period of 8 s, sweeping from 20 Hz to 2 kHz, with a voltage signal amplitude of ± 2 V. The waveform of the synchronization signal is shown in Figure 8. During the data acquisition process, both the current and vibration data acquisition devices received the synchronization signal from the signal generator simultaneously. The oscilloscope had a sampling frequency of 1 MHz, while the NI9223 acquisition card used by the NI cRIO9068 had a sampling frequency of 100 kHz.

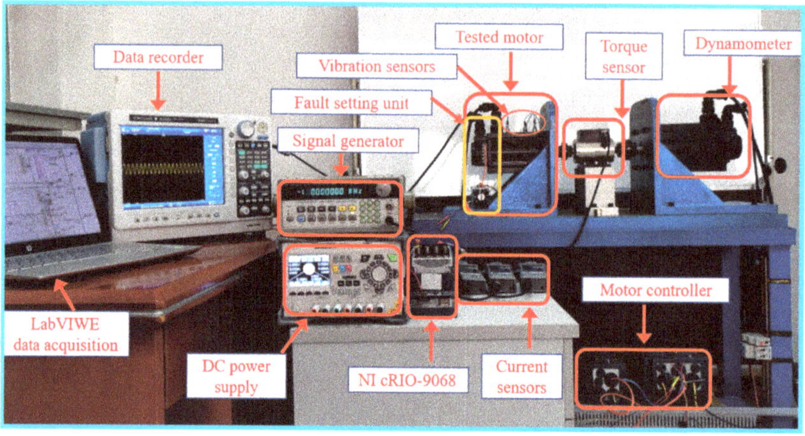

Figure 7. Fault-diagnosis test bench for PMSM.

Table 1. Specifications of the PMSM.

Parameters	Values	Parameters	Values
Rated Power	2.3 kW	Line-line resistance	1.1 Ω
Rated torque	15 Nm	Line-line inductance	4.45 mH
Rated current	9.5 A	Number of turns per phase	108
Rated speed	1500 rpm	Number of coils per phase	12
Pole pairs	4	Voltage constant	114 V/1000 r/min

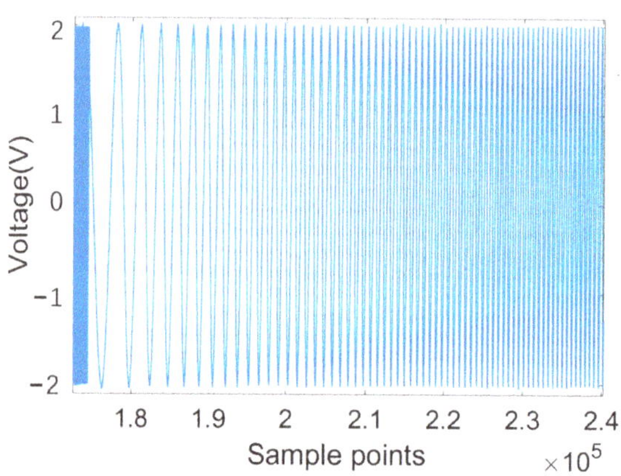

Figure 8. Synchronous signal waveform diagram.

In conducting simulation tests of ITSC faults in motors, two key factors need to be fully considered: on one hand, the establishment of early ITSC faults with varying degrees of severity; on the other hand, the operational conditions of the motor should be as comprehensive as possible. Based on the previous analysis, the severity of ITSC faults is determined by the number of shorted turns and the fault resistance. To simulate different degrees of ITSC faults, the test motor has been modified accordingly, as shown in Figure 9.

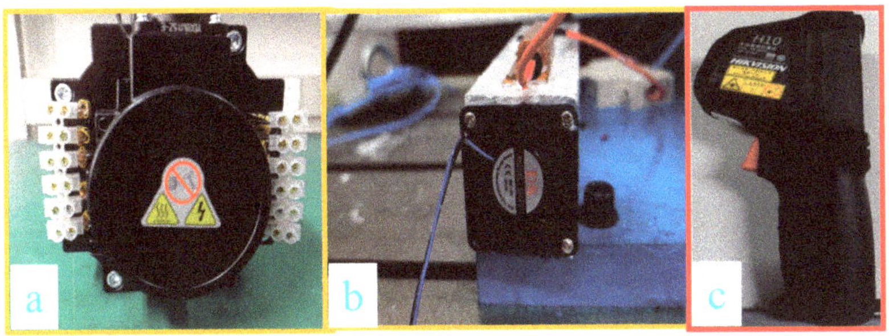

Figure 9. (**a**) The tested faulty motor. (**b**) The fault resistance its heat dissipation device. (**c**) The temperature measurement device.

Figure 9a displays the tested faulty motor, with terminals on both sides representing the lead wires of windings with different turn counts, allowing for varying degrees of short circuit simulation through paired connections. Figure 9b shows the fault resistance and its heat dissipation device; the fault resistance can be replaced to simulate different levels of insulation damage. The combination of these two parameters facilitates the simulation of varying degrees of ITSC faults. Figure 9c illustrates the temperature measurement device, which continuously monitors the temperature of the faulty motor and its fault resistance throughout the experiment to prevent damage due to excessive temperature.

The operational conditions for the motor during the fault simulation tests are presented in Table 2. To simulate the operating conditions of agricultural machinery during acceleration, deceleration, and constant speed, a total of 8 constant speed scenarios and 2 acceleration and deceleration scenarios are included. The settings for variable speed conditions are shown in Figure 10.

Table 2. Operating conditions of the PMSM to be tested.

Case	Constant				Dynamic
	8				2
Speed (rpm)	150	450	900	1350	850~1550~850
Torque (Nm)	3.0/7.5	3.0/7.5	3.0/7.5	3.0/7.5	3.0/7.5

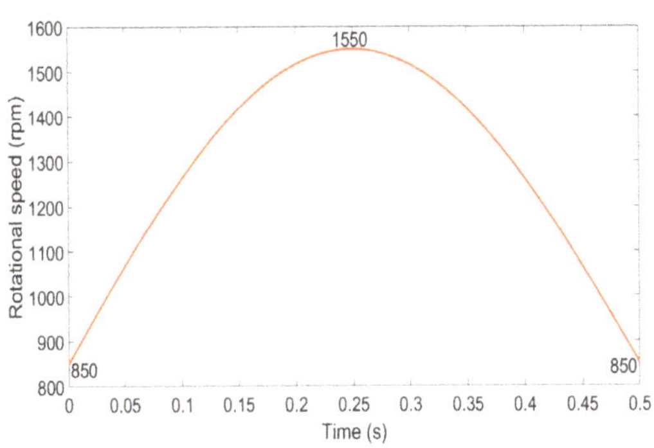

Figure 10. Schematic diagram of speed variation.

After completing the signal acquisition, a series of data preprocessing operations were required, including signal synchronization, filtering, normalization, downsampling, slicing, and grouping, to organize the current and vibration signals into a dataset suitable for multi-source data-fusion models.

The oscilloscope can record signal sampling for 10 s at a time, while the cRIO9068 has a longer signal sampling duration. Therefore, during the data acquisition phase, the cRIO9068 must be turned on first for data collection, followed by the activation of the oscilloscope. In the end, the oscilloscope is stopped first, followed by the cessation of recording on the cRIO9068. The data synchronization process between the two signals involves using the time segment occupied by the data recorded by the oscilloscope to slice the corresponding data recorded by the cRIO9068 within that time frame. Throughout this process, the synchronized signals recorded by both devices serve as timestamps to determine the start and end times of the signals. To maximize the length of the synchronized signal, the synchronization signal from the oscilloscope is downsampled to 100 kHz, denoted as y, while the synchronization signal from the cRIO9068 is denoted as x. The expression for the cross-correlation of the two signals is as follows:

$$\hat{R}_{xy}(m) = \sum_{n=0}^{N-m-1} x_{n+m} y_n, \quad m \geq 0 \tag{13}$$

where, $\hat{R}_{xy}$ represents the cross-correlation index between the two signals, m denotes the offset of the signal in the NI cRIO9068, with a range of $(0, N-1)$, and N is the length of signal x. When m takes on different values, a series of cross-correlation indices can be obtained for signal x relative to signal y after shifting m points. The maximum value in this sequence indicates the highest cross-correlation index for the two signals at the corresponding offset. Since both signals are derived from the same synchronous signal generated by the signal generator, if the two signals coincide, the offset for the corresponding cross-correlation index represents the starting point for truncating signal x using signal y, while the endpoint is determined by the number of sampling points contained in signal y.

Figure 11 illustrates the results of signal synchronization achieved through cross-correlation analysis of synchronous signals collected by two devices. To compare the effectiveness of the signal synchronization, the synchronous signals of equal amplitude have been slightly offset in the diagram. Signal x is the downsampled synchronous signal recorded by the cRIO9068, while signal y is the synchronous signal captured by the oscilloscope. Signal x′ is derived from signal y, and signal y′ is consistent with signal y. It can be observed from the figure that the length of signal x is significantly greater than that of signal y, which aligns with the setup during the experiment where the cRIO9068 was initiated earlier and stopped later than the oscilloscope. The starting point of signal x′ is determined by the maximum cross-correlation offset between signals x and y, while the endpoint of signal x is determined by the length of signal y. The locally enlarged area in the figure indicates the positions of the frequency sweep cycle transitions for the four synchronous signals. It is evident that the transition positions for these four synchronous signals are consistent, and the synchronization error is within one sampling period, indicating that the synchronization algorithm used is effective.

From the above analysis, it can be seen that during the signal synchronization process of the two devices, the synchronization signal collected by the oscilloscope has not changed. Therefore, the fundamental purpose of synchronizing the signals from these two devices is to use the synchronized signal as a timestamp to extract the corresponding time range data from the cRIO9068, thereby eliminating data that is out of sync. Due to the different sampling frequencies used by the two devices during signal acquisition, there will still be time synchronization errors when using the synchronized signal as a timestamp for data extraction. A schematic of this error is shown in Figure 12.

In Figure 12, the horizontal axis (X) represents the sampling time, and the vertical axis (Y) represents the signal amplitude. Assuming the point $(6.7 \times 10^{-5}, 3)$ in the figure is the calculated starting point of the synchronized signals from the two devices, the synchronization error between the synchronization signal and the vibration signal is t1, while the synchronization error between the synchronization signal and the current signal is t2. It is known that the algorithm ensures the synchronization error between the two signals is within one sampling period, so the upper limit of synchronization error between the current signal and the vibration signal is the sampling period of the vibration signal, which is 9.77×10^{-5} s, significantly less than 0.1 milliseconds.

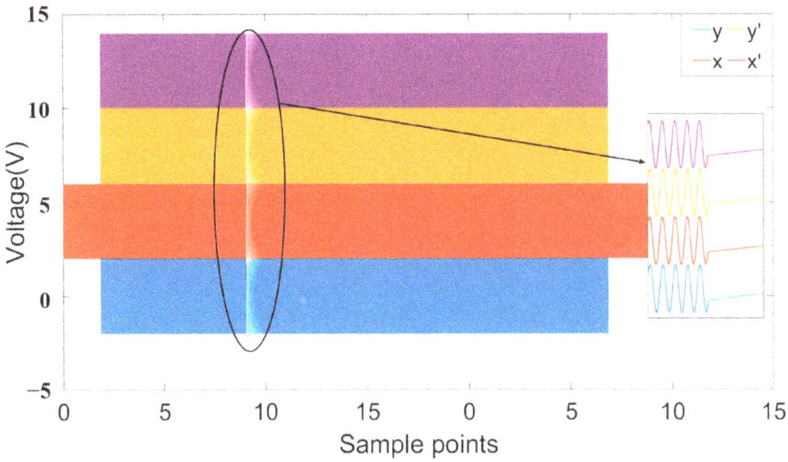

Figure 11. Schematic diagram of signal synchronization results.

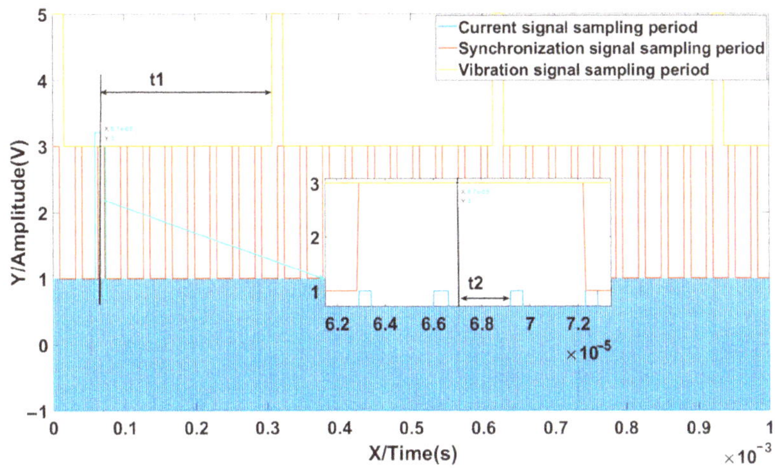

Figure 12. Schematic diagram of synchronous signal error.

During data acquisition, each group of data is collected for 10 s, and the time length of each data slice when constructing the dataset is 0.2 s. Therefore, this maximum time error represents a very small proportion of each data slice's time length, and its impact on the sample set constructed for signal synchronization in this paper can be considered negligible.

After synchronizing the current and vibration signals, the subsequent data preprocessing operations include filtering, normalization, downsampling, slicing, and grouping. The description of the resulting synchronized signal dataset is shown in Table 3. From the table, it can be observed that the dataset consists of synchronized current and vibration signals, containing 17 fault-severity labels. Among these, "HL" represents data collected under normal motor conditions, while "A*R*" indicates data collected under different ITSC fault severity. The severity of faults in the ITSC is determined by varying the combinations of shorted turn ratios and fault resistances. Different shorted turn ratios are created by connecting two specific points from the multiple lead wires, each corresponding to different turns of the first coil in phase A. Fault resistance, on the other hand, can take any value within the range of 0 to 1M ohms when an ITSC fault occurs [15]. If the fault resistance is extremely high, the fault current flowing through it will be minimal, effectively resulting in a healthy state. Conversely, if the fault resistance is too low, the fault current will be large enough to potentially cause irreversible damage to the test rig [7]. Since this paper focuses on diagnosing early-stage faults in the ITSC, fault resistance is classified into three cases based on the above analysis. First, when the fault resistance is significantly greater than the impedance of the shorted wire. Second, when the fault resistance is slightly larger than the impedance of the shorted wire, resulting in a higher fault current and more noticeable impact on the motor. Third, when the fault resistance is close to the impedance of the shorted wire, leading to a large fault current that could potentially damage the experimental setup if the test is prolonged. Based on these considerations and experimental experience, only the ITSC scenario involving a single coil is considered when determining the number of shorted turns. The fault resistances are then set within the range of 5 Ω to 0.1 Ω, ensuring that the fault currents are clearly detectable while preventing any irreversible damage to the experimental setup. The severity of faults is arranged in ascending order based on the calculation from Equation (9).

Table 3. Dataset description.

Label	Fault Setting		Current		Vibration		Total
	Fault Resistance (Ω)	Shorted Turn Ratio (%)	Training	Testing	Training	Testing	
HL	Inf	0	840	360	840	360	2400
A2R5	5	4.6	840	360	840	360	2400
A4R5	5	8.3	840	360	840	360	2400
A5R5	5	10.2	840	360	840	360	2400
A6R5	5	13.9	840	360	840	360	2400
A2R1	1	4.6	840	360	840	360	2400
A4R1	1	8.3	840	360	840	360	2400
A2R0.5	0.5	4.6	840	360	840	360	2400
A5R1	1	10.2	840	360	840	360	2400
A6R1	1	13.9	840	360	840	360	2400
A4R0.5	0.5	8.3	840	360	840	360	2400
A5R0.5	0.5	10.2	840	360	840	360	2400
A6R0.5	0.5	13.9	840	360	840	360	2400
A2R0.1	0.1	4.6	840	360	840	360	2400
A4R0.1	0.1	8.3	840	360	840	360	2400
A5R0.1	0.1	10.2	840	360	840	360	2400
A6R0.1	0.1	13.9	840	360	840	360	2400

Under each fault-severity label, both types of signals contain 1200 data samples, with 840 samples designated for training and 360 for testing. Figure 13 displays the waveform of the current signal from a single sample in the constructed dataset, along with the corresponding vibration signal for that period. The horizontal axis represents the number of sampling points, and the vertical axis represents the normalized waveform amplitude. It is evident from the figure that each set of current signals has 3000 sampling points, while the vibration signals consist of 2048 sampling points. The time duration for both sets of signals matches the specified data slicing duration, which is 0.2 s.

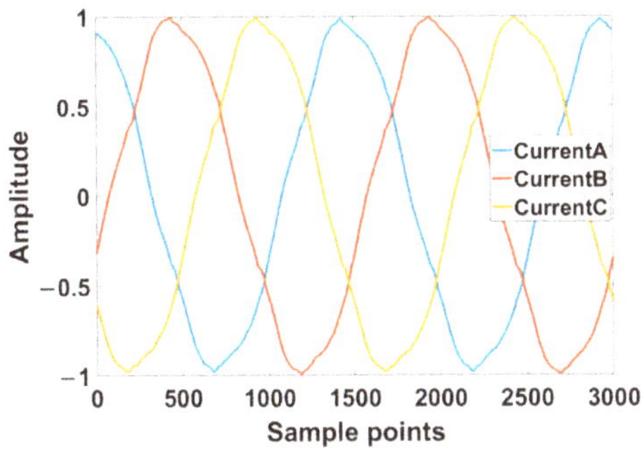

(a) Current signal slice

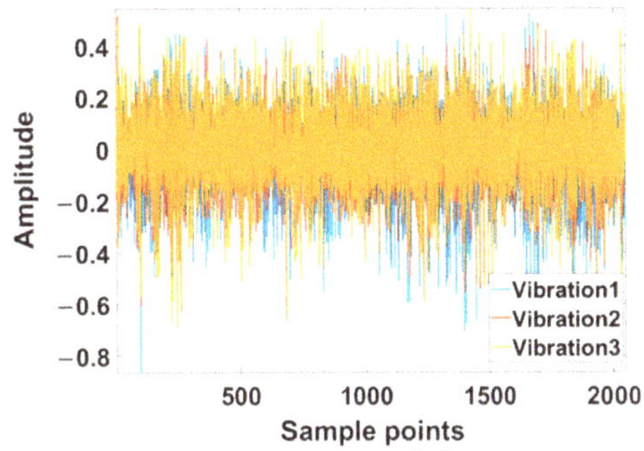

(b) Vibration signal slice

Figure 13. Waveform diagram of current signals and vibration signals over the same period in the dataset.

5. Discussion

5.1. Results and Comparisonsns

After completing the simulation tests for ITSC faults and constructing the dataset, the proposed multi-source data-fusion fault-diagnosis model is applied to analyze the synchronized current signal dataset and the vibration signal dataset. The hyperparameters that need optimization in the model include the initial learning rate, gradient optimization parameters, L2 regularization coefficient, and the dropout rate in the dropout layer, represented by the symbols L_{init}, G_1, $L2R$, and P, respectively. During the model training process, the number of Bayesian optimization iterations is set to 60, and the number of model training epochs in each optimization is set to 40. The corresponding parameter search space, data types, and the optimal hyperparameter combinations for the model are shown in Table 4.

Table 4. Hyperparameters to be optimized.

Hyperparameters	Search Intervals	Data Types	Transform	MC	MS
$Linit$	$[1 \times 10^{-5}\ 1]$	real	log	2.7586×10^{-4}	1.0001×10^{-4}
G_1	[0.5 1]	real	log	0.9037	0.8948
$L2R$	$[1 \times 10^{-10}\ 1 \times 10^{-2}]$	real	log	0.0097	0.0091
P	$[1 \times 10^{-5}\ 1]$	real	log	2.488×10^{-5}	4.7663×10^{-5}

The proposed multi-source data-fusion model is a multi-stream high-level feature fusion (MS) model, as illustrated in Figure 14a. The synchronized current and vibration signals are processed through different CNN branches for fault-feature extraction, with feature fusion occurring at a higher level. To compare the impact of different fusion methods on the diagnosis results of ITSC faults, a multi-channel information fusion model was also constructed, with its structure illustrated in Figure 14b.

The MS model is composed of three main components. The first part is the input layer. The second part handles feature extraction, which is implemented through two parallel CNN branches. Each branch consists of several stacked convolutional blocks, with each block containing a dilated convolution layer, a batch normalization layer, a ReLU layer, and a dropout layer. The third part is the classification and output layer, which includes a fully connected layer, a softmax layer, and the output layer with severity labels as outputs. In this architecture, three-phase current segments and vibration segments serve as the input signals. Each current segment has a fixed size of $1 \times 3000 \times 3$, while each vibration segment has a fixed size of $1 \times 2048 \times 3$. The feature extraction part consists of three convolutional layers, with depths of 5, 9, and 6, and widths of 18, 66, and 38, respectively. The kernel size is set to 1×3, with a dilated factor (d) of 2. The initial learning rate (Linit) is 2.7586 × 10^{-4}, Momentum (M) is 0.9037, the batch is set to 32, L2 regularization (L2R) is 0.0097, and dropout probability (P) is 2.488 × 10^{-5}.

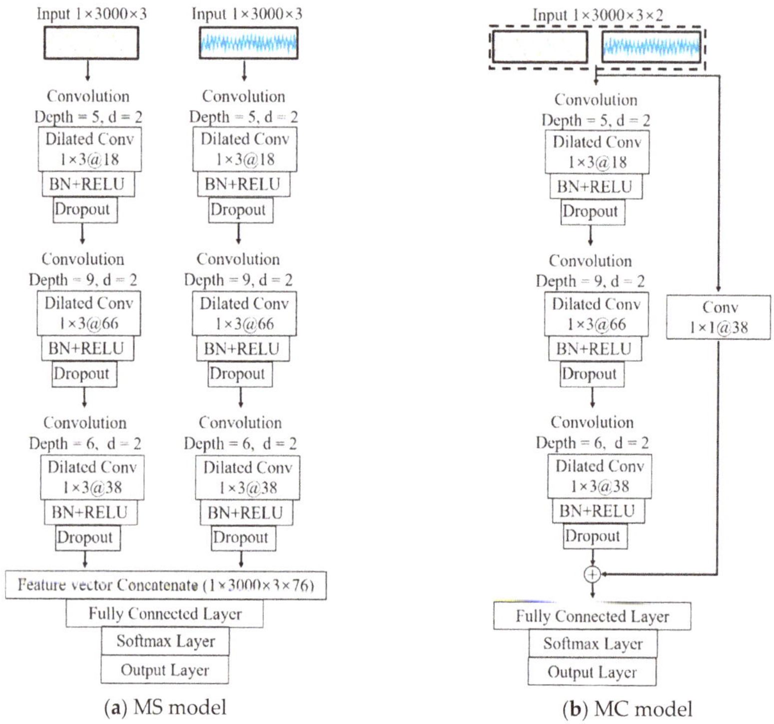

Figure 14. Schematic diagram of the multi-source data-fusion model structure.

In this model, the synchronized current and vibration signals are concatenated directly and passed through the same CNN for fault feature extraction, where the network structure parameters of the feature-extraction layer are consistent with those of the individual branches in the MS model. Furthermore, to compare the effects of fused signals versus single signals, the results of the MS model were compared with those from a residual network (Res) model using three-phase current signal, as well as an improved model that integrates a multi-scale network structure with an attention mechanism (MK-SE-Res model). The comparison results of the four models are shown in Figure 15.

Figure 15a presents a comparative trend of validation accuracy for four models as training epochs progress. From the figure, it is evident that the MS model exhibits the fastest ascent in accuracy, ultimately achieving a validation accuracy of 98.99%. The MK-SE-Res model follows closely, with a relatively rapid ascent and a final validation accuracy of 98.25%. In comparison, the Res model outperforms the MC model in its ascent rate, achieving a final validation accuracy of 97.35%, while the MC model reaches a final validation accuracy of 96.32%. Notably, the MC model's final validation accuracy is 1.03% lower than that of the Res model, which was trained solely on current signal data. In contrast, the MS model improves its final validation accuracy by 0.74% compared to the MK-SE-Res model, which performs best with current signal data. For the MC model, the introduction of multi-source fusion data not only leads to a decrease in final validation accuracy but also slows down the training efficiency throughout the training process, resulting in significant fluctuations in its curve. Conversely, the MS model, with the introduction of multi-source data, not only enhances the final validation accuracy but also accelerates the rate of change in validation accuracy concerning training epochs, leading to a notable improvement in overall stability during training. Figure 15b illustrates the trend of error loss for the four models as training epochs progress. It can be observed that in the early stages of training, both the MC and MK-SE-Res models exhibit significant error loss. As the training epochs increase, the error loss for all four models rapidly decreases, with the MS model achieving the lowest final error loss, demonstrating its superior generalization ability.

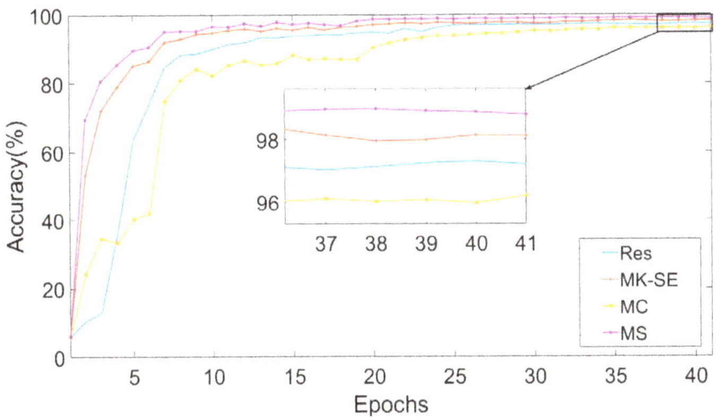

(**a**) Validation accuracy

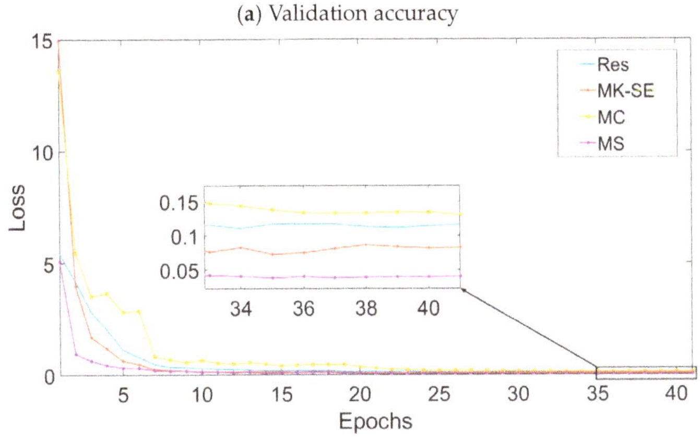

(**b**) Error loss

Figure 15. Trends of validation accuracy and error loss for four models as training epochs change.

To comprehensively analyze the performance of the MS fault-diagnosis model and the MC model on samples with varying fault levels in the validation set, the confusion matrices of these models are plotted and compared with those of the Res model and the MK-SE-Res model. The results are shown in Figures 16–19, where Figure 16 represents the confusion matrix for the Res model, Figure 17 for the MK-SE-Res model, Figure 18 for the MC model, and Figure 19 for the MS model. In the confusion matrices, the left side lists the actual fault level labels of the test set data, while the bottom displays the predicted fault level labels according to the respective fault-diagnosis models used. The values along the diagonal of the confusion matrix indicate the number of samples where the actual and predicted labels match, while the other positions represent the number of samples where the predicted labels do not correspond to the actual labels.

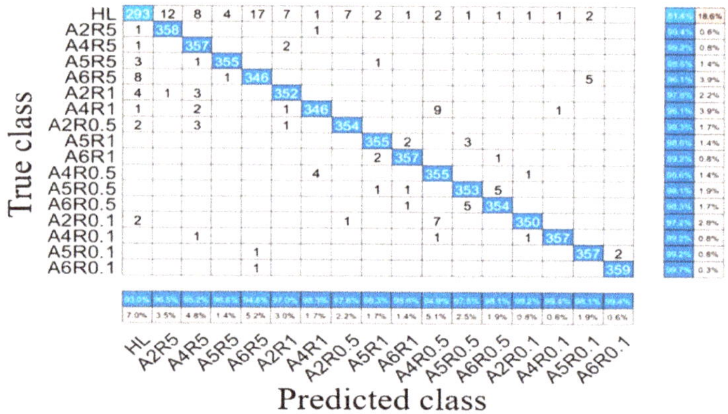

Figure 16. The confusion matrix of the Res model.

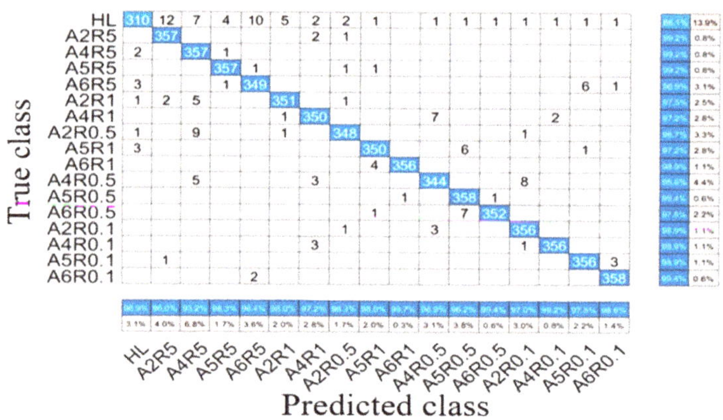

Figure 17. The confusion matrix of the MK-SE-Res model.

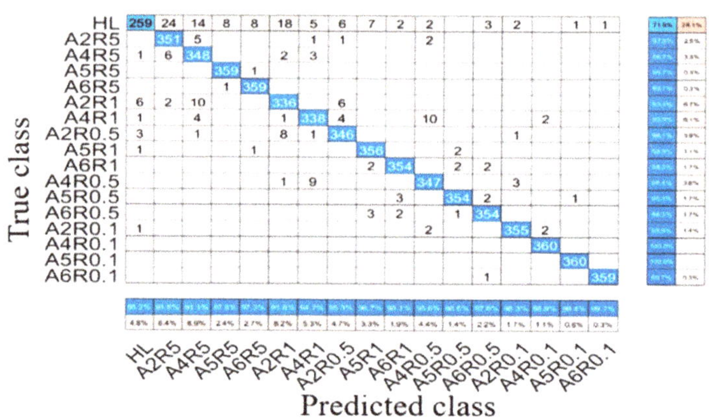

Figure 18. The confusion matrix of the MC model.

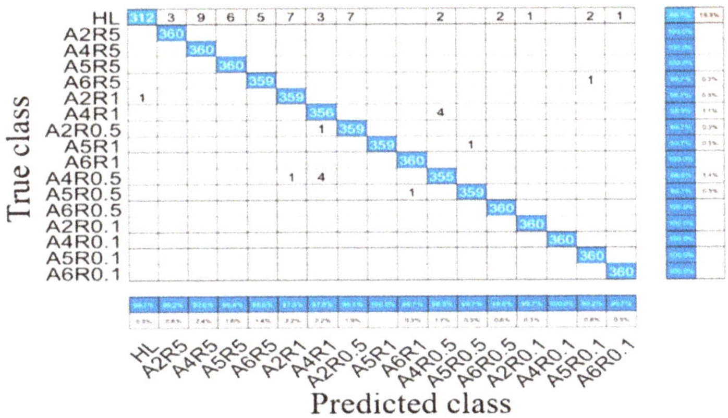

Figure 19. The confusion matrix of the MS model.

The performance of the models across different fault labels is related to two parameters: precision (p) and recall (r). Precision refers to the proportion of correctly predicted labels among the predicted labels in the model's column vector, as indicated by the bottom row of the confusion matrix; recall, on the other hand, indicates the proportion of accurately predicted samples to the total number of samples in that row. There is an inherent trade-off between precision and recall. To balance these two metrics, the F1 score was introduced as a parameter to evaluate the model's performance across various classification labels. The F1 score is the harmonic mean of precision and recall, which allows for a comprehensive consideration of the impact of both parameters, expressed as follows:

$$\mathrm{F1} = \frac{2p \times r}{p + r} \tag{14}$$

To further compare the performance of the five models across different fault-severity classification labels, the F1 scores and overall validation accuracy are calculated based on the precision and recall for each fault-severity label in the confusion matrix of the four models on the validation set. The relevant results are shown in Table 5.

Table 5. Hyperparameters to be optimized.

Label	Res (%)	MK-SE-Res (%)	MC (%)	MS (%)
Acc	97.35	98.25	96.32	98.99
HL	86.81	91.38	81.96	92.72
A2R5	97.95	97.69	94.44	99.59
A4R5	97.14	97.67	93.80	98.77
A5R5	98.61	99.31	98.63	99.17
A6R5	95.45	97.22	98.49	99.17
A2R1	97.37	99.31	92.56	98.76
A4R1	97.19	98.61	94.28	98.34
A2R0.5	98.06	98.75	95.71	98.90
A5R1	98.47	98.74	97.80	99.86
A6R1	98.89	99.44	98.20	99.86
A4R0.5	96.73	97.91	95.99	98.47
A5R0.5	97.78	99.17	98.47	99.72
A6R0.5	98.20	99.58	98.06	99.72
A2R0.1	98.18	98.04	98.47	99.86
A4R0.1	99.30	99.44	99.45	100
A5R0.1	98.62	98.62	99.72	99.59
A6R0.1	99.58	99.44	99.72	99.86

By combining the confusion matrix and Table 5, it is clear that the MS model, having the highest overall validation accuracy, has the fewest misclassified samples. Compared to the other three models, the MS model shows significant improvements in both the "false positive rate" and "false negative rate", with misclassified samples primarily concentrated in the fault-severity labels "HL", "A4R1", and "A4R0.5". In contrast, the MC model has the highest number of misclassified samples among the four models, mainly in the lighter fault-severity labels, and the number of misclassified samples decreases as fault severity increases. As shown in Table 5, the proposed model performs best in F1 scores across 14 out of 17 different fault types.

The four deep learning models mentioned above are all black box models. After training, these models are capable of extracting the necessary features for identifying the degree of ITSC faults from the given dataset. To visually illustrate the features learned by these deep learning models from the ITSC dataset samples, this paper compares the features of the input raw signal dataset with those learned at the final output layer. Since the features at each layer of the models consist of high-dimensional vectors, dimensionality reduction is necessary before visualization. This paper employs the t-SNE (t-distributed stochastic neighbor embedding) algorithm to map high-dimensional features or data samples into two-dimensional space for visualization purposes.

Figure 20 presents a comparison of the two-dimensional feature distributions of the input layer and the output layers of the four models. The feature map uses 17 colors, with each color representing a different fault degree label, and each point corresponding to a data sample. Figure 20a shows the feature distribution of the input layers for each model, revealing a chaotic distribution where sample points of different colors overlap significantly, making it difficult to discern the fault degree based solely on the input data. Figure 20b,e depict the feature distributions of the final output layers for the four ITSC fault-diagnosis models. A comparison indicates that the feature distributions in the output layers of the four models exhibit a marked intra-class clustering characteristic compared to the input layer's feature distribution, with samples of the same fault degree label being more concentrated. Notably, in the feature distribution of the MS model, there are fewer misclassified sample points, and the boundaries between different fault degree labels are clearer, with greater distances between samples, demonstrating good class separation characteristics and indicating a stronger feature learning ability.

Through the analysis of the results, it is evident that compared to the fault-diagnosis model that uses only current signals, the MS model employing multi-stream high-level feature fusion of current and vibration signals demonstrates superior performance in terms of final validation accuracy, error loss, F1 score across various fault-severity classifications, and the model's feature learning capability. This indicates that combining current and vibration signals for multi-stream high-level feature fusion can more effectively leverage the unique characteristics and fault-feature information of different signals, facilitating comprehensive collection of the motor's operating status and thereby enhancing the overall performance of the ITSC fault-diagnosis model.

5.2. Limitation Discussion

From the above analysis, it can be concluded that the algorithm focuses primarily on improving the accuracy of identifying the severity of early ITSC faults under simulated operating conditions that reflect those likely to occur in actual motors. The algorithm assumes that the training data used for the model is sufficiently large and evenly distributed. However, in real-world applications, insufficient or imbalanced data is a common issue, which poses a challenge to the practical deployment of the proposed algorithm. Although the literature suggests that transfer learning has the potential to address this problem, further research is needed to determine how it can be effectively applied in this context. Another practical challenge is that signals may experience significant interference during the data-acquisition process, including electromagnetic interference, inherent vibration noise from the equipment, vibration disturbances, and transmission noise. These factors can limit the performance of the algorithm in real-world applications. Therefore, further research is needed to improve the algorithm's robustness under these interference conditions.

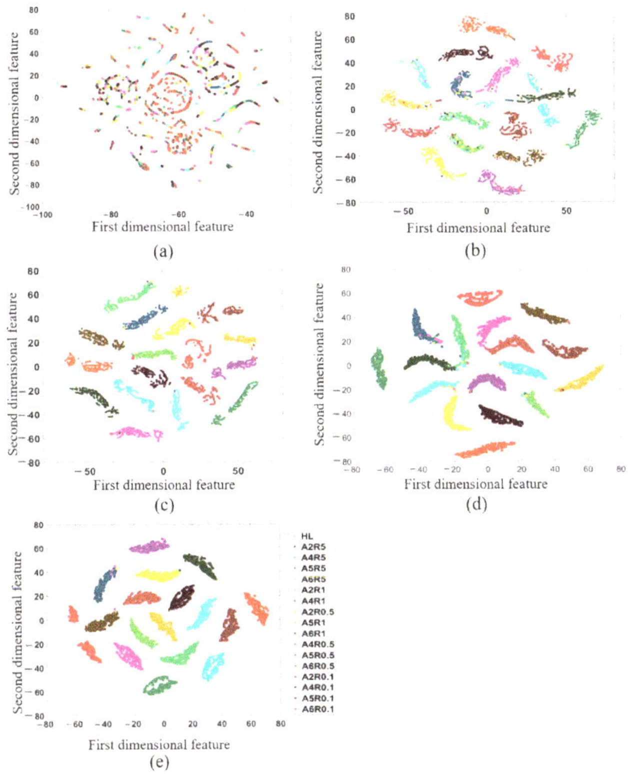

Figure 20. Comparison chart of features extracted by the four models. (**a**) Feature map of the validation dataset. (**b**) Feature map extracted by the Res model. (**c**) Feature map extracted by the MK-SE-Res model. (**d**) Feature map extracted by the MC model. (**e**) Feature map extracted by the MS model.

6. Conclusions

In this article, a novel multi-source data-fusion algorithm was proposed for the ITSC fault diagnosis. The results indicate that, compared to using only current signals, the ITSC fault-diagnosis model that combines current and vibration signals for feature fusion performs better in terms of validation accuracy, error loss, F1 score, and feature learning capability. The following conclusions can be drawn. First, an indicator suitable for early-stage ITSC fault-severity analysis is derived from the equivalent circuit. Second, a feature-level multi-source data-fusion algorithm based on CNN has been introduced. This algorithm utilizes Bayesian optimization for hyperparameter tuning and integrates current and vibration signal features, enhancing the richness of the fault-feature space, and thereby improving the accuracy of ITSC fault-severity identification. To achieve synchronization of multi-source signals during experiments, a signal synchronization method has been proposed to construct a synchronized dataset of current and vibration signals. By calculating the maximum cross-correlation of the synchronized signals collected from two devices, successful synchronization of the current and vibration signals was achieved. Finally, the experimental results indicate that, among the four comparative models, the proposed MS model achieved the highest final validation accuracy, close to 98.99%, with the smallest error loss, below 0.04. In the F1 scores for 17 classification tasks, the MS model outperformed the others in 14 cases, demonstrating the strongest feature learning capability, which validates the effectiveness of the proposed multi-source data-fusion model.

Unfortunately, the proposed algorithm primarily focuses on improving the accuracy of identifying the severity of early ITSC faults under experimental conditions, without addressing issues such as signal interference, and insufficient or imbalanced sample sizes that may arise in real-world applications. To enhance the algorithm's performance in practical scenarios, the next step will be to investigate methods for overcoming these challenges.

Author Contributions: Conceptualization, M.W., Q.S. and W.L.; methodology, M.W. and W.L.; software, M.W. and W.L.; validation, M.W.; formal analysis, M.W. and W.L.; investigation, M.W., Q.S. and W.L.; resources, Q.S.; data curation, M.W. and W.L.; writing—original draft preparation, M.W.; writing—review and editing, M.W., H.Z., Y.L. and W.L.; visualization, M.W.; supervision, Q.S.; project administration, M.W. and Q.S.; funding acquisition, Q.S. All authors have read and agreed to the published version of the manuscript.

Funding: This research received no external funding.

Data Availability Statement: The raw data supporting the conclusions of this study are available from the corresponding author upon reasonable request. (wangmingsheng_08@163.com).

Conflicts of Interest: The authors declare no conflicts of interest.

References

1. Xie, F.; Wang, Y.; Wang, G.; Sun, E.; Fan, Q.; Song, M. Fault Diagnosis of Rolling Bearings in Agricultural Machines Using SVD-EDS-GST and ResViT. *Agriculture* **2024**, *14*, 1286. [CrossRef]
2. Xie, F.; Sun, E.; Wang, L.; Wang, G.; Xiao, Q. Rolling Bearing Fault Diagnosis in Agricultural Machinery Based on Multi-Source Locally Adaptive Graph Convolution. *Agriculture* **2024**, *14*, 1333. [CrossRef]
3. Wang, J.; Lu, Z.; Wang, G.; Hussain, G.; Zhao, S.; Zhang, H.; Xiao, M. Research on Fault Diagnosis of HMCVT Shift Hydraulic System Based on Optimized BPNN and CNN. *Agriculture* **2023**, *13*, 461. [CrossRef]
4. Lagarde, Q.; Beillard, B.; Marcuzzi, D.; Mazen, S.; Leylavergne, J. Stray Currents in Livestock Farming: Electrical Diagnosis in Farms. *Agriculture* **2023**, *13*, 2010. [CrossRef]
5. Wen, C.; Zhang, J.; Zheng, K.; Li, H.; Ling, L.; Meng, Z.; Fu, W.; Yan, B. Accelerated Verification Method for the Reliability of the Motor Drive Mechanism of the Corn Precision Seed-Metering Device. *Comput. Electron. Agric.* **2023**, *212*, 108163. [CrossRef]
6. Sewiolo, M.; Mystkowski, A. Agriculture Rotary Tedder Fault Diagnosis Based on Evolutionary Convolutional Neural Network with Genetic Algorithm Optimization. In Proceedings of the 2023 27th International Conference on Methods and Models in Automation and Robotics (MMAR), Międzyzdroje, Poland, 22–25 August 2023; IEEE: New York, NY, USA, 2023; pp. 87–92.
7. Zafarani, M.; Bostanci, E.; Qi, Y.; Goktas, T.; Akin, B. Interturn Short-Circuit Faults in Permanent Magnet Synchronous Machines: An Extended Review and Comprehensive Analysis. *IEEE J. Emerg. Sel. Top. Power Electron.* **2018**, *6*, 2173–2191. [CrossRef]
8. Zorig, A.; Hedayati Kia, S.; Chouder, A.; Rabhi, A. A Comparative Study for Stator Winding Inter-Turn Short-Circuit Fault Detection Based on Harmonic Analysis of Induction Machine Signatures. *Math. Comput. Simul.* **2022**, *196*, 273–288. [CrossRef]
9. Parai, M.; Srimani, S.; Ghosh, K.; Rahaman, H. Multi-Source Data Fusion Technique for Parametric Fault Diagnosis in Analog Circuits. *Integration* **2022**, *84*, 92–101. [CrossRef]
10. Singh, A.; Nawayseh, N.; Singh, H.; Dhabi, Y.K.; Samuel, S. Internet of Agriculture: Analyzing and Predicting Tractor Ride Comfort through Supervised Machine Learning. *Eng. Appl. Artif. Intell.* **2023**, *125*, 106720. [CrossRef]
11. He, Y.; Tang, H.; Ren, Y.; Kumar, A. A Deep Multi-Signal Fusion Adversarial Model Based Transfer Learning and Residual Network for Axial Piston Pump Fault Diagnosis. *Measurement* **2022**, *192*, 110889. [CrossRef]
12. Xu, L.; Zhang, G.; Zhao, S.; Wu, Y.; Xi, Z. Fault Diagnosis of Tractor Transmission System Based on Time GAN and Transformer. *IEEE Access* **2024**, *12*, 107153–107169. [CrossRef]
13. Lee, H.; Jeong, H.; Koo, G.; Ban, J.; Kim, S.W. Attention Recurrent Neural Network-Based Severity Estimation Method for Interturn Short-Circuit Fault in Permanent Magnet Synchronous Machines. *IEEE Trans. Ind. Electron.* **2021**, *68*, 3445–3453. [CrossRef]
14. Zhu, J.; Hu, T.; Jiang, B.; Yang, X. Intelligent Bearing Fault Diagnosis Using PCA–DBN Framework. *Neural Comput. Appl.* **2020**, *32*, 10773–10781. [CrossRef]
15. Zhu, Z.; Peng, G.; Chen, Y.; Gao, H. A Convolutional Neural Network Based on a Capsule Network with Strong Generalization for Bearing Fault Diagnosis. *Neurocomputing* **2019**, *323*, 62–75. [CrossRef]
16. Husari, F.; Seshadrinath, J. Stator Turn Fault Diagnosis and Severity Assessment in Converter-Fed Induction Motor Using Flat Diagnosis Structure Based on Deep Learning Approach. *IEEE J. Emerg. Sel. Top. Power Electron.* **2023**, *11*, 5649–5657. [CrossRef]
17. Husari, F.; Seshadrinath, J. Early Stator Fault Detection and Condition Identification in Induction Motor Using Novel Deep Network. *IEEE Trans. Artif. Intell.* **2022**, *3*, 809–818. [CrossRef]

18. Ortego, P.; Diez-Olivan, A.; Ser, J.D.; Veiga, F.; Penalva, M.; Sierra, B. Evolutionary LSTM-FCN Networks for Pattern Classification in Industrial Processes. *Swarm Evol. Comput.* **2020**, *54*, 100650. [CrossRef]
19. Wang, J.; Ma, Y.; Zhang, L.; Gao, R.X., Wu, D. Deep Learning for Smart Manufacturing: Methods and Applications. *J. Manuf. Syst.* **2018**, *48*, 144–156. [CrossRef]
20. Akhmetov, Y.; Nurmanova, V.; Bagheri, M.; Zollanvari, A.; Gharehpetian, G.B. A New Diagnostic Technique for Reliable Decision-Making on Transformer FRA Data in Interturn Short-Circuit Condition. *IEEE Trans. Ind. Inf.* **2021**, *17*, 3020–3031. [CrossRef]
21. Yuan, X.-H.; He, Y.-L.; Liu, M.-Y.; Wang, H.; Wan, S.-T.; Vakil, G. Impact of the Field Winding Interturn Short-Circuit Position on Rotor Vibration Properties in Synchronous Generators. *Math. Probl. Eng.* **2021**, *2021*, 9236726. [CrossRef]
22. Lanciotti, N.; Ojeda, J.; Gabsi, M.; Boukhobza, T. Detection and Localization of Interturn Short-Circuit Fault by Analysis of Stator Accelerations Spectrum in Five-Phase Flux Switching Machine for HEV Application. In Proceedings of the 2020 Fifteenth International Conference on Ecological Vehicles and Renewable Energies (EVER), Monte-Carlo, Monaco, 10–12 September 2020; IEEE: New York, NY, USA, 2020; pp. 1–10.
23. He, Y.-L.; Wang, T.; Sun, K.; Wang, X.-L.; Peng, B.; Wan, S.-T. Enhanced Characteristic Vibration Signal Detection of Generator Based on Time-Wavelet Energy Spectrum and Multipoint Optimal Minimum Entropy Deconvolution Adjusted Method. *Math. Probl. Eng.* **2020**, *2020*, 6916289. [CrossRef]
24. He, Y.-L.; Xu, M.-X.; Zhang, W.; Wang, X.-L.; Lu, P.; Gerada, C.; Gerada, D. Impact of Stator Interturn Short Circuit Position on End Winding Vibration in Synchronous Generators. *IEEE Trans. Energy Convers.* **2021**, *36*, 713–724. [CrossRef]
25. He, Y.-L.; Deng, W.-Q.; Peng, B.; Ke, M.-Q.; Tang, G.-J.; Wan, S.-T.; Liu, X.-Y. Stator Vibration Characteristic Identification of Turbogenerator among Single and Composite Faults Composed of Static Air-Gap Eccentricity and Rotor Interturn Short Circuit. *Shock. Vib.* **2016**, *2016*, 5971081. [CrossRef]
26. Obeid, N.H.; Boileau, T.; Nahid-Mobarakeh, B. Modeling and Diagnostic of Incipient Interturn Faults for a Three-Phase Permanent Magnet Synchronous Motor. *IEEE Trans. Ind. Appl.* **2016**, *52*, 4426–4434. [CrossRef]
27. Seshadrinath, J.; Singh, B.; Panigrahi, B.K. Vibration Analysis Based Interturn Fault Diagnosis in Induction Machines. *IEEE Trans. Ind. Inf.* **2014**, *10*, 340–350. [CrossRef]
28. Qi, Y.; Bostanci, E.; Zafarani, M.; Akin, B. Severity Estimation of Interturn Short Circuit Fault for PMSM. *IEEE Trans. Ind. Electron.* **2019**, *66*, 7260–7269. [CrossRef]
29. Qi, Y.; Bostanci, E.; Gurusamy, V.; Akin, B. A Comprehensive Analysis of Short-Circuit Current Behavior in PMSM Interturn Short-Circuit Faults. *IEEE Trans. Power Electron.* **2018**, *33*, 10784–10793. [CrossRef]
30. Hang, J.; Zhang, J.; Cheng, M.; Huang, J. Online Interturn Fault Diagnosis of Permanent Magnet Synchronous Machine Using Zero-Sequence Components. *IEEE Trans. Power Electron.* **2015**, *30*, 6731–6741. [CrossRef]
31. Xie, T.; Huang, X.; Choi, S.-K. Intelligent Mechanical Fault Diagnosis Using Multisensor Fusion and Convolution Neural Network. *IEEE Trans. Ind. Inf.* **2022**, *18*, 3213–3223. [CrossRef]
32. Xia, M.; Li, T.; Xu, L.; Liu, L.; de Silva, C.W. Fault Diagnosis for Rotating Machinery Using Multiple Sensors and Convolutional Neural Networks. *IEEE/ASME Trans. Mechatron.* **2018**, *23*, 101–110. [CrossRef]
33. Chen, H.; Hu, N.; Cheng, Z.; Zhang, L.; Zhang, Y. A Deep Convolutional Neural Network Based Fusion Method of Two-Direction Vibration Signal Data for Health State Identification of Planetary Gearboxes. *Measurement* **2019**, *146*, 268–278. [CrossRef]
34. Azamfar, M.; Jia, X.; Pandhare, V.; Singh, J.; Davari, H.; Lee, J. Detection and Diagnosis of Bottle Capping Failures Based on Motor Current Signature Analysis. *Procedia Manuf.* **2019**, *34*, 840–846. [CrossRef]
35. Xiao, M.; Yang, B.; Wang, S.; Zhang, Z.; Tang, X.; Kang, L. A Feature Fusion Enhanced Multiscale CNN with Attention Mechanism for Spot-Welding Surface Appearance Recognition. *Comput. Ind.* **2022**, *135*, 103583. [CrossRef]
36. Jin, Y.; Chen, C.; Zhao, S. Multisource Data Fusion Diagnosis Method of Rolling Bearings Based on Improved Multiscale CNN. *J. Sens.* **2021**, *2021*, 2251530. [CrossRef]
37. Li, Z.; Wang, Y.; Ma, J. Fault Diagnosis of Motor Bearings Based on a Convolutional Long Short-Term Memory Network of Bayesian Optimization. *IEEE Access* **2021**, *9*, 97546–97556. [CrossRef]
38. Bai, S.; Kolter, J.Z.; Koltun, V. An Empirical Evaluation of Generic Convolutional and Recurrent Networks for Sequence Modeling. *arXiv* **2018**. [CrossRef]
39. Hu, J.; Shen, L.; Albanie, S.; Sun, G.; Wu, E. Squeeze-and-Excitation Networks. *IEEE Trans. Pattern Anal. Mach. Intell.* **2019**, *42*, 2011–2023. [CrossRef]
40. Zhang, Y.; Qu, J.; Fang, X.; Luo, G. Motor Bearing Fault Diagnosis Based on Multi-Feature Fusion and PSO-BP. In Proceedings of the 2021 IEEE 4th Student Conference on Electric Machines and Systems (SCEMS), Huzhou, China, 1–3 December 2021; IEEE: New York, NY, USA, 2021; pp. 1–5.
41. Gelbart, M.A.; Snoek, J.; Adams, R.P. Bayesian Optimization with Unknown Constraints. *arXiv* **2014**. [CrossRef]

agriculture

MDPI

Article

Fault Diagnosis of Rolling Bearings in Agricultural Machines Using SVD-EDS-GST and ResViT

Fengyun Xie [1,2,*], Yang Wang [1], Gan Wang [1], Enguang Sun [1], Qiuyang Fan [1] and Minghua Song [1]

[1] School of Mechanical Electrical and Vehicle Engineering, East China Jiaotong University, Nanchang 330013, China; wwyy0130@163.com (Y.W.); wanggan813@163.com (G.W.); sngzm999@163.com (E.S.); fqy18532671715@163.com (Q.F.); 13321775564@163.com (M.S.)

[2] State Key Laboratory of Performance Monitoring Protecting of Rail Transit Infrastructure, East China Jiaotong University, Nanchang 330013, China

* Correspondence: 1493@ecjtu.edu.cn

Abstract: In the complex and harsh environment of agriculture, rolling bearings, as the key transmission components in agricultural machinery, are very prone to failure, so research on the intelligent fault diagnosis of agricultural machinery components is critical. Therefore, this paper proposes a new method based on SVD-EDS-GST and ResNet-Vision Transformer (ResViT) for the fault diagnosis of rolling bearings in agricultural machines. Firstly, an experimental platform for rolling bearing failure in agricultural machinery is built, and one-dimensional vibration signals are obtained using acceleration sensors. Next, the signal is preprocessed for noise reduction using singular value decomposition (SVD) combined with the energy difference spectrum (EDS) to solve for the interference of complex noise and redundant components in the vibration signal. Secondly, generalized S-transform (GST) is used to process vibration signals into images. Then, the ResViT model is proposed, where the ResNet34 network is used to replace the image chunking mechanism in the original Vision Transformer model for feature extraction. Finally, an improved Vision Transformer (ViT) is utilized to synthesize global and local information for fault classification. The experimental results show that the proposed method's average accuracy in rolling bearing fault classification for agricultural machinery reaches 99.08%. In addition, compared with SVD-EDS-GST-CNN, SVD-EDS-GST-LSTM, STFT-ViT, GST-ViT, and SVD-EDS-GST-ViT, the accuracy rate was improved by 3.5%, 3.84%, 4.8%, 8.02%, and 0.56%, and the standard deviation was also minimized.

Keywords: agricultural machinery rolling bearing; fault diagnosis; SVD-EDS; GST; ResViT

1. Introduction

Today, the world is undergoing rapid changes, including population growth and changes in the global climate and ecology, so it is necessary to immediately provide sustainable and safe intelligent solutions for food production. Intelligent fault diagnosis research on agricultural machinery components is critical in this context [1]. Agricultural machinery is widely used in farming, sowing, fertilization, harvesting, and other aspects of modern agricultural production. Given the major role of this machinery, its efficient operation directly affects the efficiency and output of agricultural production [2]. Rolling bearings are a key transmission component in agricultural machinery, responsible for supporting rotating shafts and reducing friction, thereby improving mechanical efficiency and reducing energy loss. However, the operating conditions of agricultural machinery and other mechanical environments are very different. The operating environment of agricultural machinery is particularly harsh and changeable, including exposure to dust, moisture, and changing loads, and these conditions very easily produce fatigue problems such as increased strain, wear, and insufficient lubrication, which lead to the deterioration of bearing performance and failure [3]. The failure of a rolling bearing not only leads to the shutdown of agricultural machinery but also may cause more serious mechanical damage, resulting in economic losses. Therefore, it is very important to diagnose faults in agricultural machinery rolling bearings [4].

Traditional fault diagnosis methods mainly rely on practical experience and regular maintenance, but this method is low in efficiency, high in cost, and cannot easily provide early warnings [4,5]. With the advancements in artificial intelligence, more and more researchers are using intelligent diagnostic methods to perform fault detections on mechanical equipment [6]. Xie et al. [7] applied an improved SMOTE method to the early fault diagnosis of rolling bearings in agricultural machinery, enhancing the effectiveness of diagnosis under imbalanced data conditions. Wei et al. [8] applied improved DE and improved VMD algorithms for fault diagnosis in combine harvesters, successfully identifying faults in rolling bearings. Liu et al. [9] used ABC-VMD to decompose the signal of rolling bearing faults in a corn harvester, selected the best decomposed signal, and then proposed an improved EfficientNet model for diagnosis, realizing the accurate identification of rolling bearing fault conditions in a heavy noise environment. Fan et al. [10] proposed an algorithm combining FUDL and SC and applied it to tractor fault diagnosis, successfully identifying bearing and gearbox faults. Mystkowski et al. [11] applied different machine learning methods to create a model based on MLP for the fault measurement and diagnosis of rotary hay stands. Luo et al. [12] proposed a learning vector quantization neural network and a CNN VGG-16 network for tractor state recognition, providing theoretical support for angle detection error correction. Wang et al. [13] proposed a kind of infrared thermal imaging based on DCNN-SR for the early fault diagnosis of diesel engines, which showed a good ability to resist interference from temperature fluctuations.

The above methods show the application potential of artificial intelligence algorithms in fault analysis, but they also show that it is difficult to achieve both rapidity and accuracy in a model. In 2017, Google proposed the Transformer model. Traditional neural networks are not easy to parallelize, but the Transformer model shows great potential in solving this problem with its unique self-attention mechanism. Dosovitskiy et al. [14] introduced it into the field of image recognition and built the Vision Transformer (ViT) network structure, which does not rely on recurrent neural networks or convolutional neural networks. Its recognition accuracy and efficiency were found to be significantly improved, compared with those of traditional neural network models, when processing large-scale data sets [15]. This model has been widely applied. For example, Salamai et al. [16] used a ViT network for the classification and detection of agricultural rice diseases. Jamil et al. [17] used ViT for the medical classification and detection of heart valve diseases. Wang et al. [18] combined EfficientNet and Vision Transformer for cross-domain bearing fault diagnosis under mixed working conditions and achieved good results. To sum up, Vision Transformer has been proven to be superior to convolutional networks in many fields. However, when processing classification tasks, the image segmentation method of ViT may lead to the loss of local information in the curve. Therefore, in order to complete the task of fault classification more effectively, ViT needs to be improved.

At present, the vast majority of agricultural machinery fault diagnosis methods based on vibration signals still use acceleration sensors to obtain one-dimensional vibration signals of rolling bearings for analysis and processing, followed by traditional machine learning methods and deep learning methods for pattern recognition and classification [19]. Because the working environment of agricultural machinery is very complicated, the vibration signal is often subject to various types of interference, which complicates fault feature extraction. Converting a one-dimensional sequence signal into an image has potential advantages for feature extraction and differentiation in fault diagnosis. Deep learning also has better diagnostic capabilities than traditional machine learning methods. For example, Wang et al. [20] applied an optimized BPNN and CNN method to the fault diagnosis of a tractor's hydraulic continuously variable transmission shift system, converted the one-dimensional signal into a two-dimensional gray-level map, and adopted a convolutional neural network for diagnosis, which greatly improved the diagnosis accuracy when compared with the one-dimensional signal and BPNN.

In order to solve the above problems, a new fault diagnosis method based on SVD-EDS-GST and ResViT for rolling bearings in agricultural machinery is proposed in this paper. SVD-EDS-GST technology is used to enhance the feature extraction process by breaking down the vibration signals and identifying the most important components that represent fault-related information. Then, the proposed ResViT is used for feature extraction and pattern recognition on the GST fault images. The main contents and innovations of this paper are as follows:

1. An experimental platform for the fault diagnosis of rolling bearings in agricultural machinery is built, and a new fault diagnosis method for such bearings with SVD-EDS-GST and ResViT is proposed.
2. SVD combined with the EDS is proposed to reduce the noise in vibration signals in order to remove the interference of complex noise and redundant components. The vibration signal noise reduction is realized via matrix reconstruction.
3. GST is applied to the vibration signal of rolling bearings in agricultural machinery after noise removal; thus, the one-dimensional vibration signal is converted into a two-dimensional time–frequency image, and a fault data set is established.
4. An improved ViT model combined with ResNet34 (ResViT) is proposed. This model uses a ResNet34 network to replace the image segmentation mechanism in the original Vision Transformer model for feature extraction, which makes the training more efficient and has strong robustness to small perturbations. At the same time, considering that the attention mechanism is not sensitive to the element position when processing global information, relative position coding is used instead of absolute position coding in order to better retain spatial information.

Comparison experiments with the popular CNN, LSTM, and Transformer deep learning models verify that the proposed fault diagnosis model combining SVD-EDS-GST with ResViT has good test accuracy and stability.

The framework of the rest of this paper is as follows: The second section introduces relevant theories, including SVD, EDS, GST, and ViT. The third section introduces the fault diagnosis model based on SVD-EDS-GST and ResViT for rolling bearings in agricultural machinery. The fourth section describes the setup of the experimental platform for rolling bearing failure in agricultural machinery and introduces the process of data acquisition. In Section 5, the experimental results are analyzed from many aspects and compared with other models. The sixth section presents the conclusions.

2. Basic Principles

2.1. SVD-EDS

As a key matrix decomposition technique, SVD plays an important role in numerical analysis and linear algebra. In particular, in the processing of rolling bearing vibration signals, SVD technology greatly retains the fault information of the original vibration signal by removing noise [21]. However, the methods commonly used to determine the effective order of SVD often rely on the user's experience, resulting in their unsatisfactory noise reduction performance when applied to rolling bearing vibration signals [22]. To overcome this challenge, a new strategy, singular value decomposition combined with the energy difference spectrum (EDS), is proposed in this paper to determine the order of SVD more precisely, so as to improve the noise reduction effect on the vibration signals of agricultural machinery rolling bearings.

The one-dimensional signal $A = \{a_1, a_2, \cdots, a_N\}$ is transformed into a two-dimensional form using the Hankel matrix [23], as shown in Equation (1):

$$A_{m\times n} = \begin{vmatrix} a(1) & \cdots & a(n) \\ \vdots & \ddots & \vdots \\ a(m) & \cdots & a(N) \end{vmatrix} = D_{m\times n} + W_{m\times n} \tag{1}$$

Here, $A_{m\times n}$ represents the Hankel matrix, the noise signal is $W_{m\times n}$, and the processed signal is $D_{m\times n}$. When $m = N/2$, that is, when m is half the length of the original fault signal, the noise reduction effect of the Hankel matrix is generally more obvious, and the fault characteristics of the original signal can be retained [21].

In this study, the energy difference spectrum (EDS) is used to determine the singular value order, as shown in Equation (2):

$$E = \sum_{i=1}^{q} \sigma_i^2 \tag{2}$$

Here, the energy is represented by E, and the singular value is represented by σ_i. Using Equation (3) eliminates the influence of energy differences between different signals:

$$p(i) = \frac{\sigma_i^2 - \sigma_{i+1}^2}{E} \tag{3}$$

Here, the sequence $p(i)$ represents the energy difference spectrum. The energy change in the singular values in adjacent order can help to identify the useful signal and the useless signal. By finding the corresponding singular value in the energy difference spectrum, the noise signal of the rolling bearing can be effectively removed.

2.2. GST

The generalized S-transform method combines time–domain and frequency–domain analysis to obtain instantaneous frequency information from a signal [24]. The GST is expressed as shown in Equation (4):

$$GST_x(t,f) = \int_{-\infty}^{+\infty} x(\tau)w(t-\tau)e^{-\frac{(t-\tau)^2 f^2}{2}} e^{-j2\pi f\tau} d\tau \tag{4}$$

In Formula (4), $GST_x(t,f)$ represents the S-transform, $x(\tau)$ represents the signal to be analyzed, and the translation quantity is represented by τ. $w(t)$ represents the Gaussian window function, $w(t) = \frac{1}{\sigma(f)\sqrt{2\pi}} \exp\left(\frac{-t^2}{2\sigma(f)^2}\right)$, and $\sigma(f) = 1/|f|$. By introducing parameter m into the S-transform formula to adjust the width of the Gaussian window, GST can be significantly improved [25]. The improved GST is expressed as shown in Equation (5):

$$GST_x(t,f)_2 = \int_{-\infty}^{+\infty} x(t-\tau)w(t)e^{-\frac{(t-\tau)^2 f^2}{2}} e^{-j2\pi f\tau} d\tau \tag{5}$$

Here, $w(t) = \frac{1}{\sigma_m(f)\sqrt{2\pi}} \exp\left(\frac{-t^2}{2\sigma_m(f)^2}\right)$, and $\sigma_m(f) = 1/|f|^m$.

2.3. ResNet

ResNet (Residual Network) is a deep convolutional neural network architecture proposed by Microsoft Research Asia [26]. By introducing residual learning, very deep networks can be trained. The ResNet model is an improvement on VGG due to the introduction of residual network units, which use skip connections to directly add inputs to outputs, thus propagating information more effectively and making deep network structures easier to optimize and train. The most common residual blocks include Basic Block and Bottleneck Block [27], as shown in Figure 1a,b.

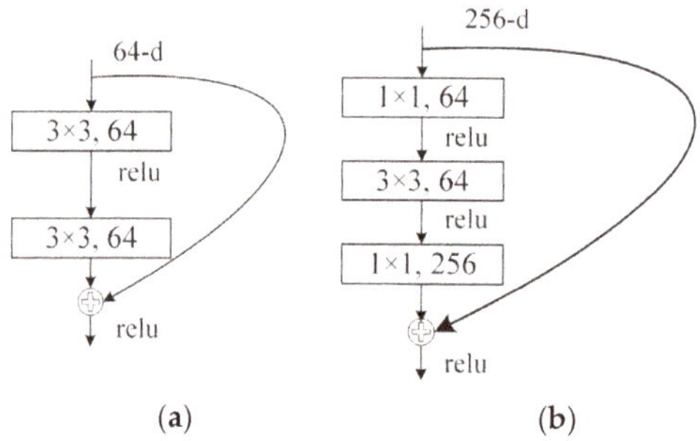

Figure 1. Residual structures: (**a**) Basic Block; (**b**) Bottleneck Block.

Basic Block consists of two convolution layers containing 3 × 3 convolution cores and a skip connection that adds the input directly to the output. Bottleneck Block contains a 1 × 1 convolution layer (to reduce the input dimensions), a 3 × 3 convolution layer, and a re-use 1 × 1 convolution layer (to restore dimensions). This design can significantly reduce the number of parameters and network computations, improving network performance. There are many different choices for network depth in ResNet [26].

2.4. Vision Transformer

In 2020, Dosovitskiy et al. proposed an image classification model with Vision Transformer (ViT) architecture [14]. Its structure is shown in Figure 2.

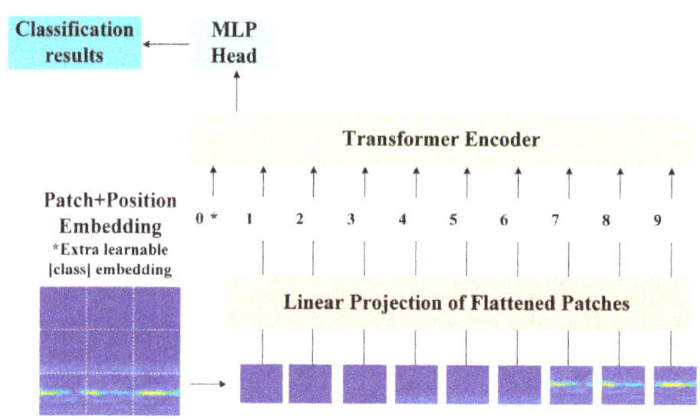

Figure 2. Vision Transformer model structure.

1. Patch and Position Embedding module. First, the GST two-dimensional fault image is divided into image blocks of the same size, and then these image blocks are linearly transformed and projected into a low-dimensional space, and the position encoding containing spatial information is added as input to the Transformer Encoder layer [28].
2. Transformer Encoder module. Transformer Encoder can efficiently extract the features of input data [29]. Its structure is shown in Figure 3. Input data from the Transformer Encoder are first normalized to improve the training stability. The data capture the in-sequence dependencies through a multi-head self-attention mechanism, and dropout is applied to prevent overfitting. After that, the processed data are fused with the original input using a residual connection and are normalized again. Finally, the data are fed into the MLP block for feature transformation. The original information is retained through residual connections [30].

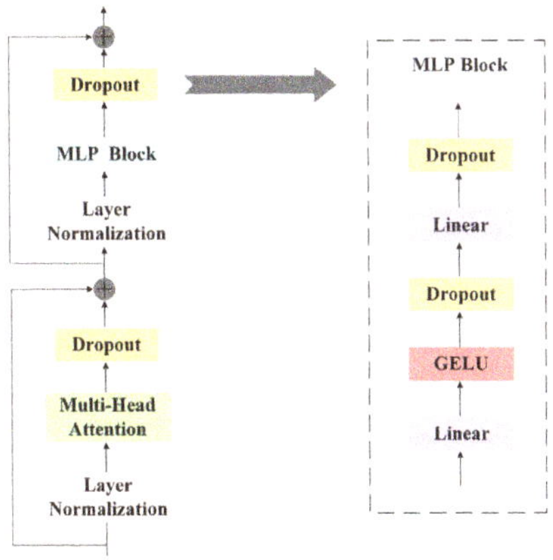

Figure 3. Transformer Encoder model structure.

3. Categorization Component. The features output by the Transformer Encoder are fed into the MLP head for classification, resulting in fault classification outcomes.

3. Fault Diagnosis Model

3.1. ResViT Model

In image processing, the first problem to be solved is to transform the two-dimensional vibration image into a two-dimensional matrix that the model can receive as input. Vision Transformer divides the input two-dimensional image into image blocks, then transforms these image blocks into one-dimensional sequences and uses the self-attention mechanism to learn the relationship between the position and sequence of each image block. Finally, the learned image data features are used in classification tasks. When expanding into a one-dimensional sequence form, a 16 × 16 two-dimensional convolution with a step size of 16 is usually used. If the step size and convolution kernel are too large, they may not be effective in capturing local information in the image [31].

In order to solve the above problems and improve the feature extraction performance of the model, a ResNet network was used to replace the Vision Transformer model for feature extraction. The ResNet network gradually extracts the features of images through cascading convolutional and pooling layers, which can capture low-level to high-level feature information. Compared with Vision Transformer, in which the image needs to be chunked and a self-attention mechanism is introduced, ResNet can show better computational performance and speed when dealing with a large amount of image information.

After a comprehensive analysis, a specific Residual Network model, ResNet34, was used. It has certain advantages over both smaller (such as ResNet18) and larger (such as ResNet50 or larger) ResNet models [27]. ResNet34 may offer a good balance of keeping the computational complexity low while achieving high performance. ResNet34 is shown in Figure 4.

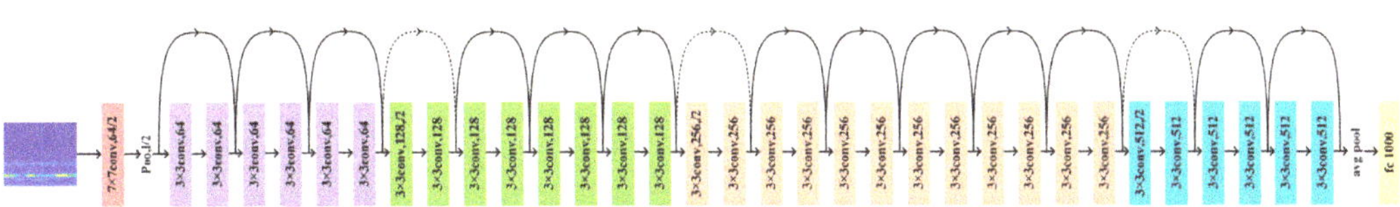

Figure 4. Resnet34.

At the end of each ResNet34 base block, there is a convolution layer that adjusts the dimensions to ensure that the input has the same dimensions as the output of the block. At the beginning of each stage, there is a subsampled convolution layer that halves the size of the feature map. These downsampling steps (stride = 2) result in the size of each stage being reduced to half its original size [27]. To sum up, when an image passes through the ResNet34 model, the total downsampling multiple is the (2 + 2 + 2 + 2)th power of 2—that is, 16-times downsampling. The specific process is shown in Figure 5.

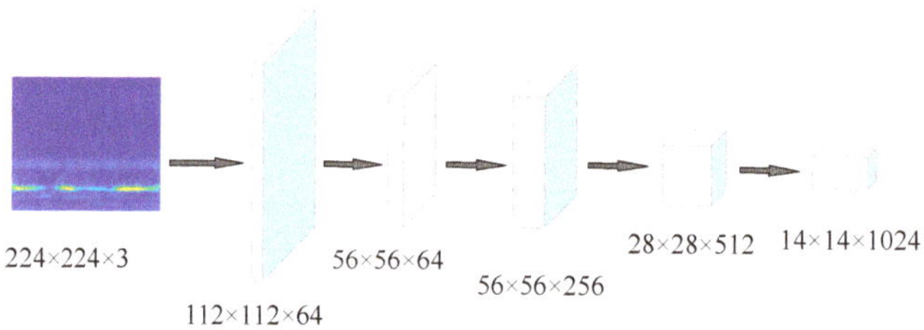

Figure 5. The ResNet process.

Considering that the attention mechanism is not sensitive to the element position when processing global information, relative position coding was used instead of absolute position coding in order to better retain spatial information. The parameters used in the Vision Transformer model are shown in Table 1.

Table 1. Improved ViT parameters.

Patch Size	Layers	Hidden Size D	MLP Size	Heads	Params
14 × 14	12	768	3072	12	86M

In the traditional Vision Transformer model, a larger patch size is usually used, such as 16 × 16 or 32 × 32. Here, in order to match the ResNet processed image data input, a patch size of 14 × 14 was chosen. The use of a smaller patch size (e.g., 14 × 14) also increases the model's sensitivity to image layers.

3.2. Proposed SVD-EDS-GST and ResViT Fault Diagnosis Model

Fault diagnosis of rolling bearings in agricultural machines using SVD-EDS-GST and ResViT is shown in Figure 6.

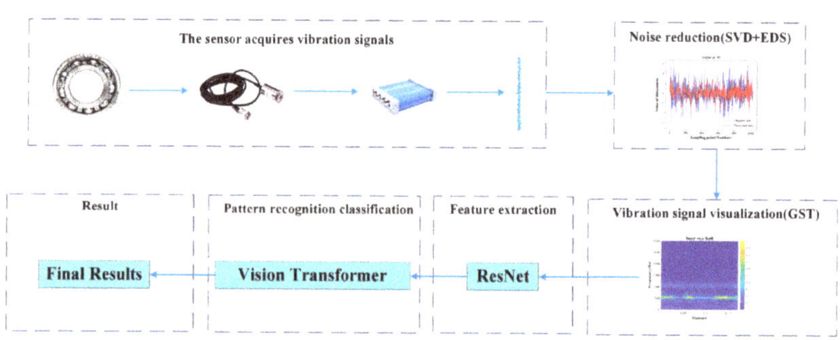

Figure 6. Overall model of SVD-EDS-GST and ResViT.

The model flow is divided into the following parts: (1) Firstly, a piezoelectric acceleration sensor is used to obtain the vibration signal of a rolling bearing in agricultural machinery. The analog signal from the sensor is received by the data acquisition card and converted into a digital signal, which is then transmitted to the computer through the computer interface and processed by YE7600 software to convert the file into a .mat format.

(2) The combined SVD and energy difference spectrum method is used to reduce the noise in the signal. (3) GST is used to process vibration signals into images. (4) A ResNet34 model is used to replace the image segmentation mechanism in the original Vision Transformer model for feature extraction, and considering that the attention mechanism is not sensitive to the element position when processing global information, relative position coding was used instead of absolute position coding. (5) A Vision Transformer model is used to recognize patterns in the image data and output the training results. (6) After the training is completed, the trained model is used to conduct a rolling bearing fault classification test.

4. Experimental Platform for Rolling Bearing Failure in Agricultural Machinery

The rolling bearing failure experimental platform constructed in this paper mainly consisted of a three-phase asynchronous motor (Model:YE3-100L2-4, Manufacturer: Zhejiang Jinsu Electric Motor Co., Ltd., Taizhou, China), gearbox reducer (Model:JZQ250, Manufacturer: Shandong Zibo Xinyuan Machinery Co., Ltd., Zibo, China), data acquisition card (Model:YE6231, Manufacturer: Jiangsu Lianeng Electronics Technology Co., Ltd., Yangzhou, China), frequency converter (Model:G7R5/P011T4, Manufacturer: Zhejiang Xintuo New Energy Co., Ltd., Yueqing, China), piezoelectric acceleration sensor (Model:CAYD051V, Manufacturer: Xiamen Nair Electronics Co., Ltd., Xiamen, China), and other components [23]. The flow of the test program is shown in Figure 7. Following the flow chart, the rolling bearing fault diagnosis experimental platform was constructed, and a photograph of the completed fault diagnosis platform is shown in Figure 8.

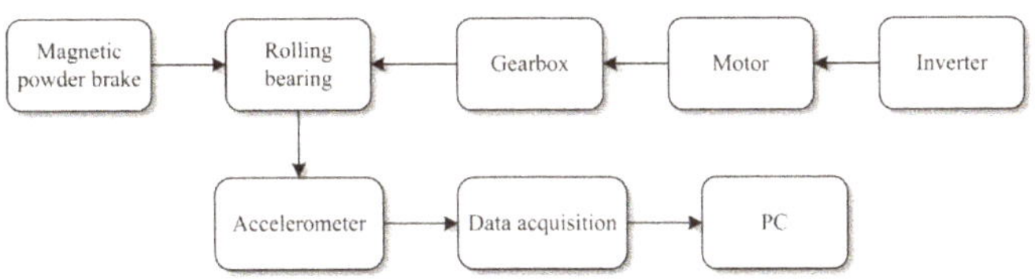

Figure 7. Fault diagnosis experiment flowchart.

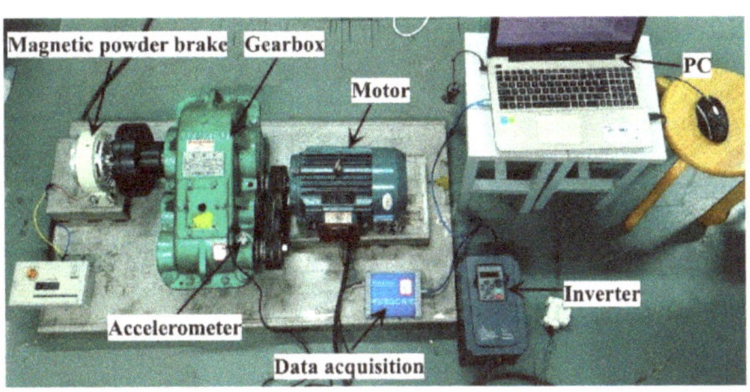

Figure 8. Fault diagnosis experimental platform.

A total of four fault states and one normal state were set on the rolling bearing fault test platform, namely, the inner ring fault, cage fracture fault, outer ring fault, rolling element fault, and a normal state. The fault form of each state is shown in Figure 9. The data settings are shown in Table 2.

For each of the five rolling bearing states, the sampling frequency was set to 6 kHz, with 1000 groups of samples, for a total of 5000 groups. Each group of samples had 1024 sampling signal points. The sample groups were divided into a training set, validation set, and test set in a ratio of 7:2:1, i.e., 3500 groups for the training set, 1000 groups for the validation set, and 500 groups for the test set. The specific division is shown in Table 3.

Figure 9. Failure pictures: (**a**) inner raceway fault (A defect in the inner track where the bearing balls run); (**b**) ball fault (A defect in the bearing ball itself); (**c**) cage fracture (A break in the cage that holds the bearing balls in place); (**d**) normal (No faults detected); (**e**) outer raceway fault (A defect in the outer track where the bearing balls run).

Table 2. Data settings.

Label	Rolling Bearing Condition	Motor Speed (r/min)	Length	Number of Data Sets	Sampling Frequency (Hz)
1	Cage fracture	900	1024	1000	6K
2	Normal	900	1024	1000	6K
3	Inner raceway fault	900	1024	1000	6K
4	Ball fault	900	1024	1000	6K
5	Outer raceway fault	900	1024	1000	6K

Table 3. Data segmentation preprocessing.

Rolling Bearing Condition	Training Set	Validation Set	Test Set	Label
Cage fracture	700	200	100	0
Normal	700	200	100	1
Inner raceway fault	700	200	100	2
Ball fault	700	200	100	3
Outer raceway fault	700	200	100	4
Total Array	3500	1000	500	

5. Results and Discussion

5.1. Vibration Signal Denoising Analysis

This study uses the SVD-EDS method to denoise. A specific analysis follows. As shown in Figure 10, the singular-value curve changed significantly after the 30th order, and the decline rate slowed down significantly. This transition was subtle and therefore required further analysis by means of energy difference spectroscopy. As shown in Figure 11, the energy difference spectrum provided a more explicit order identification. At the 30th order, a distinct peak signal can be observed, indicating a sudden change. This indicates that the useful signal and noise were separated at this point. The singular values before order 30 correspond to useful signals, and the singular values after that are mainly noise.

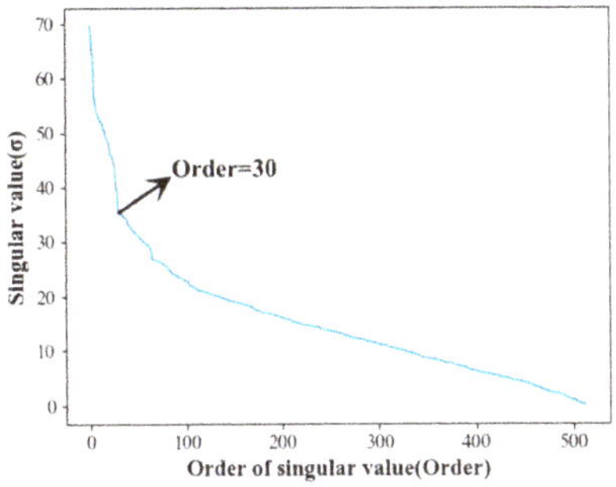

Figure 10. Vibration signal singular-value distribution curve.

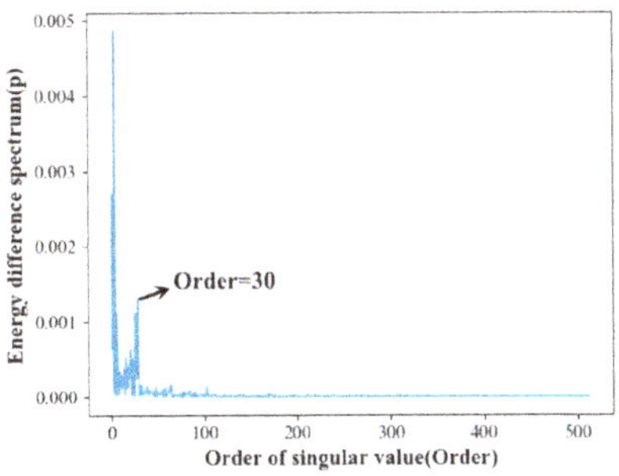

Figure 11. Vibration signal EDS distribution curve.

As shown in Figure 12. Signals reconstructed using the first 30 singular values show better periodicity and readability. The processed data (shown in red) have a clearer signal and less noise than the original data (shown in blue). This enhancement validates the effectiveness of using order 30 for signal reconstruction.

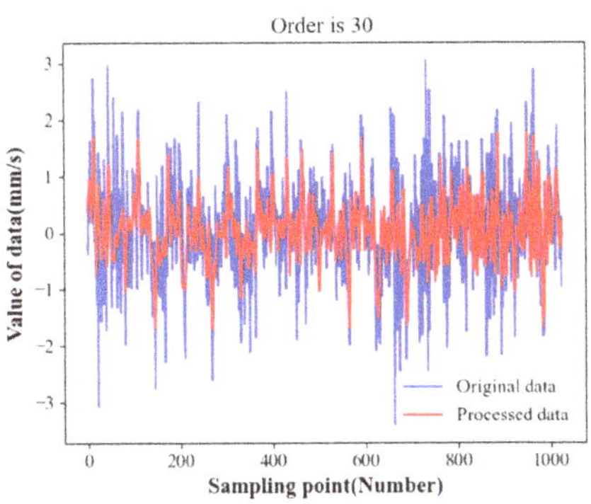

Figure 12. Denoising effect diagram.

In summary, using the SVD-EDS, the optimal order of vibration signal denoising for rolling bearings was determined to be 30. This significantly improved the clarity of useful signals and was a reliable signal-processing method.

5.2. GST 2D Graph Conversion

GST was used to convert the vibration signals denoised via SVD and EDS into two-dimensional time–frequency graphs, as shown in Figure 13.

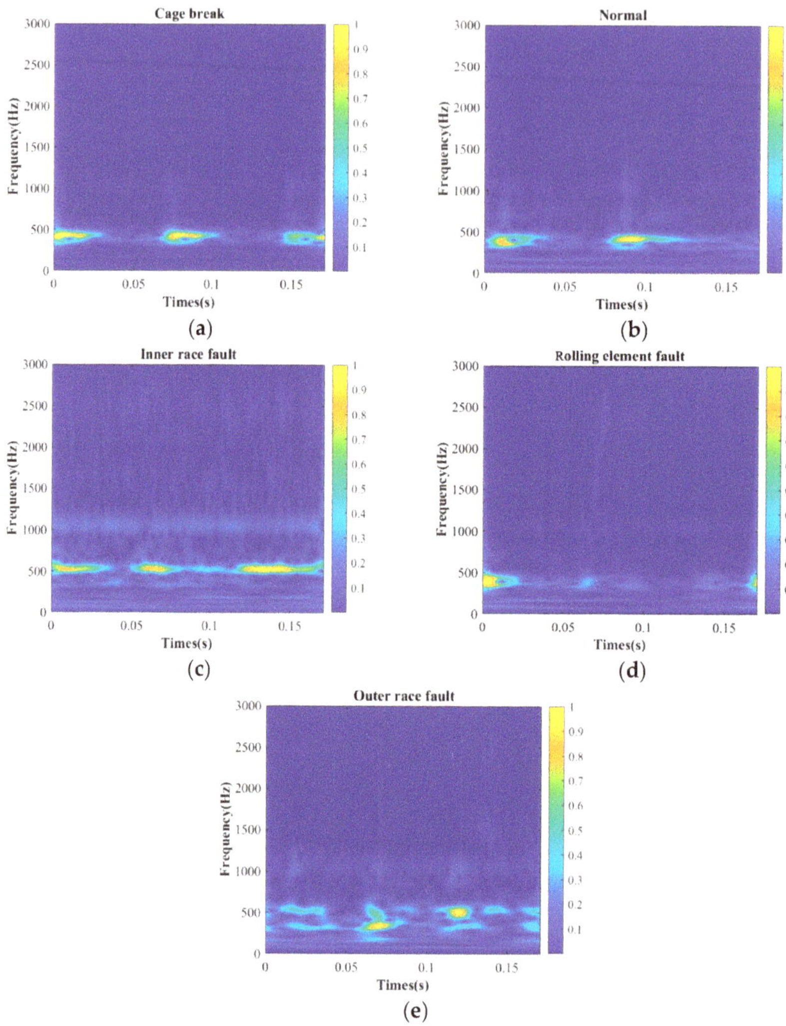

Figure 13. GST diagrams of rolling bearings in different states: (**a**) cage fracture; (**b**) normal; (**c**) inner raceway fault; (**d**) ball fault; (**e**) outer raceway fault.

5.3. Fault Diagnosis Results

5.3.1. Fault Diagnosis Model Analysis

The method used in this study was tested on the Pytorch deep learning framework on a Windows 10 system. The optimizer selected is AdamW, and cosine annealing was chosen as the learning rate adjustment strategy. Figure 14 shows the iterative changes in the accuracy and loss of the proposed method on the training and validation sets.

First, observe the change curves for the loss values in Figure 14a. As can be seen, the loss values on the training and validation samples decreased continuously with an increase in the number of epochs and finally tended to be relatively stable. The loss values on the training and validation samples decreased very rapidly within the first 10 iterations, and the decline rate of the two curves was synchronized. The loss values decreased from about 1.75 to about 0.25, indicating that the model was rapidly converging. Beyond 10 epochs, the training loss and validation loss continue to decrease gradually, approaching near-zero values by epoch 80. This suggests the model is effectively learning and fitting the training data. The curve converges very well and there is no overfitting.

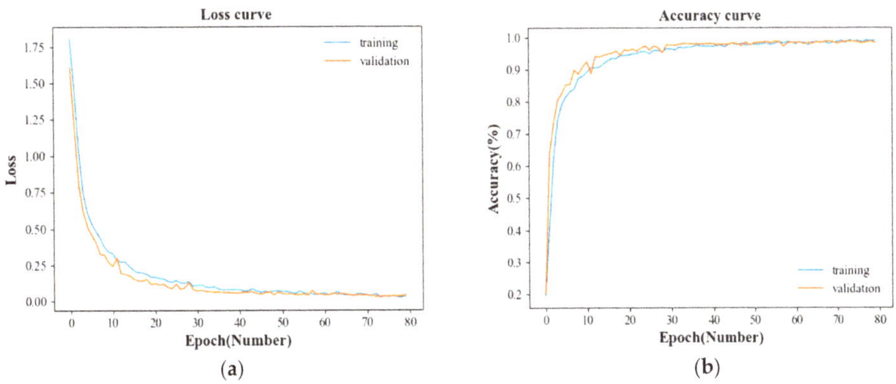

Figure 14. Iterative curves: (**a**) loss value change curves for the training and validation sets; (**b**) accuracy change curves for the training and validation sets.

Observe the change curves for the accuracy in Figure 14a. As can be seen, the training accuracy and validation accuracy start low and increases rapidly within the first 20 epochs. Beyond 20 epochs, the accuracy continues to improve gradually. Within 10 to 40 iterations, the accuracy of the training and verification samples rose relatively steadily and slowly, from about 80% to about 98%. After 40 iterations, the accuracy of the training and verifications sample almost coincided and basically remained unchanged. The curves were smooth on the whole, and there were no large line fluctuations, which indicates that the network model was trained; it also proves that the model had fast convergence and high fault diagnosis classification accuracy.

5.3.2. Feature Visualization Analysis

The features of the training result layer output are visualized in three dimensions using T-SNE [32], as shown in Figure 15.

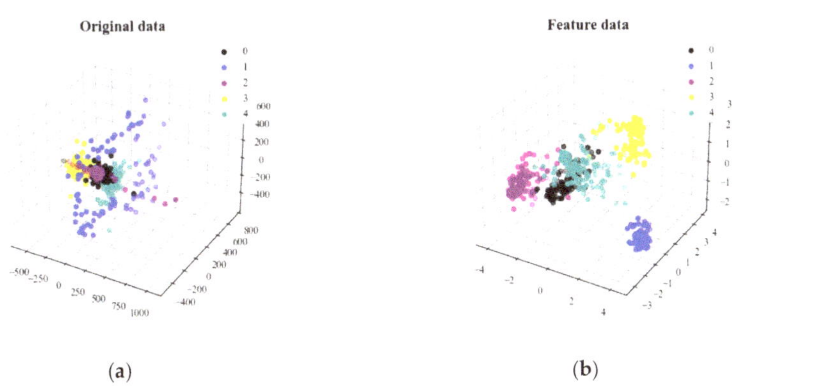

Figure 15. T-SNE visualization: (**a**) 3D distribution of the original data; (**b**) 3D distribution of feature data.

From the three-dimensional graph (a), it can be seen that the fault characteristics of each category of the original data are relatively scattered and there is a lot of overlap between different categories. It is impossible to accurately identify various faults. After feature extraction and training with the method proposed in this paper, it can be seen from the three-dimensional graph (b) that the various dispersion and overlap degrees of the T-SNE diagrams were significantly improved; thus, the accuracy and stability of the data processing were relatively good. This further proves that the model proposed in this paper is effective and advanced as a new fault diagnosis method.

5.3.3. Result Analysis

The confusion matrix [23] of the classification results of the test samples is shown in Figure 16.

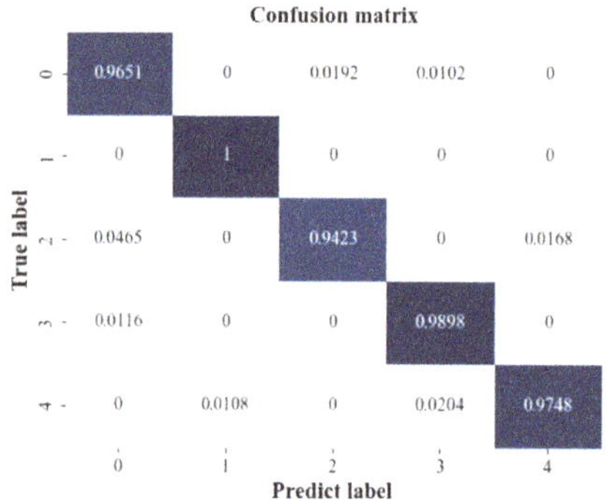

Figure 16. Confusion matrix.

As can be seen from the sample test results in Figure 16, the recognition rate of the model for label 1 (normal state) in this recognition process reached 100%. This shows that the model can identify the normal state of a rolling bearing very accurately. For label 0 (cage breakage), label 3 (rolling element failure), and label 4 (outer ring failure), the recognition rates reached 96.51%, 98.98%, and 97.48%, respectively, and the overall performance was very good. For label 2 (inner ring failure), the recognition rate was relatively low, reaching 94.23%. In all, 4.65% of the inner ring failure conditions were mistaken for bearing cage breakage, and 1.68% were mistaken for outer ring failure. Therefore, the proposed method has satisfactory diagnostic accuracy.

The results of repeated experiments were obtained and are shown in Figure 17. When the fault diagnosis model based on SVD-EDS-GST combined with ResViT was run ten times, the sixth time had the highest recognition rate of 99.88%, and the third time had the lowest recognition rate of 97.63%. The average accuracy of 10 runs is 99.08%, indicating that the proposed method has satisfactory diagnostic accuracy and stability.

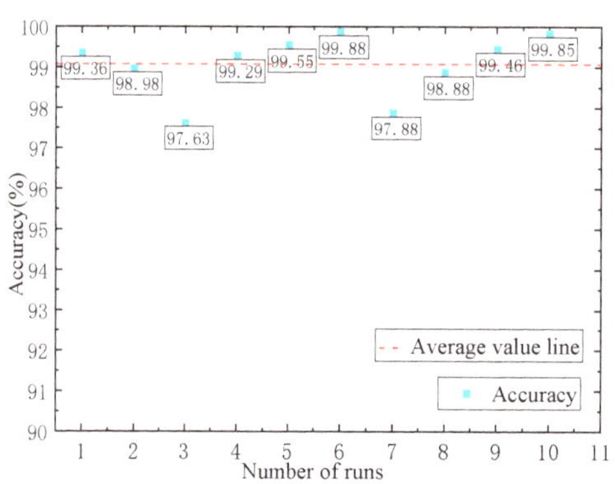

Figure 17. Display of recognition accuracy results from 10 runs.

5.4. Comparative Analysis

In order to illustrate the advantages of the SVD-EDS-GST/ResViT method proposed in this paper in fault diagnosis, it is compared with several CNN, LSTM, and Transformer deep learning models on the same diagnosis case. All models were set with the same parameters and data preprocessing to enable fair comparison. To reduce the impact of randomness, the diagnostic results obtained from 10 repeated experiments are averaged, and the standard deviation of each diagnostic result is calculated, as shown in Table 4.

Table 4. Diagnostic results of different methods.

Method	Average Accuracy (%)	Standard Deviation (%)
STFT-ViT	95.58	0.7236
SVD-EDS-GST-2DCNN	95.24	1.2933
SVD-EDS-GST-LSTM	94.28	1.7863
GST-ViT	91.06	0.9834
SVD-EDS-GST-ViT	98.52	0.4266
SVD-EDS-GST-ResViT	99.08	0.4128

As shown in Table 4, the identification accuracy of the fault diagnosis model proposed in this paper, based on SVD-EDS-GST combined with ResViT, reached 99.076%. Compared with STFT-ViT, SVD-EDS-GST-2DCNN, SVD-EDS-GST-LSTM, GST-ViT, and SVD-EDS-GST-ViT, the accuracy rate was improved by 3.5%, 3.84%, 4.8%, 8.02%, and 0.56%, respectively. The model proposed in this study has the smallest standard deviation, which is 0.4128. It proves that the method proposed in this paper has good accuracy and stability.

6. Conclusions

In the complex and harsh environment of agriculture, research on the intelligent fault diagnosis of agricultural machinery components is very important. The Vision Transformer model has high accuracy in image classification, but it processes image blocks instead of traditional pixel-level features, which may affect the acquisition of local information in some cases. To solve this problem, this study proposed a method based on SVD-EDS-GST combined with ResViT for the fault diagnosis of rolling bearings in agricultural machinery. Through comprehensive experimental verification and performance evaluation, the following conclusions were drawn:

(1) Singular value decomposition combined with the energy difference spectrum method can be used to accurately reconstruct signals and significantly improve the quality and clarity of useful signals, effectively de-noising rolling bearing vibration signals.

(2) The ResNet-Vision Transformer method proposed in this paper integrates the ResNet34 architecture and retains the relative relationships between locations through relative position coding to enhance the model's generalization ability. The experimental results showed that the average accuracy of the fault diagnosis model used in this paper was 99.08% for different fault states of rolling bearings, with higher accuracy and stability.

(3) The proposed fault diagnosis model based on SVD-EDS-GST combined with ResViT provided a more significant improvement when compared to STFT-ViT, SVD-EDS-GST-2DCNN, SVD-EDS-GST-LSTM, GST-ViT, and SVD-EDS-GST-ViT, with improvements in accuracy of 3.5%, 3.84%, 4.8%, 8.02%, and 0.56%, respectively. The standard deviation was also minimized, again proving that the proposed model has good accuracy and stability.

Although the diagnostic results of this study are very good, there are still some limitations. The method we proposed is currently only studied and applied under laboratory offline conditions, and cannot be used for online real-time monitoring. In the future, we will monitor the rolling bearing status online in real time and realize early prediction of faults through deep learning models. Agricultural machinery and equipment often operate under variable working conditions, and the introduction of transfer learning can be considered in future work. It is also hoped that the experimental theory can be applied to engineering practice.

Author Contributions: Conceptualization, F.X. and Y.W.; methodology, F.X. and G.W.; validation, F.X. and Y.W.; investigation, F.X., E.S. and Q.F.; writing—original draft preparation, F.X. and Y.W.; writing—review and editing, F.X. and M.S. All authors have read and agreed to the published version of the manuscript.

Funding: This research was funded by the National Natural Science Foundation of China (52265068), the Natural Science Foundation of Jiangxi Province (20224BAB204050), Equipment Key Laboratory Project of the Ministry of Education (KLCEZ2022-02), and the Project of Jiangxi Provincial Department of Education (GJJ2200627).

Institutional Review Board Statement: Not applicable.

Data Availability Statement: The data used to support the finding of this study are available from the corresponding author upon request.

Conflicts of Interest: The authors declare no conflicts of interest.

References

1. Zhao, S.; Li, T.; Wang, G.; Zhang, Y. Adjustment of Meat Consumption Structure under the Dual Goals of Food Security and Carbon Reduction in China. *Agriculture* **2023**, *13*, 2242. [CrossRef]
2. Li, C.; Wu, J.; Pan, X.; Dou, H.; Zhao, X.; Gao, Y.; Yang, S.; Zhai, C. Design and Experiment of a Breakpoint Continuous Spraying System for Automatic-Guidance Boom Sprayers. *Agriculture* **2023**, *13*, 2203. [CrossRef]
3. Fargnoli, M.; Lombardi, M. Safety Vision of Agricultural Tractors: An Engineering Perspective Based on Recent Studies (2009–2019). *Safety* **2020**, *6*, 1. [CrossRef]
4. Liu, Z.C.; Fang, L.L.; Jiang, D.; Qu, R.H. A machine-learning-based fault diagnosis method with adaptive secondary sampling for multiphase drive systems. *IEEE. Trans. Power Electron.* **2022**, *37*, 8767–8772. [CrossRef]
5. Cao, H.R.; Shao, H.D.; Zhang, X.; Deng, Q.W.; Yang, X.K.; Xuan, J.P. Unsupervised domain-share CNN for machine fault transfer diagnosis from steady speeds to time-varying speeds. *J. Manuf. Syst.* **2022**, *62*, 186–198. [CrossRef]
6. Rui, L.; Ding, X.X.; Wu, S.S.; Wu, Q.H.; Shao, Y.M. Signal processing collaborated with deep learning: An interpretable FIRNet for industrial intelligent diagnosis. *Mech. Syst. Signal Process.* **2024**, *212*, 111314. [CrossRef]
7. Xie, F.Y.; Li, G.; Liu, H.; Sun, E.G.; Wang, Y. Advancing Early Fault Diagnosis for Multi-Domain Agricultural Machinery Rolling Bearings through Data Enhancement. *Agriculture* **2024**, *14*, 112. [CrossRef]
8. Jiang, W.; Shan, Y.H.; Xue, X.M.; Ma, J.P.; Chen, Z.; Zhang, N. Fault Diagnosis for Rolling Bearing of Combine Harvester Based on Composite-Scale-Variable Dispersion Entropy and Self-Optimization Variational Mode Decomposition Algorithm. *Entropy* **2023**, *25*, 1111. [CrossRef] [PubMed]
9. Liu, Z.Y.; Sun, W.L.; Chang, S.K.; Zhang, K.N.; Ba, Y.J.; Jiang, R.B. Corn Harvester Bearing fault diagnosis based on ABC-VMD and optimized EfficientNet. *Entropy* **2023**, *25*, 1273. [CrossRef]
10. Fan, W.; Yang, C.X.; Chen, C.; He, C.B.; Yuan, Y.; Li, Y. Adaptive feature-oriented dictionary learning and sparse classification framework for bearing compound fault diagnosis. *IEEE Trans. Instrum. Meas.* **2024**, *73*, 1–10. [CrossRef]
11. Mystkowski, A.; Wolniakowski, A.; Idzkowski, A.; Ciężkowski, M.; Ostaszewski, M.; Kociszewski, R.; Kotowski, A.; Kulesza, Z.; Kulesza, S.; Miastkowski, K. Measurement and diagnostic system for detecting and classifying faults in the rotary hay tedder using multilayer perceptron neural networks. *Eng. Appl. Artif. Intell.* **2024**, *133*, 108513. [CrossRef]
12. Luo, Y.H.; Li, C.; Jiang, P.; Shi, Y.X.; Li, B.; Hu, W.W. Research on Tractor Condition Recognition Based on Neural Networks. *Agriculture* **2024**, *14*, 584. [CrossRef]
13. Wang, R.C.; Jia, X.S.; Liu, Z.C.; Dong, E.Z.; Li, S.Y.; Chen, Z.H. Conditional generative adversarial network based data augmentation for fault diagnosis of diesel engines applied with infrared thermography and deep convolutional neural network. *Eksploat. I Niezawodn.* **2024**, *26*, 175291. [CrossRef]
14. Dosovitskiy, A.; Beyer, L.; Kolesnikov, A.; Weissenborn, D.; Zhai, X.; Unterthiner, T.; Dehghani, M.; Minderer, M.; Heigold, G.; Gelly, S.; et al. An image is worth 16x16 words: Transformers for image recognition at scale. *arXiv* **2010**, arXiv:2010.11929.
15. Acheampong, F.A.; Nunoo-Mensah, H.; Chen, W.Y. Transformer models for text-based emotion detection: A review of BERT-based approaches. *Artif. Intell. Rev.* **2021**, *54*, 5789–5829. [CrossRef]
16. Salamai, A.A.; Ajabnoor, N.; Khalid, W.E.; Ali, M.M.; Murayr, A.A. Lesion-aware visual transformer network for Paddy diseases detection in precision agriculture. *Eur. J. Agron.* **2023**, *148*, 126884. [CrossRef]
17. Jamil, S.; Roy, A.M. An efficient and robust phonocardiography (pcg)-based valvular heart diseases (vhd) detection framework using vision transformer (vit). *Comput. Biol. Med.* **2023**, *158*, 106734. [CrossRef] [PubMed]
18. Wang, Y.G.; Zhou, J. Research on bearing fault diagnosis under mixed working condition based on Vision Transformer hybrid model. In Proceedings of the International Conference on Computer Graphics, Artificial Intelligence, and Data Processing (ICCAID 2023), Qingdao, China, 1–3 December 2023.
19. Wang, J.Y.; Mo, Z.L.; Zhang, H.; Miao, Q. A deep learning method for bearing fault diagnosis based on time-frequency image. *IEEE Access* **2019**, *7*, 42373–42383. [CrossRef]

20. Wang, J.B.; Lu, Z.X.; Wang, G.M.; Hussain, G.; Zhao, S.H.; Zhang, H.J.; Xiao, M.H. Research on fault diagnosis of HMCVT shift hydraulic system based on optimized BPNN and CNN. *Agriculture* **2023**, *13*, 461. [CrossRef]
21. Thi, T.X.C. Singular value decomposition and applications in data processing and artificial intelligence. *HPU2 J. Sci. Nat. Sci. Technol.* **2023**, *2*, 34–41.
22. Lv, Y.; Zhang, Q.X.; Yuan, R.; Dang, Z.; Ge, M. Local lowest-rank dynamic mode decomposition for transient feature extraction of rolling bearings. *ISA Trans.* **2023**, *133*, 539–558. [CrossRef] [PubMed]
23. Xie, F.Y.; Wang, G.; Zhu, H.Y.; Sun, E.G.; Fan, Q.Y.; Wang, Y. Rolling bearing fault diagnosis based on SVD-GST combined with vision transformer. *Electronics* **2023**, *12*, 3515. [CrossRef]
24. Wang, H.W.; Fang, Z.W.; Wang, H.L.; Li, Y.A.; Geng, Y.D.; Chen, L.; Chang, X. A novel time-frequency analysis method for fault diagnosis based on generalized S-transform and synchroextracting transform. *Meas. Sci. Technol.* **2023**, *35*, 36101. [CrossRef]
25. Cheng, H.; Zhang, Y.; Lu, W.; Yang, Z. A bearing fault diagnosis method based on VMD-SVD and Fuzzy clustering. *Int. J. Pattern Recognit. Artif. Intell.* **2019**, *33*, 1950018. [CrossRef]
26. He, C.B.; Cao, Y.J.; Yang, Y.; Liu, Y.B.; Liu, X.Z.; Cao, Z. Fault diagnosis of rotating machinery based on the improved multidimensional normalization ResNet. *IEEE Trans. Instrum. Meas.* **2023**, *72*, 1–11. [CrossRef]
27. Wen, L.; Li, X.Y.; Gao, L. A transfer convolutional neural network for fault diagnosis based on ResNet-50. *Neural Comput. Appl.* **2020**, *32*, 6111–6124. [CrossRef]
28. Xu, Z.B.; Tang, X.Y.; Wang, Z.G. A multi-information fusion ViT model and its application to the fault diagnosis of bearing with small data samples. *Machines* **2023**, *11*, 277. [CrossRef]
29. Liang, P.F.; Yu, Z.Z.; Wang, B.; Xu, X.F.; Tian, J.Y. Fault transfer diagnosis of rolling bearings across multiple working conditions via subdomain adaptation and improved vision transformer network. *Adv. Eng. Inform.* **2023**, *57*, 102075. [CrossRef]
30. Hou, Y.D.; Wang, J.J.; Chen, Z.G.; Ma, J.L.; Li, T.Z. Diagnosisformer: An efficient rolling bearing fault diagnosis method based on improved Transformer. *Eng. Appl. Artif. Intell.* **2023**, *124*, 106507. [CrossRef]
31. Diao, N.K.; Wang, Z.C.; Ma, H.X.; Yang, W.B. Fault diagnosis of rolling bearing under variable working conditions based on CWT and T-ResNet. *J. Vib. Eng. Technol.* **2023**, *11*, 3747–3757. [CrossRef]
32. Liang, Z.G.; Zhang, L.J.; Wang, X.Z. A novel intelligent method for fault diagnosis of steam turbines based on T-SNE and XGBoost. *Algorithms* **2023**, *16*, 98. [CrossRef]

agriculture

MDPI

Article

Precision Corn Pest Detection: Two-Step Transfer Learning for Beetles (Coleoptera) with MobileNet-SSD

Edmond Maican [1,*], Adrian Iosif [1] and Sanda Maican [2]

1 Faculty of Biotechnical Systems Engineering, National University of Science and Technology Politehnica Bucharest, RO-060042 Bucharest, Romania; adrian.iosif0104@upb.ro
2 Institute of Biology Bucharest, Romanian Academy, RO-060031 Bucharest, Romania; sanda.maican@ibiol.ro
* Correspondence: edmond.maican@upb.ro

Abstract: Using neural networks on low-power mobile systems can aid in controlling pests while preserving beneficial species for crops. However, low-power devices require simplified neural networks, which may lead to reduced performance. This study was focused on developing an optimized deep-learning model for mobile devices for detecting corn pests. We propose a two-step transfer learning approach to enhance the accuracy of two versions of the MobileNet SSD network. Five beetle species (Coleoptera), including four harmful to corn crops (belonging to genera *Anoxia*, *Diabrotica*, *Opatrum* and *Zabrus*), and one beneficial (*Coccinella* sp.), were selected for preliminary testing. We employed two datasets. One for the first transfer learning procedure comprises 2605 images with general dataset classes 'Beetle' and 'Ladybug'. It was used to recalibrate the networks' trainable parameters for these two broader classes. Furthermore, the models were retrained on a second dataset of 2648 images of the five selected species. Performance was compared with a baseline model in terms of average accuracy per class and mean average precision (mAP). MobileNet-SSD-v2-Lite achieved an mAP of 0.8923, ranking second but close to the highest mAP (0.908) obtained by MobileNet-SSD-v1 and outperforming the baseline mAP by 6.06%. It demonstrated the highest accuracy for *Opatrum* (0.9514) and *Diabrotica* (0.8066). *Anoxia* it reached a third-place accuracy (0.9851), close to the top value of 0.9912. *Zabrus* achieved the second position (0.9053), while *Coccinella* was reliably distinguished from all other species, with an accuracy of 0.8939 and zero false positives; moreover, no pest species were mistakenly identified as *Coccinella*. Analyzing the errors in the MobileNet-SSD-v2-Lite model revealed good overall accuracy despite the reduced size of the training set, with one misclassification, 33 non-identifications, 7 double identifications and 1 false positive across the 266 images from the test set, yielding an overall relative error rate of 0.1579. The preliminary findings validated the two-step transfer learning procedure and placed the MobileNet-SSD-v2-Lite in the first place, showing high potential for using neural networks on real-time pest control while protecting beneficial species.

Keywords: pest control; Coleoptera; smart agriculture; neural network; MobileNet; transfer learning; iNaturalist

1. Introduction

The ever-growing global demand for food production poses significant challenges to farmers in safeguarding their crops from harmful pests. Pest management, or the control and mitigation of pest populations, plays a crucial role in modern agriculture. To ensure food and health security in agricultural production systems, pest management plays a key role [1]. To achieve effective pest control, various strategies have been developed, encompassing chemical, biological and mechanical methods [2]. The conventional methods of pest control, which rely heavily on chemical pesticides, have raised environmental concerns and faced growing resistance from pest populations [3,4], rendering them less effective over time. Moreover, the indiscriminate use of pesticides can harm beneficial insects, disrupt ecological balance and potentially compromise human health through the food chain [5–9]. As a result, there is a pressing need to shift towards more eco-friendly and sustainable pest management approaches.

The agribusiness sector encounters notable obstacles in crop protection, marked by the considerable expenses linked to ineffective approaches and by the complexities involved in detecting pest infestations [10]. Moreover, the increasing environmental worries and the public's call for minimized reliance on toxic insecticides add further complexity to pest management practices [11].

In recent years, the integration of artificial intelligence (AI) has shown promise in revolutionizing not only irrigation optimization, yield prediction, and precision livestock farming but also weed and pest management practices [12–14]. By harnessing the power of artificial intelligence, agricultural processes can be optimized and adapted to specific and ever-changing agricultural issues, rendering it the most suitable answer to tackling these challenges [15,16]. Advanced algorithms, particularly neural networks, can enable accurate, precise and timely identification and control of pest species, thus facilitating targeted interventions and reducing reliance on broad-spectrum chemicals.

In recent agricultural research, these approaches have showed substantial promise. For instance, in disease management for sunflower crops [17], deep-learning models, including AlexNet, VGG16, InceptionV3, MobileNetV3 and EfficientNet, exhibited high precision, recall, F1-score and accuracy in classifying sunflower diseases, demonstrating the potential for early disease detection. In the horticulture industry [18], an automated system employing transfer learning in MobileNetV2, termed TL-MobileNetV2, significantly outperformed traditional models like AlexNet, VGG16, InceptionV3 and ResNet in fruit classification. The research not only achieved an impressive 99% accuracy but also highlighted the role of transfer learning in enhancing model performance. Furthermore, in citrus fruit disease detection [19], a systematic review revealed the efficacy of advanced algorithms, such as support vector machines (SVMs) in machine learning, convolutional neural networks (CNNs) in deep learning and linear discriminant analysis (LDA) in statistical techniques. These studies collectively demonstrate the adaptability and effectiveness of advanced algorithms, particularly neural networks, in diverse agricultural applications, ranging from crop disease identification to fruit classification and disease detection in citrus crops.

In terms of insect pest detection and identification, Table 1 provides an overview of various research studies focused on utilizing AI-driven techniques. Diverse datasets, including pest trap images, greenhouse images and natural environments, were used in various research activities that covered a wide range of pest species, from fruit flies and beetles to stink bugs and pine scale insects.

Table 1. Overview of studies on identification of insects with neural networks.

Study	Model/Backbone	Dataset	Performance Metrics
Kalfas et al. [20]	YOLO v5	731 sticky plates containing 74,616 bounding boxes	mAP = {0.76 across all dataset; 0.73 wooly aphid; 0.86 chicory leaf-miner; 0.61 grass-fly; 0.67 wasp}
Yang et al. [21]	YOLOv7 with insertion of CSPResNeXt-50 and VoVGSCSP modules	4533 annotated images; 13 maize pests	mAP@0.5 = 76.3%; mAP@0.5:0.95 = 51.2; recall = 77.3%; F1 = 75.2%
Wu et al. [22]	Faster R-CNN, FPN, SSD300, RefineDet, YOLOv3	IP102 dataset (18,983 images)	FPN with ResNet-50: AP@0.5 = 54.93%, YOLOv3: AP@0.5 = 50.64%
Albanese et al. [23]	Modified LeNet-5, VGG16, MobileNetV2	4400 images; 2 classes: codling moth and general insect	LeNet-5: Acc. 96.1%, Prec. 99.6% VGG16: Acc. 97.9%, Prec. 99.6% MobileNetV2: Acc. 95.1%, Prec. 98.5%

Table 1. *Cont.*

Study	Model/Backbone	Dataset	Performance Metrics
Wang et al. [24]	Faster R-CNN-FPN with ResNet-50 or ResNet-101, YOLOv5, YOLOv7	4865 images; seven species of coccinellids	YOLOv7: AP@0.50 up to 97.31 and AP up to 74.5; YOLOv5: AP@0.50 up to 96 and AP up to 73.8; Faster R-CNN: AP@0.50 up to 94.3 and AP up to 65.6;
Salamut et al. [25]	Faster R-CNN, YOLOv5 [26]	1600 annotated images of cherry fruit flies	Faster R-CNN: AP@0.50 = 0.88%; YOLOv5: AP@0.50 = 0.76%
Rustia et al. (2021) [26]	Multi-stage deep learning method	Greenhouse images	F1-scores up to 0.92
Wang et al. [27]	Faster R-CNN/VGG-16, Cascade R-CNN/ResNet-50-FPN, YOLOv3/Darknet-53	Pest24 dataset; 24 field pests; 25,000 images	YOLOv3: mAP@0.50 = 59.79%; Cascade R-CNN: mAP@0.50 = 57.23%; Faster R-CNN: mAP@0.50 = 51.10%
Li et al. [28]	Faster R-CNN (COCO pre-trained)	1500 sticky trap images; 2 classes: whitefly and thrips	Faster R-CNN (pre-trained): NA (More accurate than direct training)
Hong et al. (2021) [29]	AI-based pest counting method	Black pine bast scale images	Counting accuracy = 95%
Wang et al. [30]	Improved Faster R-CNN/Attention	AgriPest21 dataset; 21 types of pests; 25,000 images	Improved Faster R-CNN: mAP = 78.7%
Jiao et al. [31]	Faster R-CNN/ResNet50	AgriPest21 dataset 21 types of pests; 25,000 images	Faster R-CNN: mAP = 77.4%
Zhang et al. [32]	YOLO models with attention mechanism	Pest24 dataset; 25,000 images of small pests	AgriPest-YOLO: mAP@0.50 = 71.3%
Sava et al. [33]	YOLOv5m	Dataset from Maryland Biodiversity Project	YOLOv5m: mAP = 99.2%
Takimoto et al. [34]	Faster R-CNN	Web and field-collected images of herbivorous beetles	Faster R-CNN: NA
Ozdemir and Kunduraci [35]	Faster R-CNN (Inception-v3)	25,820 training images of various insect orders	Faster R-CNN: NA
Butera et al. [36]	Faster R-CNN (MobileNet-v3)	36,000 web images of Beetle-type pests and non-harmful beetles	Faster R-CNN: mAP = 92.66%
Ahmad et al. [37]	YOLO models	7046 images with 23 pests	YOLOv5-X: mAP@0.5 = 98.3%, mAP@0.05:0.95 = 79.8%
Ratnayake et al. [38]	YOLOv2 and hybrid approach	22,260 video frames with honeybees in wildflower clusters	HyDaT: Detection rate = 86.6%, YOLOv2: Detection rate = 60.7%
Bjerge et al. [39]	YOLOv5	29,960 beneficial insects	YOLOv5: mAP@0.50:0.05:0.95 = 0.592, F1-score = 0.932
Spanier [40]	YOLOv5 variant	17,000 pollinator insect images	YOLOv5 variant: Accuracy = 0.9294, F1-score = 0.9294
Bjerge et al. [41]	YOLOv5, Faster R-CNN	100,000 annotated images of small insects	YOLOv5: mAP@0.50 = 0.924, Faster R-CNN: mAP@0.50 = 0.900
Venegas et al. [42]	Deep CNN and traditional methods	2300 coccinellid beetle images	CNN model AUC = 0.977
Vega et al. [43]	CNN with weighted Hausdorff distance	2633 beetle images	Mean accuracy = 94.30%

These studies employed a **broad spectrum of neural network models**, architectures and methodologies. The choice of backbone network, such as ResNet, VGG, MobileNet and Darknet, significantly impacts performance. Wang et al. [27] demonstrated the effectiveness of YOLOv3, achieving a substantial mean average precision (mAP) of 59.79% on the Pest24 dataset, outperforming both Cascade R-CNN and Faster R-CNN. Sava et al. [33] achieved notable results with YOLOv5m, attaining an mAP of 99.2% for the detection of brown marmorated stink bugs. Additionally, Rustia et al. (2021) [26] introduced a multi-stage deep learning method for greenhouse insect detection, with F1-scores reaching up to 0.92.

Several studies also explored **model adaptations** to improve performance. Wang et al. [30] presented an enhanced Faster R CNN model integrated with attention mechanisms, resulting in enhanced detection of small pests and an mAP of 78.7%. Zhang et al. [32] and Jiao et al. [31] both investigated the integration of attention mechanisms into YOLO models, yielding significant improvements in detecting small pests and achieving mAPs of 71.3% and 77.4%, respectively. These three investigations were carried out using AgriPest21, a dataset containing insects with very small sizes. It encompasses 21 types of pests, with 21,970 training images and 2442 testing images. The original images of crop pests with a format of 2596 × 1944 pixels, were adjusted to 800 × 600 for better efficiency. The average dimensions (width and height) for all pest classes do not exceed 70 pixels. The average relative scale for all categories is less than 1%, with the smallest average relative scale measuring only 0.1129% [20]. The primary focus of these inquiries was to address the constraints of deep learning methodologies in capturing adequately detailed features for small and very small objects within an image.

In terms of **dataset size and performance**, Ahmad et al. [37] worked with a dataset of only 7046 images containing 23 types of pests. YOLOv5-X achieved an mAP@0.5 of 98.3% and an mAP@0.05:0.95 of 79.8%, showcasing reliable performance on a moderately sized dataset.

A small number of studies also extended their focus to the **identification of beneficial insects** (pollinators, natural enemies, food and food insects), biological controls of agricultural pests, or food sources to humans and other animals [44]) using AI as a tool for preserving ecosystems. Bjerge et al. [39] and Spanier [40] both demonstrated successful applications of YOLO models in the identification of various beneficial insects, highlighting the adaptability of these models for broader insect classification tasks. YOLOv5 proved top performance, with an mAP@0.50:0.05:0.95 of 0.592 and high accuracy.

Comparative analyses were also conducted. The investigation of Wu et al. [22] involved a comparison of deep learning-based object detection methods, on a dataset that contains more than 75,000 images belonging to 102 categories. The results showed that the combination of FPN (Feature Pyramid Network) with ResNet-50 architecture achieved the highest average precision (AP) of 54.93%. Following closely was YOLOv3, with an average precision of 50.64%.

Not many studies, however, have explored the application of **MobileNet networks in insect pest detection**. Albanese et al. [23] utilized modified LeNet-5, VGG16 and MobileNetV2 on a dataset comprising 4400 images, specifically focusing on classifying the codling moth from other insects. The results indicated competitive accuracy levels, with LeNet-5 achieving 96.1%, VGG16 reaching 97.9% and MobileNetV2 demonstrating an accuracy of 95.1% and a precision of 98.5%. Similarly, Butera et al. [36] employed a Faster R-CNN with a MobileNet-v3 backbone for the detection of beetle-type crop pests and non-harmful beetles, leveraging a substantial dataset of 36,000 web-sourced images. The model exhibited strong performance, achieving a mean average precision (mAP) of 92.66%. Both studies showcased competitive accuracy in insect detection tasks using MobileNet models.

An **innovative approach to detection** has been explored in a study conducted by Ratnayake et al. [38], wherein hybrid methodologies were introduced by combining YOLOv2 with background subtraction techniques for honeybee tracking. The results demonstrated that hybrid models can provide competitive detection rates, offering promising ways for tracking insects in their natural environments.

In the context of **model generalization and transfer learning**, Li et al. [28] employed transfer learning strategies by utilizing a pre-trained Faster R-CNN model derived from the COCO dataset. Their results demonstrated that pre-trained models could achieve a higher accuracy for detecting small pests, showcasing the benefits of generalization through transfer learning.

Overall, these diverse studies collectively highlight the evolving landscape of AI-powered pest detection and identification techniques, showcasing both innovation and continuous efforts to address challenges in agriculture and pest management.

With some exceptions, most of the research presented in Table 1, while promising in the context of pest detection or classification, may encounter certain limitations when applied to low-power devices that are used on agricultural mobile systems, such as counting pests from pest traps or selective use of pesticides by agricultural autonomous equipment. These systems necessitate lightweight neural networks to maximize battery life; however, reducing network complexity often results in decreased performance, particularly in tasks involving the detection of small objects, such as tiny pest insects. This trade-off between computational efficiency and accuracy poses a critical challenge for deploying these models on resource-constrained mobile devices.

In the context of sustainable pest control, where the preservation of beneficial species like pollinators and predators is crucial, the limitations of low-complexity neural networks become more apparent. These models may struggle to accurately differentiate between pests and beneficial insects, potentially leading to unintended harm to ecologically valuable species.

Additionally, several prior studies have relied on datasets comprised of high-quality, close-up images sourced from professional photographers. While such datasets may be suitable for controlled settings, they are not appropriate for real-world applications, where cameras of varying qualities and environmental conditions are the norm [45–51].

In this preliminary study, our main objective was to address these limitations and make such research more suitable for low-power mobile applications. We propose a two-step transfer learning approach to enhance the accuracy of two versions of pre-trained variations in SSD-MobileNet lower-complexity networks, namely, MobileNet-SSD-v1 and MobileNet-SSD-v2-Lite. The custom dataset we used comprises images of diverse resolutions, varying image quality, distinct lighting conditions and insects of varying sizes within the context of the image frame, mirroring the real-world conditions encountered in pest detection scenarios. The models trained on this dataset have a greater chance to handle a wide array of technical conditions, such as different camera models, setups and resolutions, benefitting from the capability to process images of various sizes.

We aimed to assess the efficiency of the proposed two-step transfer learning (TL) procedure. The first TL training step was performed on a dataset of 2605 images to recalibrate the networks' trainable parameters on the 'Beetles' and 'Ladybug' classes. Then, a second TL training step was performed on a distinct custom dataset of 2648 images with four classes of pests and one class for the beneficial 'Ladybug' insect, in order to fine-tune the trainable parameters for selected species of insects. The best-performing network was selected for the future research step aiming at employing a much larger and more representative custom dataset to reduce generalization error, as well as to test it onto an experimental model.

This research aligns with the principles of Agriculture 5.0, characterized by the integration of advanced technologies, such as artificial intelligence, the Internet of Things (IoT), robotics and data analytics, into traditional farming practices. The general aim is to enhance efficiency, sustainability and productivity in agriculture. In particular, research in the development of low-power mobile systems with neural networks capable of recognizing and localizing pests in corn crops can be useful in performing real-time crop monitoring, providing immediate information to farmers about potential pest infestations, thus allowing them to intervene promptly to limit damage, and to ensure higher yields and superior crop quality. It saves time and resources by eliminating the need for extensive manual monitoring and unplanned pesticide applications. By specifically identifying pests, farmers can apply precise and selective treatments, reducing the need for excessive pesticide use and minimizing environmental impact. These mobile systems can aid in the early identification of infestation outbreaks, allowing farmers to prevent the spread and multiplication of pests. They are easy to implement and can be integrated into existing farm infrastructure without requiring major investments in complex equipment.

To conclude, the primary and intermediate objectives of this research are outlined as follows:

- To introduce a new approach to transfer learning involving a two-step process that significantly enhances the accuracy of MobileNet SSD networks. As mentioned above, this procedure starts with broader dataset training, followed by fine-tuning the neural networks parameters on a specific dataset;
- To thoroughly assess in this context two versions of the MobileNet SSD network—MobileNet-SSD-v1 and MobileNet-SSD-v2-Lite—providing insights into their relative performance. This comparative analysis is important for understanding the balance between model complexity and accuracy in the context of low-power mobile systems;
- To assess the advantages in terms of practicality and feasibility of deploying the proposed neural network models on Jetson low-power devices, as a solution for real-world applications. The combination of the MobileNet SSD networks in conjunction with the NVIDIA Jetson platforms is known as a highly optimized solution in terms of computational speed and energy efficiency, making it well-suited for embedded and mobile applications;
- To apply the trained models to detect harmful beetle species, and also to distinguish them from a beneficial species (Coccinella), in order to demonstrate the potential utility of neural networks in real-time pest control. Emphasizing the preservation of beneficial species adds ecological significance to the research.

2. Materials and Methods

To address the limitations of low-complexity neural networks and make such research more suitable for low-power mobile applications, a strategic approach was adopted. To optimize both accuracy and efficiency, the Nvidia Jetson Nano Orin—a high-performance graphics processing unit (GPU) tailored for low-cost mobile applications, manufactured by Nvidia Corporation, based in Santa Clara, CA, USA—was selected as target hardware platform for deployment. It is important to note that, in this preliminary research, the more expensive Nvidia Jetson AGX Orin was employed for training purposes due to its superior performance in the training phase. In the second research stage that will follow this preliminary study, the best model will be retrained on a larger dataset, and then will be transitioned to the entry-level Jetson Nano Orin, thus aligning with our target low-cost mobile platform.

The study focused on five distinct beetle species. Among these, four are well-documented as significant crop pests, namely, *Anoxia villosa*, *Diabrotica virgifera virgifera*, *Opatrum sabulosum* and *Zabrus tenebrioides*. According to the information presented in Table 2, these species are polyphagous and cause substantial damage to a variety of important crops like corn, wheat, sunflower and beans. Their geographical distribution spans large agricultural regions throughout Europe, Asia and North America. This potential for crop damage served as the primary criterion for the selection of these species in our research. In addition to these crop pests, the dataset also includes images of *Coccinella* sp. (Coleoptera, Coccinellidae), known as beneficial arthropods and important aphid predators in agricultural crops [52].

The primary focus was on enhancing the accuracy of two pre-trained versions of the SSD-MobileNet network architecture, a well-established choice for real-time object detection on mobile and embedded devices. This architecture combines the Single-Shot MultiBox Detector (SSD-300) with the MobileNet backbone, known for its computational efficiency. To achieve improved accuracy while maintaining efficiency, a two-step transfer learning procedure was devised for fine-tuning the network on specific pest detection tasks. Two variations in the SSD-MobileNet architecture, namely, MobileNet-SSD-v1 and MobileNet-SSD-v2-Lite, were tested. The goal was to determine which version achieved the best accuracy and to assess the efficiency of the proposed two-step transfer learning procedure.

To facilitate model evaluation and deployment, the Nvidia-developed detectNet wrapper was employed. This wrapper streamlined the testing process, making it more efficient and accessible for further real-world applications.

Table 2. Pest species used in custom dataset.

Family/Species/ Common Name	Body Length (mm)	Distribution	Flight Period/ Optimum Temperature Range	Affected Plants	References
Carabidae/ *Zabrus tenebrioides* Goeze, 1777 (Corn Ground Beetle)	14–16	England, Southern Sweden, Northern Africa, Asia Minor, Cyprus, Ukraine, Moldova, Transcaucasia	May–June/20–26 °C	Winter wheat, corn, horn, rye, barley, oat	[53,54]
Tenebrionidae/ *Opatrum sabulosum* Linnaeus, 1761 (Darkling Beetle)	7–10	Western Europe, Northwestern Iran, Northwestern China; the European part of the former USSR, the Caucasus, South and Middle Siberia Kazakhstan, Central Asia	-/they feed actively at a temperature of 17–20 °C. Below 25 °C prefer dry plants, at temperature above 27 °C they eat green plants almost exclusively	Polyphagous (corn, sugar beet, flax, sunflower, tobacco, cotton, pumpkin, fennel, anise, castor-oil plan, safflower, buckwheat, bean)	[54,55]
Scarabaeidae/ *Anoxia villosa* (Fabricius, 1781) (Cockchafer, Steppe Beetle)	20–25	Europe, especially Mediterranean area	May–August, especially July/-	Polyphagous (corn, sunflower, wheat, barley, woody plants: vine, orchards, forest nurseries)	[56]
Chrysomelidae/ *Diabrotica virgifera virgifera* LeConte, 1868 (Western Corn Rootworm)	~5	Europe, North America	the end of June–the middle of October/high temperature, but not more than 30 °C	Preferred corn; white squash, alfalfa, clover, rape, bean, soybean, sunflower	[57,58]

2.1. Selected Neural Networks

A versatile object detection framework optimized for real-time inferencing on Nvidia Jetson platforms is detectNet [59], also from Nvidia. It accepts input images and, in return, yields a list of bounding box coordinates, along with class labels and confidence values. One of its features is its ability to integrate various pre-trained detection models, providing flexibility in choosing the most suitable architecture for the task at hand. Among them, MobileNet-SSD-v1 and MobileNet-SSD-v2-Lite both fit our specific research requirements.

The SSD-MobileNet model is a fusion of the SSD-300 Single Shot MultiBox Detector (SSD) and the MobileNet convolutional neural network (CNN), resulting in a lightweight architecture that is ideally suited for resource-constrained environments, such as mobile and embedded vision applications (Figure 1). This integration combines the efficiency of SSD as an object detector, and MobileNet as a feature extractor, thus providing a high-performance yet computationally efficient solution.

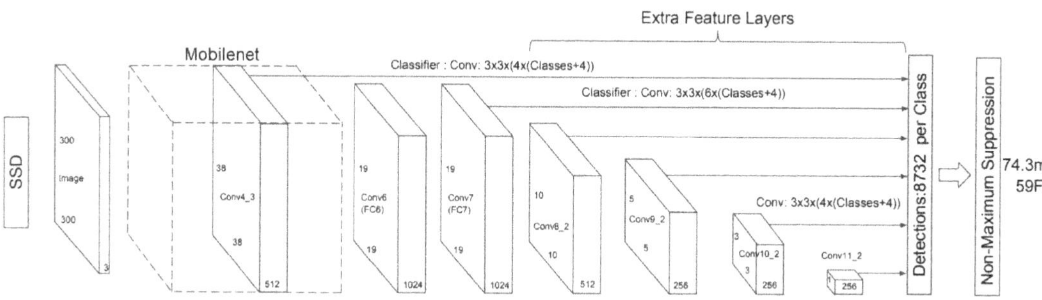

Figure 1. The SSD-MobileNet architecture (adapted by [60]).

One of the defining characteristics of MobileNet backbone is the utilization of depth-wise separable convolutions—a technique that applies a single convolution operation to each color channel individually, in order to reduce both the computational complexity and model size. It involves a depthwise convolution layer, followed by a pointwise convolution layer, thus minimizing computational demands while preserving accuracy. Subsequently, batch normalization and ReLU are employed on the output of each convolutional operation. Batch normalization fine-tunes the data by modifying learning parameters, adjusting learning rates, managing dropout ratios and preventing gradient vanishing. Feature Pyramid Network (FPN) is an integral part of SSD-MobileNet-V1 and operates by combining feature maps from different stages of the MobileNet-v1 backbone. FPN integrates feature maps from different depths of the backbone network, creating a pyramid of feature maps at various scales. This ensures that objects of different sizes, including small ones, are effectively captured in the feature representation. The SSD head is responsible for predicting object bounding boxes and class scores. The number of output channels in the final convolutional layers of the SSD head corresponds to the total number of anchor boxes and object classes to be predicted. This multi-scale approach plays an important role in accommodating objects of varying sizes, allowing the model to predict bounding boxes and categories across multiple feature maps.

The MobileNet-SSD-v2-Lite uses the MobileNet-v2-Lite architecture as its backbone, which builds upon the advancements of MobileNet-v2 by further optimizing for efficiency, thus making it even more suitable for resource-constrained environments. It prioritizes efficiency and is ideal for scenarios where real-time detection is crucial and hardware resources are limited.

The primary goal of the loss function L (Equation (1)) is to compute a weighted combination of the localization loss (L_{loc}) and the confidence loss (L_{conf}), given the number N of matched default boxes:

$$L(x, c, l, g) = \frac{1}{N}\left(L_{conf}(x, c) + \alpha L_{loc}(x, l, g)\right) \quad (1)$$

The confidence loss is calculated as the softmax loss across confidences for the following multiple classes (c):

$$L_{conf}(x, c) = -\sum_{i\in Pos}^{N} x_{ij}^{p} log\left(\hat{c}_i^p\right) - \sum_{i\in Neg} log\left(\hat{c}_i^0\right) \quad \text{where} \quad \hat{c}_i^p = \frac{\exp\left(c_i^p\right)}{\sum_p \exp\left(c_i^p\right)} \quad (2)$$

In this equation, $x_{ij}^p = \{1, 0\}$ is an indicator for matching the i-th default box with the j-th ground truth box of category p. The confidence loss evaluates how well the model predicts the presence of objects as well as their associated classes.

The localization loss measures the squared distance between the predicted coordinates, and is weighted by α, which balances these two losses and their impact on the overall loss value [61]. Using the parameters of the predicted box (l) and the ground truth box (g), one can compute the localization loss as a *Smooth* $L1$ loss [62]:

$$L_{loc}(x, l, g) = \sum_{i\in Pos}^{N} \sum_{m\in\{cx,cy,w,h\}} x_{ij}^{k} smooth_{L1}(l_i^m - \hat{g}_j^m), \quad (3)$$

where cx, cy, w and h are the coordinates of the default bounding box (d) center, its width and height. The above sum is taken over all positive samples, where positive samples are those that have a significant overlap (e.g., IoU—Intersection over Union) with ground truth objects.

2.2. The Training Platform and Framework

To further develop an efficient and optimized object detection model, for training we used the Nvidia Jetson AGX Orin platform in conjunction with PyTorch 2.0.0 for JetPack 5.1.1. The Jetson AGX Orin Nvidia Jetson AGX Orin is manufactured by Nvidia Corporation, based in Santa Clara, California, United States, and is equipped with a high-performance GPU architecture, designed for accelerating model training and inference due to parallel processing capabilities of its CUDA cores (Table 3). For increased storage capacity, we have added a 2 TB SSD (model Samsung 980 PRO Heatsink Gen.4 NVMe), with a read transfer rate of 7000 MB/s and a write transfer rate of 5100 MB/s.

Table 3. Main technical specifications of the Nvidia Jetson AGX Orin training hardware.

GPU	NVIDIA® Ampere architecture; 2048 NVIDIA CUDA cores; 64 Tensor cores
CPU	12-core Arm Cortex-A78AE v8.2 64-bit CPU; 3MB L2 + 6MB L3
DL Accelerator	2× NVDLA v2.0
Vision Accelerator	PVA v2.0
Memory	64 GB 256-bit LPDDR5; 204.8 GB/s
Storage	64 GB eMMC 5.1 + 2 TB SSD, model Samsung 980 PRO Gen.4 NVMe (added)

We choose PyTorch for JetPack—a specialized variant of the PyTorch framework that integrates with NVIDIA's CUDA technology—as it is optimized for accelerated parallel processing on Nvidia Jetson platforms. The software stack of NVIDIA JetPack is fully compatible with PyTorch for JetPack, ensuring integration and access to Nvidia's proprietary libraries and optimizations.

2.3. PyTorch Scripts

Throughout the training and testing process, the following PyTorch scripts were employed:

- train_ssd.py [63]: this script serves as a tool for training Single Shot MultiBox Detector (SSD) object detection models. It offers flexibility in configuring dataset types, network architectures and training parameters. The script incorporates data preprocessing, model training, validation and checkpointing, with support for various learning rate scheduling strategies. It also enables optional mean average precision (mAP) evaluation during validation. We used this script for re-training our SSD-MobileNet networks;
- eval_ssd.py [64]: this script is an evaluation tool for assessing the performance of trained SSD models. It allows for customizable evaluation parameters, including dataset type, model architecture and evaluation metrics, such as mean average precision (mAP). The script loads the dataset and the model pth file, performs inference and computes class-specific and/or overall accuracy and mAP values. We used this script to test the model inference performance on the test set before converting it from PyTorch to Open Neural Network Exchange (ONNX) format;
- onnx_export.py [65]: this script is designed for the conversion of PyTorch deep-learning models into the ONNX format. ONNX (Open Neural Network Exchange) is an open-standard format facilitating model interoperability and deployment across diverse inference platforms and inference environments. The script offers command-line customization options, including the choice of neural network architecture, input model checkpoint path, class labels file and output ONNX model path. It dynamically selects the inference device, loads the PyTorch model and exports it to ONNX. In order to conduct testing and real-time inference using our re-trained SSD-MobileNet models with TensorRT version 8.5.2, it was necessary to convert the PyTorch models into the ONNX format so that TensorRT can load them; TensorRT is a specialized library designed for high-performance inference on NVIDIA graphics processing units (GPUs), currently being considered the fastest way to run a trained model [66]. It automatically optimizes DNN models by fusing multiple layers and operations into a single, efficient kernel. This reduces memory bandwidth usage and minimizes the number of compute operations, resulting in faster inference. In order to further improve efficiency, TensorRT applies various graph optimizations, such as layer pruning,

to eliminate unnecessary operations. It also supports batched inference, allowing multiple inputs to be processed in parallel, which enhances throughput in real-time applications [67];

- detectnet.py [68]: this script creates an Nvidia detectNet object, and subsequently uses it to employ TensorRT for trained models inferencing on actual images and video sequences. We used 'detectnet.py' with the ONNX format of the trained models to visualize detection boxes and confidence scores on the images from the test set. This script was particularly important in the stage of errors analysis and identification of the model's strengths and limitations.

2.4. Image Datasets

As previously mentioned, the base models of MobileNet-SSD-v1 and v2-Lite come pre-trained on the PASCAL-VOC and COCO datasets, respectively, which do not include specific classes for insects. Here, the term "class" refers to the meaning used in the field of machine learning, and not in the sense of taxonomic category from the biological sciences. To adapt these models for our pest detection task we used two datasets, one for each stage of the two-step transfer learning procedure. The first dataset consisted of 2605 images, which represents a subset of Open Images [69], specifically containing classes 'Beetles' and 'Ladybug'. Maintained by Google, Open Images has 600 classes, providing a diverse range of images and has the status as one of the significant and well-regarded datasets in the field. The reason we turned to the Open Images dataset for our study consists of the following advantages:

- It has a large number of labeled images, standing as one of the most extensive sources of data available for training computer vision models;
- Each image is already annotated and comes with detailed information about the objects, which makes it readily suitable for training object detection models;
- It provides an open-licensed image sharing and use.

The second dataset consisted of 884 images representing *Coccinella* sp., *Anoxia villosa*, *Diabrotica virgifera virgifera*, *Opatrum sabulosum* and *Zabrus tenebrioides* [70–74]. They were sourced from the iNaturalist web portal via Global Biodiversity Information Facility platform (GBIF) [75], and a few from various cloud resources. iNaturalist is a publicly available platform providing free and open access to biodiversity data courtesy of the California Academy of Sciences and the National Geographic Society, with the purpose to facilitate the distribution and sharing of data and images related to plant, animal and fungal species.

Image augmentation was performed through mirroring along the horizontal and vertical axes, in order to diversify the training dataset, enhance model generalization by exposing it to a wider range of variations and real-world scenarios, and thus mitigate overfitting and improve performance. After augmentation, the dataset size increased to 2648 images. The distribution of both augmented and non-augmented images across each class is presented in Table 4.

According to Table 4, the dataset is characterized by an equitable distribution of images across species, which means that the model is trained on a representative sample of each species, preventing any single class from dominating the training process. This balance enables both models to equally learn the characteristics and features associated with each species, enhancing its ability to accurately detect and classify each of them.

The annotation process was conducted using the Computer Vision Annotation Tool (CVAT v2.9.0) software for bounding box delineation and labeling according to the Pascal VOC (Pascal Visual Object Classes) format [76]. In this format, one XML file with information such as image dimensions and bounding boxes coordinates is generated for each image.

Table 4. The distribution of images across each species in the custom dataset.

Dataset Class	Number of Images		Relative Distribution
	Before Augmentation	After Augmentation	
Anoxia	154	462	17.4%
Ladybug	173	519	19.6%
Diabrotica	206	618	23.3%
Opatrum	180	536	20.2%
Zabrus	171	513	19.5%
Total:	**884**	**2648**	**100%**

2.5. Transfer Learning Procedures

For the first transfer learning, the 2605 annotated images were downloaded from the Open Images platform. Files are structured according to the Open Images format requirements (Figure 2), namely, divided into three main directories ('train', 'validation' and 'test', with 2376, 60 and 169 images, respectively), and with csv metadata files.

```
beetles_OpenImages
      |-----train (2376 images)
      |        |------Beetles
      |                 |------ 000b3940e7d25c03.jpg
      |                 |------ 0ab44a69f16c53e3.jpg
      |                 |------ ......
      |        |------Ladybug
      |-----validation (60 images)
      |-----test (169 images)
      |-----sub-train-annotations-bbox.csv
      |-----sub-validation-annotations-bbox.csv
      |-----sub-test-annotations-bbox.csv
```

Figure 2. Dataset file structure, following the Open Images format.

Each directory is further categorized into class subdirectories ('Beetles' and 'Ladybug'), containing the corresponding class images. The csv files from the main folder provide detailed metadata, such as image identifiers, bounding box coordinates, class labels and other details.

Following the preparation of our image dataset, we employed the 'train_ssd.py' PyTorch script to retrain each of the two neural networks. The following is a sample command line for training the MobileNet-SSD-v1 network:

```
python3 train_ssd.py --dataset-type=open_images --net=mb1-ssd --data=data/
beetles_OpenImages --model-dir=models/custom_beetles --batch-size=32 --workers=12
--epochs=150 --checkpoint-folder=models/custom_beetles
```

Where the arguments used in the above command are as follows:

- `dataset-type`: the dataset format to be used; in this case, the Open Images format;
- `net`: used to select MobileNet-SSD-v1 or MobileNet-SSD-v2-Lite;
- `data`: the directory containing the training images;
- `model-dir`: the directory where the trained models are to be stored;
- `batch-size`: the number of images per batch; after experimentation on the Jetson AGX Orin platform, equipped with 64 GB of memory, we determined that a batch size of 32 provided optimal training performance in terms of training time and accuracy;

- `workers`: the number of PyTorch dataloader threads employed; in our setup, we allocated one thread per each of the 12 CPU cores of the Jetson AGX Orin platform;
- `epochs`: the number of training epochs; through iterative testing, we determined that the highest accuracy was achieved after more than 100 training epochs.

The 'train_ssd.py' script provides 29 distinct arguments for dataset balance, network architecture, loading pretrained basenet or checkpoints, fine-tuning Stochastic Gradient Descent parameters, scheduler for dynamic adjustment of the learning rate during training, and more. These arguments come with default values that proved to be appropriate for our specific training requirements. To test a potential further model refinement, a second round of transfer learning was undertaken. Our custom dataset was organized in accordance with the Pascal VOC format (Figure 3).

```
custom_dataset
      |-----Annotations
      |          |-----Anox_1.xml, Anox_2.xml, ......
      |-----ImageSets
      |          |-----Main
      |                  |-----test.txt, train.txt, trainval.txt, val.txt
      |-----JPEGImages
      |          |-----Anox_1.jpg, Anox_2.jpg, ......
      |-----labels.txt
```

Figure 3. Dataset file structure following the Pascal VOC format requirements.

The 'Annotations' directory contains XML files that characterize the images and bounding boxes. The dataset was partitioned into three subsets, with 75% of the data allocated to the training set, 15% to the validation set and the remaining 10% designated for the test set. The files 'train.txt', 'val.txt', 'test.txt' and 'trainval.txt' include the filenames of the image files, without extensions, corresponding to the training, validation and testing sets. All 2648 images in JPG format are located in the 'JPEGImages' directory. The 'labels.txt' file contains the simplified dataset class names in alphabetical order, namely, 'Anoxia', 'Diabrotica', 'Ladybug', 'Opatrum' and 'Zabrus'.

To increase the precision of statistical inference, stratified sampling was applied to create the training, testing and validation sets (Table 5). Thus, the size of the sample randomly drawn from each class is proportional to the relative size of the class in the dataset.

Table 5. Images counts for each class and subset.

Class Name	Number of Images	Relative Distribution	Train (75%)	Test (10%)	Validation (15%)
Anoxia	462	17.4%	347	46	69
Diabrotica	618	23.3%	463	63	92
Ladybug	519	19.6%	389	52	78
Opatrum	536	20.2%	402	54	80
Zabrus	513	19.5%	385	51	77
Total	2648	100%	1986	266	396

The training process is conducted using the same 'train_ssd.py' script as before, with the command line closely resembling the one presented previously, with the following exceptions:

- The Pascal VOC format is specified for the dataset, by means of the `dataset-type` argument: `--dataset-type=voc`;
- The `pretrained-ssd` argument is employed to retrain the model obtained in the previous transfer learning phase, characterized by the highest accuracy and the lowest cost

function value; for instance: `--pretrained-ssd=models/mb1-ssd-Epoch-62-Loss-0.97135413.pth`.

Following each training phase, the retrained model performance was tested. Thus, we adopted a two-step approach, involving model conversion and real-time inference. First, the 'export_onnx.py' script was employed to convert the trained model into the ONNX format, for compatibility with Nvidia TensorRT, which further optimizes the ONNX models for real-time inference as described earlier. In this respect, the 'detectnet.py' script was used to transfer the ONNX model to an Nvidia detectNet object. The following is a command line example:

```
detectnet --model=PATH_TO_ONNX_FILE -threshold=0.5 --labels=PATH_TO_CLASS_NAMES_FILE --input-blob=input_0 --output-cvg=scores --output-bbox=boxes PATH_TO_TEST_IMAGES PATH_TO_INFERENCED_IMAGES
```

In this example, the `input_blob` argument specifies the input layer of the detectNet object, and the `threshold` parameter defines the confidence threshold for object detection. detectNet employs the following two distinct output layers: the 'confidence grid' for predicting object presence likelihood across spatial locations (the `output-cvg` argument), and the 'bounding box data' for precise object localization, including coordinates and class labels (the `output-bbox` argument). This dual-output architecture enhances object detection accuracy by first identifying potential object regions and then providing detailed information about detected objects, making it very efficient for real-time object detection and tracking applications [77].

2.6. Evaluation Metrics

Mean average precision (mAP) is a very popular metric for evaluating object detection models. It considers both precision and recall across various confidence thresholds. Precision is used to measure the accuracy of positive predictions made by a model, thus assessing the model's ability to avoid false positives (instances where the model incorrectly identifies a non-existent object as positive). A high precision value indicates that, when the model predicts an object, it is usually correct. Recall measures the model's capability to avoid false negatives. In the context of object detection, false negatives refer to cases where the model fails to detect an object that actually exists in the image. A high recall value indicates that the model can effectively capture most of the relevant objects. The mAP is calculated as the mean of the average precision (AP) values for each class, where AP measures the area under the precision-recall curve. The average precision (AP) for a specific class is computed as follows:

$$AP = \frac{1}{n}\sum_{k=1}^{n} P_k \cdot \Delta R(k), \quad (4)$$

where:

- $P(k)$ represents the precision at the kth retrieved item;
- $\Delta R(k)$ is the change in recall at the kth retrieved item;
- n is the total number of retrieved items.

The mAP is then the mean of these AP values across all classes, providing an aggregated measure of the model's ability to accurately detect objects. In object detection, besides assessing if the model identifies the correct class, it is important to consider if the location of the object is correctly identified. In this respect, the AP is computed based on Intersection over Union (IoU), which is the area of overlap between the predicted bounding box and the target bounding box, divided by the area of their union [78]. In the context of mAP calculation, IoU@0.5 (Intersection over Union at a threshold of 0.5) is a common choice. It means that a predicted bounding box is considered correct if the IoU with a ground-truth box is equal to or greater than 0.5. This threshold determines whether a prediction is a true positive or a false positive.

Accuracy quantifies the model's ability to correctly classify objects into their respective classes and is an appropriate metric when the dataset is balanced, such in this case. The accuracy is calculated as the fraction of predictions that the model correctly identified as follows:

$$Accuracy = \frac{Number\ of\ Correct\ Predictions}{Total\ Number\ of\ Predictions} \tag{5}$$

These two metrics, mAP and accuracy, were employed to evaluate the retrained models, and for the subsequent selection of the model with the best performance.

3. Results and Discussion

In order to assess whether the two-step transfer learning procedure offers advantages compared to a single transfer learning procedure performed directly onto the custom dataset, the following steps were undertaken (Figure 4):

1. Initially, the original downloaded MobileNet-SSD-v1 neural network was retrained directly on the custom image dataset without augmentation. The performance of this model on the testing set served as the baseline reference level;
2. A single transfer learning step was performed, involving the retraining of the original MobileNet-SSD-v1 and MobileNet-SSD-v2-Lite models directly on the augmented custom dataset;
3. A two-step transfer learning procedure was performed. Each of the two original models was retrained twice as follows: first on the Open Images dataset (containing only the 'Ladybug' and 'Beetles' dataset classes), and then each of the resulting retrained model was further retrained on the augmented custom dataset;
4. The performances of the models obtained in steps (2) and (3) were compared in terms of accuracy and mAP, to the baseline reference model from step (1).

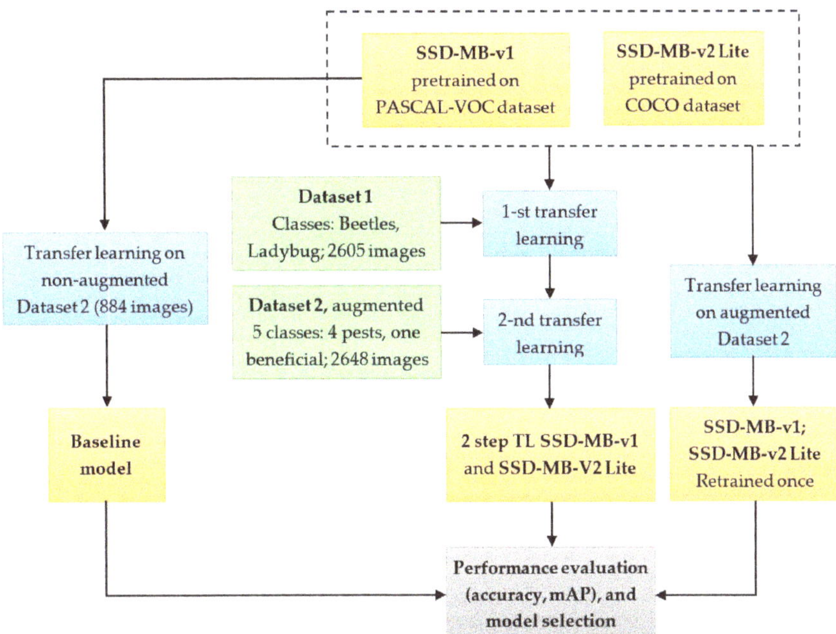

Figure 4. Assessing the performance of the two-step transfer learning procedure on SSD-MobileNet Networks; 2 step TL—two-step transfer learning; color scheme: orange-neural network models; blue-training procedures; green-datasets; grey-final results.

Figure 5 illustrates the loss variation in the MobileNet-SSD-v2-Lite network during the second retraining stage on the custom dataset. All the other training procedures exhibited a similar behavior. Continuous loss reduction indicates that the model is capturing patterns and features in the data. The stabilization of both training and validation loss values after approximately epoch 120 suggests that the model has reached a point of steady-state performance. In this phase, it has learned to represent the data adequately. The subsequent plateau and minor fluctuations in the losses show that the model is fine-tuning its parameters but not making substantial performance improvements.

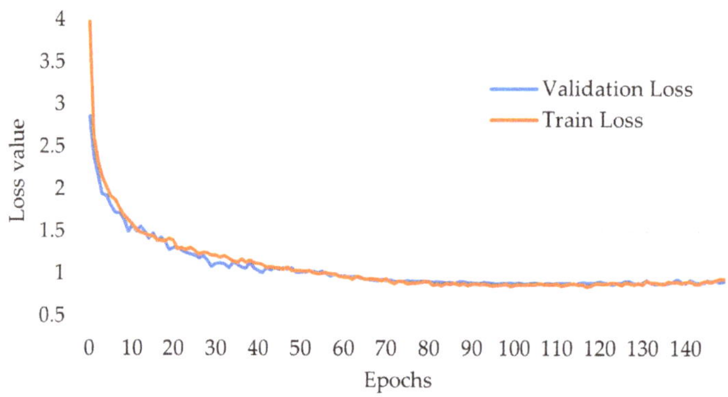

Figure 5. Loss variation in the MobileNet-SSD-v2-Lite network during the second retraining stage on the custom dataset.

The overall minimal fluctuations for both training and validation loss values indicate that the model is not overfitting to the training data. Overfitting would manifest as a later increase in validation loss relative to the training loss, which is not observed, suggesting that the model generalizes well to unseen data. The fact that both losses stabilize indicates an appropriate model complexity level. The behavior of the loss curves can also be indicative of the fact that hyperparameters, such as batch size, learning rate schedule and regularization techniques, were properly selected, thus contributing to the observed convergence and steady-state aspect.

3.1. *Evaluation Metrics and Models Ranking*

As one can notice from the results achieved on the test set and presented in Table 6 and in Figure 6, the MobileNet-SSD-v2-Lite model trained with the two-step transfer learning procedure ranks first, closely followed by the MobileNet-SSD-v1 after the same two-step transfer learning procedure.

The MobileNet-SSD-v2-Lite model achieved an mAP of 0.892, ranking second but very close to the highest mAP of 0.908, which was obtained by MobileNet-SSD-v1 after the same two-step transfer learning procedure. It outperformed the baseline mAP (0.8412) by 6.07%, and it is only 1.76% less than the highest value. Compared to the other non-baseline attempts, the two-step transfer learning trained MobileNet-SSD-v2-Lite model demonstrated the highest accuracy for *Opatrum* (0.9514, baseline excepted) and *Diabrotica* (0.8066). For *Anoxia*, it reached a third-place accuracy of 0.9851, which is only 0.6% less than the top value of 0.9912. Also, for *Zabrus*, the MobileNet-SSD-v2-Lite model achieved the second position (0.9053), surpassing the baseline model (0.803) by 12.74%, while *Coccinella* (the 'Ladybug' dataset class) was reliably differentiated from all other species, with an accuracy of 0.8939 and zero false positives. In the case of using the model for precision pest control, zero false positives ensure that the beneficial *Coccinella* species will not be mistakenly targeted or harmed, thus contributing to a more environmentally friendly and sustainable approach to agriculture.

Table 6. Accuracy and mean average precision for each type of training procedure; red font—highest value; blue font—the second highest value; TL—transfer learning.

Trained Model	Accuracy					mAP	Notes
	Anoxia	Diabrotica	Ladybug	Opatrum	Zabrus		
SSD-MB-v1 (baseline)	0.8432	0.7323	0.8276	1	0.8030	0.841	Retrained once, on non-augmented custom dataset
SSD-MB-v1 TL on custom dataset	0.9866	0.7259	0.8737	0.9240	0.9475	0.887	Retrained once, on augmented custom dataset
SSD-MB-v1 2-step TL	0.9072	0.7770	0.9436	0.9056	0.9033	0.908	Two-step transfer learning
SSD-MB-v2 Lite TL on custom dataset	0.9912	0.7273	0.8906	0.9091	0.9030	0.884	Retrained once, on augmented custom dataset
SSD-MB-v2-Lite 2-step TL	0.9851	0.8066	0.8939	0.9514	0.9053	0.892	Two-step transfer learning

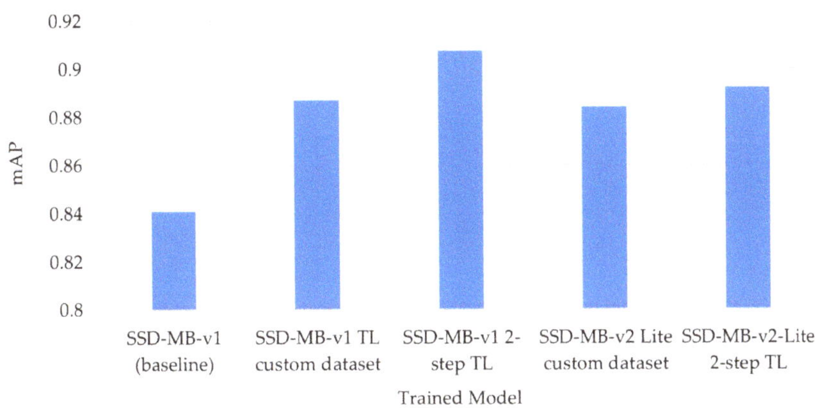

Figure 6. Neural network performances in terms of mAP for each dataset class.

Table 7 displays only the first and second-place results for mAP and accuracies across classes, clearly confirming that the two-step transfer learning procedure performs better on both MobileNet-SSD-v1 and MobileNet-SSD-v2-Lite. The MobileNet-SSD-v2-Lite model has the best overall performance. It exhibits, however, an exception in detecting the 'Anoxia' dataset class, where it ranks third but with only a marginal 0.6% difference from the first-place position.

Figure 7 shows a clear and significant performance advantage gained from the two-step transfer learning procedure applied to the MobileNet-SSD-v1-Lite model compared to the baseline model. However, one can notice from Table 7 that, in the case of the 'Opatrum' dataset class, the baseline model developed perfect performance, achieving a maximum accuracy of one. This phenomenon could be attributed to the fact that the images from the test set belonging to the 'Opatrum' class are well-separated in the feature space from other classes, thus being relatively easy to classify. Without augmentation, the number of images in the test set is small for each class, and it is possible that it does not represent the full diversity and complexity of the 'Opatrum' class. In such cases, the model might have memorized the training data and generalized well to the limited test examples.

Table 7. Ranking of accuracy and mAP, excluding the baseline model; (1)—first place; (2)—second place.

Trained Model	Ranking in Terms of Accuracy for Each Dataset Class					mAP
	Anoxia	Diabrotica	Ladybug	Opatrum	Zabrus	
SSD-MB-v1 (baseline)	-	-	-	(1)	-	-
SSD-MB-v1 TL on custom dataset	(2)	-	-	(2)	(1)	-
SSD-MB-v1 2-step TL	-	(2)	(1)	-	-	(1)
SSD-MB-v2 Lite TL on custom dataset	(1)	-	-	-	-	-
SSD-MB-v2-Lite 2-step TL	3rd place; 0.6% less than the 1st place	(1)	(2) 5.3% less than the 1st place	1st place, (except the baseline)	(2) 4.5% less than the 1st place	(2) 1.8% less than the 1st place

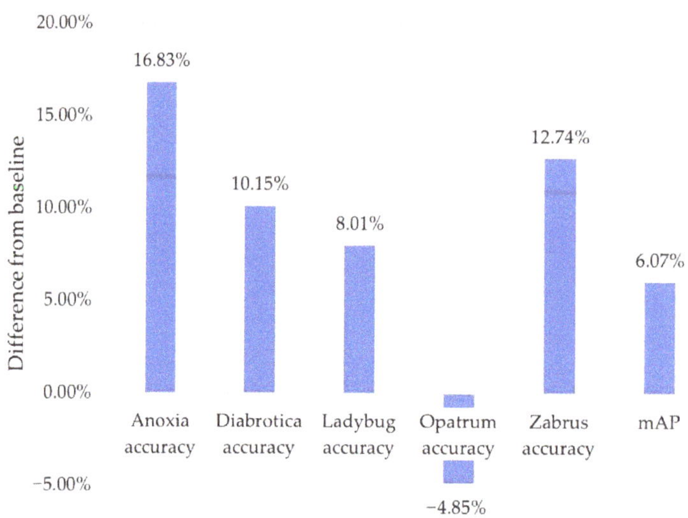

Figure 7. Performance gain of the MobileNet-SSD-v2-Lite model following the two-step transfer. learning procedure, against the baseline model.

Upon evaluating the same model trained on the augmented dataset, the accuracy decreases to 0.924. This can be explained by the fact that data augmentation introduced variations into the training data, which aids in mitigating overfitting. If the model had initially overfit to the non-augmented dataset, the addition of augmentation might have facilitated improved generalization, even if it was at the cost of some slight performance reduction.

Figure 8 shows a few instances of accurate predictions on the testing set, made by the MobileNet-SSD-v2-Lite following the two-step transfer learning procedure, some of them under challenging conditions. For example, in images (a) and (c), both *Diabrotica* and *Anoxia* exhibit small-to-medium scale in terms of image size. They are reliably detected with confidence levels of 98.5% and 99.1%, respectively. Image (b) demonstrates the correct detection of all individuals within a rich *Diabrotica* colony, some overlapping or being partially visible; moreover, in image (d) a partially visible *Zabrus* is identified with a confidence level of 95.4%.

Figure 8. Examples of accurate predictions made by MobileNet-SSD-v2-Lite, after a two-step transfer learning procedure, some of them in the following challenging conditions: (**a**,**c**): small to medium scale; (**b**) overlapping; (**d**) partially visible; (**e**) color blends with the environment; (**f**) high confidence for Ladybug; credit to the original, non-transformed images, (**a**–**f**): [79–84].

Image (e) captures the detection of *Opatrum* on the ground (92.2% confidence level), where *Opatrum*'s color blends with the environment. *Coccinella* is usually detected with high confidence levels of more than 98% (100% in image (f), for example). Most correct predictions follow the patterns presented above, underscoring a good performance, even if the model was trained on a small custom dataset.

3.2. Error Analysis

In order to provide insights into the model's strengths and limitations, and to find information for further refining its performance, we performed a detailed analysis of its predictions on the test set and identified four types of errors.

- **Misclassification errors**: Illustrated in Figure 9a, there is a singular misclassification error over the entire test set, where the model confuses *Zabrus* with *Anoxia*. This could be attributed to *Zabrus* being photographed from an unusual lateral position, a scenario represented by only two images in the training set;
- **Non-identification errors**: Errors in which the model fails to make identification, as illustrated in Figure 9b–f. Most of these misidentifications are likely due to the fact that the training set contains an extremely limited number of images with a very particular background; therefore, the model might encounter challenges when the background varies significantly from the majority of training images. Neural networks are expected to generalize across different backgrounds, but extreme variations could pose difficulties. For instance, in Figure 9b, the image has a blue sky background. In the training set, there are only four images with a blue background. Similarly, for Figure 9c, there are only two such images in the training set;
- Another circumstance involves instances where the object is considerably small relative to the overall image dimensions, as in Figure 9d, or when it is barely visible, as seen in Figure 9f. A particular scenario is presented by images featuring *Opatrum* on the

ground, where the model correctly identifies *Opatrum* in Figure 8e but fails to do so in Figure 9e, showing a nuanced performance with a confidence level of 45% in case of Figure 9e, which is just below the 50% threshold;

- **Duplication errors**: Figure 9g,h show duplication errors, where the same pest is detected twice in images with a singular specimen. Figure 9g highlights potential network confusion during the feature extraction stag, caused by the corn silk, which bears visual similarities to the legs and antennae of the pest species;
- **False Positive errors**: These are errors where a pest was detected in an area containing only background. There is a single situation over the entire test set, where a false *Diabrotica* was detected in the upper-right section of Figure 9h, with a confidence level of 62.5%.

(**a**) (**b**) (**c**) (**d**) (**e**) (**f**) (**g**) (**h**)

Figure 9. Examples of the following various types of errors: (**a**) misclassification; (**b–f**): types of failed detections; (**g**) double detection; (**h**) false and double detection. Credit to the original, non-transformed images, (**a–h**): [82,82,85,85–89].

The observed relatively high error rate of 0.1579 over the test set may be attributed, in part, to the modest size of the training set, averaging around 400 images per class after augmentation. The limited diversity in training instances could lead to challenges in the model's ability to generalize across various conditions.

4. Conclusions

To enhance accuracy while maintaining efficiency, a two-step transfer learning procedure was employed to fine-tune the network for crop pest detection tasks. Two variations in the SSD-MobileNet architecture, namely, MobileNet-SSD-v1 and MobileNet-SSD-v2-Lite, were tested, aiming to identify the version achieving the best accuracy and assess the efficiency of the proposed two-step transfer learning procedure. Five beetle species, including four harmful to corn crops and one beneficial, were selected for testing.

The results indicated that the MobileNet-SSD-v2-Lite model, trained with the two-step transfer learning procedure, exhibited the highest performance, closely followed by the MobileNet-SSD-v1. MobileNet-SSD-v2-Lite achieved an mAP of 0.892, displaying the highest accuracy for classes 'Opatrum' (excluding the baseline model) and 'Diabrotica', and the second position in terms of accuracy for classes 'Ladybug' and 'Zabrus'.

Analyzing errors in the MobileNet-SSD-v2-Lite model revealed a good overall accuracy despite the reduced size of the training set, with only 1 misclassification, 33 non-identifications, 7 double identifications and 1 false positive across the test set.

These preliminary results demonstrate the potential for real-time pest control with minimal human intervention, showing the model's capability to protect both crops and beneficial species. Considering the achieved results and the identified errors, our future research will focus on the following key areas:

- To address the issue of generalization error, we plan to employ a custom dataset comprising a minimum of 8000 images, ensuring representative sampling for real-world conditions;
- The top-performing model from current research will undergo retraining using the two-step transfer learning procedure on the new dataset. To gain a more comprehensive understanding of performance improvement, both the pretrained and untrained versions of the model will be trained on the non-augmented dataset. The outcomes will then be compared with the results obtained from training the same versions of the model on the augmented dataset;
- Our dataset enhancement strategy will involve incorporating more images that led to errors in the current study. This includes instances of small-scale beetles relative to image size, partially visible beetles, and images featuring *Opatrum* on the ground. The goal is to enhance the model's feature extraction capabilities, particularly in scenarios where the color of *Opatrum* blends with the soil color;
- Evaluation of model performance will be extended to scenarios with two or more pest species coexisting in the same image, including scenarios involving both pests and beneficial species.

In the third phase of our research, we plan to develop a prototype with an initial focus on monitoring and assessing corn crop infestation levels in real-time. The images collected during monitoring will also be used for continuous training of the model.

Author Contributions: Conceptualization, E.M., A.I. and S.M.; methodology, E.M. and A.I.; software adjustment/training/validation/testing, E.M. and A.I.; resources, E.M.; data curation, E.M. and S.M.; writing—original draft preparation, E.M. and S.M.; writing—review and editing, E.M. and S.M. All authors have read and agreed to the published version of the manuscript.

Funding: This research received no external funding.

Institutional Review Board Statement: Not applicable.

Data Availability Statement: The preprocessed data presented in this study are available on request from the corresponding author. The data are not publicly available due to self-funded nature of the research, and it was not financially supported by public sources.

Conflicts of Interest: The authors declare no conflict of interest.

References

1. Çakmakçı, R.; Salık, M.A.; Çakmakçı, S. Assessment and Principles of Environmentally Sustainable Food and Agriculture Systems. *Agriculture* **2023**, *13*, 1073. [CrossRef]
2. Flint, M.L.; Van den Bosch, R. *Introduction to Integrated Pest Management*; Springer: New York, NY, USA, 2012; pp. 1–256.
3. Jensen, S.E. Insecticide Resistance in the Western Flower Thrips, Frankliniella occidentalis. *Integr. Pest Manag. Rev.* **2000**, *5*, 131–146. [CrossRef]
4. Kranthi, K.R.; Jadhav, D.R.; Kranthi, S.; Wanjari, R.R.; Ali, S.S.; Russell, D.A. Insecticide Resistance in Five Major Insect Pests of Cotton in India. *Crop Prot.* **2002**, *21*, 449–460. [CrossRef]
5. Ngegba, P.M.; Cui, G.; Khalid, M.Z.; Zhong, G. Use of Botanical Pesticides in Agriculture as an Alternative to Synthetic Pesticides. *Agriculture* **2022**, *12*, 600. [CrossRef]
6. Krupke, C.; Holland, J.D.; Long, E.; Eitzer, B.D. Planting of Neonicotinoid-Treated Maize Poses Risks for Honey Bees and Other Non-Target Organisms Over a Wide Area without Consistent Crop Yield Benefit. *J. Appl. Ecol.* **2017**, *54*, 1449–1458. [CrossRef]
7. Krupke, C.H.; Long, E.Y. Intersections Between Neonicotinoid Seed Treatments and Honey Bees. *Curr. Opin. Insect Sci.* **2015**, *10*, 8–13. [CrossRef]
8. Bonmatin, J.M.; Giorio, C.; Girolami, V.; Goulson, D.; Kreutzweiser, D.P.; Krupke, C.; Liess, M.; Long, E.; Marzaro, M.; Mitchell, E.A.D.; et al. Environmental Fate and Exposure; Neonicotinoids and Fipronil. *Environ. Sci. Pollut. Res.* **2015**, *22*, 35–67. [CrossRef] [PubMed]
9. Sánchez-Bayo, F.; Goka, K.; Hayasaka, D. Contamination of the Aquatic Environment with Neonicotinoids and its Implication for Ecosystems. *Front. Environ. Sci.* **2016**, *4*, 71. [CrossRef]
10. Ghaderi, S.; Fathipour, Y.; Asgari, S.; Reddy, G. Economic Injury Level and Crop Loss Assessment for *Tuta absoluta* (Lepidoptera: Gelechiidae) on Different Tomato Cultivars. *J. Appl. Entomol.* **2019**, *143*, 493–507. [CrossRef]
11. Saha, T.; Chandran, N. Chemical Ecology and Pest Management: A Review. *Int. J. Chem. Stud.* **2017**, *5*, 618–621. Available online: https://www.chemijournal.com/archives/2017/vol5issue6/PartI/5-5-449-329.pdf (accessed on 7 April 2023).
12. Føre, M.; Frank, K.; Norton, T.; Svendsen, E.; Alfredsen, J.; Dempster, T.; Eguiraun, H.; Watson, W.; Stahl, A.; Sunde, L. Precision Fish Farming: A New Framework to Improve Production in Aquaculture. *Biosyst. Eng.* **2018**, *173*, 176–193. [CrossRef]
13. Eli-Chukwu, N. Applications of Artificial Intelligence in Agriculture: A Review. *Eng. Technol. Appl. Sci. Res.* **2019**, *9*, 4377–4383. [CrossRef]
14. Smith, M. Getting Value from Artificial Intelligence in Agriculture. *Anim. Prod. Sci.* **2018**, *60*, 46–54. [CrossRef]
15. Bannerjee, G.; Sarkar, U.; Das, S.; Ghosh, I. Artificial Intelligence in Agriculture: A Literature Survey. *Int. J. Sci. Res. Comput. Sci. Appl. Manag. Stud.* **2018**, *7*, 1–6.
16. Jha, K.; Doshi, A.; Patel, P.; Shah, M.A. Comprehensive Review on Automation in Agriculture using Artificial Intelligence. *Artif. Intell. Agric.* **2019**, *2*, 1–12. [CrossRef]
17. Gulzar, Y.; Ünal, Z.; Aktaş, H.; Mir, M. Harnessing the Power of Transfer Learning in Sunflower Disease Detection: A Comparative Study. *Agriculture* **2023**, *13*, 1479. [CrossRef]
18. Gulzar, Y. Fruit Image Classification Model Based on MobileNetV2 with Deep Transfer Learning Technique. *Sustainability* **2023**, *15*, 1906. [CrossRef]
19. Dhiman, P.; Kaur, A.; Balasaraswathi, V.; Gulzar, Y.; Alwan, A.; Hamid, Y. Image Acquisition, Preprocessing and Classification of Citrus Fruit Diseases: A Systematic Literature Review. *Sustainability* **2023**, *15*, 9643. [CrossRef]
20. Kalfas, I.; De Ketelaere, B.; Bunkens, K.; Saeys, W. Towards Automatic Insect Monitoring on Witloof Chicory Fields using Sticky Plate Image Analysis. *Ecol. Inf.* **2023**, *75*, 102037. [CrossRef]
21. Yang, S.; Xing, Z.; Wang, H.; Dong, X.; Gao, X.; Liu, Z.; Zhang, X.; Li, S.; Zhao, Y. Maize-YOLO: A New High-Precision and Real-Time Method for Maize Pest Detection. *Insects* **2023**, *14*, 278–291. [CrossRef]
22. Wu, X.; Zhan, C.; Lai, Y.-K.; Cheng, M.-M.; Yang, J. IP102: A Large-Scale Benchmark Dataset for Insect Pest Recognition. In Proceedings of the IEEE Conference on Computer Vision and Pattern Recognition, Long Beach, CA, USA, 15–20 June 2019.
23. Albanese, A.; Nardello, M.; Brunelli, D. Automated Pest Detection with DNN on the Edge for Precision Agriculture. *IEEE J. Emerg. Sel. Top. Circuits Syst.* **2021**, *11*, 458–467. [CrossRef]
24. Wang, C.; Grijalva, I.; Caragea, D.; McCornack, B. Detecting Common Coccinellids Found in Sorghum Using Deep Learning Models. *Sci. Rep.* **2023**, *13*, 9748. [CrossRef] [PubMed]
25. Salamut, C.; Kohnert, I.; Landwehr, N.; Pflanz, M.; Schirrmann, M.; Zare, M. Deep Learning Object Detection for Image Analysis of Cherry Fruit Fly (*Rhagoletis cerasi* L.) on Yellow Sticky Traps. *Gesunde Pflanz.* **2023**, *75*, 37–48. [CrossRef]
26. Rustia, D.J.A.; Chao, J.J.; Chiu, L.-Y.; Wu, Y.-F.; Chung, J.-Y.; Hsu, J.-C.; Lin, T.-T. Automatic Greenhouse Insect Pest Detection and Recognition Based on a Cascaded Deep Learning Classification Method. *J. Appl. Entomol.* **2020**, *145*, 206–222. [CrossRef]

27. Wang, Q.-J.; Zhang, S.-Y.; Dong, S.-F.; Zhang, G.-C.; Yang, J.; Li, R.; Wang, H.-Q. Pest24: A Large-Scale Very Small Object Data Set of Agricultural Pests for Multi-Target Detection. *Comput. Electron. Agric.* **2020**, *175*, 105585. [CrossRef]
28. Li, W.; Wang, D.; Li, M.; Gao, Y.; Wu, J.; Yang, X. Field Detection of Tiny Pests from Sticky Trap Images Using Deep Learning in Agricultural Greenhouse. *Comput. Electron. Agric.* **2021**, *183*, 106048. [CrossRef]
29. Hong, S.-J.; Nam, I.; Kim, S.-Y.; Kim, E.; Lee, C.-H.; Ahn, S.; Parl, I.-K.; Kim, G. Automatic Pest Counting from Pheromone Trap Images Using Deep Learning Object Detectors for Matsucoccus Thunbergianae Monitoring. *Insects* **2021**, *12*, 342–358. [CrossRef] [PubMed]
30. Wang, R.; Jiao, L.; Xie, C.; Chen, P.; Du, J.; Li, R. S-rpn: Sampling-Balanced Region Proposal Network for Small Crop Pest Detection. *Comput. Electron. Agric.* **2021**, *187*, 106290. [CrossRef]
31. Jiao, L.; Xie, C.; Chen, P.; Du, J.; Li, R.; Zhang, J. Adaptive Feature Fusion Pyramid Network for Multi-Classes Agricultural Pest Detection. *Comput. Electron. Agric.* **2022**, *195*, 106827. [CrossRef]
32. Zhang, W.; Huang, H.; Sun, Y.; Wu, X. Agripest-YOLO: A Rapid Light-Trap Agricultural Pest Detection Method Based on Deep Learning. *Front. Plant Sci.* **2022**, *13*, 1079384. [CrossRef]
33. Sava, A.; Ichim, L.; Popescu, D. Detection of *Halyomorpha halys* using Neural Networks. In Proceedings of the IEEE 8th International Conference on Control, Decision and Information Technologies (CoDIT), Istanbul, Turkey, 17–20 May 2022.
34. Takimoto, H.; Sato, Y.; Nagano, A.J.; Shimizu, K.K.; Kanagawa, A. Using a Two-Stage Convolutional Neural Network to Rapidly Identify Tiny Herbivorous Beetles in the Field. *Ecol. Inf.* **2021**, *66*, 101466. [CrossRef]
35. Ozdemir, D.; Kunduraci, M.S. Comparison of Deep Learning Techniques for Classification of the Insects in Order Level with Mobile Software Application. *IEEE Access* **2022**, *10*, 35675–35684. [CrossRef]
36. Butera, L.; Ferrante, A.; Jermini, M.; Prevostini, M.; Alippi, C. Precise Agriculture: Effective Deep Learning Strategies to Detect Pest Insects. *IEEE/CAA J. Autom. Sin.* **2022**, *9*, 246–258. [CrossRef]
37. Ahmad, I.; Yang, Y.; Yue, Y.; Ye, C.; Hassan, M.; Cheng, X.; Wu, Y.; Zhang, Y. Deep Learning Based Detector YOLOv5 for Identifying Insect Pests. *Appl. Sci.* **2022**, *12*, 10167. [CrossRef]
38. Ratnayake, M.N.; Dyer, A.G.; Dorin, A. Tracking Individual Honeybees Among Wildflower Clusters with Computer Vision-Facilitated Pollinator Monitoring. *PLoS ONE* **2021**, *16*, e0239504. [CrossRef] [PubMed]
39. Bjerge, K.; Alison, J.; Dyrmann, M.; Frigaard, C.E.; Mann, H.M.R.; Høye, T.T. Accurate Detection and Identification of Insects from Camera Trap Images with Deep Learning. *PLOS Sustain. Transform.* **2023**, *2*, e0000051. [CrossRef]
40. Spanier, R. Pollination AI: Deep Learning Approach to Identify Pollinators and Their Taxa Using the YOLO Architecture. Ph.D. Thesis, RWTHAachen University, Aachen, Germany, 2022.
41. Bjerge, K.; Frigaard, C.; Karstoft, H. Motion Informed Object Detection of Small Insects in Time-lapse Camera Recordings. *Sensors* **2023**, *23*, 7242. [CrossRef]
42. Venegas, P.; Calderon, F.; Riofrío, D.; Benítez, D.; Ramón, G.; Cisneros-Heredia, D.; Coimbra, M.; Rojo-Álvarez, J.-L.; Perez, N. Automatic Ladybird Beetle Detection Using Deep-Learning Models. *PLoS ONE* **2021**, *16*, e0253027. [CrossRef]
43. Vega, M.; Benitez, D.; Perez, N.P.; Riofrio, D.; Ramón-Cabrera, G.; Cisneros-Heredia, D. Coccinellidae Beetle Specimen Detection Using Convolutional Neural Networks. In Proceedings of the IEEE Colombian Conference on Applications of Computational Intelligence (ColCACI), Cali, Colombia, 26–28 May 2021.
44. Amarathunga, D.C.; Grundy, J.; Parry, H.; Dorin, A. Methods of Insect Image Capture and Classification: A Systematic Literature Review. *Smart Agric. Technol.* **2021**, *1*, 100023. [CrossRef]
45. Cheng, X.; Zhang, Y.; Chen, Y.; Wu, Y.; Yue, Y. Pest Identification via Deep Residual Learning in Complex Background. *Comput. Electron. Agric.* **2017**, *141*, 351–356. [CrossRef]
46. Kasinathan, T.; Singaraju, D.; Uyyala, S.R. Insect Classification and Detection in Field Crops using Modern Machine Learning Techniques. *Inf. Proc. Agric.* **2021**, *8*, 446–457. [CrossRef]
47. Li, Y.; Yang, J. Few-Shot Cotton Pest Recognition and Terminal Realization. *Comput. Electron. Agric.* **2020**, *169*, 105240. [CrossRef]
48. Nanni, L.; Maguolo, G.; Pancino, F. Insect Pest Image Detection and Recognition Based on Bio-Inspired Methods. *Ecol. Inf.* **2020**, *57*, 101089. [CrossRef]
49. Pattnaik, G.; Shrivastava, V.K.; Parvathi, K. Transfer Learning-Based Framework for Classification of Pest in Tomato Plants. *Appl. Artif. Intell.* **2020**, *34*, 981–993. [CrossRef]
50. Wang, R.J.; Zhang, J.; Dong, W.; Yu, J.; Xie, C.J.; Li, R.; Chen, T.J.; Chen, H.B. Crop Pests Image Classification Algorithm Based on Deep Convolutional Neural Network. *Telkomnika* **2017**, *15*, 1239–1246. [CrossRef]
51. Wang, J.; Li, Y.; Feng, H.; Ren, L.; Du, X.; Wu, J. Common Pests Image Recognition Based on Deep Convolutional Neural Network. *Comput. Electron. Agric.* **2020**, *179*, 105834. [CrossRef]
52. You, Y.; Zeng, Z.; Zheng, J.; Zhao, J.; Luo, F.; Chen, Y.; Xie, M.; Liu, X.; Wei, H. The Toxicity Response of *Coccinella septempunctata* L. (Coleoptera: Coccinellidae) after Exposure to Sublethal Concentrations of Acetamiprid. *Agriculture* **2022**, *12*, 1642. [CrossRef]
53. Ovsyannikova, E.I. Zabrus tenebrioides Goeze-Corn Ground Beetle. 2008. Available online: http://agroatlas.ru/en/content/pests/Zabrus_tenebrioides/index.html (accessed on 7 April 2023).
54. Afonin, A.N.; Greene, S.L.; Dzyubenko, N.I.; Frolov, A.N. Interactive Agricultural Ecological Atlas of Russia and Neighboring Countries. Economic Plants and their Diseases, Pests and Weeds. 2008. Available online: http://www.agroatlas.ru (accessed on 7 April 2023).

55. Ovsyannikova, E.I.; Grichanov, I.Y. *Opatrum sabulosum* (L.)-Darkling Beetle. 2008. Available online: http://agroatlas.ru/en/content/pests/Opatrum_sabulosum/index.html (accessed on 7 April 2023).
56. Fătu, A.-C.; Dinu, M.M.; Andrei, A.M. Susceptibility of some melolonthine scarab species to entomopathogenic fungus *Beauveria brongniartii* (Sacc.) Petch and *Metarhizium anisopliae* (Metsch.). *Sci. Bull. Ser. F Biotech.* **2018**, *22*, 42–49.
57. Grozea, I.; Trusca, R.; Virteiu, A.M.; Stef, R.; Butnariu, M. Interaction between *Diabrotica virgifera virgifera* and host plants determined by feeding behavior and chemical composition. *Rom. Agric. Res.* **2017**, *34*, 329–337.
58. CABI. *Diabrotica virgifera virgifera* (Western Corn Rootworm). 2021. Available online: https://www.cabidigitallibrary.org/doi/full/10.1079/cabicompendium.18637 (accessed on 24 May 2023).
59. Franklin, D. NVIDIA: DNN Vision Library (Jetson-Inference): detectNet. 2023. Available online: https://rawgit.com/dusty-nv/jetson-inference/master/docs/html/group__detectNet.html (accessed on 14 July 2023).
60. Liu, W.; Anguelov, D.; Erhan, D.; Szegedy, C.; Reed, S.; Fu, C.-Y.; Berg, A. SSD: Single Shot MultiBox Detector. In Proceedings of the 14th European Conference on Computer Vision–ECCV 2016, Amsterdam, The Netherlands, 11–14 October 2016.
61. Teng, T.W.; Veerajagadheswar, P.; Ramalingam, B.; Yin, J.; Mohan, R.E.; Gómez, B.F. Vision Based Wall Following Framework: A Case Study with HSR Robot for Cleaning Application. *Sensors* **2020**, *20*, 3298. [CrossRef]
62. Girshick, R. Fast R-CNN. In Proceedings of the IEEE International Conference on Computer Vision (ICCV), Santiago, Chile, 7–13 December 2015.
63. Franklin, D. SSD-Based Object Detection in PyTorch: Model Training. 2023. Available online: https://github.com/dusty-nv/pytorch-ssd/blob/master/train_ssd.py (accessed on 14 July 2023).
64. Franklin, D. SSD-Based Object Detection in PyTorch: Model Evaluation. 2023. Available online: https://github.com/dusty-nv/pytorch-ssd/blob/master/eval_ssd.py (accessed on 14 July 2023).
65. Franklin, D. SSD-Based Object Detection in PyTorch: Export ONNX. 2023. Available online: https://github.com/dusty-nv/pytorch-ssd/blob/master/onnx_export.py (accessed on 14 July 2023).
66. Nelson, J. What is TensorRT. 2021. Available online: https://blog.roboflow.com/what-is-tensorrt/ (accessed on 21 June 2021).
67. NVIDIA TensorRT. 2023. Available online: https://docs.nvidia.com/deeplearning/tensorrt/pdf/TensorRT-Developer-Guide.pdf (accessed on 21 June 2023).
68. Franklin, D. SSD-Based Object Detection in PyTorch: Detectnet. 2023. Available online: https://github.com/dusty-nv/jetson-inference/blob/master/python/examples/detectnet.py (accessed on 21 July 2023).
69. Open Images Dataset V7 and Extensions. 2022. Available online: https://storage.googleapis.com/openimages/web/factsfigures_v7.html (accessed on 21 June 2023).
70. *Coccinella* Linnaeus, 1758 in GBIF Secretariat. GBIF Backbone Taxonomy. Checklist Dataset accessed via GBIF.org. Available online: https://www.gbif.org/search?q=Coccinella%20sp. (accessed on 14 July 2023).
71. *Anoxia villosa* (Fabricius, 1781) in GBIF Secretariat. GBIF Backbone Taxonomy. Checklist Dataset accessed via GBIF.org. Available online: https://www.gbif.org/species/1054733 (accessed on 14 July 2023).
72. *Diabrotica virgifera* LeConte, 1868 in GBIF Secretariat. GBIF Backbone Taxonomy. Checklist Dataset accessed via GBIF.org. Available online: https://www.gbif.org/species/1048497 (accessed on 14 July 2023).
73. *Opatrum sabulosum* (Linnaeus, 1761) in GBIF Secretariat. GBIF Backbone Taxonomy. Checklist Dataset accessed via GBIF.org. Available online: https://www.gbif.org/species/4454749 (accessed on 14 July 2023).
74. *Zabrus tenebrioides* (Goeze, 1777) in GBIF Secretariat. GBIF Backbone Taxonomy. Checklist Dataset accessed via GBIF.org. Available online: https://www.gbif.org/species/4473277 (accessed on 14 July 2023).
75. GBIF.org, GBIF Home Page. 2023. Available online: https://www.gbif.org (accessed on 14 July 2023).
76. Everingham, M.; Ali Eslami, S.M.; Van Gool, L.; Williams, C.K.I.; Winn, J.; Zisserman, A. The PASCAL Visual Object Classes Challenge: A Retrospective. *Int. J. Comput. Vision* **2014**, *111*, 98–136. [CrossRef]
77. NVIDIA Transfer Learning Toolkit for Intelligent Video Analytics-Getting Started Guide. 2020. Available online: https://docs.nvidia.com/metropolis/TLT/archive/tlt-10/pdf/Transfer-Learning-Toolkit-Getting-Started-Guide-IVA.pdf (accessed on 24 May 2023).
78. Geron, A. *Hands-on Machine Learning with Scikit-Learn, Keras, and TensorFlow*, 2nd ed.; O'Reilly Media: Sebastopol, CA, USA, 2019; p. 491.
79. BBirgit, iNaturalist Contributors, iNaturalist (2023). iNaturalist Research-Grade Observations. iNaturalist.org. Occurrence dataset accessed via GBIF.org. 2023. Available online: https://www.gbif.org/occurrence/3338144902 (accessed on 30 July 2023).
80. Miquet, A. iNaturalist Contributors, iNaturalist (2023). iNaturalist Research-Grade Observations. iNaturalist.org. Occurrence Dataset accessed via GBIF.org. 2023. Available online: https://www.gbif.org/occurrence/4039229776 (accessed on 30 July 2023).
81. Ferreira, R. iNaturalist Contributors, iNaturalist (2023). iNaturalist Research-Grade Observations. iNaturalist.org. Occurrence Dataset accessed via GBIF.org. 2023. Available online: https://www.gbif.org/occurrence/4121193187 (accessed on 30 July 2023).
82. Mobbini. iNaturalist Contributors, iNaturalist (2023). iNaturalist Research-Grade Observations. iNaturalist.org. Occurrence Dataset accessed via GBIF.org. 2020. Available online: https://www.gbif.org/occurrence/2901580832 (accessed on 30 July 2023).
83. Jeltov, P. iNaturalist Contributors, iNaturalist (2023). iNaturalist Research-Grade Observations. iNaturalist.org. Occurrence Dataset accessed via GBIF.org. 2023. Available online: https://www.gbif.org/occurrence/4075854369 (accessed on 30 July 2023).
84. Le Mao, P. iNaturalist Contributors, iNaturalist (2023). iNaturalist Research-Grade Observations. iNaturalist.org. Occurrence Dataset accessed via GBIF.org. 2023. Available online: https://www.gbif.org/occurrence/4018220177 (accessed on 30 July 2023).

85. Levon, A. iNaturalist Contributors, iNaturalist (2023). iNaturalist Research-Grade Observations. iNaturalist.org. Occurrence Dataset accessed via GBIF.org. 2022. Available online: https://www.gbif.org/occurrence/3903140984 (accessed on 30 July 2023).
86. Barileva, N. iNaturalist Contributors, iNaturalist (2023). iNaturalist Research-grade Observations. iNaturalist.org. Occurrence Dataset accessed via GBIF.org. 2023. Available online: https://www.gbif.org/occurrence/4014953025 (accessed on 30 July 2023).
87. Danielle. iNaturalist Contributors, iNaturalist (2023). iNaturalist Research-grade Observations. iNaturalist.org. Occurrence Dataset accessed via GBIF.org. 2023. Available online: https://www.gbif.org/occurrence/4018183044 (accessed on 30 July 2023).
88. Mednii, A. iNaturalist Contributors, iNaturalist (2023). iNaturalist Research-grade Observations. iNaturalist.org. Occurrence Dataset accessed via GBIF.org. 2023. Available online: https://www.gbif.org/occurrence/4091424606 (accessed on 30 July 2023).
89. Fogliato, S. iNaturalist Contributors, iNaturalist (2023). iNaturalist Research-grade Observations. iNaturalist.org. Occurrence Dataset accessed via GBIF.org. 2023. Available online: https://www.gbif.org/occurrence/3874204663 (accessed on 30 July 2023).

agriculture

MDPI

Article

Efficient and Lightweight Automatic Wheat Counting Method with Observation-Centric SORT for Real-Time Unmanned Aerial Vehicle Surveillance

Jie Chen [1], Xiaochun Hu [2], Jiahao Lu [1], Yan Chen [1,3,*] and Xin Huang [4]

1 School of Computer and Electronic Information, Guangxi University, Nanning 530004, China; 2213393002@st.gxu.edu.cn (J.C.); 2113591012@st.gxu.edu.cn (J.L.)
2 School of Big Data and Artificial Intelligence, Guangxi University of Finance and Economics, Nanning 530003, China; hxch@gxufe.edu.cn
3 Guangxi Intelligent Digital Services Research Center of Engineering Technology, Nanning 530004, China
4 College of Information Engineering, Guangxi Vocational University of Agriculture, Nanning 530007, China; huangxin543@163.com
* Correspondence: cy@gxu.edu.cn

Abstract: The number of wheat ears per unit area is crucial for assessing wheat yield, but automated wheat ear counting still faces significant challenges due to factors like lighting, orientation, and density variations. Departing from most static image analysis methodologies, this study introduces Wheat-FasterYOLO, an efficient real-time model designed to detect, track, and count wheat ears in video sequences. This model uses FasterNet as its foundational feature extraction network, significantly reducing the model's parameter count and improving the model's inference speed. We also incorporate deformable convolutions and dynamic sparse attention into the feature extraction network to enhance its ability to capture wheat ear features while reducing the effects of intricate environmental conditions. To address information loss during up-sampling and strengthen the model's capacity to extract wheat ear features across varying feature map scales, we integrate a path aggregation network (PAN) with the content-aware reassembly of features (CARAFE) up-sampling operator. Furthermore, the incorporation of the Kalman filter-based target-tracking algorithm, Observation-centric SORT (OC-SORT), enables real-time tracking and counting of wheat ears within expansive field settings. Experimental results demonstrate that Wheat-FasterYOLO achieves a mean average precision (mAP) score of 94.01% with a small memory usage of 2.87MB, surpassing popular detectors such as YOLOX and YOLOv7-Tiny. With the integration of OC-SORT, the composite higher order tracking accuracy (HOTA) and counting accuracy reached 60.52% and 91.88%, respectively, while maintaining a frame rate of 92 frames per second (FPS). This technology has promising applications in wheat ear counting tasks.

Keywords: object detection; deep learning; wheat ears counting; Kalman filter; lightweight model

1. Introduction

Wheat is one of the world's primary staple crops, playing a crucial role in meeting global food demands. Its production and quality control are critical factors in ensuring global food security [1], with the wheat ear as the reproductive organ of the wheat plant, directly influencing both the yield and quality of wheat [2]. Research indicates that the number of ears per unit area is the dominant yield component, and negative correlations were obtained between most of the parameters affecting the yield per plant and the yield per unit area [3]. Plant breeding experts utilize the information on wheat ear quantities from different varieties, combining these data with genetic and environmental factors. Through hybridization experiments involving a large number of wheat varieties, they select strains suited to various growing conditions, developing more resistant and higher-yielding varieties [4]. Effective detection and counting of wheat ears are essential for wheat yield prediction and ensuring food security [5]. Therefore, building a fast and efficient automatic counting method for wheat ears is of great significance, and wheat ear detection technology enables precise assessment of wheat ear quantities in fields, providing vital support for agricultural production, field management, and food trade [6].

Due to the high planting density of wheat, accurate counting of wheat ears is a challenging task. In the past, wheat yield estimation relied primarily on labor-intensive manual counting [7,8] and expert visual estimation [9]. The former approach is not only time-consuming and inefficient but also struggles to acquire accurate data in large-scale farmland settings. The latter method is subjective, demanding a high level of expertise in agriculture and resulting in difficulties in scientifically and accurately estimating the correct yield. As a result, the aforementioned methods are unable to rapidly and precisely estimate the wheat yield in large-scale wheat fields.

With the advancement of computer vision technology, image processing techniques have been widely applied in agricultural production. Fernandez-Gallego et al. proposed a method for automatic wheat ear counting using RGB drone images. This method utilizes techniques such as frequency filtering, segmentation, and feature extraction to achieve efficient and accurate wheat ear counting [10]. Tan et al. introduced a rapid identification method for field wheat ears based on superpixel segmentation algorithms and digital images. This approach involves image classification based on color feature parameters and analysis of wheat ear morphology, demonstrating both speed and accuracy [11]. Bao et al. presented a wheat ear counting method based on frequency domain decomposition. They employed multiscale support value filtering (MSVF) in combination with improved sampling contour transformation (ISCT) for frequency domain decomposition of wheat ear images. At last, the wheat ear images are segmented and counted [12]. Fang et al. proposed an automatic wheat tiller counting algorithm based on ground LiDAR data. This algorithm utilizes adaptive hierarchical and hierarchical clustering algorithms to comprehensively leverage 3D crop information in field environments, successfully counting wheat tillers of different varieties, nitrogen levels, planting densities, and ecological conditions [13].

In recent years, an increasing number of researchers have begun using deep learning techniques to use in agriculture. Compared to traditional image processing methods, deep learning offers higher adaptability, accuracy, generalization, and scalability, thus demonstrating better performance in processing large-scale image datasets and handling complex image tasks. Pérez-Porras et al. proposed a method for early on-ground image-based detection of poppies (Papaver rhoeas) in wheat using the YOLO architecture. Their research findings demonstrate that the deep-learning-based object detection strategy can accurately identify poppies at an early stage, providing precise information for the development of accurate wheat weed management [14]. Yang et al. proposed a deep-learning-based cross-platform model for wheat ear counting. This model combines a collaborative attention mechanism, achieving high-density counting of wheat ears while maintaining high counting accuracy and a reduced number of model parameters [15]. Zaji et al. introduced automatic object level augmentation (AutoOLA), which decouples different objects in wheat images and generates augmented images through random combinations, significantly reducing the required training sample size for the wheat ear model [16]. Alkhudaydi et al. introduced SpikeCount, a density-based method for wheat ear counting. This approach automatically extracts useful features from images using a fully convolutional neural network and utilizes transfer learning to optimize model training [17]. Qiu et al. proposed an unsupervised learning method that automatically detects and labels wheat ears from wheat ear images. They established a dataset to train a deep convolutional neural network model for accurate detection and counting of wheat ears [18].

The Internet and computer technology are finding broader applications in agriculture [19]. Modern agriculture is increasingly demanding efficient and precise intelligent solutions. Despite the considerable research efforts focused on wheat ear counting [20], challenges persist due to variations in wheat plants across growth stages and environmental conditions, as well as the diversity of wheat ears in images. Achieving accurate and efficient wheat ear counting remains a complex task. Furthermore, previous studies on wheat ear counting, whether based on image processing or deep learning methods, predominantly employed static counting approaches. This involves counting wheat ears in acquired images and then summing up the counts from each image to obtain the total count. However, this approach lacks real-time capability and involves time-consuming and intricate data preparation processes. To avoid repetitive counting of the same wheat ears, data collectors need to precisely control the shooting range while capturing images of wheat ears. This necessity introduces challenges such as difficulties in field operation, time consumption, and inefficiency during data collection. Wu et al. used YOLOv7 and DeepSORT to train on a subset of the GWHD dataset for real-time wheat ear counting [21]. However, the model has too many parameters and slow inference speed. On a high-end GPU like 3090TI, it only achieved 14 FPS, making it unsuitable for real-time counting in large wheat fields. This approach also demands high computational resources, increasing hardware costs and hindering model deployment on mobile devices in the future.

In response to the challenges outlined above and to address the complex issues related to efficient automatic wheat ear counting, our main objectives were as follows: (i) to propose a novel lightweight wheat ear counting model, introducing an innovative and efficient real-time wheat ear counting method based on applying leading-edge artificial intelligence (AI) and Internet technology (IT) solutions, and (ii) to further advance the globally important agriculture practices in wheat monitoring and production. The new method implemented by our model is intended for accurate identifying and counting of wheat ears in real time under unmanned aerial vehicle (UAV) conditions, thereby significantly reducing the manual labor, and effectively auto-calculating the number of wheat ears, thereby conducting a preliminary evaluation of wheat yield in the field to help agriculture management and decision-making processes. We pursue our main objectives by addressing our computational research hypotheses in the background of the following proposed technological approaches and specific objectives:

(1) To enhance the robustness of our model performance, various data augmentation methods were applied to the acquired dataset to ensure it would perform well under diverse conditions, such as different contrast, lighting, and environments.
(2) To improve the computational efficiency of our model, FasterNet [22] was utilized as the primary backbone for feature extraction. A specific objective was to enhance computational efficiency while minimizing the number of parameters, thereby making the model easily deployable on mobile devices.
(3) To enhance the backbone network, dynamic sparse attention and deformable convolution models were integrated into the model. A specific objective was to mitigate the influence of intricate environmental factors, such as the stickiness of wheat ears, while improving the model's capability to efficiently extract wheat ear features.
(4) To comprehensively capture fine details and context characteristics, feature pyramid network (FPN) [23] and lightweight upsampling operators were integrated into the PAN [24]. A specific objective was to enhance the capability of the proposed model to detect various sizes of wheat ears by optimal extraction of multi-scale features while minimizing the information loss during the upsampling process.
(5) To further build upon the wheat ear detection algorithms, the Kalman filter-based tracking algorithm was incorporated into our model. A specific objective was to overcome the limitations of traditional image-based counting methods by achieving accurate motion prediction, and thereby avoid repeated counting in the continuous sequence by analyzing the context of video frames. Another objective was to significantly decrease the amount of manual work for wheat ear counting in the field.

2. Materials and Methods

2.1. Data Acquisition and Processing

To train and validate Wheat-FasterYOLO, we prepared two different categories of datasets. These include a wheat ear image dataset for the object detection training phase and a wheat ear video dataset for the object tracking evaluation phase. Specifically, we employed the Global Wheat Head Detection (GWHD) [25] dataset for training and validating our wheat ear detection model. Furthermore, we utilized wheat ear video data captured by UAV to assess the model's performance in practical applications.

2.1.1. Source of Image Dataset

In order to obtain a universal and reliable wheat ears detection model, we utilized the GWHD dataset for model training. This dataset is the world's first large-scale and diverse collection of labeled wheat ear images, originating from the collaboration of nine research institutions across seven countries. GWHD encompasses high-resolution images of wheat ears at different growth stages, varieties, and cultivation conditions. These image data are sourced from multiple countries, exhibiting rich genotypic characteristics and diverse growth phases, thereby providing robust support for the methodology proposed in this study.

2.1.2. Image Data Partitioning and Augmentation

In this study, we obtained a total of 3372 annotated wheat ear image data. To ensure the validity of the experiments, we used the hold-out method to randomly partition the dataset, dividing the 3372 image data into training, testing, and validation sets in proportions of 7:2:1, respectively. During the training process, the model's hyperparameters were adjusted and optimized based on various indicators from the validation set, and the final performance of the model was evaluated on the test set.

Wheat ear images gathered from the field are impacted by factors like weather, lighting, and variety, resulting in diverse color variations and disruptions. To enhance the model's generalization ability and robustness, making it suitable for different wheat varieties, accurate counting can also be carried out, even in cases of overlapping wheat ears and interference from weed occlusion. This research utilizes the ImgAug [26] algorithm library to augment the data; after data augmentation, the dataset increased from the original 3372 images to 4899 images, comprising 3954 images in the training set, 675 images in the validation set, and 270 images in the test set. Details of augmentation techniques and outcomes are illustrated in Figure 1.

Figure 1. Data augmentation effect display. (**a**) Original figure. (**b**) Contrast adjustment. (**c**) Diffuse focus blur. (**d**) Fogging noise. (**e**) Gaussian noise. (**f**) Sharpening and hue change.

2.1.3. Video Data Collection

To validate the real-time tracking and counting effectiveness of the model in actual scenarios, we visited three different wheat cultivation locations in the Guangxi Zhuang Autonomous Region of China. We collected and processed multiple segments of wheat ear videos. The wheat ear data collection equipment used was the DJI Mavic 3, manufactured by DJI Innovations, headquartered in Shenzhen, China. This drone was equipped with a 4/3-inch CMOS Hasselblad camera, which is from a Hasselblad in Gothenburg, Sweden. We have collected a total of tens of thousands of frames of wheat ear video data. In this study, considering the actual workload, we extracted some representative data, and the basic information of the data used is shown in Table 1.

Table 1. Wheat ear video data collection and processing information.

Video Name	Wheat Variety	Collection Location	Location Coordinates	Video Length/Frames
Yangmai 17.mp4	Yangmai 17	Taodeng Town, Laibin City	Longitude 109°16′36″ E, Latitude 23°52′18″ N	3267
Huanuo No.1.mp4	Huanuo No.1	Changfu Village, Laibin City	Longitude 109°14′36″ E, Latitude 23°52′18″ N	2327
Xumai 45.mp4	Xumai 45	Shuangqiao Village, Guilin City	Longitude 111°11′3″ E, Latitude 26°4′22″ N	3264

During the data collection process, the drone operated at an altitude of 4 m, and we used a 7× telephoto camera manufactured by Zeiss, a company based in Germany. This camera features a fixed focal length of 166mm and captures an 80-degree downward view. The collected wheat ear videos had a resolution of 1920 × 1080 pixels at a frame rate of 60 frames per second. Using DJI Mavic 3's fixed-speed cruise function, the UAV autonomously collected wheat video data based on pre-planned flight routes, maintaining a constant speed of 0.1 m per second. In the collected and processed data, the drone flew a total of 14.76 m. The acquired data examples are shown in Figure 2.

(**a**) Yangmai 17 (**b**) Huanuo No.1 (**c**) Xumai 45

Figure 2. Collected data on wheat ears characteristics from different varieties.

2.1.4. Annotation of Video Data

In the annotation process of the wheat ear video dataset, we utilized DarkLabel version 2.4 [27]. A total of 8858 frames from the video sequences were annotated. Starting from frame 0, wheat ear images were marked with rectangular bounding boxes frame by frame. These manually annotated real target positions and trajectory information help us to evaluate the model's performance in practical applications. The annotation results are shown in Figure 3.

2.2. *Model Design Method*

This work focuses on real-time wheat ear tracking and counting, which is an object-tracking task. It involves accurate detection and positioning of wheat ear targets within images, followed by trajectory prediction using target-tracking algorithms. The study utilizes FasterNet as the baseline model, enhancing it through analysis of wheat ear recognition challenges and their integration with target-tracking algorithms. The outcome is the proposed Wheat-FasterYOLO, a model for real-time wheat ear counting.

Figure 3. Annotation of wheat ear video data.

2.2.1. FasterNet

In vast wheat fields, wheat ear detection and counting often require a substantial amount of computation time. Mobile and embedded devices have limited power, memory, and storage, making complex models impractical. Lightweight models like GhostNet [28], MobileNet [29–31], and ShuffleNet [32,33] utilize techniques such as group convolutions and depthwise convolutions for feature extraction. However, these methods primarily focus on reducing floating-point operations (FLOPs), which frequently result in increased memory access during this process, consequently leading to lower floating-point operations per second (FLOPS). The formula relating latency and FLOPs is as follows:

$$\text{Latency} = \frac{\text{FLOPs}}{\text{FLOPS}} \tag{1}$$

The formula implies that latency does not solely connect with FLOPs, but also with their FLOPS. In order to create a genuinely fast and efficient neural network, this study adopts the FasterNet, rooted in partial convolution (PConv) [22], as its benchmark model. Its aim is to lighten the device load for wheat ear counting tasks while achieving enhanced FPS rates. The PConv's architecture, showcased in Figure 4, engages convolutions solely with select input feature map channels, leaving the remaining channels unaltered.

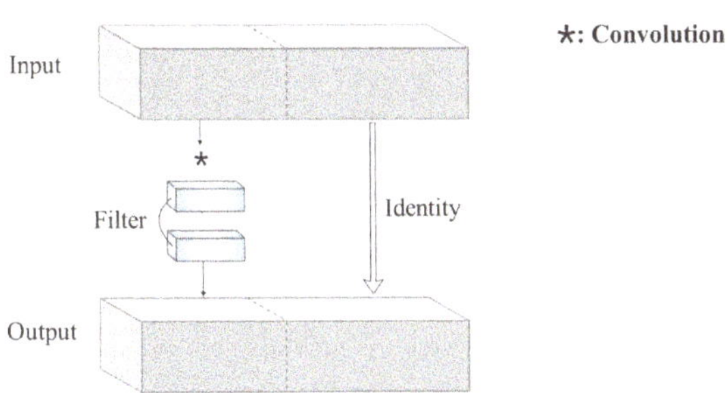

Figure 4. The structure of PConv.

2.2.2. Loss Function

The proposed wheat ear detection model in this study employs object confidence and bounding box regression losses. While conventional bounding box regression uses intersection over union (IoU) [34] for alignment, limited to area overlap, we introduce SCYLLA intersection over union (SIoU) [35]. It also includes angle, distance, and shape losses, accelerating convergence. This improves accuracy and precision in localizing wheat ears.

The formula for object confidence loss is as follows:

$$L_{obj} = w_{obj}\text{BCE}_{obj}^{sig}(p_o, p_{iou}) \tag{2}$$

where p_o represents the predicted object confidence. p_{iou} denotes the IoU of the predicted bounding box. BCE_{obj}^{sig} signifies the binary cross entropy loss function for binary classification. w_{obj} embodies the weight coefficient for the object loss.

Introducing angle loss seeks to penalize model errors in angle predictions, enhancing the precise alignment of wheat ear angles. Angle loss is defined as follows:

$$\Lambda = 1 - 2 \times \sin^2(\arcsin(\frac{c_h}{\sigma}) - \frac{\pi}{4}) \tag{3}$$

where c_h indicates a vertical disparity between the predicted and ground truth bounding box centers, while σ represents their horizontal difference.

Introducing the concept of distance loss to penalize the model for inaccuracies in distance prediction helps the model better capture spatial relationships between wheat ears. The loss is defined as follows:

$$\Delta = \sum_{t=x,y} \left(1 - \text{e}^{-\rho_t(2-\Lambda)}\right) \tag{4}$$

where

$$\rho_x = \left(\frac{b_{c_x}^{gt} - b_{cx}}{c_w}\right)^2, \rho_y = \left(\frac{b_{c_y}^{gt} - b_{cy}}{c_h}\right)^2 \tag{5}$$

where $b_{c_x}^{gt}$ is the horizontal coordinate value of the center point of the ground truth bounding box, while $b_{c_y}^{gt}$ represents their vertical coordinate value. c_w signifies the width of the minimum bounding rectangle for both the ground truth and predicted bounding boxes, while c_h represents the height of the minimum bounding rectangle.

Introducing shape loss to refine shape predictions enhances the model's grasp of wheat ear morphology. The shape loss is defined as:

$$\Omega = \sum_{t=w,h} (1 - \text{e}^{-w_t})^\theta \tag{6}$$

where

$$w_w = \frac{| w - w^{gt} |}{\max(w, w^{gt})}, w_h = \frac{| h - h^{gt} |}{\max(h, h^{gt})} \tag{7}$$

Among them, w and h represent the width and height of the predicted bounding box, respectively. w^{gt} and h^{gt} represent the width and height of the ground truth bounding box, respectively. θ indicates the degree of attention to the shape loss.

In summary, the final loss function of the Wheat-FasterYOLO is as follows:

$$\text{LOSS}_{\text{total}} = 1 - IoU + \frac{\Delta + \Omega}{2} + L_{obj} \tag{8}$$

2.2.3. BiFormer

The attention mechanism enhances the network's capability to extract wheat ear features while reducing the influence of complex backgrounds (such as weeds resembling wheat ear characteristics). However, in most cases, it faces challenges of high computational complexity and memory consumption. Therefore, in Wheat-FasterYOLO, we have incorporated BiFormer consisting of bi-level routing attention (BRA) [36]. By incorporating sparsity and query-aware adaptability, BiFormer efficiently models pivotal image regions globally. This method has moderately enhanced the precision of matching wheat ear samples amid intricate backgrounds, concurrently guaranteeing minimal memory utilization and reduced computational intricacy.

The BRA module's structure is shown in Figure 5. First, the height of the input feature map is H, the width is W, and the number of channels is C. Next, the feature map is divided into $S \times S$ distinct regions, and key-value pairs are collected for the first N regions. Each region contains $\frac{HW}{S^2}$ feature vectors, which are linearly mapped to obtain the corresponding Q, K, and V. An adjacency matrix is used to construct a directed graph to determine the participating relationships for different key-value pairs. As the routing regions are distributed across the entire feature map, it is necessary to aggregate the tensors of keys and values. The aggregated tensors of K^g and V^g are obtained. Matrix multiplication (mm) is performed between K^g and Q, followed by a softmax operation. The obtained matrix A, representing the index of region-to-region routing, is multiplied with V^g using matrix multiplication. Ultimately, a fine-grained token-to-token attention is achieved.

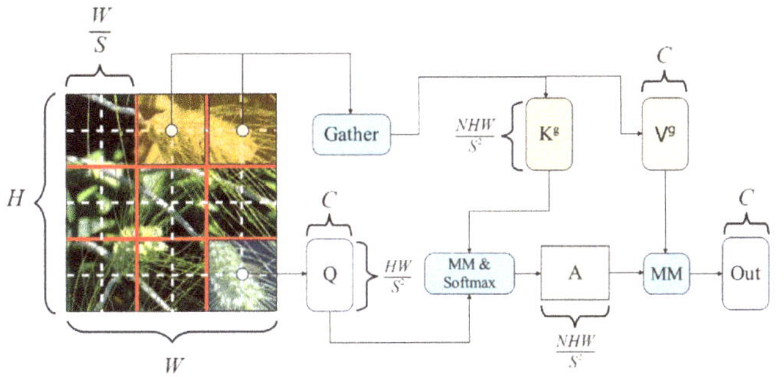

Figure 5. The structure of BRA.

2.2.4. DCNv2

Throughout the growth stages, wheat ears can exhibit diverse shapes, colors, and densities due to different growth factors. Furthermore, during UAV flights, camera motion may distort wheat ear images. The factors mentioned above can easily cause errors in wheat ear detection models. Models should adapt to changing feature distributions. Traditional convolutions ignore target deformation and position, yielding poor results for uneven features. Thus, this study introduces Deformable ConvNets v2 (DCNv2), a neural network using offsets for deformable convolutions [37]. DCNv2 dynamically adjusts kernel positions to precisely address target changes. This adaptation empowers the feature layer to adjust the receptive field and scale, resulting in improved performance in detecting wheat ear targets of different characteristics.

DcnV2's learning is shown in Figure 6. Extra convolutional layers learn offsets from the input feature map through convolutional operations, forming a $2N$ channel dimension. This dimension signifies **x** and **y** coordinate axes offsets for each pixel in an N-sized kernel. Inputs to the deformable convolutional layer are the feature map and offsets, where sampling points are shifted. Following convolutions, an output feature map is obtained.

2.2.5. Improving the PAN Architecture

The FPN acquires multi-scale contextual information by integrating feature maps from different scales. It not only captures texture information in lower-level feature maps but also extracts rich semantic information from higher-level feature maps. Building on this, the PAN further enhances performance through additional sampled fusion.

Conventional upsampling methods focus only on spatial distances among sub-pixel neighbors. However, in the task of wheat ears detection, due to the limited feature information that wheat ears can capture, this approach cannot fully grasp the hidden semantic information in the feature map. This may result in the loss of important details and contextual information. To address this issue, we introduce the lightweight upsampling operator CARAFE [38] in PAN. During the upsampling process, CARAFE aggregates extensive contextual information, enabling PAN to better capture wheat ear details and context, thus reducing feature loss during upsampling. Figure 7 illustrates the structure of CARAFE-PAN.

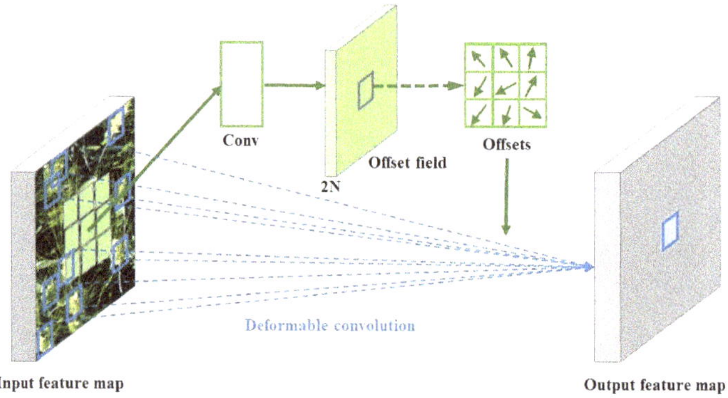

Figure 6. Illustration of deformable convolution.

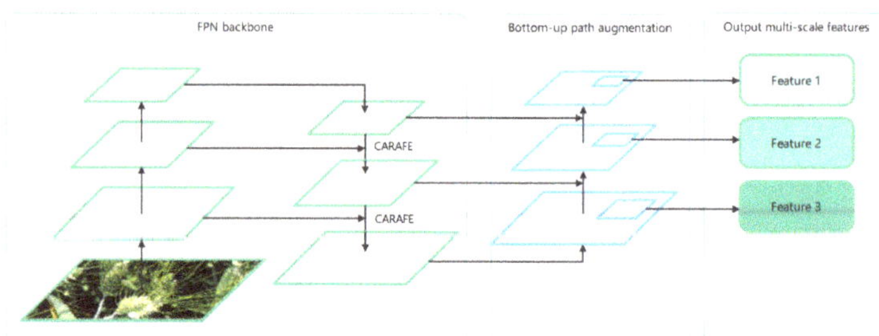

Figure 7. Illustrates the structure of CARAFE-PAN.

2.2.6. OC-SORT

To achieve real-time tracking and counting of wheat ears in a video stream, we adopt the OC-SORT [39] algorithm. This method reduces noise and enhances robustness using observation-centric techniques.

In consecutive video frames, the motion of the same target can be seen as linear. The Kalman filter is suitable for position estimation, but noise can lead to inaccurate motion direction estimation, causing target loss. OC-SORT integrates the observation-centric momentum (OCM) strategy, reducing noise impact.

During wheat ear tracking and counting tasks, to prevent the parameter bias of the Kalman filter from causing recurrent losses of detected wheat ear targets, OC-SORT employs the observation-centric re-update (ORU) strategy. When re-tracking, Kalman filter parameters adjust through virtual trajectories, reducing errors.

Natural wind can briefly obstruct wheat ear targets, and traditional tracking algorithms may mislead target loss or counting errors. OC-SORT leverages the observation-centric recovery (OCR) strategy, attempting a second association to restore trajectories.

The OC-SORT workflow is shown in Figure 8.

2.2.7. Wheat-FasterYOLO Model Structure

As depicted in Figure 9, the Wheat-FasterYOLO framework incorporates FasterNet with DCNv2 and BiFormer in its backbone. Employing PAN with CARAFE in the network's neck and three YOLO heads in the head predicts wheat ear target regression. OC-SORT enables efficient real-time tracking and updates of wheat ear positions, enabling real-time wheat ear tracking and counting.

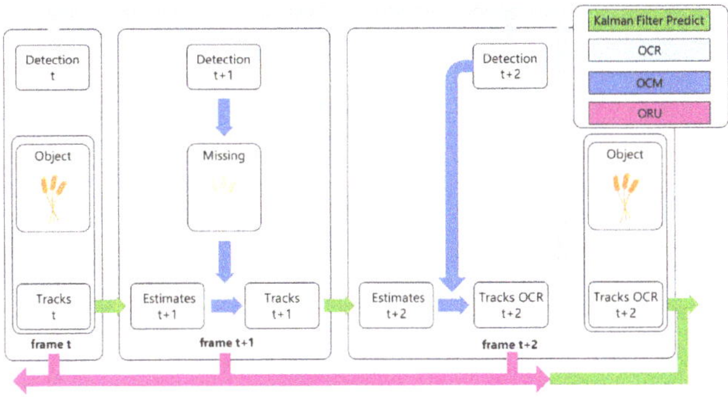

Figure 8. The pipeline of OC-SORT. During association, OCM introduces velocity consistency cost, but the target is momentarily obscured in frame t + 1 due to occlusion. However, in the subsequent frame, the target is successfully regained by referencing its observation in frame t using OCR. This re-tracking event then prompts ORU to update the parameters of its Kalman Filter from t to t + 2.

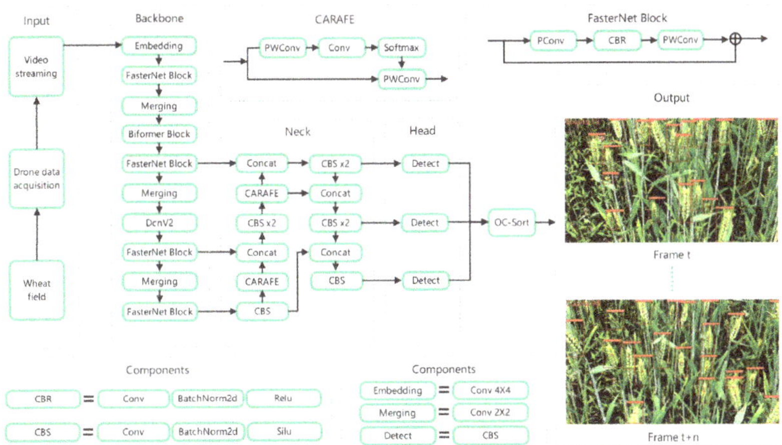

Figure 9. Comprehensive architecture of the Wheat-FasterYOLO.

2.2.8. Practical Application Process of the Model

In our research, the process of using a model for actual counting is illustrated in Figure 10. First, based on the features of the actual wheat field, manual route planning for the drone's flight path is conducted to ensure that every area is scanned by the drone. Subsequently, the automatic cruise function is initiated, and the drone will autonomously fly and capture a sequence of wheat ear videos based on the previously established flight route. Finally, when the drone begins its operation, real-time captured video data are transmitted to our mobile computing device terminal through DJI's OcuSync transmission technology. Running our models on mobile devices with the PyTorch deep learning framework, real-time detection and counting of the wheat ear video sequence are performed, providing real-time counting results.

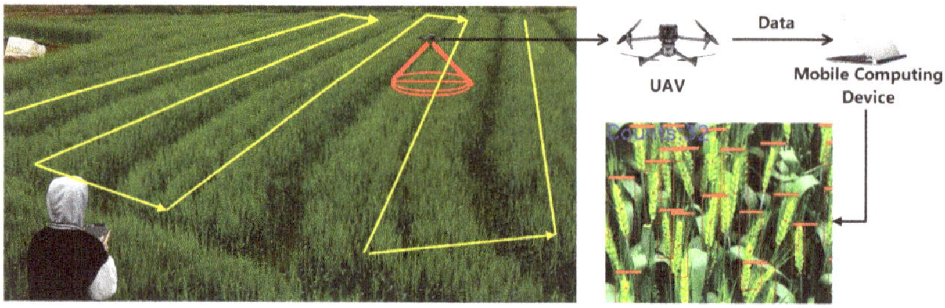

Figure 10. The process in the actual counting of wheat ears, with the yellow line in the diagram representing the planned route and the red area indicating the area being scanned by the UAV.

2.3. Evaluation Indicators

In wheat ears object detection, we employ the P, R, mAP, and $F1$ metrics to comprehensively evaluate detector effectiveness. P represents the proportion of actual wheat ears detected by the model within the detected wheat ears, while R denotes the proportion of successfully detected wheat ears by the model out of all the true wheat ears. mAP provides a comprehensive assessment of precision and recall across target categories, and $F1$ is the harmonic mean of precision and recall. The specific calculations are as follows:

$$P = \frac{TP}{TP + FP} \tag{9}$$

$$R = \frac{TP}{TP + FN} \tag{10}$$

$$mAP = \sum_{i=1}^{1} \left[\int_0^1 P(R) dR \right]_i \tag{11}$$

$$F1 = 2 \times \frac{P \times R}{P + R} \tag{12}$$

where TP is the accurate wheat ear predictions, FP is non-target samples misclassified as ears and FN is the missed number of actual wheat ear detection.

In the task of wheat ear tracking, the performance of the model in tracking targets after detecting wheat ears is measured using $DetA$, $AssA$, $DetRe$, $AssRe$, and $HOTA$ [40]. $DetA$ signifies]precise detection ratio, while $AssA$ indicates the association accuracy of tracking. $DetRe$ and $AssRe$ are the ratios of successfully detected and associated wheat ears to all true targets. $HOTA$ assesses the overall t performance in detection and tracking. The specific calculation methods are as follows:

$$DetA = \int_0^1 \frac{|TP|}{|TP| + |FN| + |FP|} d\alpha \tag{13}$$

$$AssA = \int_0^1 \frac{1}{|TP|} \sum_{c \in (\mathrm{TP})} \frac{|TPA(c)|}{|TPA(c)| + |FNA(c)| + |FPA(c)|} d\alpha \tag{14}$$

$$DetRe = \int_0^1 \frac{|TP|}{|TP| + |FN|} d\alpha \tag{15}$$

$$AssRe = \int_0^1 \frac{1}{|\ TP\ |} \sum_{c \in \{TP\}} \frac{|\ TPA(c)\ |}{|\ TPA(c)\ | + |\ FNA(c)\ |} d\alpha \tag{16}$$

$$HOTA = \int_0^1 \sqrt{DetA_\alpha \times AssA_\alpha} d\alpha \tag{17}$$

where TP, FP, and FN have the same definitions as in formulas 9 and 10. TPA, FNA, and FPA correspond to accurate positives, incorrect negatives, and incorrect positives. α is the association accuracy threshold, and c represents any point within TP.

3. Results and Discussion

This study trained and tested the model on the Ubuntu 18.04.5 LTS 64-bit operating system. The experimental environment employed an NVIDIA RTX 3090 (24G) graphics card with a CUDA 11.1 driver. Python 3.8.3 and the deep learning framework PyTorch 1.8.0 were utilized. The final set of hyperparameters is presented in Table 2.

Table 2. Training and testing hyperparameter information.

Set of Parameters	Value or Name
Batch size	16
Learning rate	0.01
Epoch	230
Image resize	640 × 640
Optimizer	SGD
Momentum	0.937
IoU-thres	0.55

During training, we employed common object detection techniques like mosaic data augmentation, cosine learning rate scheduling, and hyperparameter evolution. Figure 11 visualizes bounding box regression loss, confidence loss, precision, and recall for the Wheat-FasterYOLO model's validation set.

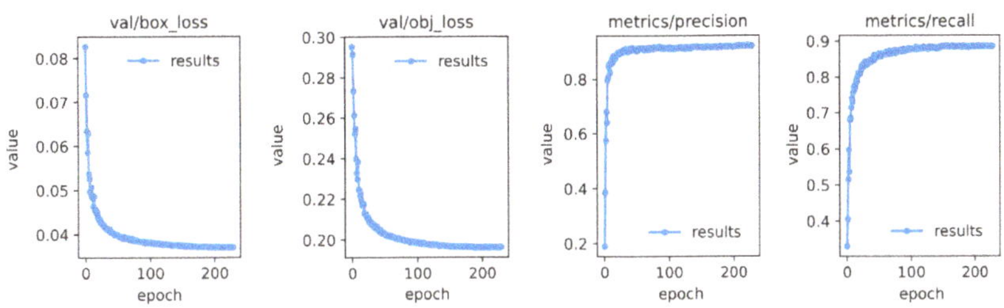

Figure 11. Visualization of Wheat-FasterYOLO detection model training data.

3.1. The Impact of Data Augmentation

Data augmentation experiments trained the baseline model on original and augmented GWHD datasets, resulting in two models. As shown in Table 3, using FasterNet as the baseline, the non-augmented model achieved mAP and F1 scores of 84.91% and 81.17%. After augmentation, scores improved to 85.66% and 81.97%, confirming data augmentation's necessity for field-derived wheat ear images.

Table 3. Comparison before and after data augmentation.

Data State	P/%	R/%	mAP/%	F1/%
Non-augmented	86.28	76.62	84.91	81.17
Augmented	**86.52**	**77.88**	**85.66**	**81.97**

3.2. Comparative Experiments with Different Attention Integrations

Attention mechanisms typically enhance model performance. In the context of the FasterNet backbone feature extraction network for global wheat ear detection datasets, we assessed various attention mechanisms' effectiveness. The comparative experimental results, showcased in Table 4, illustrate the impact of different attention mechanisms within the augmented global wheat ears detection dataset.

Table 4. Results of comparative experiments incorporating different attentional mechanisms in the data-enhanced GWHD dataset.

Attention	P/%	R/%	mAP/%	F1/%
None	86.52	77.88	85.66	81.97
SimAM	86.75	78.25	86.11	82.28
CBAM	86.78	78.85	85.81	82.63
GAM	89.99	84.02	90.49	86.9
SE	86.19	77.31	85.18	81.51
BiFormer	**90.2**	**85.35**	**91.21**	**87.71**

It can be observed that compared to the baseline model, the inclusion of SimAM [41] and CBAM [42] resulted in a slight improvement in model performance, with an increase of 0.45% and 0.15% in mAP values, and 0.31% and 0.66% in F1 values, respectively. In contrast, incorporating GAM [43] significantly boosted performance, with mAP and F1 values rising by 4.83% and 4.93%. However, SE [44] had no positive impact; instead, it led to a 0.48% and 0.46% decrease in mAP and F1 values, indicating its unsuitability for this model.

It is worth noting that BiFormer performed the best in the experiments, with mAP and F1 values reaching 91.21% and 87.71%, respectively, marking a significant improvement of 5.55% and 5.74% compared to the baseline model. Figure 12 visually presents heatmaps depicting various attention mechanisms, illustrating the model's precise targeting of wheat ear objectives. These findings reaffirm BiFormer's commendable performance in the realm of wheat ear detection tasks.

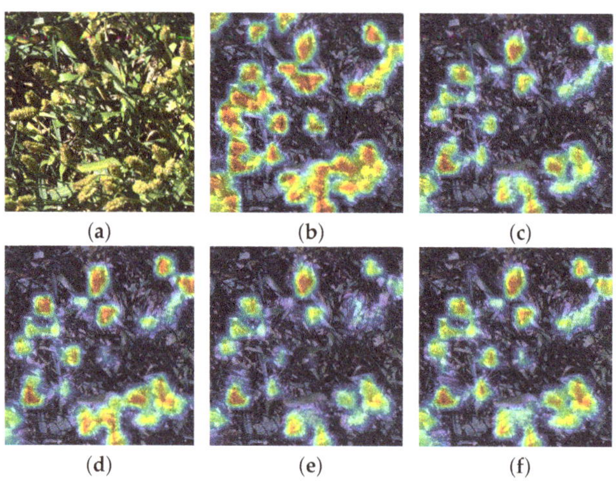

Figure 12. Heat maps of attention to the target after incorporating different attention mechanisms. (**a**) Original figure. (**b**) BiFormer. (**c**) CBAM. (**d**) GAM. (**e**) SE. (**f**) SinAM.

3.3. Ablation Experiment

To validate the effectiveness of the wheat ear detection model improvement, we conducted ablation experiments, and the results are shown in Table 5. The experimental results demonstrate a significant enhancement in the model's performance after the incorporation of BiFormer, with an increase of 5.55% in mAP and 5.74% in F1. BiFormer, with its unique sparsity and query-aware adaptability, can effectively model regions of interest across the feature maps globally.

By introducing the improved upsampling operator path aggregation network, the mAP and F1 of the wheat ear detection model improved by 2.37% and 2.65%, respectively. The path aggregation network helps the model better fuse multiscale features in the feature maps. After upsampling with the lightweight operator CARAFE, the model can capture the details and contextual information of wheat ears more effectively.

Table 5. Results of model ablation experiments.

BiFormer	CARAFE-PAN	DCNv2	P/%	R/%	mAP/%	F1/%
			86.52	77.88	85.66	81.97
√			90.2	85.35	91.21	87.71
√	√		92.32	88.47	93.58	90.36
√	√	√	**92.63**	**89.04**	**94.01**	**90.8**

With the addition of the DcnV2 module, the mAP of the wheat ear detection model increased by 0.43%, and the F1 increased by 0.44%. Deformable convolution adjusts the position information of convolution kernels dynamically, responding more accurately to the deformation and spatial positional changes in wheat ear targets.

The mAP of the improved wheat ear detection model reached 94.01%, and the F1 score reached 90.8%. Compared to the baseline model before improvement, there was an 8.35% increase in mAP and an 8.83% increase in F1 score. The proposed improvement methods in this paper have played a significant role in the wheat ear detection model, effectively enhancing its performance. The detection results of wheat ears before and after the model improvement are shown in Figure 13.

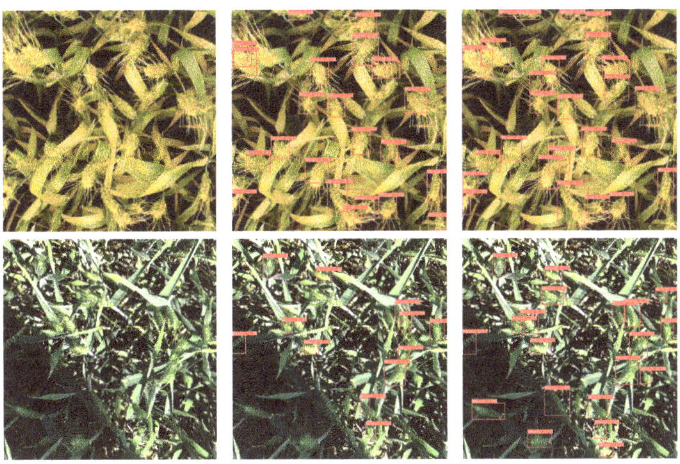

(**a**) (**b**) (**c**)

Figure 13. Comparison of the wheat ear detection model's performance before and after improvement. (**a**) Pictures of primitive wheat ears. (**b**) Before the model improvement, there were varying degrees of missed and false detections in wheat ear detection. (**c**) After the model improvement, the wheat ear detection performance has been greatly enhanced, significantly reducing missed and false detections.

3.4. Comparative Experimental Analysis of Different Detection Models

In order to evaluate the performance of our proposed wheat ear detection model, we conducted a comparative analysis with popular object detection models. We utilized the same set of parameters and dataset, and each model underwent training in the same experimental environment. The experimental findings in Table 6 reveal that our Wheat-FasterYOLO model, introduced in this study, outperformed in terms of P, R, mAP, and F1 scores, achieving high scores of 92.63%, 89.04%, 94.01%, and 90.8%, respectively. Furthermore, our model has fewer parameters and lower computational complexity, with a mere 1.34×10^6 parameters and 3.9 GFLOPs. Additionally, it demonstrated a faster speed; the frame rate reached 185 FPS.

Table 6. Performance of different object detection models in a wheat ear detection task.

Model	P/%	R/%	mAP/%	F1/%	Parameters	GFLOPs	FPS
SSD-VGG	90.94	63.96	82.59	75.1	2.36×10^7	136.6	66
SSD-MobileNet	**93.44**	71.21	88.45	80.82	3.54×10^6	**3.0**	87
Faster R-CNN	68.52	85.41	81.13	76.04	2.83×10^7	474.1	30
EfficientDet	92.43	79.01	89.69	85.19	6.56×10^6	5.7	21
YOLOX	93.04	**89.6**	93.69	**91.29**	8.04×10^6	21.6	117
YOLOv7-Tiny	92.89	88.86	93.0	90.83	6.01×10^6	13.0	125
Wheat-FasterYOLO	92.89	89.04	**94.01**	90.8	$\mathbf{1.34 \times 10^6}$	3.9	**185**

Wheat-FasterYOLO outperforms SSD-VGG [45,46], SSD-MobileNet [29], Faster R-CNN [47], and EfficientDet [48] significantly, even though Faster R-CNN and EfficientDet exhibit FPS of only 30 and 21. However, the introduction of tracking algorithms, requiring increased computational resources for Kalman filtering to estimate target motion, makes it unsuitable for real-time wheat ear tracking. Compared to SSD-MobileNet, Wheat-FasterYOLO experiences a slight 0.9 increase in GFLOPs, but its parameters are only 37.93% of SSD-MobileNet. It also achieves a 98 FPS boost, demonstrating Wheat-FasterYOLO's fast and lightweight performance despite increased GFLOPs, with superior FPS and fewer parameters.

While Wheat-FasterYOLO has a slightly lower F1 score compared to YOLOX [49] and YOLOv7-Tiny [50], it surpasses all other models in terms of mAP. This indicates that Wheat-FasterYOLO may not have the absolute best precision and recall. However, in terms of mAP, it outperforms all other models, which means it is better in overall detection accuracy across a range of confidence thresholds. It stands out for its efficiency, requiring only 16.71% of YOLOX's parameters and 22.36% of YOLOv7-Tiny's parameters, with computational demands at 18.06% of YOLOX and 30% of YOLOv7-Tiny. Combining the highest mAP with faster speed, Wheat-FasterYOLO demonstrates better overall performance, making it a suitable choice for real-time wheat ear tracking and counting tasks.

3.5. Comparative Experiments Incorporating Different Tracking Algorithms

In the real-time wheat ear tracking and counting task, the wheat ear detection model is combined with popular target-tracking algorithms. Through comparative experiments in different wheat varieties and using the TrackEval [51] evaluation, the performance differences in models incorporating different tracking algorithms in practical wheat ear counting applications are assessed. The results of the experiment are shown in Table 7.

According to the experiment data, OC-SORT achieved the best performance, with an average HOTA of 60.52%, which is 7.04% higher than ByteTrack [52] and 13.05% higher than StrongSORT [53]. When working in conjunction with the wheat ear detector, OC-SORT had a slightly lower average FPS than ByteTrack. However, its average DetA, AssA, DetRe, and AssRe were 9.36%, 3.2%, 13.54%, and 5.65% higher than ByteTrack, respectively, indicating that it outperformed ByteTrack comprehensively. StrongSORT had a higher DetRe than OC-SORT in testing. However, its HOTA metric was significantly lower than OC-SORT, indicating that StrongSORT's overall performance in real-time wheat ear counting tasks was unsatisfactory. Additionally, due to the introduction of the feature re-identification network, StrongSORT consumed a large amount of computational resources, resulting in high latency, with an average FPS of only 20, making it unsuitable for practical wheat ear tracking and counting tasks. The feature re-identification network is able to recapture a similar-looking target and confirm whether it is the same target as the previously detected one. In StrongSORT, the feature re-identification network tends to misidentify different wheat ear targets as the same target when dealing with wheat ears with highly similar appearance features, significantly affecting the counting results.

Table 7. Scores of different tracking algorithms in conjunction with an object detection model for various metrics in different varieties of wheat ears.

Tracker	Wheat Variety	DetA/%	AssA/%	DetRe/%	AssRe/%	HOTA/%	FPS
StrongSORT	Yangmai 17	60.58	38.69	74.57	75.04	48.06	23
	Huanuo No.1	58.52	42.72	74.83	72.54	49.85	17
	Xumai 45	48.58	30.71	57.0	62.12	38.49	20
	Average	55.89	37.37	**68.8**	69.33	47.47	20
ByteTrack	Yangmai 17	58.45	65.92	65.43	73.12	61.75	115
	Huanuo No.1	60.16	62.28	67.27	70.43	61.09	101
	Xumai 45	25.82	54.92	26.61	59.31	37.61	127
	Average	48.14	61.04	53.10	67.62	53.48	**114**
OC-SORT	Yangmai 17	63.0	69.25	74.41	78.75	65.64	104
	Huanuo No.1	62.79	65.14	73.88	75.95	63.82	83
	Xumai 45	46.72	58.34	51.63	65.1	52.11	90
	Average	**57.5**	**64.24**	66.64	**73.27**	**60.52**	92

In summary, the Wheat-FasterYOLO model proposed in this paper, when integrated with the OC-SORT algorithm, achieved higher HOTA and overall performance compared to ByteTrack and StrongSORT. It achieved an average FPS of 92, meeting the requirements of real-time wheat ear tracking and counting. Figure 14 shows the HOTA, DetA, AssA, DetRe, and AssRe curves of OC-SORT at different association accuracy threshold values "alpha", reflecting the variations in scores of various metrics with the threshold "alpha".

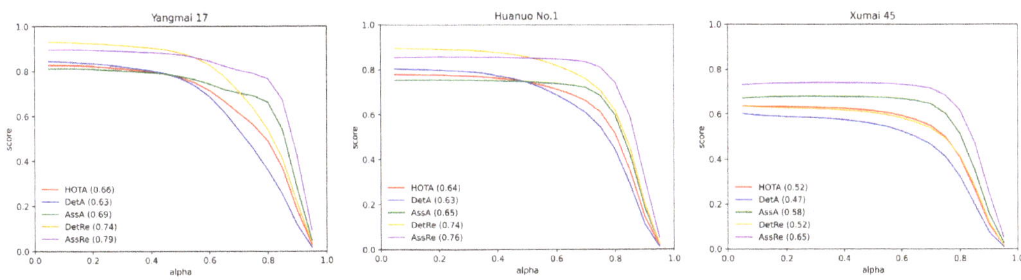

Figure 14. The variations in HOTA, DetA, AssA, DetRe, and AssRe curves under different threshold values "alpha" for OC-SORT.

3.6. *Analysis of Counting Accuracy in the Wheat-FasterYOLO Model*

When collecting data with UAV, wheat ear targets are prone to temporary loss in the detector due to motion blur or occlusion. However, by integrating target-tracking algorithms, as long as the wheat ear target is detected once in the video sequence, a unique ID can be assigned and counted. In subsequent detections, if the detector redetects the lost target and the target has not undergone significant irregular motion or severe deformation, the tracking algorithm will ensure consistent ID recognition. Wheat-FasterYOLO avoids the issue of the same target being counted repeatedly in different video sequences, as shown in Figure 15 for a specific illustration.

Table 8 shows the counting results for three different types of wheat ears. In the table, "IDs" represent the model's counting results, "GT_IDs" represent the actual number of wheat ears, and "Counting accuracy" reflects the accuracy of the model in practical wheat ear counting tasks. As shown in Figure 16, a linear regression analysis is performed between the model count results and the actual quantity over a period of time. R^2 reflects the degree of agreement between the model count values and the actual values. The closer its value is to 1, the better the fit. RMSE represents the deviation between the calculated value of the model and the actual value. It can be seen from this that there is a strong correlation between the counting results of the wheat ear counting method proposed in this study and the manual counting results, indicating that our method is practical.

In the wheat ear counting experiment, the accuracy rates for Yangmai 17, Huanuo No.1, and Xumai 45 were 91.71%, 92.66%, and 91.28%, respectively, with an average accuracy rate of 91.88%. By analyzing the detection results, it was found that there were weeds in Yangmai 17 and Huanuo No.1 with heights similar to wheat ears, leading to the model mistakenly identifying weeds as wheat ears. Additionally, in windy conditions, when the wheat ears moved only slightly in the wind, the model was able to track the wheat ear targets well. However, when strong winds caused the wheat ears to sway significantly, the model had difficulty accurately capturing the same wheat ear target, resulting in the model incorrectly considering wheat ears that moved significantly before and after as different objects, ultimately leading to an overestimation of the detected wheat ear count.

Table 8. Practical counting performance of Wheat-FasterYOLO in ears of three different wheat varieties.

Wheat Variety	IDs	GT_IDs	Counting Accuracy/%
Yangmai 17	374	343	91.71
Huanuo No.1	518	480	92.66
Xumai 45	680	745	91.28
Average	524	523	91.88

Figure 15. The demonstration of Wheat-FasterYOLO's tracking effect on the same wheat ear object in different video sequences. The rectangular box in the figure shows the location of the wheat target detected by the model, and the red dot is the center of the rectangular box. (**a**) In the first frame of the video sequence, target number 1's and target number 3's wheat ears are in a detected state, included in the total count of wheat ears. (**b**) In the 64th frame of the video sequence, target number 1's wheat ear was temporarily lost by the detector due to motion blur, and target number 3's wheat ear was excessively occluded by leaves. (**c**) In the 76th frame of the video sequence, target number 3's wheat ear was re-detected, the ID number remained unchanged, and the tracking algorithm successfully identified it as the same target, restoring the label for target number 3's wheat ear. (**d**) In the 105th frame of the video sequence, similar to target number 3's wheat ear, target number 1's wheat ear was re-detected.

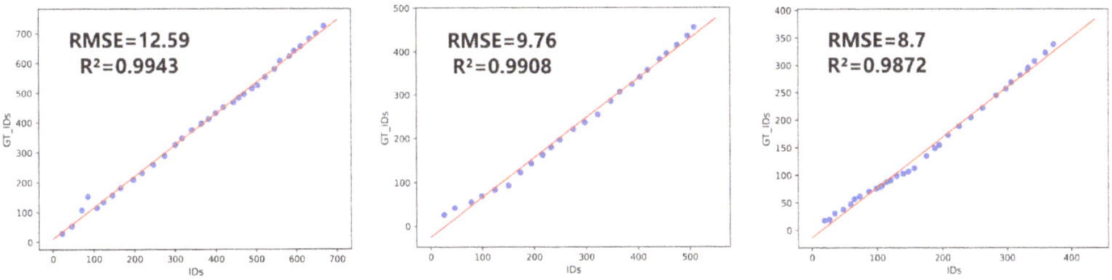

(**a**) (**b**) (**c**)

Figure 16. Scatter plot of model-based and manual counting results for three different types of wheat ears in a video sequence. (**a**) Yanmai 17.mp4. (**b**) Huanuo No.1.mp4. (**c**) Xumai 45.mp4.

In the case of Xumai 45 detection, there were no issues related to weeds with heights similar to wheat ears or interference from strong winds. However, due to the heavy overlap and occlusion of Xumai 45 wheat ears, the model erroneously identified overlapped wheat ears as a single target. Furthermore, the wheat ears of Xumai 45 had a relatively large aspect ratio, making it difficult for the model to fit the position information of the real bounding boxes. These factors led to instances of missed detections, resulting in a lower detected wheat ear count compared to the actual count.

Figure 17 shows a randomly selected frame from the three detection video sequences, illustrating the counting results of Wheat-FasterYOLO. The top-left corner displays the total number of different wheat plants detected by the model from the first frame to the currently selected frame. The information displayed above the detection boxes indicates the wheat ear's ID value, category, and confidence level.

(**a**) (**b**) (**c**)

Figure 17. Wheat-FasterYOLO shows the counting results in a video sequence of one of the three different wheat ear varieties (wheat ear counting begins from the first frame of the video sequence and continues until the last frame). (**a**) Frame 794 of Yangmai 17.mp4. At this point, the cumulative count is 98 wheat ears. (**b**) Frame 620 of Huanuo No.1.mp4. At this point, the cumulative count is 179 wheat ears. (**c**) Frame 10 of Xumai 45.mp4. At this point, the cumulative count is 32 wheat ears.

3.7. Advantages and Limitations

In this subsection, we will discuss the advantages and limitations of Wheat-FasterYOLO in detail as follows:

Firstly, we employed a combined model approach by introducing the OC-SORT algorithm based on the Kalman filter into the wheat ear detection model under study. This integration enables the model to accurately estimate the motion of wheat ear targets in UAV video sequences. By assigning a unique identification number (ID) to each wheat ear target, we achieved non-repetitive and high-precision counting. Farm owners only need to plan the drone's flight path based on real-world conditions to automatically obtain the desired wheat ear count information for a better preliminary assessment and decision-making regarding their wheat fields.

Secondly, we recognized the critical importance of GPU resource allocation in our approach. While ensuring sufficient GPU resources for effective YOLO operation, allocating a processing layer for tracking is a key consideration. In extensive tests, we found that the tracking algorithm typically consumes fewer resources than YOLO. The fast and lightweight nature of Wheat-FasterYOLO allows it to operate on a variety of devices, reducing hardware costs and enabling real-time counting in diverse environments. However, for optimal results and to prevent processing delays during counting, we recommend using a GTX 1050 or higher image processor to ensure the quality of wheat ear detection and tracking in various scenarios.

Moreover, understanding the growth stages of wheat is crucial for making informed agricultural decisions. Wheat growth can be divided into six distinct phases: germination, vegetative growth, heading, flowering, grain filling, and maturation. Our model, trained on a diverse dataset, is capable of effectively counting wheat heads during the flowering stage and beyond. This feature provides valuable insights for farmers during the mid to late stages of wheat growth, contributing to improved crop management and planning.

Finally, although our model can perform real-time counting for different wheat varieties in general, there are limitations. In some cases, wheat ears may be empty, and since our model was not trained on samples of empty wheat ears, it cannot effectively handle this specific situation. To address this limitation, we plan to collect more samples in future research and continuously enhance our model to make it more versatile.

4. Conclusions

In this study, we utilized the path planning and constant-speed cruising functions of UAV to automatically collect video sequences of wheat ears and achieved real-time tracking and counting of wheat ears in the field environment using the proposed Wheat-FasterYOLO. Compared to target detection and counting methods focused solely on static images, our approach circumvented the complexities associated with the operation, time-consuming, and low efficiency associated with data collection processes. Compared to existing real-time wheat ear counting models, our approach has fewer parameters, faster speed, and can achieve good results. In practical applications, it significantly enhanced the level of automation in wheat ear counting.

In the wheat ear detection method, we trained the Wheat-FasterYOLO model based on the GWHD dataset. Its mAP, F1 score, parameter count, GFLOPs, and FPS are 94.01%, 90.8%, 1.34×10^6, 3.9, and 185, respectively. This model combines lightweight design with speed and accuracy, demonstrating better overall performance than many popular object detection models and showing great potential for wheat ear detection tasks.

For wheat ear tracking and counting tasks, this study integrated the Kalman filter-based object tracking algorithm OC-SORT with the wheat ear detection model. We collected video sequences of three different wheat varieties using the DJI Mavic 3 and annotated them frame by frame. In multi-object tracking tests, the average HOTA reaches 60.52%, and the FPS is 92. In actual wheat ear counting scenarios, the average RMSE is 10.35, R^2 is 99.08%, and the counting accuracy is 91.88%. The lightweight design of Wheat-FasterYOLO makes it suitable for mobile edge terminals such as drones, allowing for rapid completion of wheat ear counting tasks in field environments and further advancing agricultural automation.

The effectiveness of the tracking algorithm is influenced by the detection model, as well as factors such as motion blur, image distortion generated during the drone flight, and the mutual occlusion of wheat ears, all of which can introduce certain interference into the counting results. In future research, we will continuously improve the quality of the wheat ear detection model and explore more stable counting methods to achieve efficient and accurate detection and counting of wheat ears in high-density wheat field scenarios. This will provide strong support for field management, grain trade, and agricultural production.

Author Contributions: Conceptualization, X.H. (Xiaochun Hu) and Y.C.; Data curation, J.C., X.H. (Xiaochun Hu), J.L. and X.H. (Xin Huang); Formal analysis, Y.C. and X.H. (Xin Huang); Funding acquisition, Y.C. and X.H. (Xin Huang); Methodology, J.C.; Project administration, X.H. (Xiaochun Hu) and J.L.; Resources, J.C. and X.H. (Xin Huang); Software, J.C., X.H. (Xiaochun Hu), J.L. and Y.C.; Supervision, X.H. (Xiaochun Hu), Y.C. and X.H. (Xin Huang); Validation, J.C., J.L. and Y.C.; Visualization, J.C., X.H. (Xiaochun Hu), J.L. and X.H. (Xin Huang); Writing—original draft, J.C. and Y.C.; Writing—review and editing, J.C., Y.C. and X.H. (Xin Huang). All authors have read and agreed to the published version of the manuscript.

Funding: This research was funded by the Guangxi Scientific Research and Technology Development Plan Project grant number AA20302002-3 and the Guangxi Vocational University of Agriculture Natural Science and Technology Development Program Project grant number XKJ2357.

Institutional Review Board Statement: Not applicable.

Informed Consent Statement: Not applicable.

Data Availability Statement: The data used in this study consists of two components: the GWHD Global Wheat Head Detection dataset and our self-collected video dataset for wheat ear tracking. The GWHD dataset is publicly available and can be accessed at http://www.global-wheat.com (accessed on 19 February 2023).

Conflicts of Interest: The authors declare no conflict of interest.

References

1. Zhao, J.; Yan, J.; Xue, T.; Wang, S.; Qiu, X.; Yao, X.; Tian, Y.; Zhu, Y.; Cao, W.; Zhang, X. A deep learning method for oriented and small wheat spike detection (OSWSDet) in UAV images. *Comput. Electron. Agric.* **2022**, *198*, 107087. [CrossRef]
2. Zhou, H.; Riche, A.B.; Hawkesford, M.J.; Whalley, W.R.; Atkinson, B.S.; Sturrock, C.J.; Mooney, S.J. Determination of wheat spike and spikelet architecture and grain traits using X-ray Computed Tomography imaging. *Plant Methods* **2021**, *17*, 26. [CrossRef] [PubMed]
3. Nerson, H. Effects of population density and number of ears on wheat yield and its components. *Field Crops Res.* **1980**, *3*, 225–234. [CrossRef]
4. Madec, S.; Jin, X.; Lu, H.; De Solan, B.; Liu, S.; Duyme, F.; Heritier, E.; Baret, F. Ear density estimation from high resolution RGB imagery using deep learning technique. *Agric. For. Meteorol.* **2019**, *264*, 225–234. [CrossRef]
5. Sadeghi-Tehran, P.; Virlet, N.; Ampe, E.M.; Reyns, P.; Hawkesford, M.J. DeepCount: In-field automatic quantification of wheat spikes using simple linear iterative clustering and deep convolutional neural networks. *Front. Plant Sci.* **2019**, *10*, 1176. [CrossRef]
6. Sun, J.; Yang, K.; Chen, C.; Shen, J.; Yang, Y.; Wu, X.; Norton, T. Wheat head counting in the wild by an augmented feature pyramid networks-based convolutional neural network. *Comput. Electron. Agric.* **2022**, *193*, 106705. [CrossRef]
7. Zhang, L.; Chen, Y.; Li, Y.; Ma, J.; Du, K. Detection and Counting System for winter wheat ears based on convolutional neural network. *Trans. Chin. Soc. Agric. Mach.* **2019**, *50*, 144–150.
8. Ma, J.; Li, Y.; Liu, H.; Wu, Y.; Zhang, L. Towards improved accuracy of UAV-based wheat ears counting: A transfer learning method of the ground-based fully convolutional network. *Expert Syst. Appl.* **2022**, *191*, 116226. [CrossRef]
9. Zhou, X.; Zheng, H.; Xu, X.; He, J.; Ge, X.; Yao, X.; Cheng, T.; Zhu, Y.; Cao, W.; Tian, Y. Predicting grain yield in rice using multi-temporal vegetation indices from UAV-based multispectral and digital imagery. *ISPRS J. Photogramm. Remote Sens.* **2017**, *130*, 246–255. [CrossRef]
10. Fernandez-Gallego, J.A.; Lootens, P.; Borra-Serrano, I.; Derycke, V.; Haesaert, G.; Roldán-Ruiz, I.; Araus, J.L.; Kefauver, S.C. Automatic wheat ear counting using machine learning based on RGB UAV imagery. *Plant J.* **2020**, *103*, 1603–1613. [CrossRef]
11. Tan, C.; Zhang, P.; Zhang, Y.; Zhou, X.; Wang, Z.; Du, Y.; Mao, W.; Li, W.; Wang, D.; Guo, W. Rapid recognition of field-grown wheat spikes based on a superpixel segmentation algorithm using digital images. *Front. Plant Sci.* **2020**, *11*, 259. [CrossRef]

12. Bao, W.; Lin, Z.; Hu, G.; Liang, D.; Huang, L.; Zhang, X. Method for wheat ear counting based on frequency domain decomposition of MSVF-ISCT. *Inf. Process. Agric.* **2023**, *10*, 240–255. [CrossRef]
13. Fang, Y.; Qiu, X.; Guo, T.; Wang, Y.; Cheng, T.; Zhu, Y.; Chen, Q.; Cao, W.; Yao, X.; Niu, Q.; et al. An automatic method for counting wheat tiller number in the field with terrestrial LiDAR. *Plant Methods* **2020**, *16*, 132. [CrossRef] [PubMed]
14. Pérez-Porras, F.J.; Torres-Sánchez, J.; López-Granados, F.; Mesas-Carrascosa, F.J. Early and on-ground image-based detection of poppy (Papaver rhoeas) in wheat using YOLO architectures. *Weed Sci.* **2023**, *71*, 50–58. [CrossRef]
15. Yang, B.; Pan, M.; Gao, Z.; Zhi, H.; Zhang, X. Cross-Platform Wheat Ear Counting Model Using Deep Learning for UAV and Ground Systems. *Agronomy* **2023**, *13*, 1792. [CrossRef]
16. Zaji, A.; Liu, Z.; Xiao, G.; Bhowmik, P.; Sangha, J.S.; Ruan, Y. AutoOLA: Automatic object level augmentation for wheat spikes counting. *Comput. Electron. Agric.* **2023**, *205*, 107623. [CrossRef]
17. Alkhudaydi, T.; De la Lglesia, B. Counting spikelets from infield wheat crop images using fully convolutional networks. *Neural Comput. Appl.* **2022**, *34*, 17539–17560. [CrossRef]
18. Qiu, R.; He, Y.; Zhang, M. Automatic Detection and Counting of Wheat Spikelet Using Semi-Automatic Labeling and Deep Learning. *Front. Plant Sci.* **2022**, *13*, 872555. [CrossRef]
19. Dimitrov, D.D. Internet and Computers for Agriculture. *Agriculture* **2023**, *13*, 155. [CrossRef]
20. Zaji, A.; Liu, Z.; Xiao, G.; Sangha, J.S.; Ruan, Y. A survey on deep learning applications in wheat phenotyping. *Appl. Soft Comput.* **2022**, *131*, 109761. [CrossRef]
21. Wu, T.; Zhong, S.; Chen, H.; Geng, X. Research on the Method of Counting Wheat Ears via Video Based on Improved YOLOv7 and DeepSort. *Sensors* **2023**, *23*, 4880. [CrossRef]
22. Chen, J.; Kao, S.h.; He, H.; Zhuo, W.; Wen, S.; Lee, C.H.; Chan, S.H.G. Run, Do not Walk: Chasing Higher FLOPS for Faster Neural Networks. In Proceedings of the IEEE/CVF Conference on Computer Vision and Pattern Recognition, Vancouver, BC, Canada, 18–22 June 2023; pp. 12021–12031.
23. Lin, T.Y.; Dollár, P.; Girshick, R.; He, K.; Hariharan, B.; Belongie, S. Feature pyramid networks for object detection. In Proceedings of the IEEE Conference on Computer Vision and Pattern Recognition, Honolulu, HI, USA, 21–26 July 2017; pp. 2117–2125.
24. Liu, S.; Qi, L.; Qin, H.; Shi, J.; Jia, J. Path aggregation network for instance segmentation. In Proceedings of the IEEE Conference on Computer Vision and Pattern Recognition, Salt Lake City, UT, USA, 18–23 June 2018; pp. 8759–8768.
25. David, E.; Madec, S.; Sadeghi-Tehran, P.; Aasen, H.; Zheng, B.; Liu, S.; Kirchgessner, N.; Ishikawa, G.; Nagasawa, K.; Badhon, M.A.; et al. Global Wheat Head Detection (GWHD) dataset: A large and diverse dataset of high-resolution RGB-labelled images to develop and benchmark wheat head detection methods. *Plant Phenomics* **2020**, *2020*, 3521852. [CrossRef]
26. Jung, A.B.; Wada, K.; Crall, J.; Tanaka, S.; Graving, J.; Reinders, C.; Yadav, S.; Banerjee, J.; Vecsei, G.; Kraft, A.; et al. Imgaug. Available online: https://github.com/aleju/imgaug (accessed on 5 June 2023).
27. DarkLabel. Available online: https://github.com/darkpgmr/DarkLabel (accessed on 1 June 2023).
28. Han, K.; Wang, Y.; Tian, Q.; Guo, J.; Xu, C.; Xu, C. Ghostnet: More features from cheap operations. In Proceedings of the IEEE/CVF Conference on Computer Vision and Pattern Recognition, Seattle, WA, USA, 13–19 June 2020; pp. 1580–1589.
29. Howard, A.G.; Zhu, M.; Chen, B.; Kalenichenko, D.; Wang, W.; Weyand, T.; Andreetto, M.; Adam, H. Mobilenets: Efficient convolutional neural networks for mobile vision applications. *arXiv* **2017**, arXiv:1704.04861.
30. Sandler, M.; Howard, A.; Zhu, M.; Zhmoginov, A.; Chen, L.C. Mobilenetv2: Inverted residuals and linear bottlenecks. In Proceedings of the IEEE Conference on Computer Vision and Pattern Recognition, Salt Lake City, UT, USA, 18–23 June 2018; pp. 4510–4520.
31. Howard, A.; Sandler, M.; Chu, G.; Chen, L.C.; Chen, B.; Tan, M.; Wang, W.; Zhu, Y.; Pang, R.; Vasudevan, V.; et al. Searching for mobilenetv3. In Proceedings of the IEEE/CVF International Conference on Computer Vision, Seoul, South Korea, 27 October–2 November 2019; pp. 1314–1324.
32. Zhang, X.; Zhou, X.; Lin, M.; Sun, J. Shufflenet: An extremely efficient convolutional neural network for mobile devices. In Proceedings of the IEEE Conference on Computer Vision and Pattern Recognition, Salt Lake City, UT, USA, 18–23 June 2018; pp. 6848–6856.
33. Ma, N.; Zhang, X.; Zheng, H.T.; Sun, J. Shufflenet v2: Practical guidelines for efficient cnn architecture design. In Proceedings of the European Conference on Computer Vision (ECCV), Munich, Germany, 8–14 September 2018; pp. 116–131.
34. Yu, J.; Jiang, Y.; Wang, Z.; Cao, Z.; Huang, T. Unitbox: An advanced object detection network. In Proceedings of the 24th ACM International Conference on Multimedia, Amsterdam, The Netherlands, 15–19 October 2016; pp. 516–520.
35. Gevorgyan, Z. SIoU loss: More powerful learning for bounding box regression. *arXiv* **2022**, arXiv:2205.12740.
36. Zhu, L.; Wang, X.; Ke, Z.; Zhang, W.; Lau, R.W. BiFormer: Vision Transformer with Bi-Level Routing Attention. In Proceedings of the IEEE/CVF Conference on Computer Vision and Pattern Recognition, Vancouver, BC, Canada, 18–22 June 2023; pp. 10323–10333.
37. Zhu, X.; Hu, H.; Lin, S.; Dai, J. Deformable convnets v2: More deformable, better results. In Proceedings of the IEEE/CVF Conference on Computer Vision and Pattern Recognition, Long Beach, CA, USA, 15–20 June 2019; pp. 9308–9316.
38. Wang, J.; Chen, K.; Xu, R.; Liu, Z.; Loy, C.C.; Lin, D. Carafe: Content-aware reassembly of features. In Proceedings of the IEEE/CVF International Conference on Computer Vision, Seoul, South Korea, 27 October–2 November 2019; pp. 3007–3016.

39. Cao, J.; Pang, J.; Weng, X.; Khirodkar, R.; Kitani, K. Observation-centric sort: Rethinking sort for robust multi-object tracking. In Proceedings of the IEEE/CVF Conference on Computer Vision and Pattern Recognition, Vancouver, BC, Canada, 18–22 June 2023; pp. 9686–9696.
40. Luiten, J.; Osep, A.; Dendorfer, P.; Torr, P.; Geiger, A.; Leal-Taixé, L.; Leibe, B. Hota: A higher order metric for evaluating multi-object tracking. *Int. J. Comput. Vis.* **2021**, *129*, 548–578. [CrossRef]
41. Qin, X.; Li, N.; Weng, C.; Su, D.; Li, M. Simple attention module based speaker verification with iterative noisy label detection. In Proceedings of the ICASSP 2022–2022 IEEE International Conference on Acoustics, Speech and Signal Processing (ICASSP), Singapore, 22–27 May 2022; pp. 6722–6726.
42. Woo, S.; Park, J.; Lee, J.Y.; Kweon, I.S. Cbam: Convolutional block attention module. In Proceedings of the European Conference on Computer Vision (ECCV), Munich, Germany, 8–14 September 2018; pp. 3–19.
43. Liu, Y.; Shao, Z.; Hoffmann, N. Global attention mechanism: Retain information to enhance channel-spatial interactions. *arXiv* **2021**, arXiv:2112.05561.
44. Hu, J.; Shen, L.; Sun, G. Squeeze-and-excitation networks. In Proceedings of the IEEE Conference on Computer Vision and Pattern Recognition, Salt Lake City, UT, USA, 18–23 June 2018; pp. 7132–7141.
45. Liu, W.; Anguelov, D.; Erhan, D.; Szegedy, C.; Reed, S.; Fu, C.Y.; Berg, A.C. Ssd: Single shot multibox detector. In *Proceedings of the Computer Vision–ECCV 2016: 14th European Conference, Amsterdam, The Netherlands, 11–14 October 2016, Proceedings, Part I 14*; Springer International Publishing: Cham, Switzerland, 2016; pp. 21–37.
46. Simonyan, K.; Zisserman, A. Very deep convolutional networks for large-scale image recognition. *arXiv* **2014**, arXiv:1409.1556.
47. Ren, S.; He, K.; Girshick, R.; Sun, J. Faster r-cnn: Towards real-time object detection with region proposal networks. In Proceedings of the Advances in Neural Information Processing Systems 28 (NIPS 2015), Montreal, QC, Canada, 7–12 December 2015.
48. Tan, M.; Pang, R.; Le, Q.V. Efficientdet: Scalable and efficient object detection. In Proceedings of the IEEE/CVF Conference on Computer Vision and Pattern Recognition, Seattle, WA, USA, 13–19 June 2019; pp. 10781–10790.
49. Ge, Z.; Liu, S.; Wang, F.; Li, Z.; Sun, J. Yolox: Exceeding yolo series in 2021. *arXiv* **2021**, arXiv:2107.08430.
50. Wang, C.Y.; Bochkovskiy, A.; Liao, H.Y.M. YOLOv7: Trainable bag-of-freebies sets new state-of-the-art for real-time object detectors. In Proceedings of the IEEE/CVF Conference on Computer Vision and Pattern Recognition, Vancouver, BC, Canada, 18–22 June 2023; pp. 7464–7475.
51. Jonathon, L.; Arne, H. TrackEval. Available online: https://github.com/JonathonLuiten/TrackEval (accessed on 21 June 2023).
52. Zhang, Y.; Sun, P.; Jiang, Y.; Yu, D.; Weng, F.; Yuan, Z.; Luo, P.; Liu, W.; Wang, X. Bytetrack: Multi-object tracking by associating every detection box. In *Proceedings of the European Conference on Computer Vision*; Springer Nature Switzerland: Cham, Switzerland, 2022; pp. 1–21.
53. Du, Y.; Zhao, Z.; Song, Y.; Zhao, Y.; Su, F.; Gong, T.; Meng, H. Strongsort: Make deepsort great again. *IEEE Trans. Multimed.* **2023**, *Early Access*. [CrossRef]

agriculture

MDPI

Review

Emerging Technologies for Precision Crop Management Towards Agriculture 5.0: A Comprehensive Overview

Mohamed Farag Taha [1,2,3], Hanping Mao [1,*], Zhao Zhang [4], Gamal Elmasry [5], Mohamed A. Awad [6], Alwaseela Abdalla [7], Samar Mousa [8], Abdallah Elshawadfy Elwakeel [9] and Osama Elsherbiny [1,10]

1 School of Agricultural Engineering, Jiangsu University, Zhenjiang 212013, China; 1000006483@ujs.edu.cn (M.F.T.); osama_algazeery@mans.edu.eg (O.E.)
2 College of Biosystems Engineering and Food Science, Zhejiang University, Hangzhou 310058, China
3 Department of Soil and Water Sciences, Faculty of Environmental Agricultural Sciences, Arish University, Arish 45516, Egypt
4 Key Laboratory of Smart Agriculture System Integration, Ministry of Education, China Agricultural University, Beijing 100083, China; zhaozhangcau@cau.edu.cn
5 Agricultural Engineering Department, Faculty of Agriculture, Suez Canal University, Ismailia 41522, Egypt; gamal.elmasry@agr.suez.edu.eg
6 Department of Plant Production, Faculty of Environmental Agricultural Sciences, Arish University, Arish 45516, Egypt; mohamed.awad@agri.aru.edu.eg
7 Department of Agricultural Biology, Colorado State University, Fort Collins, CO 80523, USA; alwaseela.hassan@colostate.edu
8 Agricultural Botany Department, Faculty of Agriculture, Suez Canal University, Ismailia 41522, Egypt; samarmer89@mails.ucas.ac.cn
9 Agricultural Engineering Department, Faculty of Agriculture and Natural Resources, Aswan University, Aswan 81528, Egypt; abdallah_elshawadfy@agr.aswu.edu.eg
10 Agricultural Engineering Department, Faculty of Agriculture, Mansoura University, Mansoura 35516, Egypt
* Correspondence: maohp@ujs.edu.cn; Tel.: +86-135-1169-5868

Abstract: Agriculture 5.0 (Ag5.0) represents a groundbreaking shift in agricultural practices, addressing the global food security challenge by integrating cutting-edge technologies such as artificial intelligence (AI), machine learning (ML), robotics, and big data analytics. To adopt the transition to Ag5.0, this paper comprehensively reviews the role of AI, machine learning (ML) and other emerging technologies to overcome current and future crop management challenges. Crop management has progressed significantly from early agricultural methods to the advanced capabilities of Ag5.0, marking a notable leap in precision agriculture. Emerging technologies such as collaborative robots, 6G, digital twins, the Internet of Things (IoT), blockchain, cloud computing, and quantum technologies are central to this evolution. The paper also highlights how machine learning and modern agricultural tools are improving the way we perceive, analyze, and manage crop growth. Additionally, it explores real-world case studies showcasing the application of machine learning and deep learning in crop monitoring. Innovations in smart sensors, AI-based robotics, and advanced communication systems are driving the next phase of agricultural digitalization and decision-making. The paper addresses the opportunities and challenges that come with adopting Ag5.0, emphasizing the transformative potential of these technologies in improving agricultural productivity and tackling global food security issues. Finally, as Agriculture 5.0 is the future of agriculture, we highlight future trends and research needs such as multidisciplinary approaches, regional adaptation, and advancements in AI and robotics. Ag5.0 represents a paradigm shift towards precision crop management, fostering sustainable, data-driven farming systems that optimize productivity while minimizing environmental impact.

Keywords: Agriculture 5.0 (Ag5.0); artificial intelligence (AI); autonomous mobile robots; crop monitoring; convolutional neural networks (CNNs)

1. Introduction

Population statistics indicate a substantial rise in the global population, projected to reach 10 billion by 2050. As a result, the global challenges of energy and water scarcity, food security, and climate change are expected to intensify [1,2]. Agricultural industries, as one of the primary sectors addressing these challenges, must undergo both quantitative and qualitative development. In addition to these challenges, the agricultural sector faces the migration of workers to more profitable industries, coupled with the increasing average age of farmers [3]. Consequently, there is an urgent need for scientists and agricultural experts to develop more innovative and sustainable food production systems to address these issues and meet the growing population's nutritional needs. Agricultural experts have developed innovative monitoring strategies to address these challenges, including the optimal monitoring of nutrients, remote sensing systems, global information, global positioning, and variable-rate applications. These strategies are collectively known as precision agriculture (PA) or precision farming. Precision agriculture involves applying the right actions—such as irrigation or fertilization—in the right way, at the right place, and at the right time, to increase productivity and quality while minimizing waste [4]. Furthermore, the concept of "smart agriculture" has emerged, which integrates digitization and artificial intelligence (AI) technologies into agriculture [5]. The application of precision and smart farming techniques not only enhances productivity but also improves profit margins.

AI applications, particularly those related to machine learning (ML) technologies and the IoT, have seen remarkable growth in recent years. The emergence of various epidemics and pandemics, such as COVID-19, has accelerated the global prominence of these technologies [6,7]. Machine learning refers to AI methods that enable machines to perform specific tasks and solve problems without explicit programming [6,8]. In essence, these techniques allow a machine (or computer) to be "trained" on data related to a particular problem or task, enabling it to perform or solve that problem effectively.

The significant advancements in computing power and big data have facilitated the development of a more complex and applied form of machine learning—"deep learning" [9,10]. Deep-learning (DL) applications have revolutionized fields such as object detection [11,12], powdery food identification [13], plant nutrient status detection [14], leaf area measurement [15], spore detection [16], water stress [17,18], fruit quality [19], and other complex tasks involving large-scale data analysis [20]. The performance of computer systems in these applications has reached exceptional levels, enabling them to either fully replace or significantly assist human involvement in certain operations with remarkable success. These technologies have permeated various sectors, including agriculture, where AI, ML, DL, Internet of Things (IoT), and other digital technologies have been widely adopted, giving rise to the concept of "smart agriculture" and the aspiration to provide the fifth generation of agriculture (Agriculture 5.0 (Ag5.0)) [21].

Ag5.0 is an emerging innovative agricultural version that aims to address issues that Agriculture 4.0 failed to address. Ag5.0 is a subcategory of Industry 5.0 [22], an emerging concept to adopt that entails the integration of emerging technologies such as artificial intelligence, IoT, big data analytics, digital twins, blockchain, 6G networks, robotics, and human–machine interaction to handle various agricultural processes. One of the most important agricultural practices that are expected to benefit from Agriculture 5.0 technologies is crop monitoring and management.

Tracking and monitoring crops from germination to harvest is a key aspect of smart agriculture, crucial for optimizing both the quantitative and qualitative productivity of crops. To improve the efficiency of food production systems, assessing seed germination is essential for comparing the performance of different seed batches and selecting the most effective one for sowing [23]. Furthermore, continuous monitoring of various stages of plant development is vital. The existing literature provides extensive insights into the application of AI, ML, and DL techniques in crop monitoring, including growth assessment, monitoring of nutritional status, disease detection, crop water status, and yield prediction. Accordingly, this article offers a comprehensive review of the role of AI, being one of the backbones of Ag5.0 in addressing current and future challenges related to detecting, tracking, and managing crops throughout their growth stages.

This paper provides a comprehensive exploration of the role of AI in advancing agriculture towards Ag5.0, focusing on the integration of ML, DL, and AI-driven robotics. It begins by tracing the evolution of crop detection from the traditional methods of Agriculture 1.0 to the cutting-edge technologies of smart farming and Ag5.0. The paper then reviews the enabling technologies that facilitate the transition to Ag5.0, offering insights into the critical technological infrastructure required for this shift. It further examines how AI, ML, and contemporary agricultural technologies can be leveraged to monitor, analyze, and optimize key stages of crop growth and management, highlighting their role in improving efficiency and precision. The manuscript includes a detailed evaluation of the practical applications of machine learning in crop monitoring and protection, demonstrating its tangible impact on agricultural practices. It also discusses emerging technologies, such as smart sensors, advanced devices, communication systems, and AI-based robotics, which are driving the next generation of digitalization, decision-making, and automation in agriculture. In conclusion, the paper addresses the opportunities and challenges associated with adopting Ag5.0, outlining both the potential benefits and the obstacles to its widespread implementation. Ultimately, this paper emphasizes the transformative potential of Ag5.0 and underscores the importance of embracing AI-driven technologies to promote sustainability, productivity, and food security in the agricultural sector.

To accomplish this review, we focused on international and reliable scientific databases to ensure the comprehensiveness and quality of the selected literature. The databases used included Web of Science, IEEE Xplore, MDPI, and Science Direct. Regarding the selection of literature, studies published between 2015 and 2024 were included unless we had to use a more ancient study. All the selected studies focused on the applications of agricultural technologies in agriculture and crop management, especially artificial intelligence applications, as it is one of the most critical backbones for adopting the trend towards Agriculture 5.0. Keywords such as "agriculture", "artificial intelligence", "Internet of Things", "agricultural collaborative robots", "machine learning applications in agriculture", "deep learning applications in agriculture", and so on were used to ensure the improvement of search results. The criteria for the selected studies focused on articles published in international scientific journals and conference papers that were peer-reviewed to ensure scientific accuracy and research authority, while the literature that was not directly related to the review topic or of low scientific quality, such as preprints that were not peer-reviewed, were excluded.

2. From Traditional Agriculture to Smart Farming and Agriculture 5.0

Figure 1 shows the development roadmap for the agricultural revolution from Agriculture 1.0 to Ag5.0. Over time, humans have developed various methods to manage and monitor crops during growing seasons, addressing both positive (growth, fruiting) and negative (diseases, pests, nutrient deficiencies) changes. From the 19th century until around 1950, "Agriculture 1.0" relied on manual labor and traditional tools, which limited productivity and efficiency in crop monitoring [24]. In the 1950s, the introduction of mechanized farming equipment and synthetic pesticides marked the transition to "Agriculture 2.0" [25]. This shift resulted in increased productivity and lower production costs but also led to environmental concerns. By the late 20th century, the development of embedded systems, communication technologies, and data modeling led to "Agriculture 3.0" [26]. Precision agriculture emerged, with technologies like global navigation satellite systems (GNSS), remote sensing, and geographic information systems (GIS) enabling site-specific monitoring that optimized resource use and improved sustainability. Recently, the integration of IoT, big data, and autonomous machines has given rise to "Agriculture 4.0" [27]. This phase focuses on precision crop monitoring with minimal human intervention, supported by AI, machine learning, and robotics.

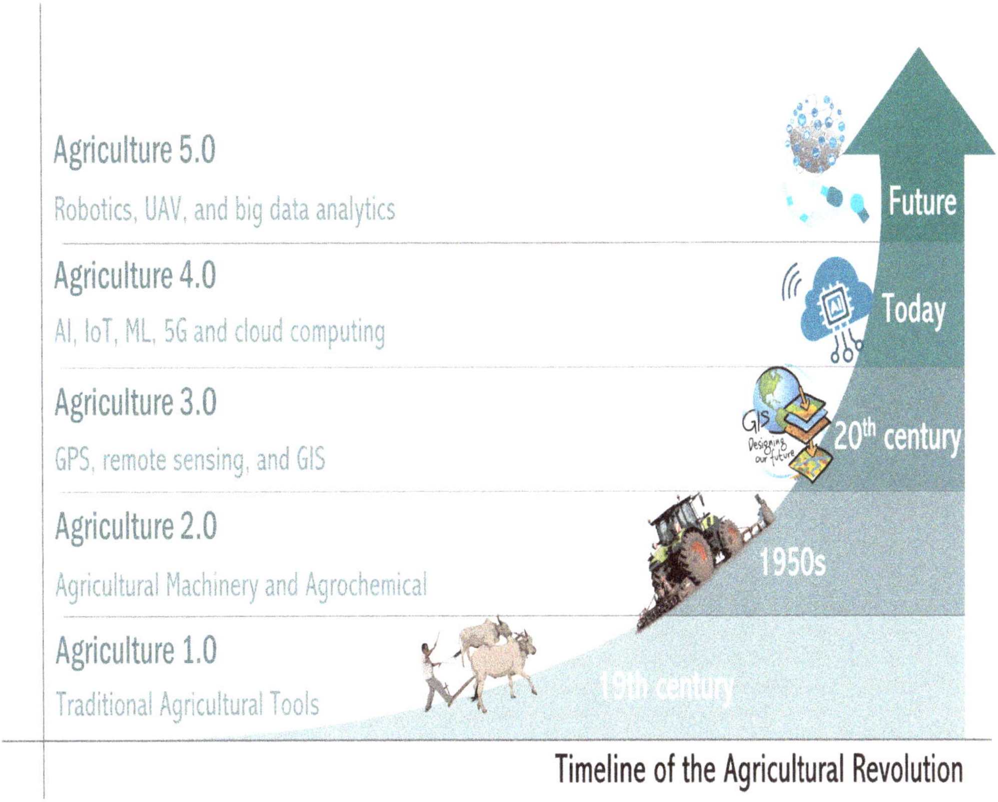

Figure 1. The development roadmap for the agricultural revolution from Agriculture 1.0 to Ag5.0.

Looking ahead, Ag5.0 is set to revolutionize crop monitoring further through autonomous systems, AI-driven decision support, and sustainable practices [28]. While its precise definition varies, Ag5.0 is widely understood as integrating AI, big data, robotics, and cloud computing to enhance crop monitoring and sustainability [29]. Hence, we can conclude that Ag5.0 includes the principles of precision agriculture and the use of AI algorithms and advanced robotic structures, including unmanned operations and autonomous decision support systems [30]. Ag5.0 aims to address emerging challenges such as climate change, irregular cropping seasons, and pest emergence [31]. With global population growth and the need for sustainable food production, AI, machine learning, and innovative technologies will be key to achieving efficient, eco-friendly crop monitoring [32]. The next phase must minimize chemical inputs and introduce new, sustainable methods to ensure long-term food security. Table 1 shows the main differences between the different agriculture versions.

Table 1. The main differences between the different agriculture versions.

Version	Features	Focus and Major Issues	Major Driving Factors	Information and Cybersecurity Issues
Agriculture 1.0	Traditional Agriculture dominated by manpower and animal forces	Human-centric, unsustainable low performance, and not resilient	N/A	N/A
Agriculture 2.0	Agriculture mechanization	Machine-focused, machinery and chemicals usage, unsustainable not resilient	Industrial revolution	N/A
Agriculture 3.0	Automatic agriculture with high-speed development	Technology-focused, Computers, programs, Unsustainable not resilient, and cybersecurity issues	Invention of computers, robotics, programming	Systems security Network security Devices security
Agriculture 4.0	Smart agriculture featured by AI and IoT	Automation-focused Smart systems/devices, renewable energies, sustainable, not resilient, efficient, and cybersecurity issues	Introduction of AI, IoT, cloud computing, and big data	Data security, systems security, network security, security devices, cloud security
Agriculture 5.0	Human-focused agriculture featured AI, IoT, robotics and Human–machine interactions	Human-centered highly sustainable resilient societal well-being cybersecurity issues	Chronic social issues. resilience need, increasing, consumer demands, and unsustainable production	Data security, systems security, network security, security devices, cloud security, and Human–machine security

3. Enabling Technologies for Agriculture 5.0

The enabling technologies of Agriculture 5.0 (Ag5.0), which build on advances in Agriculture 4.0, will contribute to addressing some of the current global challenges of producing sufficient, affordable, and healthy food while protecting ecosystems [33]. Ag5.0 integrates a range of advanced technologies and biotechnologies to promote sustainability, enhance human–machine collaboration, and develop circular agricultural production systems. Among the key enabling technologies is collaborative robotics (cobots), which are designed to perform labor-intensive, repetitive, and hazardous tasks. By automating these processes, cobots not only improve production efficiency but also make agriculture more appealing to younger generations, helping to address labor shortages in the sector [34]. Moreover, 6G technology, with its capability of delivering up to 1 Tbps throughput, extends IoT connectivity through higher data rates, minimal latency, and broader coverage. This allows for enhanced AI-based monitoring and precision farming techniques, improving the management of agricultural resources [34]. Digital twin technologies integrate data from weather, farm operations, environmental conditions, and supply chains, facilitating predictive analysis. This data integration supports informed decision-making, ultimately fostering more sustainable and resilient agricultural systems [35]. The Internet of Things (IoT) plays a central role in driving data-driven, automated farming practices. By optimizing resource utilization, reducing waste, and boosting productivity, IoT technologies contribute significantly to the efficiency and sustainability of modern farming practices [33,36]. AI further enhances agriculture by supporting intelligent automation, enabling rapid decision-making, ensuring quality assurance, and improving operational efficiency across various stages of farming. Big data analytics aid in real-time monitoring, enhancing decision-making capabilities and enabling the development of tailored solutions to meet specific agricultural needs. Edge computing reduces latency and enhances cybersecurity, ensuring seamless data flow and improving the interoperability of systems. Meanwhile, blockchain technologies ensure secure and transparent monitoring of IoT systems, allowing for traceability and accountability in agricultural practices. Cloud computing enables improved data storage and monitoring, supports collaboration among stakeholders, and offers cost-effective solutions to manage agricultural operations on scale. Finally, quantum technologies, with their potential for ultra-fast data transmission and enhanced security, promise to push the boundaries of data processing and encryption, offering unprecedented capabilities for Ag5.0 applications. These technologies enable Ag5.0 to revolutionize sustainable agricultural practices through advanced automation, data integration, and intelligent decision-making [22].

4. Perception, Analysis, and Actuation of Precision Crop Monitoring

Figure 2 shows the perception, analysis, and actuation stages of precision crop monitoring. Emerging crop monitoring technologies aim to identify plant conditions and symptoms during different growth stages, enabling site-specific chemical or mechanical control actions [37]. This process involves three main stages: Perception Stage: Crop-related data are collected through sensors and cameras, either on-ground or via remote sensing platforms like UAVs. UAVs, equipped with digital, multispectral, hyperspectral, thermal, light detection and ranging (LiDAR), radar, or sonar sensors, enable large-scale, rapid crop monitoring. These technologies detect symptoms and disorders at different stages of growth, including diseases and nutrient deficiencies. Analysis Stage: Acting as the critical link between perception and actuation, this stage involves processing crop data using advanced AI, machine learning, and deep-learning algorithms. These techniques evaluate plant conditions in real-time, such as growth stages, nutrient deficiencies, diseases, and environmental factors like weather and soil properties [38]. Machine learning leverages high-level data features to build predictive models, classify images, and propose solutions, with CPU, graphics processing units (GPU), and tensor processing units (TPU) technology advancements enabling more complex analyses. Actuation Stage: This stage involves applying precision crop monitoring strategies based on insights from the analysis phase. Smart devices with GNSS support execute site-specific actions, including autonomous crop service machines such as smart sprayers and agricultural robots. Prescription maps further enhance accuracy in crop management [39]. These technologies have revolutionized crop monitoring and management, providing effective solutions for detecting and addressing crop conditions, optimizing resource use, and enhancing precision agriculture.

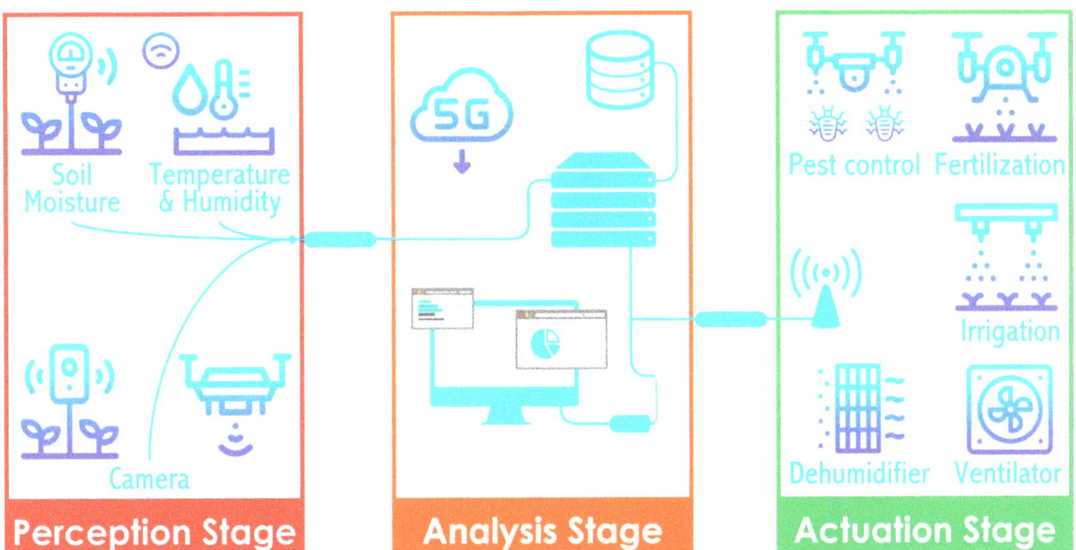

Figure 2. Perception, analysis, and actuation of precision crop monitoring.

5. Machine Learning Applications for Precision Crop Monitoring

AI is integral to Ag5.0, with its key subfields, ML and DL, playing critical roles. ML enables machines to learn automatically from data without explicit programming. ML is crucial for Ag5.0, aiding in crop yield prediction, plant disease monitoring, nutritional and water status assessment, weed detection, and forecasting adverse climatic conditions like rainfall, drought, and wind [40]. DL, a subset of both AI and ML, uses artificial neural networks (ANNs) inspired by human brain function for complex computations. DL methods such as convolutional neural networks (CNNs), recurrent neural networks (RNNs), autoencoders, and generative adversarial networks (GANs) are applied to specific agricultural challenges. CNNs, the most common DL architecture in precision agriculture, excel in computer vision tasks like image classification and object detection for crop protection [41]. Unlike traditional ML, CNNs autonomously extract features from image data, improving accuracy. Their performance depends on network depth, computational resources, and optimization algorithms like gradient descent and ReLU activation [42]. CNNs consist of three core layers. The input layer processes the raw data, which is split for training and testing [43,44]. The second layer includes convolutional, pooling, and fully connected layers. The convolutional layer extracts data and optimizes weights, while the pooling layer filters and selects the maximum value for the next layer. The fully connected layer processes the output from previous layers, applying complex functions to produce the final result. The output layer then converts this information into an equivalent score using statistical methods. Figure 3 illustrates the general architecture of CNN.

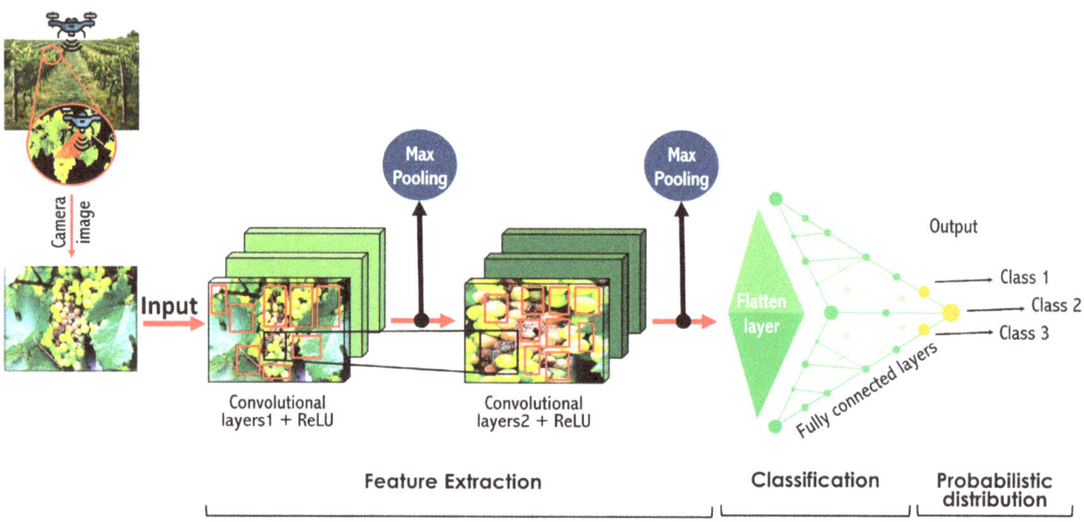

Figure 3. Illustration of general CNN architecture.

5.1. Germination Assessment

Precision crop monitoring begins at the germination stage, as seed germination assessment is essential for determining seed quality and predicting the crop's initial yield. Traditional manual evaluation is time-consuming and prone to errors [45,46]. ML and computer vision techniques have recently been applied to automate seed monitoring. El-Masry et al. presented a computer-integrated multichannel spectral imaging system as a high-throughput phenotyping tool for the analysis of individual cowpea seeds harvested at different developmental stages. The developed linear discriminant analysis (LDA) was robust in classifying the seeds based on their germination capacity, with overall correct classification of 96.33% and 95.67% in the training and validation datasets, respectively [47]. Peng et al. developed an automated system for seed germination assessment using deep learning, incorporating a modified thermostat, digital camera, monitoring software, and a dense small-target detection algorithm (DDST-CenterNet). Their system demonstrated high efficiency, unaffected by seed background, lighting, or environmental factors, with strong scalability [48]. Genze et al. employed CNNs to evaluate maize, rye, and pearl millet germination, achieving predictive efficiencies of over 94% for rye and pearl millet and 97.9% for maize [45]. Similarly, Awty-Carroll et al. used the K-Nearest Neighbors (K-NN) algorithm to analyze diverse germination patterns in Miscanthus sinensis. K-NN achieved a validity score of 0.69–0.7 and an optimized ROC area of 0.89, demonstrating its reliability compared to traditional scoring methods [49]. Despite promising applications of machine learning and computer vision in seed germination assessment, significant research gaps remain. These include limited exploration of AI techniques and the need for more robust models capable of handling diverse environmental conditions and seed types. To advance with Ag5.0, the scientific community must prioritize developing scalable, efficient smart seed monitoring systems for broader agricultural use.

5.2. Diseases Detection and Crop Protection

Plant diseases cause significant global economic losses, accounting for 13% of crop yield reductions and 20–40% of food production losses [50,51]. Innovative methods for monitoring and diagnosing plant diseases are crucial to mitigate these losses. ML has recently emerged as a powerful tool for detecting plant diseases and safeguarding crops from threats like weeds and epidemics [52]. ML methodologies analyze plant biochemical changes and sensory data related to pathogen spread. Ground-based measurements using digital or spectral cameras and spectral fingerprint analysis are commonly employed. Classification and regression algorithms, combined with dimensionality reduction techniques like principal component analysis (PCA), are used to distinguish healthy plants from diseased ones and predict spectral responses of infected plants. Notable ML applications include Abdalla et al.'s integration of CNNs with Bidirectional Long Short-Term Memory (BiLSTM) networks using UAV images, achieving 89.7% accuracy with the ResNet101-BiLSTM model [53]. Ahmed and Yadav developed an automated system employing CNNs and support vector machines (SVMs) to identify nine plant diseases, including bacterial, viral, and fungal infections [54]. Orchi et al. conducted a comparative study between conventional machine-learning algorithms (SVM, LDA, KNN, CART, RF, and NB) and deep-learning algorithms (VGG16, VGG19, InceptionV3, ResNet50, and CNN) using PlantVillage Dataset to classify diseased and healthy leaves of crops. InceptionV3 achieved the best performance and the highest classification accuracy of 98.01% [55]. Wang et al. proposed a hybrid CNN-LSTM model to detect cucumber downy mildew, achieving a high predictive accuracy (R^2 = 91%) (Wang et al., 2024a) [16]. Nagachandrika et al. introduced a multi-scale fused feature approach using VGG16, Variational Autoencoders, and Visual Transformers, optimizing parameters with the Enhanced Gannet Optimization Algorithm, achieving over 94% classification accuracy for leaf diseases [56]. Guerrero-Ibañez and Reyes-Muñoz combined CNNs with generative adversarial networks to identify tomato leaf diseases, avoiding overfitting [57]. Tamilvizhi et al. introduced a quantum-behaved particle swarm optimization based on deep transfer learning (QBPSO-DTL) to detect and classify sugarcane leaf diseases with high accuracy, achieving 99.48% accuracy in validation and 89.65% accuracy in training. The integration of quantum computing and deep learning significantly improves disease classification accuracy [58]. For pest monitoring, Zhao et al. designed an automated system using deep learning and cameras, achieving a detection quality of 0.93 [59]. Lee et al. implemented a web application with the EfficientNet-b0 model for citrus insect classification, achieving 97% accuracy [60]. Dai et al. introduced a digital twin (DT) system for managing aphid pests through the integration of data and models. The framework provides a comprehensive approach to aphid management throughout the crop growth cycle. A predictive digital twin model was developed using a random forest algorithm optimized with a genetic algorithm to forecast aphid populations. Based on the twin data, a search strategy was employed to design pest control interventions, and decision optimization was applied to improve management efficiency. The system demonstrated a prediction accuracy of 85.73% when evaluated on a test dataset [61]. Despite significant progress in machine learning and deep learning for plant disease detection, several key challenges remain. These include limited data availability, difficulties in generalizing models across different crops and environments, and the integration of diverse data sources. To meet the demands of Ag5.0, the scientific community must focus on developing more adaptable, scalable, and data-efficient models. Overcoming these challenges will unlock the full potential of AI technologies for creating sustainable and efficient agricultural systems. Table 2 summarizes key research advancements in ML applications for disease detection and crop protection.

Table 2. An overview of the key research on ML applications in disease detection and crop protection.

Application	ML Algorithm	Dataset Type	Acc, %	Ref.
Coton diseases	CNN-BiLSTM	RGB	89.7	[53]
Detect diseases	CNN	RGB	94	[56]
Diagnosing plant diseases	CNN	RGB	90	[62]
Adzuki bean rust disease	CNN, ResNet-ViT, RMT	Multi-source	99	[63]
Cucumber downy mildew prediction	CNN-LSTM	RGB	91	[64]
Diagnosing plant diseases	SVM and CNN	RGB	99	[54]
Detection of plant diseases	ML and DL models	RGB	98	[55]
Rice diseases	GA and CNN	RGB	95	[65]
Tomato leaf disease	CNN	RGB	99	[57]
Tomato gray mold, cucumber downy mold, and cucumber powdery mildew spores	SVM	Fingerprint characteristics of diffraction–polarized	95	[66]
Detect brown spots in rice	CNN	RGB	97	[43]
Detection of apple diseases	CNN	RGB	97	[67]
Detection of cassava diseases	CNN	RGB	98	[68]
Wheat diseases	CNN	RGB	97	[69]
Detection of fusarium head blight disease	CNN	Hyperspectral	75	[60]
Tomato spotted wilt virus	CNN	Hyperspectral	96	[60]
Diseases and pests in tomatoes	ANN for regression, SVM	Spectral	99	[70]
Powdery mildew in wheat	PLSR, SVM, RF	Hyperspectral	85	[71]
Northern leaf blight in maize	CNNs	RGB	96	[72]
Tomato water stress	DNNs	RGB	Performed well	[73]
Tomato diseases and pests	Faster R-CNN, R-FCN, SSD, ResNet	RGB	90	[74]
Disease in Eggplant	CNN	RGB	99	[75]
Plant disease identification	CNN	RGB	99	[76]

Table 2. *Cont.*

Application	ML Algorithm	Dataset Type	Acc, %	Ref.
Detection of grapevine esca disease	SIFT encoding and CNN	RGB	90	[77]
Grapevine yellows symptoms	CNN	RGB	99	[78]
Plant disease	CNN	RGB	99	[79]
Late blight in potato	RF and PLS-DA	Spectral	83	[80]
Laurel wilt	DT and MLP	Spectral	100	[81]
Bacterial spots in tomato	PCA and k-NN	Spectral	100	[82]
Anthracnose crown rot in strawberry	FDA, SDA, and kNN	Spectral	73	[83]
Rhizoctonia root and crown rot in sugar beet	PLS, RF, k-NN, and SVM	Hyperspectral	72	[84]
Potato virus y	CNN	Hyperspectral	88	[85]
Tobacco mosaic virus in tobacco	PLS-DA, RF, SVM, BPNN, and ELM	Hyperspectral	95	[86]
Bacterial blight in coffee	RF, SVM, and Naïve Bayes	Multi-spectral and thermal	75	[87]
Xylella fastidiosa infection in olive	LDA, SVM, RBF, and ensemble classifier	Hyperspectral and thermal	80	[88]
Fall armyworm (*Spodoptera frugiperda*) in cotton	Multiple ML algorithms	Hyperspectral	91	[89]
Pest monitoring	DPeNet, Faster R-CNN, SSD, and Yolov3	RGB	93	[59]
Citrus pest	EfficientNet-b0	RGB	97	[60]
Spotted spider mite in cotton	SVM	Multispectral	85	[90]
Plague species	CNN	RGB	75	[91]
Plague species in insect images	CNN	RGB	89	[92]

5.3. Weeds Detection

Weeds account for over 40% of global crop yield losses annually [93]. Managing weeds on a large scale is crucial to minimizing environmental and economic impacts, making automated weed monitoring systems essential for enhancing global food production. Recent advancements in automated systems have focused on differentiating crops from weeds. Almalky and Ahmed developed a deep-learning-based approach to classify the growth stages of Consolida regalis weeds, using YOLOv5, RetinaNet, and Faster R-CNN models. Their YOLOv5-small model achieved the best performance with a recall of 0.794 [94]. Similarly, Mu et al. hybridized ResNeXt and Faster R-CNN, achieving 95% accuracy in distinguishing between crop seedlings and weeds [95]. Zhu et al. introduced a YOLOX CNN-based weeding robot employing a blue laser, achieving 88.94% weed identification accuracy [96]. Drones are emerging as an efficient alternative to robotic and satellite imaging for weed monitoring, offering cost-effective, high-resolution field data collection. Islam et al. utilized drones equipped with RGB cameras to capture images and calculate vegetation indices (e.g., normalized red, green, and blue bands), mitigating the effects of varying lighting conditions [97]. Agricultural robots play a pivotal role in modern farming, performing tasks like laser weeding [98], pesticide spraying [99], spot picking [100], and fertilizer application. These robots range from modified tractors to aerial and small service robots [99]. Although automated weed monitoring systems have advanced significantly, important scientific gaps remain. These include the development of more robust models capable of handling diverse environmental conditions and crop variations, as well as the integration of various sensor technologies for real-time decision-making. To align with

Ag5.0, the scientific community must enhance the accuracy, efficiency, and scalability of weed management systems, enabling the full potential of AI and robotics to boost global food production. Table 3 summarizes key applications of machine learning and robotics in weed monitoring.

Table 3. An overview of the most significant studies conducted on crop weed detection using ML.

Crop	ML Algorithm	Dataset Type	Acc, %	Ref.
Peanuts	YOLOv4-Tiny	RGB	96.7	[101]
Sunflower	U-Net	Multispectral	90	[102]
Soybean	ML	Thermal	82	[103]
Tobacco, tomato, and sorghum	PlantNet	High Precision 3D Laser	>95	[104]
Carrot	ANN with 15 units in ensemble	Multispectral	83.5	[105]
Chilli	RF and SVM	RGB	96 and 94	[97]
Rice	SVM	RGB	73	[106]
Wheat	Wheat-V2	Spectral	>96.7	[107]
Tobacco	Faster R-CNN and YOLOv5	RGB	98.43 and 94.45	[108]
Pea and Strawberry	Faster R-CNN	RGB	95.3	[109]
Sugar Beet and Oilseed	An encoder–decoder network with atrous separable convolution	RGB	96.12	[110]
Soybean	CNN	RGB and spectral	99.66	[111]
Soybean	CNNLVQ	RGB	99.44	[112]
Bean and Spinach	RF	RGB	96.99	[113]
Carrot	ANN	Multispectral	83.5	[114]

5.4. Nutrient Stress Detection and Chlorophyll Estimation

Accurate assessment of crop nutritional status and nutrient requirements is critical for effective farm monitoring, influencing both environmental sustainability and economic viability [115]. Nutrient excess or deficiency can result in yield losses, inefficient resource use, reduced soil organic carbon, and associated challenges [116]. Precise diagnosis enhances crop yields, optimizes fertilizer use, and increases revenue. Traditional methods like visual assessments and chemical analyses are often destructive, costly, and labor-intensive. Recent ML advancements have enabled automated diagnostic systems that effectively monitor and manage plant nutrients. For instance, Bera et al. proposed PND-Net, a hybrid graph convolutional network (GNN) and CNN for classifying plant nutrient deficiencies and diseases across multiple crops, achieving 90.54% accuracy for coffee and 96.18% for potato diseases [117]. To efficiently assess the nutritional status of oilseed rape, Abdalla et al. encoded the spatiotemporal information of plants in a single time-series model. They conducted a comparative study between SVM and deep-learning models (VGG16, VGG19, InceptionV3, ResNet50) hybridized with LSTM. The hybridization of InceptionV3 and LSTM achieved the highest diagnostic accuracy of 95% [118]. Similarly, Taha et al. achieved over 95% accuracy in assessing nitrogen, phosphorus, potassium, and chlorophyll levels in aquaponic plants using deep models [119,120]. Kou et al. used UAV-based digital cameras and CNNs to estimate nitrogen levels in cotton canopies, achieving a prediction accuracy of $R^2 = 0.80$ [121]. Yu et al. combined hyperspectral imaging and deep-learning techniques, such as CNN regression and stacked autoencoders (SAE), to estimate nitrogen levels in canola leaves with $R^2 = 0.903$ [122]. Anami et al. utilized VGG-16 to detect biotic and abiotic stresses in rice, achieving classification accuracy exceeding 95% [123]. The integration of sensors, smart irrigation systems, and remote sensing techniques has generated large datasets that, combined with ML advancements, enable precise nutrient monitoring. These breakthroughs enhance decision-making, support sustainable agricultural practices, and improve crop yields, contributing to global food security [124,125]. Table 4 summarizes significant research in nutrient stress detection and chlorophyll estimation using ML.

Table 4. An overview of the most significant research conducted in the field of nutrient stress detection and chlorophyll estimation using ML.

Crop	Dataset	ML Algorithm	Detected Nutrient	Ref.
Banana, coffee and potatoes	RGB	CNN—Graph convolutional networks (GCN	Br, Ca, Fe, Mn, Mg, N, K, P, and more deficiencies	[117]
oilseed rape	RGB	CNN-LSTM	N-P-K	[118]
Coton	RGB	CNN-based regression	Nitrogen	[121]
Lettuce	RGB	CNN	NPK	[119]
Lettuce	Spectral data and RGB	SVM, PLSR, BPNN, RF, and AutoML	Chlorophyll content	[120]
Tomato	RGB	Pre-trained deep-learning model	Ca and Mg	[126]
Wheat	RGB	BP-ANN and KNN—stepwise-based ridge regression (SBRR)	Chlorophyll content	[127]
Rice	RGB	Ensembling of various Transfer Learning (TL) architectures	Multiple deficiencies	[128]
Soybean	RGB	Deep CNN Model framework	Multiple stress, and potassium deficiency	[129]
Rice	RGB	CNN, pre-eminent classifier-SVM	Nitrogen	[130]
Black gram	RGB	Image Segmentation and CNN	Multiple deficiencies	[131]
Paddy	RGB	Deep CNN with pre-trained VGG 16	Various classes of Biotic and Abiotic stress	[123]
Rice	RGB	CNN and using Edge as a service	Biotic stress	[132]
Muskmelon	RGB	CNN, BPNN, DCNN, LSTM	Nitrogen	[133]
Guava, Groundnut	RGB	CNN, RCNN	N-P-K	[134]
Sorghum Plant	RGB	Multilayered Deep Learning	Nitrogen	[135]
Sugar beet	RGB	CNN using RGB images	N, P, K, Calcium and fertilization status	[136]
Lettuce	RGB	CNN	Nitrogen	[137]

5.5. *Water Status*

Droughts, exacerbated by global climate change, are increasingly impacting agriculture and raising concerns about food security. Irregular rainfall limits water availability, causing water stress in crops—one of the most critical abiotic stresses affecting plant growth. Early detection of water stress is essential for sustainable agricultural productivity. Akbari et al. developed a neural network-based GenPhenML approach to predict barley resistance to drought, achieving over 97% classification accuracy [138]. Gupta et al. evaluated nine machine-learning models for automated wheat water stress detection, with Random Forest (RF) achieving the highest accuracy of over 91% [139]. Okyere et al. combined vegetation indices (VIs) from hyperspectral images with ML models, finding RF regression most effective for monitoring wheat drought stress [140]. Banerjee et al. used the Vision Transformer classifier to analyze NIR reflectance in maize, achieving 85% accuracy for early water stress detection. For rice water condition prediction, Wu et al. utilized thermal imaging-based temperature indicators with RF, achieving an $R^2 = 0.78$ [141]. Jin et al. evaluated multiple deep-learning models to predict cotton water status using 5200 thermal images under sub-membrane drip irrigation. MobilenetV3 outperformed other models, achieving an F1 score of 0.9990 [142]. Although machine learning has made considerable strides in early water stress detection, significant gaps remain in the research. Key challenges include the need for more adaptable models that can generalize across various crops and environmental contexts, along with the integration of diverse sensor data to refine prediction accuracy. To align with Ag5.0, the scientific community must prioritize enhancing the scalability, flexibility, and precision of these models to drive sustainable agricultural productivity amid escalating drought risks. Table 5 summarizes key studies on monitoring crop water status using ML applications.

Table 5. An overview of the most significant research conducted in the field of water status of crops using ML.

Crop	ML Algorithm	Dataset Type	Acc.	Ref.
Barley	ANN	Phenotype and genotype features	$R^2 = 0.99$	[138]
Wheat	RF	Fluorescence	91%	[139]
Maize	ResNet50	RGB	98%	[143]
Wheat	SVM, RF, and DNN	Hyperspectral	94%	[140]
Rice	RF	Thermal	$R^2 = 0.78$	[141]
Maize	CNN+SVM	RGB	94%	[144]
Rice	ML models	Spectral	$R^2 = 0.87$	[145]
Cotton	Multi CNN models	Thermal	F1 = 0.99	[142]
Tomato	VGG-16 and Resnet-50	RGB, NIR and depth images	99.18%	[146]

5.6. *Prediction of Crop Yield*

The technological revolution, particularly deep learning, has significantly improved agricultural yield forecasting by overcoming the limitations of traditional methods. Deep learning enables predictive models that are highly accurate and adaptable to dynamic agricultural conditions, addressing the growing demand for precise yield predictions. Li et al. used UAV multispectral data to construct vegetation indices for predicting winter wheat yield across growth stages. Their CNN model outperformed other algorithms, achieving the highest prediction accuracy [147]. Similarly, Li and Wu developed a lightweight CNN model (SqueezeNet) for tiger nut yield estimation, achieving $R^2 = 0.78$ using multispectral UAV data [148]. Tanaka et al. employed CNNs with over 22,000 RGB images from six countries to estimate rice yields, explaining 68% yield variation with a root mean square error of 0.22 [149]. Mia et al. combined UAV imagery and meteorological data in a multimodal deep-learning model for rice yield forecasting, achieving 86% accuracy [150]. Zhou et al. compared CNN, ConvLSTM, and CNN-LSTM models for rice yield prediction using remote sensing and geographical heterogeneity variables, with the CNN-LSTM model demonstrating the best performance [151]. Sarr and Sultan used SVM, Random Forest, and Neural Networks to predict staple crop yields in Senegal, finding optimal performance when combining vegetation and climate data, although soil conditions were not considered [152]. Surana and Khandelwal applied a three-layer ANN with ReLU activation to predict crop yields in Maharashtra, India, achieving 82% accuracy [153]. Kuradusenge et al. used Random Forest models to predict potato and maize yields in Rwanda, with R^2 scores of 0.875 and 0.817, though the study omitted key meteorological variables [154]. In smart agriculture, quantum-enhanced AI technologies optimize crop yields, improve fertilizer efficiency, and support pest control through advanced robotics and automation [155]. Despite progress, challenges remain in standardizing input data, integrating findings across crop types, and evaluating the comparative efficacy of deep-learning architectures. Table 6 summarizes key research in crop yield prediction using ML.

Table 6. An overview of the most significant research conducted in the field of crop yield prediction using ML.

Crop	ML Algorithm	Dataset Type	Acc. %	Ref.
Wheat	Traditional ML and 1D-CNN	Multispectral data	$R^2 = 0.703$	[147]
Tiger nut	SqueezeNet	Multispectral	78	[148]
Rice	CNN	RGB	68	[149]
Rice	CNN (AlexNet)	Meteorological	86	[150]
Rice	CNN-LSTM	Meteorological	93.4	[151]
Multiple	LR, NB, and RF	Meteorological	92.81	[156]
Peanut, maize, millet and sorghum	SVM, RF, and ANN	Meteorological	$R^2 \geq 0.50$	[152]
Multiple	RF, Adaboost, Gradient Boost, and (SVM)	Meteorological	82	[153]
Irish potatoes and Maize	RF and SVM	Meteorological	87.5	[154]
Wheat	LSTM-RF	Multispectral	71	[157]
Soybean	PLSR, RF, SVM, DNN-F1 and DNN (DNN-F2)	RGB, multispectral and thermal	72	[158]
Wheat	OLS and LASSO, SVM, RF, AdaBoost, and DNN	Satellite images, climate data, soil maps, and historical yield records.	84	[159]
Wheat	RF, DNN, 1D-CNN and LSTM	Climate, satellite, soil properties, and spatial information data	90	[160]
Corn	1D-CNN	Hyperspectral	75.50	[161]
Wheat and Barley	CNN	RGB	MAPE = 8.8	[162]
Almond	CNN (U-Net)	RGB	78	[163]

6. Innovative Technologies Associated with Ag5.0 for Precision Crop Monitoring

Crop monitoring is increasingly adopting emerging technologies such as AI and ML to automate tasks and make data-driven decisions, aligning with precision agriculture and Agriculture 4.0. Recent advancements in ML have accelerated AI applications in agriculture, enabling the development of innovative approaches suitable for Ag5.0. Key challenges in crop monitoring include monitoring nutritional deficiencies, irrigation issues, weeds, diseases, and pests. Emerging technologies focus on two objectives: (1) enabling rapid, early crop monitoring to implement proactive strategies before conditions deteriorate, and (2) developing autonomous, real-time multi-tasking systems. These goals rely on integrating diverse data sources, such as climate data, sensors, soil information, and farm monitoring systems, to improve early detection and monitoring capabilities [68,164,165]. Autonomous systems aim to perform three critical crop monitoring phases in real-time: monitoring crop conditions (e.g., nutrient stress, irrigation issues, pests), analyzing collected data, and making informed decisions. Ag5.0 technologies are expected to address these challenges by leveraging ML, hardware, drones, robotics, and telecommunications advancements.

6.1. Innovative Hardware-Based Crop Monitoring

The industrial revolution in microprocessors has driven the development of innovative electronic devices, such as smart sensors, IoT technologies, and multi-core embedded systems, transforming agriculture through digitalization. Modern smart sensors, like Sony's IMX500 and IMX501, integrate AI processing capabilities directly onto image sensors, enabling real-time data analysis with minimal latency, reduced power consumption, and enhanced privacy. These sensors process pixel data through an image signal processor (ISP) and perform AI tasks at the logic chip stage, outputting metadata for efficient crop monitoring. They facilitate the detection and diagnosis of disturbances (e.g., nutrient deficiencies, diseases, pests) and provide data formats like ISP images and region-specific outputs for optimized crop monitoring (Sony Group Corporation, Tokyo, Japan). The pixel chip captures the signals and then processes them through the ISP (Image Signal Processor) and AI processing is performed in the processing step on the logical chip and the information is output as metadata, which reduces the size and amount of processed data. Users can specify the image output format according to the application requirements, including ISP (YUV/RGB) output images and ROI (region of interest) region extraction images, as shown in Figure 4.

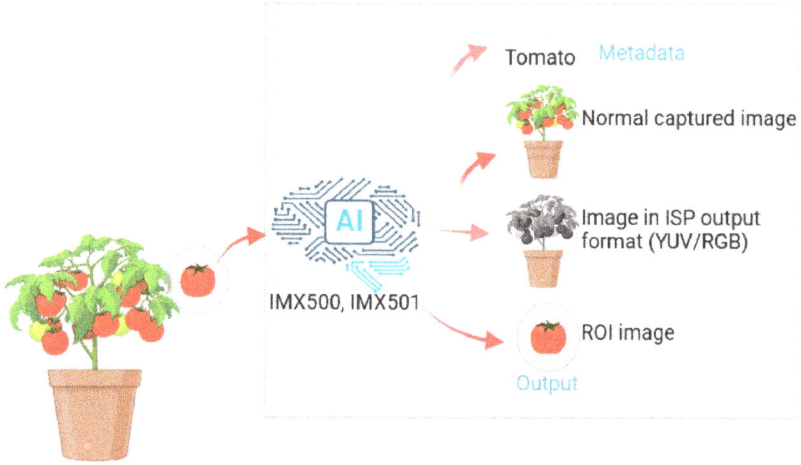

Figure 4. Data output format selectable to meet various needs using IMX500 and IMX501 sensors.

Advancements in GPU, TPU, Radeon DNA (RDNA), and protocols like non-volatile memory express (NVMe) have enhanced programmability, enabling Ag5.0 applications such as virtual modeling, digital twins, and supercomputer-based crop monitoring analysis. Supercomputers improve CNN model training by accelerating data augmentation and reducing computational time, enabling precise crop monitoring and classification. field programmable gate arrays (FPGAs) further enhance CNN-based designs, supporting multi-core embedded systems and mobile-compatible Ag-IoT crop detectors [166,167]. IoT devices pose security risks, addressed through application-specific integrated circuits (ASICs) for edge gateways, essential for secure Ag-IoT systems [168,169]. ML-optimized gateways improve task efficiency under resource constraints, reducing latency and enhancing privacy in Ag5.0 edge computing systems.

In the medium term, hardware and software advancements, particularly memristor-based devices, are poised to revolutionize agriculture. Memristors integrate storage and processing capabilities, mimicking neuronal synapses and enabling neuromorphic computing with spiking neural networks (SNNs). SNNs outperform traditional artificial neural networks

(ANNs) with lower latency, reduced training times, higher accuracy, and lower energy consumption, making them critical for future Ag5.0 systems [170–172]. Neuromorphic computing and SNNs will drive innovative computational systems with significant impacts on Ag5.0.

6.2. Crop Monitoring Through Communication Technology

Precision crop monitoring increasingly utilizes a system-of-systems approach, integrating various interconnected technologies to establish effective crop monitoring strategies. Essential technologies, including on-ground, proximal, and remote sensing, monitor factors affecting crop health, while telecommunications connect devices such as platforms, processors, and actuators to facilitate data transfer. This interconnected framework supports improved data processing, pest forecasting, and decision-making. Wireless Sensor Networks (WSNs) (Figure 5) play a central role in agricultural communication systems, employing technologies like Bluetooth and Zigbee. Bluetooth offers faster data transfer rates (1–24 Mbps) but a shorter range (<10 m), whereas Zigbee supports a longer range (up to 100 m) but slower transfer speeds (40–240 Kbps) [173]. Wi-Fi extends operational range to 50–100 m or more, while WiMAX covers distances up to 50 km. IoT and low-power wide area networks (LPWAN) have enabled technologies like LoRa and LoRaWAN, which are well-suited for rural agriculture due to their long-range (dozens of kilometers), low power consumption, and secure connectivity [174]. LoRaWAN uses the ISM frequency bands, varying regionally [175].

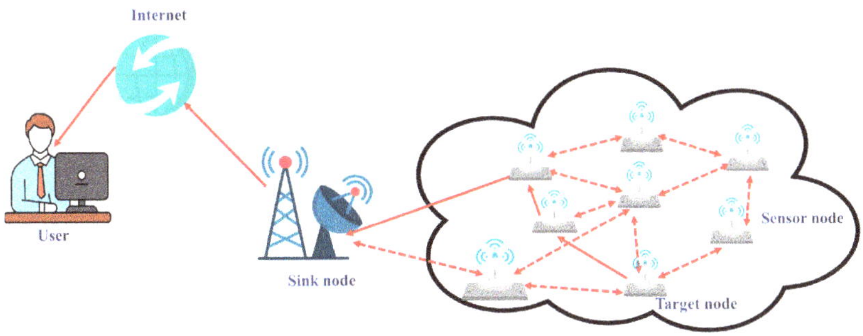

Figure 5. Wireless Sensor Network (WSN).

Efficient telecommunications are vital for real-time operations on actuator platforms such as unmanned aerial vehicles (UAVs), unmanned ground vehicles (UGVs), and self-propelled sprayers, designed to analyze and respond to pest occurrences in milliseconds. Cloud computing, edge computing, and edge AI are key technologies enabling real-time precision crop monitoring in the context of Ag5.0. Cloud computing, supported by companies like Amazon and Microsoft, processes sensor data through machine learning and sends prescriptions back to devices. These operations require low-latency networks like 5G, with advancements toward 6G supporting UAV-based precision applications [176]. Undoubtedly, Ag5.0 will significantly benefit from the development of 6G wireless networks as they are expected to bring several features such as increased network speed and improved data throughput while enhancing time efficiency. Moreover, 6G networks are also expected to serve as the backbone for the development of connectivity in wireless network architecture and IoT sensors. The adoption of 6G networks will significantly facilitate the deployment of smart agricultural applications. These applications include real-time crop monitoring, disease and nutrient deficiency diagnosis in plants, as well as reliable and decentralized agricultural operations [177]. The integration of IoT sensors with blockchain-based systems enables new types of decentralized applications. IoT sensors collect agricultural information about plants such as monitoring plant diseases, insects, nutritional disorders, and water status, and then store and analyze it with blockchain-enabled devices and its trustless, decentralized environment, enabling new types of business processes across more transparent organizations. Figure 6 illustrates the integration of blockchain and IoT in agricultural operations.

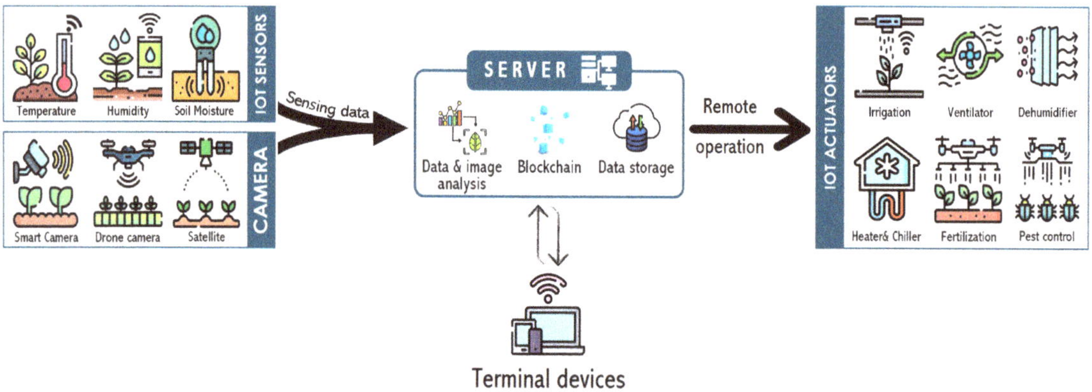

Figure 6. IoT and blockchain in agriculture.

Security and privacy remain critical challenges in cloud computing, prompting interest in decentralized fog computing to reduce latency and enhance bandwidth efficiency [178,179]. Edge computing further addresses these challenges by enabling localized data processing near the source, reducing communication costs, and improving privacy [180]. The next step in this evolution is edge AI, which integrates machine learning algorithms into embedded systems for autonomous crop monitoring. While still in early research [181], edge AI and AI chips capable of executing CNNs show significant promise for advancing agricultural robotics and automation [182]. This multi-layered technological framework ensures robust, efficient, and secure solutions for modern agriculture, emphasizing real-time precision and autonomous operations.

6.3. Advancements in Robotics Towards Ag5.0

Autonomous mobile robots (AMRs) hold great potential to significantly boost industry productivity, particularly within the context of Ag5.0, which focuses on improving precision crop monitoring. These agricultural robots are purpose-built for farming tasks [183]. As integral members of the robotics family, they are designed with advanced perceptual capabilities, autonomous decision-making, control systems, and precise execution abilities, enabling them to operate effectively in complex and hazardous environments. The growing demand for labor efficiency and enhanced agricultural production has led to an expansion in the types and applications of agricultural robots.

Agricultural robots can generally be categorized into three main types: field robots, fruit and vegetable robots, and livestock robots [183]. Research primarily focuses on field robots and those used for harvesting fruits and vegetables. Despite the differences in their specific applications, these robots share common technological features, including stable mobile platforms, multi-sensor integration, advanced image processing, sophisticated algorithms, and flexible locomotion control. Figure 7 shows the core technologies included in agricultural robotic applications. A summary of various agricultural robots, their applications, and key tasks is provided in Table 7. This evolution in agricultural robotics promises greater automation and efficiency in farming practices, contributing to more precise and sustainable crop monitoring.

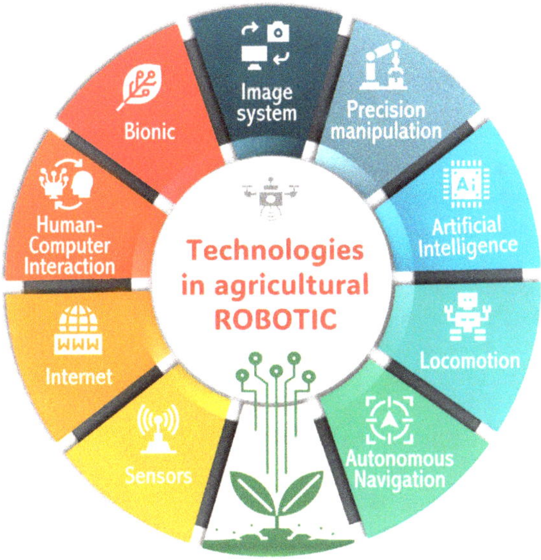

Figure 7. Core technologies are involved in agricultural robotic applications.

Table 7. An overview of some agricultural robots developed to perform specific tasks.

Application	Accuracy	Ref.
Tilling	Performed very well	[184]
Tractor	Performed very well	[185]
Wheat Precision Seeding	The qualified rates of seeding exceed 93%	[186]
Spraying fertilizers and pesticides	Performed very well	[187]
Spraying fertilizers and pesticides	The system can detect lines in plantations and can be used to retrofit conventional boom sprayers.	[188]
Planting	Performed well	[189]
Plant protection	Path planning accuracy is up to 97.8%	[190]
Strawberry sensing and harvesting	Trials showed an overall success rate of 78% in dealing with harvestable strawberries with a 23% damage rate	[191]
Inspect the presence of pests and diseases	A detection rate of 66.4% was obtained for images obtained in the laboratory and 59.8% for images obtained in the field.	[192]
Harvesting paddy	Harvesting process improved	[193]
Detecting and classifying weed	The results showed that the development deployed automatically on AgBot II was effective in controlling all weeds.	[194]
Tomato treatment and harvesting	Performed well	[195]
Tomato harvesting	Performed well	[196]

The future of Ag5.0 will increasingly rely on collaborative robots that improve agricultural practices by enhancing ergonomics and sharing workspace [197]. These robots will play a key role in organic food production, particularly in non-chemical pest control using mechanical methods [198], as well as in efficient harvesting systems [199], which will improve organic crop monitoring effectiveness [200].

As Ag5.0 technologies advance, it will facilitate the integration of UGVs and UAVs into coordinated operations under a single control system, leading to the development of multi-robot fleet systems (MFS). These fleets, consisting of smaller robots, can perform tasks similar to those of larger machines while offering greater precision in positioning [201]. Technological progress has also enabled UAVs to develop sensory-motor, reactive, and cognitive autonomy [202], making them highly effective for crop monitoring when equipped with RGB, multispectral, and hyperspectral sensors.

In Ag5.0, ML algorithms will be integrated into UAVs' systems, enabling decentralized, real-time monitoring of autonomous vehicle fleets (both UAVs and UGVs). This will support advanced navigation systems, such as the redundant navigation system developed by Belhajem et al., which combines artificial neural networks, genetic algorithms, and the Extended Kalman Filter to estimate vehicle positions in real time, even without GPS [203]. Autonomous vehicle fleets will enable real-time crop monitoring, allowing for quick detection of issues affecting crop health [204]. Ultimately, this technology will reduce production costs, improve economic returns, enhance sustainability, and minimize the environmental impact of traditional farming practices.

7. Opportunities and Challenges Towards Ag5.0

The transition to Ag5.0 is expected to improve human well-being and security by creating numerous community opportunities, particularly in addressing unemployment. A significant challenge in agriculture today is the decline in productivity, which has contributed to many countries' food security issues and social crises. This decline is partly due to the negative perception of the sector among younger generations [205] and the aging agricultural workforce [206]. Ag5.0 technologies, such as AI and collaborative robots, will generate new, innovative job opportunities that improve job satisfaction and shift the perception of agriculture from a low-status, unskilled field to one requiring technical expertise [207]. This requires a highly skilled workforce to enhance productivity and compete in the global market.

Ag5.0 can potentially reduce unemployment, alleviate poverty, decrease social tensions, and potentially reduce terrorism. It will also allow consumers to choose products tailored to their preferences, increasing customer satisfaction, offering more specialized services, and reducing waste [34]. Integrating human workers with intelligent machines will improve both the quality and quantity of agricultural production, optimizing resource use and maximizing profits. Additionally, Ag5.0 will enhance data monitoring and control through advanced data collection, analysis, decision-making, and monitoring technologies while promoting safer, organic, and sustainable production practices [33].

However, the transition to Ag5.0 faces several challenges, particularly cybersecurity. Emerging technologies may introduce public safety and privacy risks, including data security, access control, communication networks, and human–machine interactions [22]. One of the main challenges in adopting Ag5.0 is the need for more expertise and infrastructure. Ag5.0 technologies such as blockchain, big data analytics, and quantum computing are relatively new technological developments and need to be better understood by people in all sectors of the global economy, including agriculture [155]. As a result, technology adoption in the specific context of smart agriculture may be higher because most industry stakeholders need the necessary competencies to use them. The scope of most Ag5.0 technologies, such as blockchain and quantum computing in smart agriculture, is far-reaching, meaning that it is difficult for most people to learn and use the technology in all its various applications [155]. Although Ag5.0 holds great promise, it remains an emerging field with limited availability of hardware and software worldwide, making it difficult to test and optimize quantum algorithms in a variety of areas, including smart agriculture [208]. The lack of standardization makes it difficult for farmers and agricultural organizations to effectively use Ag5.0 technologies, such as quantum computing, in their operations, as they may need to create solutions tailored to their specific needs. In this regard, the perceived scarcity of hardware and software may result from the fact that it is difficult for different stakeholders in agriculture to find the specific quantum computing options they need, as, in most cases, the technology must be customized for each user. Thus, unifying hardware and software for the agricultural sector can overcome the scarcity of technology resources. Cybersecurity risks are a major challenge facing the adoption of Agriculture 5.0, leading to negative impacts on society such as security risks and privacy issues [22]. The issues involved include agricultural data security, access control issues, network and communication security, security related to human–machine interactions, and safe machine operations [22]. Ag5.0 technologies, such as quantum computing, can break traditional encryption methods, making sensitive data vulnerable to attack. The reliance on Ag5.0 on state-of-the-art technology has made it very easy to hack most software-based security systems, increasing the risk of unauthorized individuals, such as hackers, accessing essential data about organizations in various contexts, including smart agriculture. In this regard, stakeholders in the agricultural sector may be reluctant to adopt the technology for fear of being exposed to security breaches by malicious parties [155,209]. Another challenge in adopting Ag5.0 is requiring specialized quantum sensors and devices. Quantum sensors are essential for the growth of quantum computing technology in smart agriculture [209]. These sensors are adept at measuring physical properties, making soil and water quality monitoring possible. Despite the many benefits, the development and implementation of sensors suitable for Ag5.0 adoption is still in its early stages, and many unsolved technical challenges need to be addressed.

8. Future Trends and Research Needs

Agriculture 5.0 is the next revolution in the agricultural sector thanks to the integration of advanced technologies such as IoT, AI, quantum computing, robotics, and smart grids, which opens new horizons for improving sustainability, efficiency, and productivity, in addition to addressing global challenges such as labor shortages and food insecurity.

The success of Ag5.0 depends on integrating technology with agricultural science, which contributes to modernizing agricultural operations and creating high-skilled jobs. Using smart machines, such as collaborative robots, will enhance productivity and empower farmers [206]. In addition, the collaboration between human expertise and smart technologies enhances data management and allows farmers to control their data and generate additional income through knowledge exchange [35]. Regional adaptation is crucial to the successful implementation of Ag5.0, as technologies such as AI and IoT can solve unemployment problems and contribute to improving food production in areas with labor shortages [207]. Designing agricultural solutions tailored to local needs also boosts the local economy and increases customer satisfaction [205]. On the other hand, advances in AI and robotics, such as neuromorphic computing and evolving neural networks, offer great potential to improve the efficiency of agricultural robots in performing precise tasks such as harvesting and soil monitoring [206]. Combining these technologies with quantum computing will enable advanced agricultural data analysis, enhancing farm management and contributing to waste reduction and more sustainable agricultural practices [33].

9. Conclusions

This article presents a framework for the future of precision crop Management, emphasizing the scientific, agronomic, and industrial applications of traditional ML algorithms, alongside recent advancements in artificial neural network (ANN) models. Between 2015 and 2024, various algorithms have been identified and applied across multiple disciplines to address growth assessment, water status, nutritional stress, crop diseases, weeds, pests, and prediction of yields. These algorithms aim to tackle tasks involving classification, regression, clustering, anomaly detection, dimensionality reduction, and association rule learning, thereby advancing precision crop detection in alignment with the emerging concept of Ag5.0. The transition to Ag5.0 will require innovations and tailored hardware, telecommunications, and robotics solutions—some already being implemented in agriculture, while others remain in development. This article highlights these emerging technologies. The shift from current Agriculture 4.0 strategies to future Ag5.0 approaches in precision crop detection will depend largely on the focus and level of automation involved. Ag5.0 will usher in a new era of intelligent crop monitoring, with an emphasis on resolving complex detection challenges, such as the early identification of crop pests, while improving overall monitoring practices, including autonomous real-time multitasking. A central focus will be on automated decision-making processes, unmanned operations, and a progressively reduced need for human intervention supported by cutting-edge AI systems, advanced robotics, and robust ML algorithms.

Author Contributions: Conceptualization, M.F.T. and H.M.; investigation, M.F.T., Z.Z. and G.E.; writing—original draft preparation, M.F.T., S.M. and M.A.A.; writing—review and editing, M.F.T., S.M., M.A.A., G.E., Z.Z., O.E., A.E.E. and A.A.; visualization, M.A.A., A.E.E. and Z.Z.; supervision, H.M.; project administration, H.M.; funding acquisition, H.M. All authors have read and agreed to the published version of the manuscript.

Funding: This work was financially supported by the Jiangsu Funding Program for Excellent Postdoctoral Talent (No. 2024ZB876).

Institutional Review Board Statement: Not applicable.

Data Availability Statement: The original contributions presented in this study are included in the article.

Conflicts of Interest: The authors declare no conflicts of interest.

Abbreviations

The following abbreviations are used in this manuscript:

Ag5.0	Agriculture 5.0
AI	Artificial intelligence
ML	Machine Learning
IoT	Internet of Things
cobots	Collaborative Robots
GNSS	Global Navigation Satellite Systems
GIS	Geographic Information Systems
LiDAR	Light Detection and Ranging
RNN	Recurrent Neural Network
GANs	Generative Adversarial Networks
PCA	Principal Component Analysis
SVMs	Support Vector Machines
GNN	Graph Convolutional Network
ISP	Image Signal Processor
GPU	Graphics Processing Units
TPU	Tensor Processing Units
NVMe	Non-Volatile Memory Express
FPGAs	Field Programmable Gate Arrays
ASICs	Application-Specific Integrated Circuits
ANNs	Artificial Neural Networks
SNNs	Spiking Neural Networks
WSNs	Wireless Sensor Networks
LPWAN	Low-Power Wide Area Networks
AMRs	Autonomous Mobile Robots
UGVs	Unmanned Ground Vehicles
UAV	Unmanned Aerial Vehicles
MFS	Multirobot Fleet Systems

References

1. Eltohamy, K.M.; Taha, M.F. Use of Inductively Coupled Plasma Mass Spectrometry (ICP-MS) to Assess the Levels of Phosphorus and Cadmium in Lettuce. In *Plant Chemical Compositions and Bioactivities*; Springer: New York, NY, USA, 2024; pp. 231–248.
2. Maja, M.M.; Ayano, S.F. The Impact of Population Growth on Natural Resources and Farmers' Capacity to Adapt to Climate Change in Low-Income Countries. *Earth Syst. Environ.* **2021**, *5*, 271–283. [CrossRef]
3. Miyake, Y.; Kimoto, S.; Uchiyama, Y.; Kohsaka, R. Income Change and Inter-Farmer Relations through Conservation Agriculture in Ishikawa Prefecture, Japan: Empirical Analysis of Economic and Behavioral Factors. *Land* **2022**, *11*, 245. [CrossRef]
4. Nowak, B. Precision Agriculture: Where Do We Stand? A Review of the Adoption of Precision Agriculture Technologies on Field Crops Farms in Developed Countries. *Agric. Res.* **2021**, *10*, 515–522. [CrossRef]
5. Zinke-Wehlmann, C.; Charvát, K. Introduction of Smart Agriculture. In *Big Data in Bioeconomy*; Springer International Publishing: Cham, Switzerland, 2021; pp. 187–190.
6. Ramachandran, K.K.; Apsara, A.; Hawladar, S.; Asokk, D.; Bhaskar, B.; Pitroda, J.R. Machine Learning and Role of Artificial Intelligence in Optimizing Work Performance and Employee Behavior. *Mater. Today Proc.* **2022**, *51*, 2327–2331. [CrossRef]
7. Rajendra, P.; Kumari, M.; Rani, S.; Dogra, N.; Boadh, R.; Kumar, A.; Dahiya, M. Impact of Artificial Intelligence on Civilization: Future Perspectives. *Mater. Today Proc.* **2022**, *56*, 252–256. [CrossRef]
8. Bensoussan, A.; Li, Y.; Nguyen, D.P.C.; Tran, M.-B.; Yam, S.C.P.; Zhou, X. Machine Learning and Control Theory. In *Handbook of Numerical Analysis*; Elsevier: Amsterdam, The Netherlands, 2022; pp. 531–558.
9. Mustafa, M.A.A.; Alshaibi, A.J.; Kostyuchenko, E.; Shelupanov, A. A Review of Artificial Intelligence Based Malware Detection Using Deep Learning. *Mater. Today Proc.* **2023**, *80*, 2678–2683. [CrossRef]
10. Song, J.; Rondao, D.; Aouf, N. Deep Learning-Based Spacecraft Relative Navigation Methods: A Survey. *Acta Astronaut.* **2022**, *191*, 22–40. [CrossRef]
11. Liang, N.; Sun, S.; Zhou, L.; Zhao, N.; Taha, M.F.; He, Y.; Qiu, Z. High-Throughput Instance Segmentation and Shape Restoration of Overlapping Vegetable Seeds Based on Sim2real Method. *Measurement* **2023**, *207*, 112414. [CrossRef]
12. Liang, N.; Sun, S.; Yu, J.; Farag Taha, M.; He, Y.; Qiu, Z. Novel Segmentation Method and Measurement System for Various Grains with Complex Touching. *Comput. Electron. Agric.* **2022**, *202*, 107351. [CrossRef]
13. Zhou, L.; Wang, X.; Zhang, C.; Zhao, N.; Taha, M.F.; He, Y.; Qiu, Z. Powdery Food Identification Using NIR Spectroscopy and Extensible Deep Learning Model. *Food Bioprocess Technol.* **2022**, *15*, 2354–2362. [CrossRef]
14. Taha, M.F.; Mao, H.; Mousa, S.; Zhou, L.; Wang, Y.; Elmasry, G.; Al-Rejaie, S.; Elwakeel, A.E.; Wei, Y.; Qiu, Z. Deep Learning-Enabled Dynamic Model for Nutrient Status Detection of Aquaponically Grown Plants. *Agronomy* **2024**, *14*, 2290. [CrossRef]
15. Niu, Z.; Huang, T.; Xu, C.; Sun, X.; Taha, M.F.; He, Y.; Qiu, Z. A Novel Approach to Optimize Key Limitations of Azure Kinect DK for Efficient and Precise Leaf Area Measurement. *Agriculture* **2025**, *15*, 173. [CrossRef]
16. Wang, Y.; Yang, N.; Ma, G.; Farag Taha, M.; Mao, H.; Zhang, X.; Shi, Q. Detection of Spores Using Polarization Image Features and BP Neural Network. *Int. J. Agric. Biol. Eng.* **2024**, *17*, 213–221. [CrossRef]
17. Elsherbiny, O.; Fan, Y.; Zhou, L.; Qiu, Z. Fusion of Feature Selection Methods and Regression Algorithms for Predicting the Canopy Water Content of Rice Based on Hyperspectral Data. *Agriculture* **2021**, *11*, 51. [CrossRef]
18. Elsherbiny, O.; Zhou, L.; He, Y.; Qiu, Z. A Novel Hybrid Deep Network for Diagnosing Water Status in Wheat Crop Using IoT-Based Multimodal Data. *Comput. Electron. Agric.* **2022**, *203*, 107453. [CrossRef]
19. Galal, H.; Elsayed, S.; Elsherbiny, O.; Allam, A.; Farouk, M. Using RGB Imaging, Optimized Three-Band Spectral Indices, and a Decision Tree Model to Assess Orange Fruit Quality. *Agriculture* **2022**, *12*, 1558. [CrossRef]
20. Fan, H. The Digital Asset Value and Currency Supervision under Deep Learning and Blockchain Technology. *J. Comput. Appl. Math.* **2022**, *407*, 114061. [CrossRef]
21. Doshi, M.; Varghese, A. Smart Agriculture Using Renewable Energy and AI-Powered IoT. In *AI, Edge and IoT-Based Smart Agriculture*; Elsevier: Amsterdam, The Netherlands, 2022; pp. 205–225.
22. Mourtzis, D.; Angelopoulos, J.; Panopoulos, N. A Literature Review of the Challenges and Opportunities of the Transition from Industry 4.0 to Society 5.0. *Energies* **2022**, *15*, 6276. [CrossRef]
23. King, T.; Cole, M.; Farber, J.M.; Eisenbrand, G.; Zabaras, D.; Fox, E.M.; Hill, J.P. Food Safety for Food Security: Relationship between Global Megatrends and Developments in Food Safety. *Trends Food Sci. Technol.* **2017**, *68*, 160–175. [CrossRef]
24. Rapela, M.A. *Fostering Innovation for Agriculture 4.0*; Springer International Publishing: Cham, Switzerland, 2019; ISBN 978-3-030-32492-6.
25. Liu, Y.; Ma, X.; Shu, L.; Hancke, G.P.; Abu-Mahfouz, A.M. From Industry 4.0 to Agriculture 4.0: Current Status, Enabling Technologies, and Research Challenges. *IEEE Trans. Ind. Inform.* **2021**, *17*, 4322–4334. [CrossRef]
26. Mavridou, E.; Vrochidou, E.; Papakostas, G.A.; Pachidis, T.; Kaburlasos, V.G. Machine Vision Systems in Precision Agriculture for Crop Farming. *J. Imaging* **2019**, *5*, 89. [CrossRef] [PubMed]
27. Zhai, Z.; Martínez, J.F.; Beltran, V.; Martínez, N.L. Decision Support Systems for Agriculture 4.0: Survey and Challenges. *Comput. Electron. Agric.* **2020**, *170*, 105256. [CrossRef]

28. Saiz-Rubio, V.; Rovira-Más, F. From Smart Farming towards Agriculture 5.0: A Review on Crop Data Management. *Agronomy* **2020**, *10*, 207. [CrossRef]
29. Ragazou, K.; Garefalakis, A.; Zafeiriou, E.; Passas, I. Agriculture 5.0: A New Strategic Management Mode for a Cut Cost and an Energy Efficient Agriculture Sector. *Energies* **2022**, *15*, 3113. [CrossRef]
30. Zambon, I.; Cecchini, M.; Egidi, G.; Saporito, M.G.; Colantoni, A. Revolution 4.0: Industry vs. Agriculture in a Future Development for SMEs. *Processes* **2019**, *7*, 36. [CrossRef]
31. Mulla, S.; Singh, S.K.; Singh, K.K.; Praveen, B. Climate Change and Agriculture: A Review of Crop Models. In *Global Climate Change and Environmental Policy*; Springer: Singapore, 2020; pp. 423–435.
32. van Dijk, M.; Gramberger, M.; Laborde, D.; Mandryk, M.; Shutes, L.; Stehfest, E.; Valin, H.; Faradsch, K. Stakeholder-Designed Scenarios for Global Food Security Assessments. *Glob. Food Secur.* **2020**, *24*, 100352. [CrossRef]
33. Fraser, E.D.G.; Campbell, M. Agriculture 5.0: Reconciling Production with Planetary Health. *One Earth* **2019**, *1*, 278–280. [CrossRef]
34. Maddikunta, P.K.R.; Pham, Q.-V.; Prabadevi, B.; Deepa, N.; Dev, K.; Gadekallu, T.R.; Ruby, R.; Liyanage, M. Industry 5.0: A Survey on Enabling Technologies and Potential Applications. *J. Ind. Inf. Integr.* **2022**, *26*, 100257. [CrossRef]
35. Cesco, S.; Sambo, P.; Borin, M.; Basso, B.; Orzes, G.; Mazzetto, F. Smart Agriculture and Digital Twins: Applications and Challenges in a Vision of Sustainability. *Eur. J. Agron.* **2023**, *146*, 126809. [CrossRef]
36. Pandrea, V.-A.; Ciocoiu, A.-O.; Machedon-Pisu, M. IoT-Based Irrigation System for Agriculture 5.0. In Proceedings of the 2023 17th International Conference on Engineering of Modern Electric Systems (EMES), Oradea, Romania, 9–10 June 2023; IEEE: Piscataway, NJ, USA, 2023; pp. 1–4.
37. Behmann, J.; Mahlein, A.-K.; Rumpf, T.; Römer, C.; Plümer, L. A Review of Advanced Machine Learning Methods for the Detection of Biotic Stress in Precision Crop Protection. *Precis. Agric.* **2015**, *16*, 239–260. [CrossRef]
38. Pätzold, S.; Hbirkou, C.; Dicke, D.; Gerhards, R.; Welp, G. Linking Weed Patterns with Soil Properties: A Long-Term Case Study. *Precis. Agric.* **2020**, *21*, 569–588. [CrossRef]
39. Shafi, U.; Mumtaz, R.; García-Nieto, J.; Hassan, S.A.; Zaidi, S.A.R.; Iqbal, N. Precision Agriculture Techniques and Practices: From Considerations to Applications. *Sensors* **2019**, *19*, 3796. [CrossRef]
40. Sarker, I.H. Deep Learning: A Comprehensive Overview on Techniques, Taxonomy, Applications and Research Directions. *SN Comput. Sci.* **2021**, *2*, 420. [CrossRef]
41. Hong, S.-J.; Kim, S.-Y.; Kim, E.; Lee, C.-H.; Lee, J.-S.; Lee, D.-S.; Bang, J.; Kim, G. Moth Detection from Pheromone Trap Images Using Deep Learning Object Detectors. *Agriculture* **2020**, *10*, 170. [CrossRef]
42. Yamashita, R.; Nishio, M.; Do, R.K.G.; Togashi, K. Convolutional Neural Networks: An Overview and Application in Radiology. *Insights Imaging* **2018**, *9*, 611–629. [CrossRef]
43. Debnath, O.; Saha, H.N. An IoT-Based Intelligent Farming Using CNN for Early Disease Detection in Rice Paddy. *Microprocess. Microsyst.* **2022**, *94*, 104631. [CrossRef]
44. Wang, Y.; Wang, H.; Peng, Z. Rice Diseases Detection and Classification Using Attention Based Neural Network and Bayesian Optimization. *Expert Syst. Appl.* **2021**, *178*, 114770. [CrossRef]
45. Genze, N.; Bharti, R.; Grieb, M.; Schultheiss, S.J.; Grimm, D.G. Accurate Machine Learning-Based Germination Detection, Prediction and Quality Assessment of Three Grain Crops. *Plant Methods* **2020**, *16*, 157. [CrossRef]
46. ElMasry, G.; Mandour, N.; Al-Rejaie, S.; Belin, E.; Rousseau, D. Recent Applications of Multispectral Imaging in Seed Phenotyping and Quality Monitoring—An Overview. *Sensors* **2019**, *19*, 1090. [CrossRef]
47. ElMasry, G.; Mandour, N.; Ejeez, Y.; Demilly, D.; Al-Rejaie, S.; Verdier, J.; Belin, E.; Rousseau, D. Multichannel Imaging for Monitoring Chemical Composition and Germination Capacity of Cowpea (*Vigna unguiculata*) Seeds during Development and Maturation. *Crop J.* **2022**, *10*, 1399–1411. [CrossRef]
48. Peng, Q.; Tu, L.; Wu, Y.; Yu, Z.; Tang, G.; Song, W. Automatic Monitoring System for Seed Germination Test Based on Deep Learning. *J. Electr. Comput. Eng.* **2022**, *2022*, 4678316. [CrossRef]
49. Awty-Carroll, D.; Clifton-Brown, J.; Robson, P. Using K-NN to Analyse Images of Diverse Germination Phenotypes and Detect Single Seed Germination in Miscanthus Sinensis. *Plant Methods* **2018**, *14*, 5. [CrossRef] [PubMed]
50. Ahmad, A.; Saraswat, D.; El Gamal, A. A Survey on Using Deep Learning Techniques for Plant Disease Diagnosis and Recommendations for Development of Appropriate Tools. *Smart Agric. Technol.* **2023**, *3*, 100083. [CrossRef]
51. Oerke, E.-C.; Dehne, H.-W. Safeguarding Production—Losses in Major Crops and the Role of Crop Protection. *Crop Prot.* **2004**, *23*, 275–285. [CrossRef]
52. Saleem, M.H.; Potgieter, J.; Arif, K.M. Automation in Agriculture by Machine and Deep Learning Techniques: A Review of Recent Developments. *Precis. Agric.* **2021**, *22*, 2053–2091. [CrossRef]
53. Abdalla, A.; Wheeler, T.A.; Dever, J.; Lin, Z.; Arce, J.; Guo, W. Assessing Fusarium Oxysporum Disease Severity in Cotton Using Unmanned Aerial System Images and a Hybrid Domain Adaptation Deep Learning Time Series Model. *Biosyst. Eng.* **2024**, *237*, 220–231. [CrossRef]

54. Ahmed, I.; Yadav, P.K. A Systematic Analysis of Machine Learning and Deep Learning Based Approaches for Identifying and Diagnosing Plant Diseases. *Sustain. Oper. Comput.* **2023**, *4*, 96–104. [CrossRef]
55. Orchi, H.; Sadik, M.; Khaldoun, M.; Sabir, E. Automation of Crop Disease Detection through Conventional Machine Learning and Deep Transfer Learning Approaches. *Agriculture* **2023**, *13*, 352. [CrossRef]
56. Nagachandrika, B.; Prasath, R.; Praveen Joe, I.R. An Automatic Classification Framework for Identifying Type of Plant Leaf Diseases Using Multi-Scale Feature Fusion-Based Adaptive Deep Network. *Biomed. Signal Process. Control* **2024**, *95*, 106316. [CrossRef]
57. Guerrero-Ibañez, A.; Reyes-Muñoz, A. Monitoring Tomato Leaf Disease through Convolutional Neural Networks. *Electronics* **2023**, *12*, 229. [CrossRef]
58. Tamilvizhi, T.; Surendran, R.; Anbazhagan, K.; Rajkumar, K. Quantum Behaved Particle Swarm Optimization-Based Deep Transfer Learning Model for Sugarcane Leaf Disease Detection and Classification. *Math. Probl. Eng.* **2022**, *2022*, 3452413. [CrossRef]
59. Zhao, N.; Zhou, L.; Huang, T.; Taha, M.F.; He, Y.; Qiu, Z. Development of an Automatic Pest Monitoring System Using a Deep Learning Model of DPeNet. *Measurement* **2022**, *203*, 111970. [CrossRef]
60. Lee, S.; Choi, G.; Park, H.-C.; Choi, C. Automatic Classification Service System for Citrus Pest Recognition Based on Deep Learning. *Sensors* **2022**, *22*, 8911. [CrossRef] [PubMed]
61. Dai, M.; Shen, Y.; Li, X.; Liu, J.; Zhang, S.; Miao, H. Digital Twin System of Pest Management Driven by Data and Model Fusion. *Agriculture* **2024**, *14*, 1099. [CrossRef]
62. Routis, G.; Michailidis, M.; Roussaki, I. Plant Disease Identification Using Machine Learning Algorithms on Single-Board Computers in IoT Environments. *Electronics* **2024**, *13*, 1010. [CrossRef]
63. Ma, X.; Zhang, X.; Guan, H.; Wang, L. Recognition Method of Crop Disease Based on Image Fusion and Deep Learning Model. *Agronomy* **2024**, *14*, 1518. [CrossRef]
64. Wang, Y.; Li, T.; Chen, T.; Zhang, X.; Taha, M.F.; Yang, N.; Mao, H.; Shi, Q. Cucumber Downy Mildew Disease Prediction Using a CNN-LSTM Approach. *Agriculture* **2024**, *14*, 1155. [CrossRef]
65. Balaji, V.; Anushkannan, N.K.; Narahari, S.C.; Rattan, P.; Verma, D.; Awasthi, D.K.; Pandian, A.A.; Veeramanickam, M.R.M.; Mulat, M.B. Deep Transfer Learning Technique for Multimodal Disease Classification in Plant Images. *Contrast Media Mol. Imaging* **2023**, *2023*, 5644727. [CrossRef]
66. Wang, Y.; Zhang, X.; Taha, M.F.; Chen, T.; Yang, N.; Zhang, J.; Mao, H. Detection Method of Fungal Spores Based on Fingerprint Characteristics of Diffraction–Polarization Images. *J. Fungi* **2023**, *9*, 1131. [CrossRef]
67. Liu, B.; Zhang, Y.; He, D.; Li, Y. Identification of Apple Leaf Diseases Based on Deep Convolutional Neural Networks. *Symmetry* **2017**, *10*, 11. [CrossRef]
68. Ramcharan, A.; Baranowski, K.; McCloskey, P.; Ahmed, B.; Legg, J.; Hughes, D.P. Deep Learning for Image-Based Cassava Disease Detection. *Front. Plant Sci.* **2017**, *8*, 1852. [CrossRef] [PubMed]
69. Lu, J.; Hu, J.; Zhao, G.; Mei, F.; Zhang, C. An In-Field Automatic Wheat Disease Diagnosis System. *Comput. Electron. Agric.* **2017**, *142*, 369–379. [CrossRef]
70. Fuentes, S.; Tongson, E.; Unnithan, R.R.; Gonzalez Viejo, C. Early Detection of Aphid Infestation and Insect-Plant Interaction Assessment in Wheat Using a Low-Cost Electronic Nose (E-Nose), Near-Infrared Spectroscopy and Machine Learning Modeling. *Sensors* **2021**, *21*, 5948. [CrossRef] [PubMed]
71. Feng, Z.-H.; Wang, L.-Y.; Yang, Z.-Q.; Zhang, Y.-Y.; Li, X.; Song, L.; He, L.; Duan, J.-Z.; Feng, W. Hyperspectral Monitoring of Powdery Mildew Disease Severity in Wheat Based on Machine Learning. *Front. Plant Sci.* **2022**, *13*, 828454. [CrossRef]
72. DeChant, C.; Wiesner-Hanks, T.; Chen, S.; Stewart, E.L.; Yosinski, J.; Gore, M.A.; Nelson, R.J.; Lipson, H. Automated Identification of Northern Leaf Blight-Infected Maize Plants from Field Imagery Using Deep Learning. *Phytopathology* **2017**, *107*, 1426–1432. [CrossRef] [PubMed]
73. Kaneda, Y.; Shibata, S.; Mineno, H. Multi-Modal Sliding Window-Based Support Vector Regression for Predicting Plant Water Stress. *Knowl. Based Syst.* **2017**, *134*, 135–148. [CrossRef]
74. Fuentes, A.; Yoon, S.; Kim, S.; Park, D. A Robust Deep-Learning-Based Detector for Real-Time Tomato Plant Diseases and Pests Recognition. *Sensors* **2017**, *17*, 2022. [CrossRef]
75. Krishnaswamy Rangarajan, A.; Purushothaman, R. Disease Classification in Eggplant Using Pre-Trained VGG16 and MSVM. *Sci. Rep.* **2020**, *10*, 2322. [CrossRef]
76. Too, E.C.; Yujian, L.; Njuki, S.; Yingchun, L. A Comparative Study of Fine-Tuning Deep Learning Models for Plant Disease Identification. *Comput. Electron. Agric.* **2019**, *161*, 272–279. [CrossRef]
77. Rançon, F.; Bombrun, L.; Keresztes, B.; Germain, C. Comparison of SIFT Encoded and Deep Learning Features for the Classification and Detection of Esca Disease in Bordeaux Vineyards. *Remote Sens.* **2018**, *11*, 1. [CrossRef]
78. Cruz, A.; Ampatzidis, Y.; Pierro, R.; Materazzi, A.; Panattoni, A.; De Bellis, L.; Luvisi, A. Detection of Grapevine Yellows Symptoms in *Vitis Vinifera* L. with Artificial Intelligence. *Comput. Electron. Agric.* **2019**, *157*, 63–76. [CrossRef]

79. Liang, Q.; Xiang, S.; Hu, Y.; Coppola, G.; Zhang, D.; Sun, W. PD2SE-Net: Computer-Assisted Plant Disease Diagnosis and Severity Estimation Network. *Comput. Electron. Agric.* **2019**, *157*, 518–529. [CrossRef]
80. Gold, K.M.; Townsend, P.A.; Herrmann, I.; Gevens, A.J. Investigating Potato Late Blight Physiological Differences across Potato Cultivars with Spectroscopy and Machine Learning. *Plant Sci.* **2020**, *295*, 110316. [CrossRef]
81. Abdulridha, J.; Ampatzidis, Y.; Ehsani, R.; de Castro, A.I. Evaluating the Performance of Spectral Features and Multivariate Analysis Tools to Detect Laurel Wilt Disease and Nutritional Deficiency in Avocado. *Comput. Electron. Agric.* **2018**, *155*, 203–211. [CrossRef]
82. Lu, J.; Ehsani, R.; Shi, Y.; de Castro, A.I.; Wang, S. Detection of Multi-Tomato Leaf Diseases (Late Blight, Target and Bacterial Spots) in Different Stages by Using a Spectral-Based Sensor. *Sci. Rep.* **2018**, *8*, 2793. [CrossRef] [PubMed]
83. Lu, J.; Ehsani, R.; Shi, Y.; Abdulridha, J.; de Castro, A.I.; Xu, Y. Field Detection of Anthracnose Crown Rot in Strawberry Using Spectroscopy Technology. *Comput. Electron. Agric.* **2017**, *135*, 289–299. [CrossRef]
84. Barreto, A.; Paulus, S.; Varrelmann, M.; Mahlein, A.-K. Hyperspectral Imaging of Symptoms Induced by Rhizoctonia Solani in Sugar Beet: Comparison of Input Data and Different Machine Learning Algorithms. *J. Plant Dis. Prot.* **2020**, *127*, 441–451. [CrossRef]
85. Polder, G.; Blok, P.M.; de Villiers, H.A.C.; van der Wolf, J.M.; Kamp, J. Potato Virus Y Detection in Seed Potatoes Using Deep Learning on Hyperspectral Images. *Front. Plant Sci.* **2019**, *10*, 209. [CrossRef]
86. Zhu, H.; Chu, B.; Zhang, C.; Liu, F.; Jiang, L.; He, Y. Hyperspectral Imaging for Presymptomatic Detection of Tobacco Disease with Successive Projections Algorithm and Machine-Learning Classifiers. *Sci. Rep.* **2017**, *7*, 4125. [CrossRef]
87. de Carvalho Alves, M.; Pozza, E.A.; Sanches, L.; Belan, L.L.; de Oliveira Freitas, M.L. Insights for Improving Bacterial Blight Management in Coffee Field Using Spatial Big Data and Machine Learning. *Trop. Plant Pathol.* **2021**, *47*, 118–139. [CrossRef]
88. Zarco-Tejada, P.J.; Camino, C.; Beck, P.S.A.; Calderon, R.; Hornero, A.; Hernández-Clemente, R.; Kattenborn, T.; Montes-Borrego, M.; Susca, L.; Morelli, M.; et al. Previsual Symptoms of Xylella Fastidiosa Infection Revealed in Spectral Plant-Trait Alterations. *Nat. Plants* **2018**, *4*, 432–439. [CrossRef]
89. Ramos, A.P.M.; Gomes, F.D.G.; Pinheiro, M.M.F.; Furuya, D.E.G.; Gonçalvez, W.N.; Junior, J.M.; Michereff, M.F.F.; Blassioli-Moraes, M.C.; Borges, M.; Alaumann, R.A.; et al. Detecting the Attack of the Fall Armyworm (*Spodoptera frugiperda*) in Cotton Plants with Machine Learning and Spectral Measurements. *Precis. Agric.* **2022**, *23*, 470–491. [CrossRef]
90. Garcia-Ruiz, F.; Sankaran, S.; Maja, J.M.; Lee, W.S.; Rasmussen, J.; Ehsani, R. Comparison of Two Aerial Imaging Platforms for Identification of Huanglongbing-Infected Citrus Trees. *Comput. Electron. Agric.* **2013**, *91*, 106–115. [CrossRef]
91. Liu, L.; Wang, R.; Xie, C.; Yang, P.; Wang, F.; Sudirman, S.; Liu, W. PestNet: An End-to-End Deep Learning Approach for Large-Scale Multi-Class Pest Detection and Classification. *IEEE Access* **2019**, *7*, 45301–45312. [CrossRef]
92. Xia, D.; Chen, P.; Wang, B.; Zhang, J.; Xie, C. Insect Detection and Classification Based on an Improved Convolutional Neural Network. *Sensors* **2018**, *18*, 4169. [CrossRef] [PubMed]
93. Ding, Y.; Jiang, C.; Song, L.; Liu, F.; Tao, Y. RVDR-YOLOv8: A Weed Target Detection Model Based on Improved YOLOv8. *Electronics* **2024**, *13*, 2182. [CrossRef]
94. Almalky, A.M.; Ahmed, K.R. Deep Learning for Detecting and Classifying the Growth Stages of Consolida Regalis Weeds on Fields. *Agronomy* **2023**, *13*, 934. [CrossRef]
95. Mu, Y.; Feng, R.; Ni, R.; Li, J.; Luo, T.; Liu, T.; Li, X.; Gong, H.; Guo, Y.; Sun, Y.; et al. A Faster R-CNN-Based Model for the Identification of Weed Seedling. *Agronomy* **2022**, *12*, 2867. [CrossRef]
96. Zhu, H.; Zhang, Y.; Mu, D.; Bai, L.; Zhuang, H.; Li, H. YOLOX-Based Blue Laser Weeding Robot in Corn Field. *Front. Plant Sci.* **2022**, *13*, 1017803. [CrossRef]
97. Islam, N.; Rashid, M.M.; Wibowo, S.; Xu, C.-Y.; Morshed, A.; Wasimi, S.A.; Moore, S.; Rahman, S.M. Early Weed Detection Using Image Processing and Machine Learning Techniques in an Australian Chilli Farm. *Agriculture* **2021**, *11*, 387. [CrossRef]
98. Tran, D.; Schouteten, J.J.; Degieter, M.; Krupanek, J.; Jarosz, W.; Areta, A.; Emmi, L.; De Steur, H.; Gellynck, X. European Stakeholders' Perspectives on Implementation Potential of Precision Weed Control: The Case of Autonomous Vehicles with Laser Treatment. *Precis. Agric.* **2023**, *24*, 2200–2222. [CrossRef] [PubMed]
99. Aravind, K.R.; Raja, P.; Pérez-Ruiz, M. Task-Based Agricultural Mobile Robots in Arable Farming: A Review. *Span. J. Agric. Res.* **2017**, *15*, e02R01. [CrossRef]
100. Jiang, W.; Quan, L.; Wei, G.; Chang, C.; Geng, T. A Conceptual Evaluation of a Weed Control Method with Post-Damage Application of Herbicides: A Composite Intelligent Intra-Row Weeding Robot. *Soil Tillage Res.* **2023**, *234*, 105837. [CrossRef]
101. Zhang, L.; Zhang, Z.; Wu, C.; Sun, L. Segmentation Algorithm for Overlap Recognition of Seedling Lettuce and Weeds Based on SVM and Image Blocking. *Comput. Electron. Agric.* **2022**, *201*, 107284. [CrossRef]
102. Guo, X.; Ge, Y.; Liu, F.; Yang, J. Identification of Maize and Wheat Seedlings and Weeds Based on Deep Learning. *Front. Earth Sci.* **2023**, *11*, 1146558. [CrossRef]
103. Eide, A.; Koparan, C.; Zhang, Y.; Ostlie, M.; Howatt, K.; Sun, X. UAV-Assisted Thermal Infrared and Multispectral Imaging of Weed Canopies for Glyphosate Resistance Detection. *Remote Sens.* **2021**, *13*, 4606. [CrossRef]

104. Li, D.; Shi, G.; Li, J.; Chen, Y.; Zhang, S.; Xiang, S.; Jin, S. PlantNet: A Dual-Function Point Cloud Segmentation Network for Multiple Plant Species. *ISPRS J. Photogramm. Remote Sens.* **2022**, *184*, 243–263. [CrossRef]
105. Fawakherji, M.; Potena, C.; Pretto, A.; Bloisi, D.D.; Nardi, D. Multi-Spectral Image Synthesis for Crop/Weed Segmentation in Precision Farming. *Robot. Auton. Syst.* **2021**, *146*, 103861. [CrossRef]
106. Ashraf, T.; Khan, Y.N. Weed Density Classification in Rice Crop Using Computer Vision. *Comput. Electron. Agric.* **2020**, *175*, 105590. [CrossRef]
107. Shen, Y.; Yin, Y.; Li, B.; Zhao, C.; Li, G. Detection of Impurities in Wheat Using Terahertz Spectral Imaging and Convolutional Neural Networks. *Comput. Electron. Agric.* **2021**, *181*, 105931. [CrossRef]
108. Alam, M.S.; Alam, M.; Tufail, M.; Khan, M.U.; Güneş, A.; Salah, B.; Nasir, F.E.; Saleem, W.; Khan, M.T. TobSet: A New Tobacco Crop and Weeds Image Dataset and Its Utilization for Vision-Based Spraying by Agricultural Robots. *Appl. Sci.* **2022**, *12*, 1308. [CrossRef]
109. Khan, S.; Tufail, M.; Khan, M.T.; Khan, Z.A.; Anwar, S. Deep Learning-Based Identification System of Weeds and Crops in Strawberry and Pea Fields for a Precision Agriculture Sprayer. *Precis. Agric.* **2021**, *22*, 1711–1727. [CrossRef]
110. Ramirez, W.; Achanccaray, P.; Mendoza, L.F.; Pacheco, M.A.C. Deep Convolutional Neural Networks for Weed Detection in Agricultural Crops Using Optical Aerial Images. In Proceedings of the 2020 IEEE Latin American GRSS & ISPRS Remote Sensing Conference (LAGIRS), Santiago, Chile, 22–26 March 2020; IEEE: Piscataway, NJ, USA, 2020; pp. 133–137.
111. Milioto, A.; Lottes, P.; Stachniss, C. Real-Time Blob-Wise Sugar Beets Vs Weeds Classification For Monitoring Fields Using Convolutional Neural Networks. *ISPRS Ann. Photogramm. Remote Sens. Spat. Inf. Sci.* **2017**, *4*, 41–48. [CrossRef]
112. Anul Haq, M. CNN Based Automated Weed Detection System Using UAV Imagery. *Comput. Syst. Sci. Eng.* **2022**, *42*, 837–849. [CrossRef]
113. Bah, M.D.; Hafiane, A.; Canals, R. Deep Learning with Unsupervised Data Labeling for Weed Detection in Line Crops in UAV Images. *Remote Sens.* **2018**, *10*, 1690. [CrossRef]
114. Andy Lease, B.; Wong, W.; Gopal, L.; Chiong, W.R. Weed Pixel Level Classification Based on Evolving Feature Selection on Local Binary Pattern With Shallow Network Classifier. *IOP Conf. Ser. Mater. Sci. Eng.* **2020**, *943*, 012001. [CrossRef]
115. Dhal, S.B.; Bagavathiannan, M.; Braga-Neto, U.; Kalafatis, S. Nutrient Optimization for Plant Growth in Aquaponic Irrigation Using Machine Learning for Small Training Datasets. *Artif. Intell. Agric.* **2022**, *6*, 68–76. [CrossRef]
116. Ennaji, O.; Vergütz, L.; El Allali, A. Machine Learning in Nutrient Management: A Review. *Artif. Intell. Agric.* **2023**, *9*, 1–11. [CrossRef]
117. Bera, A.; Bhattacharjee, D.; Krejcar, O. PND-Net: Plant Nutrition Deficiency and Disease Classification Using Graph Convolutional Network. *Sci. Rep.* **2024**, *14*, 15537. [CrossRef]
118. Abdalla, A.; Cen, H.; Wan, L.; Mehmood, K.; He, Y. Nutrient Status Diagnosis of Infield Oilseed Rape via Deep Learning-Enabled Dynamic Model. *IEEE Trans. Ind. Inform.* **2021**, *17*, 4379–4389. [CrossRef]
119. Taha, M.F.; Abdalla, A.; ElMasry, G.; Gouda, M.; Zhou, L.; Zhao, N.; Liang, N.; Niu, Z.; Hassanein, A.; Al-Rejaie, S.; et al. Using Deep Convolutional Neural Network for Image-Based Diagnosis of Nutrient Deficiencies in Plants Grown in Aquaponics. *Chemosensors* **2022**, *10*, 45. [CrossRef]
120. Taha, M.F.; Mao, H.; Wang, Y.; ElManawy, A.I.; Elmasry, G.; Wu, L.; Memon, M.S.; Niu, Z.; Huang, T.; Qiu, Z. High-Throughput Analysis of Leaf Chlorophyll Content in Aquaponically Grown Lettuce Using Hyperspectral Reflectance and RGB Images. *Plants* **2024**, *13*, 392. [CrossRef] [PubMed]
121. Kou, J.; Duan, L.; Yin, C.; Ma, L.; Chen, X.; Gao, P.; Lv, X. Predicting Leaf Nitrogen Content in Cotton with UAV RGB Images. *Sustainability* **2022**, *14*, 9259. [CrossRef]
122. Yu, X.; Lu, H.; Liu, Q. Deep-Learning-Based Regression Model and Hyperspectral Imaging for Rapid Detection of Nitrogen Concentration in Oilseed Rape (*Brassica napus* L.) Leaf. *Chemom. Intell. Lab. Syst.* **2018**, *172*, 188–193. [CrossRef]
123. Anami, B.S.; Malvade, N.N.; Palaiah, S. Deep Learning Approach for Recognition and Classification of Yield Affecting Paddy Crop Stresses Using Field Images. *Artif. Intell. Agric.* **2020**, *4*, 12–20. [CrossRef]
124. Thompson, L.J.; Ferguson, R.B.; Kitchen, N.; Frazen, D.W.; Mamo, M.; Yang, H.; Schepers, J.S. Model and Sensor-Based Recommendation Approaches for In-Season Nitrogen Management in Corn. *Agron. J.* **2015**, *107*, 2020–2030. [CrossRef]
125. Chlingaryan, A.; Sukkarieh, S.; Whelan, B. Machine Learning Approaches for Crop Yield Prediction and Nitrogen Status Estimation in Precision Agriculture: A Review. *Comput. Electron. Agric.* **2018**, *151*, 61–69. [CrossRef]
126. Kusanur, V.; Chakravarthi, V.S. Using Transfer Learning for Nutrient Deficiency Prediction and Classification in Tomato Plant. *Int. J. Adv. Comput. Sci. Appl.* **2021**, *12*, 784–790. [CrossRef]
127. Song, Y.; Teng, G.; Yuan, Y.; Liu, T.; Sun, Z. Assessment of Wheat Chlorophyll Content by the Multiple Linear Regression of Leaf Image Features. *Inf. Process. Agric.* **2021**, *8*, 232–243. [CrossRef]
128. Sharma, M.; Nath, K.; Sharma, R.K.; Kumar, C.J.; Chaudhary, A. Ensemble Averaging of Transfer Learning Models for Identification of Nutritional Deficiency in Rice Plant. *Electronics* **2022**, *11*, 148. [CrossRef]

129. Ghosal, S.; Blystone, D.; Singh, A.K.; Ganapathysubramanian, B.; Singh, A.; Sarkar, S. An Explainable Deep Machine Vision Framework for Plant Stress Phenotyping. *Proc. Natl. Acad. Sci. USA* **2018**, *115*, 4613–4618. [CrossRef]
130. Sethy, P.K.; Barpanda, N.K.; Rath, A.K.; Behera, S.K. Nitrogen Deficiency Prediction of Rice Crop Based on Convolutional Neural Network. *J. Ambient Intell. Humaniz. Comput.* **2020**, *11*, 5703–5711. [CrossRef]
131. Watchareeruetai, U.; Noinongyao, P.; Wattanapaiboonsuk, C.; Khantiviriya, P.; Duangsrisai, S. Identification of Plant Nutrient Deficiencies Using Convolutional Neural Networks. In Proceedings of the 2018 International Electrical Engineering Congress (iEECON), Krabi, Thailand, 7–9 March 2018; IEEE: Piscataway, NJ, USA, 2018; pp. 1–4.
132. Joshi, P.; Das, D.; Udutalapally, V.; Pradhan, M.K.; Misra, S. RiceBioS: Identification of Biotic Stress in Rice Crops Using Edge-as-a-Service. *IEEE Sens. J.* **2022**, *22*, 4616–4624. [CrossRef]
133. Chang, L.; Li, D.; Hameed, M.K.; Yin, Y.; Huang, D.; Niu, Q. Using a Hybrid Neural Network Model DCNN–LSTM for Image-Based Nitrogen Nutrition Diagnosis in Muskmelon. *Horticulturae* **2021**, *7*, 489. [CrossRef]
134. Manoharan, S.; Sariffodeen, B.; Ramasinghe, K.T.; Rajaratne, L.H.; Kasthurirathna, D.; Wijekoon, J.L. Smart Plant Disorder Identification Using Computer Vision Technology. In Proceedings of the 2020 11th IEEE Annual Information Technology, Electronics and Mobile Communication Conference (IEMCON), Vancouver, BC, Canada, 4–7 November 2020; IEEE: Piscataway, NJ, USA, 2020; pp. 0445–0451.
135. Azimi, S.; Kaur, T.; Gandhi, T.K. A Deep Learning Approach to Measure Stress Level in Plants Due to Nitrogen Deficiency. *Measurement* **2021**, *173*, 108650. [CrossRef]
136. Yi, J.; Krusenbaum, L.; Unger, P.; Hüging, H.; Seidel, S.J.; Schaaf, G.; Gall, J. Deep Learning for Non-Invasive Diagnosis of Nutrient Deficiencies in Sugar Beet Using RGB Images. *Sensors* **2020**, *20*, 5893. [CrossRef]
137. Ahsan, M.; Eshkabilov, S.; Cemek, B.; Küçüktopcu, E.; Lee, C.W.; Simsek, H. Deep Learning Models to Determine Nutrient Concentration in Hydroponically Grown Lettuce Cultivars (*Lactuca sativa* L.). *Sustainability* **2021**, *14*, 416. [CrossRef]
138. Akbari, M.; Sabouri, H.; Sajadi, S.J.; Yarahmadi, S.; Ahangar, L. Classification and Prediction of Drought and Salinity Stress Tolerance in Barley Using GenPhenML. *Sci. Rep.* **2024**, *14*, 17420. [CrossRef]
139. Gupta, A.; Kaur, L.; Kaur, G. Drought Stress Detection Technique for Wheat Crop Using Machine Learning. *PeerJ Comput. Sci.* **2023**, *9*, e1268. [CrossRef]
140. Okyere, F.G.; Cudjoe, D.K.; Virlet, N.; Castle, M.; Riche, A.B.; Greche, L.; Mohareb, F.; Simms, D.; Mhada, M.; Hawkesford, M.J. Hyperspectral Imaging for Phenotyping Plant Drought Stress and Nitrogen Interactions Using Multivariate Modeling and Machine Learning Techniques in Wheat. *Remote Sens.* **2024**, *16*, 3446. [CrossRef]
141. Wu, Y.; Jiang, J.; Zhang, X.; Zhang, J.; Cao, Q.; Tian, Y.; Zhu, Y.; Cao, W.; Liu, X. Combining Machine Learning Algorithm and Multi-Temporal Temperature Indices to Estimate the Water Status of Rice. *Agric. Water Manag.* **2023**, *289*, 108521. [CrossRef]
142. Jin, K.; Zhang, J.; Wang, Z.; Zhang, J.; Liu, N.; Li, M.; Ma, Z. Application of Deep Learning Based on Thermal Images to Identify the Water Stress in Cotton under Film-Mulched Drip Irrigation. *Agric. Water Manag.* **2024**, *299*, 108901. [CrossRef]
143. An, J.; Li, W.; Li, M.; Cui, S.; Yue, H. Identification and Classification of Maize Drought Stress Using Deep Convolutional Neural Network. *Symmetry* **2019**, *11*, 256. [CrossRef]
144. Zhuang, S.; Wang, P.; Jiang, B.; Li, M. Learned Features of Leaf Phenotype to Monitor Maize Water Status in the Fields. *Comput. Electron. Agric.* **2020**, *172*, 105347. [CrossRef]
145. Sun, H.; Feng, M.; Xiao, L.; Yang, W.; Wang, C.; Jia, X.; Zhao, Y.; Zhao, C.; Muhammad, S.K.; Li, D. Assessment of Plant Water Status in Winter Wheat (*Triticum aestivum* L.) Based on Canopy Spectral Indices. *PLoS ONE* **2019**, *14*, e0216890. [CrossRef]
146. Zuo, Z.; Mu, J.; Li, W.; Bu, Q.; Mao, H.; Zhang, X.; Han, L.; Ni, J. Study on the Detection of Water Status of Tomato (*Solanum lycopersicum* L.) by Multimodal Deep Learning. *Front. Plant Sci.* **2023**, *14*, 1094142. [CrossRef]
147. Li, Z.; Chen, Z.; Cheng, Q.; Fei, S.; Zhou, X. Deep Learning Models Outperform Generalized Machine Learning Models in Predicting Winter Wheat Yield Based on Multispectral Data from Drones. *Drones* **2023**, *7*, 505. [CrossRef]
148. Li, D.; Wu, X. Individualized Indicators and Estimation Methods for Tiger Nut (*Cyperus esculentus* L.) Tubers Yield Using Light Multispectral UAV and Lightweight CNN Structure. *Drones* **2023**, *7*, 432. [CrossRef]
149. Tanaka, Y.; Watanabe, T.; Katsura, K.; Tsujimoto, Y.; Takai, T.; Tanaka, T.S.T.; Kawamura, K.; Saito, H.; Homma, K.; Mairoua, S.G.; et al. Deep Learning Enables Instant and Versatile Estimation of Rice Yield Using Ground-Based RGB Images. *Plant Phenomics* **2023**, *5*, 0073. [CrossRef] [PubMed]
150. Mia, M.S.; Tanabe, R.; Habibi, L.N.; Hashimoto, N.; Homma, K.; Maki, M.; Matsui, T.; Tanaka, T.S.T. Multimodal Deep Learning for Rice Yield Prediction Using UAV-Based Multispectral Imagery and Weather Data. *Remote Sens.* **2023**, *15*, 2511. [CrossRef]
151. Zhou, S.; Xu, L.; Chen, N. Rice Yield Prediction in Hubei Province Based on Deep Learning and the Effect of Spatial Heterogeneity. *Remote Sens.* **2023**, *15*, 1361. [CrossRef]
152. Sarr, A.B.; Sultan, B. Predicting Crop Yields in Senegal Using Machine Learning Methods. *Int. J. Climatol.* **2023**, *43*, 1817–1838. [CrossRef]

153. Surana, R.; Khandelwal, R. Crop Yield Prediction Using Machine Learning: A Pragmatic Approach 2024. Available online: https://www.researchgate.net/publication/381910719_Crop_Yield_Prediction_Using_Machine_Learning_A_Pragmatic_Approach (accessed on 10 June 2024).
154. Kuradusenge, M.; Hitimana, E.; Hanyurwimfura, D.; Rukundo, P.; Mtonga, K.; Mukasine, A.; Uwitonze, C.; Ngabonziza, J.; Uwamahoro, A. Crop Yield Prediction Using Machine Learning Models: Case of Irish Potato and Maize. *Agriculture* **2023**, *13*, 225. [CrossRef]
155. Maraveas, C.; Konar, D.; Michopoulos, D.K.; Arvanitis, K.G.; Peppas, K.P. Harnessing Quantum Computing for Smart Agriculture: Empowering Sustainable Crop Management and Yield Optimization. *Comput. Electron. Agric.* **2024**, *218*, 108680. [CrossRef]
156. Nigam, A.; Garg, S.; Agrawal, A.; Agrawal, P. Crop Yield Prediction Using Machine Learning Algorithms. In Proceedings of the 2019 Fifth International Conference on Image Information Processing (ICIIP), Shimla, India, 15–17 November 2019; IEEE: Piscataway, NJ, USA, 2019; pp. 125–130.
157. Shen, Y.; Mercatoris, B.; Cao, Z.; Kwan, P.; Guo, L.; Yao, H.; Cheng, Q. Improving Wheat Yield Prediction Accuracy Using LSTM-RF Framework Based on UAV Thermal Infrared and Multispectral Imagery. *Agriculture* **2022**, *12*, 892. [CrossRef]
158. Maimaitijiang, M.; Sagan, V.; Sidike, P.; Hartling, S.; Esposito, F.; Fritschi, F.B. Soybean Yield Prediction from UAV Using Multimodal Data Fusion and Deep Learning. *Remote Sens. Environ.* **2020**, *237*, 111599. [CrossRef]
159. Wang, Y.; Zhang, Z.; Feng, L.; Du, Q.; Runge, T. Combining Multi-Source Data and Machine Learning Approaches to Predict Winter Wheat Yield in the Conterminous United States. *Remote Sens.* **2020**, *12*, 1232. [CrossRef]
160. Cao, J.; Zhang, Z.; Luo, Y.; Zhang, L.; Zhang, J.; Li, Z.; Tao, F. Wheat Yield Predictions at a County and Field Scale with Deep Learning, Machine Learning, and Google Earth Engine. *Eur. J. Agron.* **2021**, *123*, 126204. [CrossRef]
161. Yang, W.; Nigon, T.; Hao, Z.; Dias Paiao, G.; Fernández, F.G.; Mulla, D.; Yang, C. Estimation of Corn Yield Based on Hyperspectral Imagery and Convolutional Neural Network. *Comput. Electron. Agric.* **2021**, *184*, 106092. [CrossRef]
162. Nevavuori, P.; Narra, N.; Lipping, T. Crop Yield Prediction with Deep Convolutional Neural Networks. *Comput. Electron. Agric.* **2019**, *163*, 104859. [CrossRef]
163. Chakraborty, M.; Pourreza, A.; Zhang, X.; Jafarbiglu, H.; Shackel, K.A.; DeJong, T. Early Almond Yield Forecasting by Bloom Mapping Using Aerial Imagery and Deep Learning. *Comput. Electron. Agric.* **2023**, *212*, 108063. [CrossRef]
164. Shankar, P.; Johnen, A.; Liwicki, M. Data Fusion and Artificial Neural Networks for Modelling Crop Disease Severity. In Proceedings of the 2020 IEEE 23rd International Conference on Information Fusion (FUSION), Rustenburg, South Africa, 6–9 July 2020; IEEE: Piscataway, NJ, USA, 2020; pp. 1–8.
165. Picon, A.; Alvarez-Gila, A.; Seitz, M.; Ortiz-Barredo, A.; Echazarra, J.; Johannes, A. Deep Convolutional Neural Networks for Mobile Capture Device-Based Crop Disease Classification in the Wild. *Comput. Electron. Agric.* **2019**, *161*, 280–290. [CrossRef]
166. Qiu, J.; Wang, J.; Yao, S.; Guo, K.; Li, B.; Zhou, E.; Yu, J.; Tang, T.; Xu, N.; Song, S.; et al. Going Deeper with Embedded FPGA Platform for Convolutional Neural Network. In Proceedings of the 2016 ACM/SIGDA International Symposium on Field-Programmable Gate Arrays, Monterey, CA, USA, 21–23 February 2016; ACM: New York, NY, USA, 2016; pp. 26–35.
167. Shawahna, A.; Sait, S.M.; El-Maleh, A. FPGA-Based Accelerators of Deep Learning Networks for Learning and Classification: A Review. *IEEE Access* **2019**, *7*, 7823–7859. [CrossRef]
168. Ibrahim, A.; Gebali, F. Compact Hardware Accelerator for Field Multipliers Suitable for Use in Ultra-Low Power IoT Edge Devices. *Alex. Eng. J.* **2022**, *61*, 13079–13087. [CrossRef]
169. Oñate, W.; Sanz, R. Analysis of Architectures Implemented for IIoT. *Heliyon* **2023**, *9*, e12868. [CrossRef]
170. Strukov, D.B.; Snider, G.S.; Stewart, D.R.; Williams, R.S. The Missing Memristor Found. *Nature* **2008**, *453*, 80–83. [CrossRef]
171. Zidan, M.A.; Strachan, J.P.; Lu, W.D. The Future of Electronics Based on Memristive Systems. *Nat. Electron.* **2018**, *1*, 22–29. [CrossRef]
172. Esser, S.K.; Merolla, P.A.; Arthur, J.V.; Cassidy, A.S.; Appuswamy, R.; Andreopoulos, A.; Berg, D.J.; McKinstry, J.L.; Melano, T.; Barch, D.R.; et al. Convolutional Networks for Fast, Energy-Efficient Neuromorphic Computing. *Proc. Natl. Acad. Sci. USA* **2016**, *113*, 11441–11446. [CrossRef]
173. Zeadally, S.; Siddiqui, F.; Baig, Z. 25 Years of Bluetooth Technology. *Future Internet* **2019**, *11*, 194. [CrossRef]
174. Castro, S.; Iñacasha, J.; Mesias, G.; Oñate, W. Prototype Based on a LoRaWAN Network for Storing Multivariable Data, Oriented to Agriculture with Limited Resources. In Proceedings of the Seventh International Congress on Information and Communication Technology, London, UK, 21–24 February 2022; Springer Nature: Singapore, 2023; pp. 245–255.
175. Lavric, A. LoRa (Long-Range) High-Density Sensors for Internet of Things. *J. Sens.* **2019**, *2019*, 3502987. [CrossRef]
176. Ullah, Z.; Al-Turjman, F.; Mostarda, L. Cognition in UAV-Aided 5G and Beyond Communications: A Survey. *IEEE Trans. Cogn. Commun. Netw.* **2020**, *6*, 872–891. [CrossRef]
177. Yang, Y.; Lin, M.; Lin, Y.; Zhang, C.; Wu, C. A Survey of Blockchain Applications for Management in Agriculture and Livestock Internet of Things. *Future Internet* **2025**, *17*, 40. [CrossRef]
178. Maniah; Abdurachman, E.; Gaol, F.L.; Soewito, B. Survey on Threats and Risks in the Cloud Computing Environment. *Procedia Comput. Sci.* **2019**, *161*, 1325–1332. [CrossRef]

179. Sun, P. Security and Privacy Protection in Cloud Computing: Discussions and Challenges. *J. Netw. Comput. Appl.* **2020**, *160*, 102642. [CrossRef]
180. García-Valls, M.; Escribano-Barreno, J.; García-Muñoz, J. An Extensible Collaborative Framework for Monitoring Software Quality in Critical Systems. *Inf. Softw. Technol.* **2019**, *107*, 3–17. [CrossRef]
181. Zhou, Z.; Chen, X.; Li, E.; Zeng, L.; Luo, K.; Zhang, J. Edge Intelligence: Paving the Last Mile of Artificial Intelligence With Edge Computing. *Proc. IEEE* **2019**, *107*, 1738–1762. [CrossRef]
182. Gao, W.; Zhou, P. Customized High Performance and Energy Efficient Communication Networks for AI Chips. *IEEE Access* **2019**, *7*, 69434–69446. [CrossRef]
183. Cheng, C.; Fu, J.; Su, H.; Ren, L. Recent Advancements in Agriculture Robots: Benefits and Challenges. *Machines* **2023**, *11*, 48. [CrossRef]
184. Panarin, R.N.; Khvorova, L.A. Software Development for Agricultural Tillage Robot Based on Technologies of Machine Intelligence. In Proceedings of the International Conference on High-Performance Computing Systems and Technologies in Scientific Research, Automation of Control and Production, Barnaul, Russia, 20–21 May 2022; Springer: Cham, Switzerland, 2022; pp. 354–367.
185. Backman, J.; Linkolehto, R.; Lemsalu, M.; Kaivosoja, J. Building a Robot Tractor Using Commercial Components and Widely Used Standards. *IFAC-PapersOnLine* **2022**, *55*, 6–11. [CrossRef]
186. Haibo, L.; Shuliang, D.; Zunmin, L.; Chuijie, Y. Study and Experiment on a Wheat Precision Seeding Robot. *J. Robot.* **2015**, *2015*, 696301. [CrossRef]
187. Ghafar, A.S.A.; Hajjaj, S.S.H.; Gsangaya, K.R.; Sultan, M.T.H.; Mail, M.F.; Hua, L.S. Design and Development of a Robot for Spraying Fertilizers and Pesticides for Agriculture. *Mater. Today Proc.* **2023**, *81*, 242–248. [CrossRef]
188. Terra, F.P.; Nascimento, G.H.D.; Duarte, G.A.; Drews, P.L.J., Jr. Autonomous Agricultural Sprayer Using Machine Vision and Nozzle Control. *J. Intell. Robot. Syst.* **2021**, *102*, 38. [CrossRef]
189. Putu Devira Ayu Martini, N.; Tamami, N.; Husein Alasiry, A. Design and Development of Automatic Plant Robots with Scheduling System. In Proceedings of the 2020 International Electronics Symposium (IES), Surabaya, Indonesia, 29–30 September 2020; IEEE: Piscataway, NJ, USA, 2020; pp. 302–307.
190. An, Z.; Wang, C.; Raj, B.; Eswaran, S.; Raffik, R.; Debnath, S.; Rahin, S.A. Application of New Technology of Intelligent Robot Plant Protection in Ecological Agriculture. *J. Food Qual.* **2022**, *2022*, 1257015. [CrossRef]
191. Ren, G.; Wu, T.; Lin, T.; Yang, L.; Chowdhary, G.; Ting, K.C.; Ying, Y. Mobile Robotics Platform for Strawberry Sensing and Harvesting within Precision Indoor Farming Systems. *J. Field Robot.* **2024**, *41*, 2047–2065. [CrossRef]
192. Cubero, S.; Marco-Noales, E.; Aleixos, N.; Barbé, S.; Blasco, J. RobHortic: A Field Robot to Detect Pests and Diseases in Horticultural Crops by Proximal Sensing. *Agriculture* **2020**, *10*, 276. [CrossRef]
193. Pooranam, N.; Vignesh, T. A Swarm Robot for Harvesting a Paddy Field. In *Nature-Inspired Algorithms Applications*; Wiley: Hoboken, NJ, USA, 2021; pp. 137–156.
194. McCool, C.S.; Beattie, J.; Firn, J.; Lehnert, C.; Kulk, J.; Bawden, O.; Russell, R.; Perez, T. Efficacy of Mechanical Weeding Tools: A Study into Alternative Weed Management Strategies Enabled by Robotics. *IEEE Robot. Autom. Lett.* **2018**, *3*, 1184–1190. [CrossRef]
195. Sembiring, A.; Budiman, A.; Lestari, Y.D. Design And Control Of Agricultural Robot For Tomato Plants Treatment And Harvesting. *J. Phys. Conf. Ser.* **2017**, *930*, 012019. [CrossRef]
196. Casseem, M.S.I.S.; Venkannah, S.; Bissessur, Y. Design of a Tomato Harvesting Robot for Agricultural Small and Medium Enterprises (SMEs). In Proceedings of the 2022 IST-Africa Conference (IST-Africa), Virtual, 16–20 May 2022; IEEE: Piscataway, NJ, USA, 2022; pp. 1–8.
197. Maurice, P.; Malaisé, A.; Amiot, C.; Paris, N.; Richard, G.-J.; Rochel, O.; Ivaldi, S. Human Movement and Ergonomics: An Industry-Oriented Dataset for Collaborative Robotics. *Int. J. Robot. Res.* **2019**, *38*, 1529–1537. [CrossRef]
198. Machleb, J.; Peteinatos, G.G.; Kollenda, B.L.; Andújar, D.; Gerhards, R. Sensor-Based Mechanical Weed Control: Present State and Prospects. *Comput. Electron. Agric.* **2020**, *176*, 105638. [CrossRef]
199. Zhang, K.; Lammers, K.; Chu, P.; Li, Z.; Lu, R. System Design and Control of an Apple Harvesting Robot. *Mechatronics* **2021**, *79*, 102644. [CrossRef]
200. Giampieri, F.; Mazzoni, L.; Cianciosi, D.; Alvarez-Suarez, J.M.; Regolo, L.; Sánchez-González, C.; Capocasa, F.; Xiao, J.; Mezzetti, B.; Battino, M. Organic vs. Conventional Plant-Based Foods: A Review. *Food Chem.* **2022**, *383*, 132352. [CrossRef] [PubMed]
201. Gonzalez-de-Santos, P.; Ribeiro, A.; Fernandez-Quintanilla, C.; Lopez-Granados, F.; Brandstoetter, M.; Tomic, S.; Pedrazzi, S.; Peruzzi, A.; Pajares, G.; Kaplanis, G.; et al. Fleets of Robots for Environmentally-Safe Pest Control in Agriculture. *Precis. Agric.* **2017**, *18*, 574–614. [CrossRef]
202. Floreano, D.; Wood, R.J. Science, Technology and the Future of Small Autonomous Drones. *Nature* **2015**, *521*, 460–466. [CrossRef] [PubMed]

203. Belhajem, I.; Ben Maissa, Y.; Tamtaoui, A. A Robust Low Cost Approach for Real Time Car Positioning in a Smart City Using Extended Kalman Filter and Evolutionary Machine Learning. In Proceedings of the 2016 4th IEEE International Colloquium on Information Science and Technology (CiSt), Tangier, Morocco, 24–26 October 2016; IEEE: Piscataway, NJ, USA, 2016; pp. 806–811.
204. Emmi, L.; Gonzalez-de-Soto, M.; Pajares, G.; Gonzalez-de-Santos, P. New Trends in Robotics for Agriculture: Integration and Assessment of a Real Fleet of Robots. *Sci. World J.* **2014**, *2014*, 404059. [CrossRef] [PubMed]
205. Concepcion, R.; Josh Ramirez, T.; Alejandrino, J.; Janairo, A.G.; Jahara Baun, J.; Francisco, K.; Relano, R.-J.; Enriquez, M.L.; Grace Bautista, M.; Vicerra, R.R.; et al. A Look at the Near Future: Industry 5.0 Boosts the Potential of Sustainable Space Agriculture. In Proceedings of the 2022 IEEE 14th International Conference on Humanoid, Nanotechnology, Information Technology, Communication and Control, Environment, and Management (HNICEM), Boracay Island, Philippines, 1–4 December 2022; IEEE: Piscataway, NJ, USA, 2022; pp. 1–6.
206. Polymeni, S.; Plastras, S.; Skoutas, D.N.; Kormentzas, G.; Skianis, C. The Impact of 6G-IoT Technologies on the Development of Agriculture 5.0: A Review. *Electronics* **2023**, *12*, 2651. [CrossRef]
207. Humayun, M. Industrial Revolution 5.0 and the Role of Cutting Edge Technologies. *Int. J. Adv. Comput. Sci. Appl.* **2021**, *12*, 605. [CrossRef]
208. Rietsche, R.; Dremel, C.; Bosch, S.; Steinacker, L.; Meckel, M.; Leimeister, J.-M. Quantum Computing. *Electron. Mark.* **2022**, *32*, 2525–2536. [CrossRef]
209. Siddiquee, K.N.-A.; Islam, M.S.; Singh, N.; Gunjan, V.K.; Yong, W.H.; Huda, M.N.; Naik, D.S.B. Development of Algorithms for an IoT-Based Smart Agriculture Monitoring System. *Wirel. Commun. Mob. Comput.* **2022**, *2022*, 7372053. [CrossRef]

MDPI AG
Grosspeteranlage 5
4052 Basel
Switzerland
Tel.: +41 61 683 77 34

Agriculture Editorial Office
E-mail: agriculture@mdpi.com
www.mdpi.com/journal/agriculture

www.ingramcontent.com/pod-product-compliance
Lightning Source LLC
Chambersburg PA
CBHW061555220926
51850CB00005B/345
* 9 7 8 3 7 2 5 8 5 2 8 9 5 *